AB INITIO MOLECULAR DYNAMICS:
BASIC THEORY AND ADVANCED METHODS

Ab initio molecular dynamics revolutionized the field of realistic computer simulation of complex molecular systems and processes, including chemical reactions, by unifying molecular dynamics and electronic structure theory. This book provides the first coherent presentation of this rapidly growing field, covering a vast range of methods and their applications, from basic theory to advanced methods.

This fascinating text for graduate students and researchers contains systematic derivations of various *ab initio* molecular dynamics techniques to enable readers to understand and assess the merits and drawbacks of commonly used methods. It also discusses the special features of the widely used Car–Parrinello approach, correcting various misconceptions currently found in the research literature.

The book contains pseudo-code and program layout for typical plane wave electronic structure codes, allowing newcomers to the field to understand commonly used program packages, and enabling developers to improve and add new features in their code.

DOMINIK MARX is Chair of Theoretical Chemistry at Ruhr-Universität Bochum, Germany. His main areas of research are in studying the dynamics and reactions of complex molecular many-body systems and the development of novel *ab initio* simulation techniques.

JÜRG HUTTER is a Professor at the Physical Chemistry Institute at the University of Zürich in Switzerland, where he researches problems in theoretical chemistry, in particular, methods for large-scale density functional calculations.

D1500778

AB INITIO MOLECULAR DYNAMICS: BASIC THEORY AND ADVANCED METHODS

DOMINIK MARX

Ruhr-Universität Bochum

and

JÜRG HUTTER

University of Zürich

CAMBRIDGE
UNIVERSITY PRESS

CAMBRIDGE UNIVERSITY PRESS
Cambridge, New York, Melbourne, Madrid, Cape Town,
Singapore, São Paulo, Delhi, Mexico City

Cambridge University Press
The Edinburgh Building, Cambridge CB2 8RU, UK

Published in the United States of America by Cambridge University Press, New York

www.cambridge.org
Information on this title: www.cambridge.org/9781107663534

First published 2009
Reprinted 2010
First paperback edition 2012

Printed and bound in the United Kingdom by the MPG Books Group

A catalogue record for this publication is available from the British Library

ISBN 978-0-521-89863-8 Hardback
ISBN 978-1-107-66353-4 Paperback

Contents

Preface   *page* viii

1   **Setting the stage: why *ab initio* molecular dynamics?**   1

     Part I   Basic techniques   9

2   **Getting started: unifying MD and electronic structure**   11
     2.1   Deriving classical molecular dynamics   11
     2.2   Ehrenfest molecular dynamics   22
     2.3   Born–Oppenheimer molecular dynamics   24
     2.4   Car–Parrinello molecular dynamics   27
     2.5   What about Hellmann–Feynman forces?   51
     2.6   Which method to choose?   56
     2.7   Electronic structure methods   67
     2.8   Basis sets   75

3   **Implementation: using the plane wave basis set**   85
     3.1   Introduction and basic definitions   85
     3.2   Electrostatic energy   93
     3.3   Exchange and correlation energy   99
     3.4   Total energy, gradients, and stress tensor   104
     3.5   Energy and force calculations in practice   109
     3.6   Optimizing the Kohn–Sham orbitals   111
     3.7   Molecular dynamics   119
     3.8   Program organization and layout   128

4   **Atoms with plane waves: accurate pseudopotentials**   136
     4.1   Why pseudopotentials?   137
     4.2   Norm-conserving pseudopotentials   138
     4.3   Pseudopotentials in the plane wave basis   152
     4.4   Dual-space Gaussian pseudopotentials   157

4.5	Nonlinear core correction	160
4.6	Pseudopotential transferability	162
4.7	Example: pseudopotentials for carbon	167

Part II   Advanced techniques — 175

5	**Beyond standard *ab initio* molecular dynamics**	**177**
5.1	Introduction	177
5.2	Beyond microcanonics: thermostats, barostats, meta-dynamics	178
5.3	Beyond ground states: ROKS, surface hopping, FEMD, TDDFT	194
5.4	Beyond classical nuclei: path integrals and quantum corrections	233
5.5	Hybrid QM/MM molecular dynamics	267
6	**Beyond norm-conserving pseudopotentials**	**286**
6.1	Introduction	286
6.2	The PAW transformation	287
6.3	Expectation values	290
6.4	Ultrasoft pseudopotentials	292
6.5	PAW energy expression	296
6.6	Integrating the Car–Parrinello equations	297
7	**Computing properties**	**309**
7.1	Perturbation theory: Hessian, polarizability, NMR	309
7.2	Wannier functions: dipole moments, IR spectra, atomic charges	327
8	**Parallel computing**	**350**
8.1	Introduction	350
8.2	Data structures	352
8.3	Computational kernels	354
8.4	Massively parallel processing	359

Part III   Applications — 369

9	**From materials to biomolecules**	**371**
9.1	Introduction	371
9.2	Solids, minerals, materials, and polymers	372
9.3	Interfaces	376
9.4	Mechanochemistry and molecular electronics	380
9.5	Water and aqueous solutions	382

9.6   Non-aqueous liquids and solutions   385
9.7   Glasses and amorphous systems   389
9.8   Matter at extreme conditions   390
9.9   Clusters, fullerenes, and nanotubes   392
9.10   Complex and fluxional molecules   394
9.11   Chemical reactions and transformations   396
9.12   Homogeneous catalysis and zeolites   399
9.13   Photophysics and photochemistry   400
9.14   Biophysics and biochemistry   403

10   Properties from *ab initio* simulations   **407**
10.1   Introduction   407
10.2   Electronic structure analyses   407
10.3   Infrared spectroscopy   410
10.4   Magnetism, NMR and EPR spectroscopy   411
10.5   Electronic spectroscopy and redox properties   412
10.6   X-ray diffraction and Compton scattering   413
10.7   External electric fields   414

11   Outlook   **416**
Bibliography   419
Index   550

Preface

In this book we develop the rapidly growing field of *ab initio* molecular dynamics computer simulations from the underlying basic ideas up to the latest techniques, from the most straightforward implementation up to multilevel parallel algorithms. Since the seminal contributions of Roberto Car and Michele Parrinello starting in the mid-1980s, the unification of molecular dynamics and electronic structure theory, often dubbed "Car–Parrinello molecular dynamics" or just "CP", widened the scope and power of *both* approaches considerably. The forces are described at the level of the many-body problem of interacting electrons and nuclei, which form atoms and molecules as described in the framework of quantum mechanics, whereas the dynamics is captured in terms of classical dynamics and statistical mechanics. Due to its inherent virtues, *ab initio* molecular dynamics is currently an extremely popular and ever-expanding computational tool employed to study physical, chemical, and biological phenomena in a very broad sense. In particular, it is the basis of what could be called a "virtual laboratory approach" used to study complex processes at the molecular level, including the difficult task of the breaking and making of chemical bonds, by means of purely theoretical methods. In a nutshell, *ab initio* molecular dynamics allows one to tackle vastly different systems such as amorphous silicon, Ziegler-Natta heterogeneous catalysis, and wet DNA using the same computational approach, thus opening avenues to deal with molecular phenomena in physics, chemistry, and biology in a unified framework.

We now feel that the time has come to summarize the impressive developments of the last 20 years in this field within a unified framework at the level of an advanced text. Currently, any newcomer in the field has to face the problem of first working through the many excellent and largely complementary review articles or Lecture Notes that are widespread. Even worse, much of the significant development of the last few years is not even accessible at that level. Thus, our aim here is to provide not only an introduction to the beginner such as graduate students, but also as far as possible a comprehensive and up-to-date overview of the entire field including its prospects and limitations. Both aspects are also of value to the increasing number of those scientists who wish only to apply *ab initio* molecular dynamics as a powerful problem-solving tool in their daily research, without having to bother too much about the technical aspects, let alone about method development. This is indeed possible, in principle, since several rather easy-to-use program packages are now on the market, mostly for free or at low cost for academic users.

In particular, different flavors of *ab initio* molecular dynamics methods are explained and compared in the first part of this book at an introductory level, the focus being on the efficient extended Lagrangian approach as introduced by Car and Parrinello in 1985. But in the meantime, a wealth of techniques that go far beyond what we call here the "standard approach", that is microcanonical molecular dynamics in the electronic ground state using classical nuclei and norm-conserving pseudopotentials, have been devised. These advanced techniques are outlined in Part II and include methods that allow us to work in other ensembles, to enhance sampling, to include excited electronic states and nonadiabatic effects, to deal with quantum effects on nuclei, and to treat complex biomolecular systems in terms of mixed quantum/classical approaches. Most important for the practitioner is the computation of properties during the simulations, such as optical, IR, Raman, or NMR properties, mostly in the context of linear response theory or the analysis of the dynamical electronic structure in terms of fragment dipole moments, localized orbitals, or effective atomic charges. Finally in Part III, we provide a glimpse of the wide range of applications, which not only demonstrate the enormous potential of *ab initio* molecular dynamics for both explaining and predicting properties of matter, but also serve as a compilation of pertinent literature for future reference and upcoming applications.

In addition to all these aspects we also want to provide a solid basis of technical knowledge for the younger generation such as graduate students, postdocs, and junior researchers beginning their career in a nowadays well-established field. For this very reason we also decided to include, as far as possible, specific references in the text to the original literature as well as to review articles. To achieve this, the very popular approach of solving the electronic structure problem in the framework of Kohn–Sham density functional theory as formulated in terms of plane waves and pseudopotentials is described in detail in Part I. Although a host of "tricks" can already be presented at that stage, specific aspects can only be made clear when discussing them at the level of implementation. Here, the widely used and ever-expanding program package CPMD serves as our main reference, but we stress that the techniques and paradigms introduced apply analogously to many other available codes that are in extensive use. This needs to be supplemented with an introduction to the concept of norm-conserving pseudopotentials, including definitions of various widely used pseudopotential types. In Part I, the norm-conserving pseudopotentials are explained, whereas in Part II, the reader will be exposed to the powerful projector augmented-wave transformation and ultrasoft pseudopotentials. A crucial aspect for large-scale applications, given the current computer architectures and the foreseeable future developments, is how to deal with parallel platforms. We account for this sustainable trend by devoting special attention in Part II to parallel programming, explaining a very powerful hierarchical multilevel scheme. This paradigm allows one to use not only the ubiquitous Beowulf clusters efficiently, but also the largest machines available, viz. clustered shared-memory parallel servers and ultra-dense massively parallel computers.

Overall, our hope is that this book will contribute not only to strengthen applications of *ab initio* molecular dynamics in both academia and industry, but also to foster further technical development of this family of computer simulation methods. In the spirit of this idea, we will maintain the site www.theochem.rub.de/go/aimd-book.html where corrections and additions to this book will be collected and provided in an open access mode. We thus encourage all readers to send us information about possible errors, which are definitively hidden at many places despite our investment of much care in preparing this manuscript.

Last but not least we would like to stress that our knowledge of *ab initio* molecular dynamics has grown slowly within the realms of a fruitful and longstanding collaboration with Michele Parrinello, initially at IBM Zurich Research Laboratory in Rüschlikon and later at the Max-Planck-Institut für Festkörperforschung in Stuttgart, which we gratefully acknowledge on this occasion. In addition, we profited enormously from pleasant cooperations with too many friends and colleagues to be named here.

1

Setting the stage: why *ab initio* molecular dynamics?

Classical molecular dynamics using predefined potentials, force fields, either based on empirical data or on independent electronic structure calculations, is well established as a powerful tool serving to investigate many-body condensed matter systems, including biomolecular assemblies. The broadness, diversity, and level of sophistication of this technique are documented in several books as well as review articles, conference proceedings, lecture notes, and special issues [25, 120, 136, 272, 398, 468, 577, 726, 1189, 1449, 1504, 1538, 1539]. At the very heart of any molecular dynamics scheme is the question of how to describe – that is in practice how to approximate – the interatomic interactions. The traditional route followed in molecular dynamics is to determine these potentials in advance. Typically, the full interaction is broken up into two-body and many-body contributions, long-range and short-range terms, electrostatic and non-electrostatic interactions, etc., which have to be represented by suitable functional forms, see Refs. [550, 1405] for detailed accounts. After decades of intense research, very elaborate interaction models, including the nontrivial aspect of representing these potentials analytically, were devised [550, 1280, 1380, 1405, 1539].

Despite their overwhelming success – which will, however, not be praised in this book – the need to devise a fixed predefined potential implies serious drawbacks [1123, 1209]. Among the most significant are systems in which (i) many different atom or molecule types give rise to a myriad of different interatomic interactions that have to be parameterized and/or (ii) the electronic structure and thus the chemical bonding pattern changes qualitatively during the course of the simulation. Such systems are termed here "chemically complex". An additional aspect (iii) is of a more practical nature: once a specific system is understood after elaborate development of satisfactory potentials, changing a single species provokes typically enormous efforts to

1

parameterize the new potentials needed. As a result, systematic studies are a *tour de force* if no suitable set of consistent potentials is already available.

The reign of traditional molecular dynamics *and* electronic structure methods was extended greatly by a family of techniques that is referred to here as "*ab initio* molecular dynamics" (AIMD). Apart from the widely used general notion of "Car–Parrinello" or just "CP simulations" as defined in the *Physics and Astronomy Classification Scheme*, PACS [1093], other names including common abbreviations that are currently in use for such methods are for instance first principles (FPMD), on-the-fly, direct, extended Lagrangian (ELMD), density functional (DFMD), quantum chemical, Hellmann–Feynman, Fock-matrix, potential-free, or just quantum (QMD) molecular dynamics amongst others. The basic idea underlying every *ab initio* molecular dynamics method is to compute the forces acting on the nuclei from electronic structure calculations that are performed "on-the-fly" as the molecular dynamics trajectory is generated, see Fig. 1.1 for a simplifying scheme. In this way, the electronic variables are not integrated out beforehand and represented by fixed interaction potentials, rather they are considered to be active and explicit degrees of freedom in the course of the simulation. This implies that, given a suitable approximate solution of the many-electron problem, also "chemically complex" systems, or those where the electronic structure changes drastically during the dynamics, can be handled easily by molecular dynamics. But this also implies that the approximation is shifted from the level of devising an interaction potential to the level of selecting a particular approximation for solving the Schrödinger equation, since it cannot be solved exactly for the typical problems at hand.

Applications of *ab initio* molecular dynamics are particularly widespread in physics, chemistry, and more recently also in biology, where the aforementioned difficulties (i)-(iii) are particularly severe [39, 934]. A collection of problems that have already been tackled by *ab initio* molecular dynamics, including the pertinent references, can be found in Chapter 9 of Part III. The power of this novel family of techniques led to an explosion of activity in this field in terms of the number of published papers, see the squares in Fig. 1.2 that can be interpreted as a measure of the activity in the area of *ab initio* molecular dynamics. This rapid increase in activity started in the mid to late 1980s. As a matter of fact the time evolution of the number of citations of a particular paper, the one by Car and Parrinello from 1985 entitled "Unified approach for molecular dynamics and density-functional theory" [222, 1216], initially parallels the growth trend of the entire field, see the circles in Fig. 1.2. Thus, the resonance evoked by this publication and, at its very heart, the introduction of the Car–Parrinello

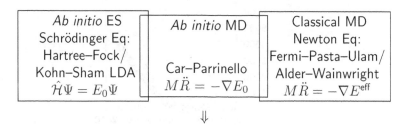

Statics and Dynamics
Electrons and Nuclei

Molecules, Clusters, Complexes
Liquids, Solids, Surfaces
Composites
⋮
at $T \geq 0$

Fig. 1.1. *Ab initio* molecular dynamics unifies approximate *ab initio* electronic structure theory (i.e. solving Schrödinger's wave equation numerically using, for instance, Hartree–Fock theory or the local density approximation within Kohn–Sham theory) and classical molecular dynamics (i.e. solving Newton's equation of motion numerically for a given interaction potential as reported by Fermi, Pasta, Ulam, and Tsingou for a one-dimensional anharmonic chain model of solids [409] and published by Alder and Wainwright for the three-dimensional hard-sphere model of fluids [19]; see Refs. [33, 272, 308, 453, 652] for historic perspectives on these early molecular dynamics studies).

"Lagrangean" [995], has gone hand in hand with the popularity of the entire field over the last decade. Incidentally, the 1985 paper by Car and Parrinello is the last one included in the section "Trends and Prospects" in the reprint collection of "key papers" from the field of atomistic computer simulations [272]. Evidence that the entire field of *ab initio* molecular dynamics has matured is also provided by the separate PACS classification number ("71.15.Pd - Electronic Structure: Molecular dynamics calculations

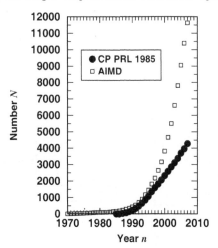

Fig. 1.2. Publication and citation analysis up to the year 2007. Squares: number of publications N which appeared up to the year n containing the keyword "ab initio molecular dynamics" (or synonyms such as "first principles MD", "Car–Parrinello simulations" etc.) in title, abstract or keyword list. Circles: number of publications N which appeared up to the year n citing the 1985 paper by Car and Parrinello [222] (including misspellings of the bibliographic reference). Self-citations and self-papers are excluded, i.e. citations of Ref. [222] in their own papers and papers coauthored by R. Car and/or M. Parrinello are *not* considered in the respective statistics; note that this, together with the correction for misspellings, is probably the main reason for a slightly different citation number up to the year 2002 as given here compared to that (2819 citations) reported in Ref. [1216]. The analysis is based on Thomson/ISI Web of Science (WoS), literature file CAPLUS of the Chemical Abstracts Service (CAS), and INSPEC file (Physics Abstracts) as accessible under the database provider STN International. Earlier reports of these statistics [933, 934, 943] are updated as of March 13, 2008; the authors are most grateful to Dr. Werner Marx (Information Service for the Institutes of the Chemical Physical Technical Section of the Max Planck Society) for carrying out these analyses.

(Car–Parrinello) and other numerical simulations") introduced in 1996 into the *Physics and Astronomy Classification Scheme* [1093].

Despite its obvious advantages, it is evident that a price has to be payed for putting molecular dynamics onto an *ab initio* foundation: the correlation lengths and relaxation times that are accessible are much smaller than what is affordable in the framework of standard molecular dynamics. More recently, this discrepancy was counterbalanced by the ever-increasing power of available computing resources, in particular massively parallel platforms [661, 662], which shifted many problems in the physical sciences right into the realm of *ab initio* molecular dynamics. Another appealing feature of standard molecular dynamics is less evident, namely the experimental aspect of "playing with the potential". Thus, tracing back the properties of a given system to a simple physical picture or mechanism is much harder in *ab initio* molecular dynamics, where certain interactions cannot easily be "switched off" like in standard molecular dynamics. On the other hand, *ab initio* molecular dynamics has the power to eventually map phenomena onto a firm basis in terms of the underlying electronic structure and chemical bonding patterns. Most importantly, however, is the fact that new phenomena, which were not foreseen before starting the simulation, can simply happen if necessary. All this lends *ab initio* molecular dynamics a~truly predictive power.

Ab initio molecular dynamics can also be viewed from another perspective, namely from the field of classical trajectory calculations [1284, 1514]. In this approach, which has its origin in gas phase reaction dynamics, a *global* potential energy surface is constructed in a first step either empirically, semi-empirically or, more and more, based on high-level electronic structure calculations. After fitting it to a suitable analytical form in a second step (but without imposing additional approximations such as pairwise additivity, etc.), the dynamical evolution of the nuclei is generated in a third step by using classical mechanics, quantum mechanics, or semi/quasiclassical approximations of various sorts. In the case of using classical mechanics to describe the dynamics - which is the focus of the present book - the limiting step for large systems is the first one, why should this be so? There are $3N-6$ internal degrees of freedom that span the global potential energy surface of an unconstrained N-body system. Using, for simplicity, 10 discretization points per coordinate implies that of the order of 10^{3N-6} electronic structure calculations are needed in order to map such a global potential energy surface. Thus, the computational workload for the first step in the approach outlined above grows roughly like $\sim 10^N$ with increasing system size $\sim N$. This is what might be called the "curse of dimensionality" or "dimensionality bottleneck" of calculations that rely on *global* potential energy surfaces, see for instance the discussion on p. 420 in Ref. [551].

What is needed in *ab initio* molecular dynamics instead? Suppose that a useful trajectory consists of about 10^M molecular dynamics steps, i.e. 10^M electronic structure calculations are needed to generate one trajectory. Furthermore, it is assumed that 10^n independent trajectories are necessary in order to average over different initial conditions so that 10^{M+n} *ab initio* molecular dynamics steps are required in total. Finally, it is assumed that each single-point electronic structure calculation needed to devise the global potential energy surface and one *ab initio* molecular dynamics time step require roughly the same amount of CPU time. Based on this truly simplistic order of magnitude estimate, the advantage of *ab initio* molecular dynamics vs. calculations relying on the computation of a global potential energy surface amounts to about $10^{3N-6-M-n}$. The crucial point is that for a given statistical accuracy (that is for M and n fixed and independent of N) and for a given electronic structure method, the computational advantage of "on-the-fly" approaches grows like $\sim 10^N$ with system size. Thus, Car–Parrinello methods always outperform the traditional three-step approaches *if the system is sufficiently large and complex.* Conversely, computing global potential energy surfaces beforehand and running many classical trajectories afterwards without much additional cost always pays off for a given system size N like $\sim 10^{M+n}$ *if the system is small enough* so that a global potential energy surface can be computed and parameterized.

Of course, considerable progress has been achieved in accelerating the computation of global potentials by carefully selecting the discretization points and reducing their number, choosing sophisticated representations and internal coordinates, exploiting symmetry and decoupling of irrelevant modes, implementing efficient sampling and smart extrapolation techniques and so forth. Still, these improvements mostly affect the prefactor but not the overall scaling behavior, $\sim 10^N$, with the number of active degrees of freedom. Other strategies consist of, for instance, reducing the number of active degrees of freedom by constraining certain internal coordinates, representing less important ones by a (harmonic) bath or by friction forces, or building up the global potential energy surface in terms of few-body fragments. All these approaches, however, invoke approximations beyond those of the electronic structure method itself. Finally, it is evident that the computational advantage of the "on-the-fly" approaches diminishes as more and more trajectories are needed for a given (small) system. For instance, extensive averaging over many different initial conditions is required in order to calculate scattering or reactive cross-sections quantitatively. Summarizing this discussion, it can be concluded that *ab initio* molecular dynamics is the method of choice to investigate large and "chemically complex" systems.

Quite a few reviews, conference articles, lecture notes, and overviews dealing with *ab initio* molecular dynamics have appeared since the early 1990s [38, 228, 338, 460, 485, 486, 510, 563, 564, 669, 784, 933, 934, 936–938, 943, 1099, 1103, 1104, 1123, 1209, 1272, 1306, 1307, 1498, 1512, 1544] and the interested reader is referred to them for various complementary viewpoints. This book originates from the Lecture Notes [943] "*Ab initio* molecular dynamics: Theory and implementation" written by the present authors on the occasion of the *NIC Winter School 2000* titled "Modern Methods and Algorithms of Quantum Chemistry". However, it incorporates in addition many recent developments as covered in a variety of lectures, courses, and tutorials given by the authors as well as parts from our previous review and overview articles. Here, emphasis is put on both the broad extent of the approaches and the depth of the presentation as demanded from both the practitioner's and newcomer's viewpoints.

With respect to the broadness of the approaches, the discussion starts in Part I, "Basic techniques", at the coupled Schrödinger equation for electrons and nuclei. Classical, Ehrenfest, Born–Oppenheimer, and Car–Parrinello molecular dynamics are derived in Chapter 2 from the time-dependent mean-field approach that is obtained after separating the nuclear and electronic degrees of freedom. The most extensive discussion is related to the features of the standard Car–Parrinello approach, however, all three *ab initio* approaches to molecular dynamics - Car–Parrinello, Born–Oppenheimer, and Ehrenfest - are contrasted and compared. The important issue of how to obtain the correct forces in these schemes is discussed in some depth. The two most popular electronic structure theories implemented within *ab initio* molecular dynamics, Kohn–Sham density functional theory but also the Hartree–Fock approach, are only touched upon since excellent textbooks [363, 397, 625, 760, 762, 913, 985, 1102, 1423] already exist in these well-established fields. Some attention is also given to another important ingredient in *ab initio* molecular dynamics, the choice of the basis set.

As for the depth of the presentation, the focus in Part I is clearly on the implementation of the basic *ab initio* molecular dynamics schemes in terms of the powerful and widely used plane wave/pseudopotential formulation of Kohn–Sham density functional theory outlined in Chapter 3. The explicit formulae for the energies, forces, stress, pseudopotentials, boundary conditions, optimization procedures, etc. are noted for this choice of method to solve the electronic structure problem, making particular reference to the CPMD software package [696]. One should, however, keep in mind that an *increasing* number of other powerful codes able to perform *ab initio* molecular dynamics simulations are available today (for instance ABINIT [2], CASTEP [234], CONQUEST [282], CP2k [287], CP-PAW [288], DACAPO [303], FHI98md [421], NWChem [1069], ONETEP [1085], PINY [1153], PWscf [1172], SIESTA [1343], S/PHI/nX [1377], or VASP [1559] amongst others), which are partly based on very similar techniques. An important ingredient in any plane wave-based technique is the usage of pseudopotentials to represent the core electrons, therefore enabling them not to be considered explicitly. Thus, Chapter 4 of Part I introduces the norm-conserving pseudopotentials up to the point of providing an overview about the different generation schemes and functional forms that are commonly used.

In Part II devoted to "Advanced techniques", the standard *ab initio* molecular dynamics approach as outlined in Part I is extended and generalized in various directions. In Chapter 5, ensembles other than the microcanonical one are introduced and explained along with powerful techniques used to deal with large energetic barriers and rare events, and methods to treat other electronic states than the ground state, such as time-dependent density functional theory in both the frequency and time domains. The approximation of using classical nuclei is lifted by virtue of the path integral formulation of quantum statistical mechanics, including a discussion of how to approximately correct classical time-correlation functions for quantum effects. Various techniques that allow us to represent only part of the entire system in terms of an electronic structure treatment, the hybrid, quantum/classical, or "QM/MM" molecular dynamics simulation methods, are outlined, including continuum solvation models. Subsequently, advanced pseudopotential concepts such as Vanderbilt's ultrasoft pseudopotentials and Blöchl's projector augmented-wave (PAW) transformation are introduced in Chapter 6.

Modern techniques to calculate properties directly from the available electronic structure information in *ab initio* molecular dynamics, such as infrared, Raman or NMR spectra, and methods to decompose and analyze the electronic structure including its dynamical changes are discussed in Chapter 7. Last but not least, the increasingly important aspect of writing highly efficient parallel computer codes within the framework of *ab initio* molecular dynamics, which take as much advantage as possible of the parallel platforms currently available and of those in the foreseeable future, is the focus of the last section in Part II, Chapter 8.

Finally, Part III is devoted to the wealth of problems that can be addressed using state-of-the-art *ab initio* molecular dynamics techniques by referring to an extensive set of references. The problems treated are briefly outlined with respect to the broad variety of systems in Chapter 9 and to specific properties in Chapter 10. The book closes with a short outlook in Chapter 11. In addition to this printed version of the book corrections and additions will be provided at `www.theochem.rub.de/go/aimd-book.html` in an open access mode.

Part I

Basic techniques

2

Getting started: unifying molecular dynamics and electronic structure

2.1 Deriving classical molecular dynamics

The starting point of all that follows is non-relativistic quantum mechanics as formalized via the time-dependent Schrödinger equation

$$i\hbar\frac{\partial}{\partial t}\Phi(\{\mathbf{r}_i\}, \{\mathbf{R}_I\}; t) = \mathcal{H}\Phi(\{\mathbf{r}_i\}, \{\mathbf{R}_I\}; t) \qquad (2.1)$$

in its position representation in conjunction with the standard Hamiltonian

$$\mathcal{H} = -\sum_I \frac{\hbar^2}{2M_I}\nabla_I^2 - \sum_i \frac{\hbar^2}{2m_{\mathrm{e}}}\nabla_i^2$$

$$+ \frac{1}{4\pi\varepsilon_0}\sum_{i<j}\frac{e^2}{|\mathbf{r}_i - \mathbf{r}_j|} - \frac{1}{4\pi\varepsilon_0}\sum_{I,i}\frac{e^2 Z_I}{|\mathbf{R}_I - \mathbf{r}_i|} + \frac{1}{4\pi\varepsilon_0}\sum_{I<J}\frac{e^2 Z_I Z_J}{|\mathbf{R}_I - \mathbf{R}_J|}$$

$$= -\sum_I \frac{\hbar^2}{2M_I}\nabla_I^2 - \sum_i \frac{\hbar^2}{2m_{\mathrm{e}}}\nabla_i^2 + V_{\mathrm{n-e}}(\{\mathbf{r}_i\}, \{\mathbf{R}_I\})$$

$$= -\sum_I \frac{\hbar^2}{2M_I}\nabla_I^2 + \mathcal{H}_{\mathrm{e}}(\{\mathbf{r}_i\}, \{\mathbf{R}_I\}) \qquad (2.2)$$

for the electronic $\{\mathbf{r}_i\}$ and nuclear $\{\mathbf{R}_I\}$ degrees of freedom. Thus, only the bare electron-electron, electron-nuclear, and nuclear-nuclear Coulomb interactions are taken into account. Here, M_I and Z_I are mass and atomic number of the Ith nucleus, the electron mass and charge are denoted by m_{e} and $-e$, and ε_0 is the vacuum permittivity. In order to keep the current derivation as transparent as possible, the more convenient atomic units (a.u.) will be introduced only at a later stage.

The goal of this section is to derive molecular dynamics of classical point particles [25, 468, 577, 1189], that is essentially classical mechanics, starting

11

from Schrödinger's quantum-mechanical wave equation Eq. (2.1) for both electrons and nuclei. As an intermediate step to molecular dynamics based on force fields, two variants of *ab initio* molecular dynamics are derived in passing. To achieve this, two complementary derivations will be presented, both of which are not considered to constitute rigorous derivations in the spirit of mathematical physics. In the first, more traditional route [355] the starting point is to consider the electronic part of the Hamiltonian for fixed nuclei, i.e. the clamped-nuclei part $\mathcal{H}_e$ of the full Hamiltonian, Eq. (2.2). Next, it is supposed that the exact solution of the corresponding *time-independent* (stationary) electronic Schrödinger equation,

$$\mathcal{H}_e(\{\mathbf{r}_i\}; \{\mathbf{R}_I\})\Psi_k = E_k(\{\mathbf{R}_I\})\Psi_k(\{\mathbf{r}_i\}; \{\mathbf{R}_I\}) \ , \qquad (2.3)$$

is known for clamped nuclei at positions $\{\mathbf{R}_I\}$. Here, the spectrum of $\mathcal{H}_e$ is assumed to be discrete and the eigenfunctions to be orthonormalized

$$\int \Psi_k^{\star}(\{\mathbf{r}_i\}; \{\mathbf{R}_I\})\Psi_l(\{\mathbf{r}_i\}; \{\mathbf{R}_I\}) \, d\mathbf{r} = \delta_{kl} \qquad (2.4)$$

at all possible positions of the nuclei; $\int \cdots d\mathbf{r}$ refers to integration over all $i = 1, \ldots$ variables $\mathbf{r} = \{\mathbf{r}_i\}$. Knowing all these adiabatic eigenfunctions at all possible nuclear configurations, the total wave function in Eq. (2.1) can be expanded

$$\Phi(\{\mathbf{r}_i\}, \{\mathbf{R}_I\}; t) = \sum_{l=0}^{\infty} \Psi_l(\{\mathbf{r}_i\}; \{\mathbf{R}_I\})\chi_l(\{\mathbf{R}_I\}; t) \qquad (2.5)$$

in terms of the complete set of eigenfunctions $\{\Psi_l\}$ of $\mathcal{H}_e$ where the nuclear wave functions $\{\chi_l\}$ can be viewed to be time-dependent expansion coefficients. This is an *ansatz* of the total wave function, introduced by Born in 1951 [179, 811] for the time-independent problem, in order to separate systematically the light electrons from the heavy nuclei [180, 771, 811] by invoking a hierarchical viewpoint.[1]

Insertion of this ansatz Eq. (2.5) into the *time-dependent* coupled Schrödinger equation Eq. (2.1) followed by multiplication from the left by

---

[1] "The terms of the molecular spectra comprise, as is known, contributions of varying orders of magnitude; the largest contribution originates from the electron movement around the nuclei, there then follows a contribution stemming from the nuclear vibrations, and, ultimately, the contribution arising from the nuclear rotation. The justification of the existence of such a hierarchy emanates from the magnitude of the mass of the nuclei, compared to that of the electrons." Translated by the authors from "Die Terme der Molekelspektren setzen sich bekanntlich aus Anteilen verschiedener Größenordnung zusammen; der größte Beitrag rührt von der Elektronenbewegung um die Kerne her, dann folgt ein Beitrag der Kernschwingungen, endlich die von den Kernrotationen erzeugten Anteile. Der Grund für die Möglichkeit einer solchen Ordnung liegt offensichtlich in der Größe der Masse der Kerne, verglichen mit der der Elektronen." Cited from the Introduction of the seminal paper [180] by Born and Oppenheimer from 1927.

$\Psi_k^\star(\{\mathbf{r}_i\}; \{\mathbf{R}_I\})$ and integration over all electronic coordinates $\mathbf{r}$ leads to a set of coupled differential equations

$$\left[-\sum_I \frac{\hbar^2}{2M_I} \nabla_I^2 + E_k(\{\mathbf{R}_I\}) \right] \chi_k + \sum_l C_{kl} \chi_l = i\hbar \frac{\partial}{\partial t} \chi_k \qquad (2.6)$$

where

$$C_{kl} = \int \Psi_k^\star \left[-\sum_I \frac{\hbar^2}{2M_I} \nabla_I^2 \right] \Psi_l \, d\mathbf{r}$$

$$+ \frac{1}{M_I} \sum_I \left\{ \int \Psi_k^\star \left[-i\hbar \nabla_I \right] \Psi_l \, d\mathbf{r} \right\} \left[-i\hbar \nabla_I \right] \qquad (2.7)$$

is the exact nonadiabatic coupling operator. The first term is a matrix element of the kinetic energy operator of the nuclei, whereas the second term depends on their momenta.

The diagonal contribution C_{kk} depends only on a single adiabatic wave function Ψ_k and as such represents a correction to the adiabatic eigenvalue E_k of the electronic Schrödinger equation Eq. (2.3) in this kth state. As a result, the "adiabatic approximation" to the fully nonadiabatic problem Eq. (2.6) is obtained by considering only these diagonal terms,

$$C_{kk} = -\sum_I \frac{\hbar^2}{2M_I} \int \Psi_k^\star \, \nabla_I^2 \, \Psi_k \, d\mathbf{r} \ , \qquad (2.8)$$

the second term of Eq. (2.7) being zero when the electronic wave function is real, which leads to complete decoupling

$$\left[-\sum_I \frac{\hbar^2}{2M_I} \nabla_I^2 + E_k(\{\mathbf{R}_I\}) + C_{kk}(\{\mathbf{R}_I\}) \right] \chi_k = i\hbar \frac{\partial}{\partial t} \chi_k \qquad (2.9)$$

of the fully coupled original set of differential equations Eq. (2.6). This, in turn, implies that the motion of the nuclei proceeds without changing the quantum state, k, of the electronic subsystem during time evolution. Correspondingly, the coupled wave function in Eq. (2.1) can be decoupled simply

$$\Phi(\{\mathbf{r}_i\}, \{\mathbf{R}_I\}; t) \approx \Psi_k(\{\mathbf{r}_i\}; \{\mathbf{R}_I\}) \chi_k(\{\mathbf{R}_I\}; t) \qquad (2.10)$$

into a direct product of an electronic and a nuclear wave function. Note that this amounts to taking into account only a single term in the general expansion Eq. (2.5).

The ultimate simplification consists in neglecting also the diagonal coupling terms

$$\left[-\sum_I \frac{\hbar^2}{2M_I} \nabla_I^2 + E_k(\{\mathbf{R}_I\}) \right] \chi_k = i\hbar \frac{\partial}{\partial t} \chi_k \ , \tag{2.11}$$

which defines the famous "Born–Oppenheimer approximation". Thus, both the adiabatic approximation and the Born–Oppenheimer approximation (introduced in Ref. [180] using a cumbersome perturbation expansion in powers of the mass ratio $(m_e/M_I)^{1/4}$, see also § 14 and Appendix VII in Ref. [179]) are readily derived as special cases based on the particular functional ansatz Eq. (2.5) of the total wave function. In the above simplified presentation subtleties due to Berry's geometric phase [1329] have been ignored, but the interested reader is referred to excellent reviews [168, 986, 1642] that cover this general phenomenon with a focus on molecular systems.

The next step in the derivation of molecular dynamics is the task of approximating the nuclei as classical point particles. How can this be achieved in the framework where a full quantum-mechanical wave equation, χ_k, describes the motion of all nuclei in a selected electronic state Ψ_k? In order to proceed, it is first noted that for a great number of physical situations the Born–Oppenheimer approximation can safely be applied, but see Section 5.3 for a discussion of cases where this is not the case. Based on this assumption, the following derivation will be built on Eq. (2.11) being the Born–Oppenheimer approximation to the fully coupled solution, Eq. (2.6). Secondly, a well-known route to extract semiclassical mechanics from quantum mechanics in general starts with rewriting the corresponding wave function

$$\chi_k(\{\mathbf{R}_I\}; t) = A_k(\{\mathbf{R}_I\}; t) \ \exp\left[iS_k(\{\mathbf{R}_I\}; t)/\hbar \right] \tag{2.12}$$

in terms of an amplitude factor A_k and a phase S_k which are both considered to be real and $A_k > 0$ in this polar representation, see for instance Refs. [345, 996, 1268]. After transforming the nuclear wave function in Eq. (2.11) for a chosen electronic state k accordingly and after separating the real and imaginary parts, the equations for the nuclei

$$\frac{\partial S_k}{\partial t} + \sum_I \frac{1}{2M_I} (\nabla_I S_k)^2 + E_k = \hbar^2 \sum_I \frac{1}{2M_I} \frac{\nabla_I^2 A_k}{A_k} \tag{2.13}$$

$$\frac{\partial A_k}{\partial t} + \sum_I \frac{1}{M_I} (\nabla_I A_k)(\nabla_I S_k) + \sum_I \frac{1}{2M_I} A_k (\nabla_I^2 S_k) = 0 \tag{2.14}$$

are re-expressed (exactly) in terms of the new variables S_k and A_k instead of using $\text{Re}\chi_k$ and $\text{Im}\chi_k$. It is noted in passing that this quantum fluid dynamic (or hydrodynamic, Bohmian) representation [169, 1636], Eqs. (2.13)-(2.14), can actually be used to solve the time-dependent Schrödinger equation [340, 878].

The relation for the amplitude, Eq. (2.14), may be rewritten after multiplying by $2A_k$ from the left as a continuity equation [345, 996, 1268]

$$\frac{\partial A_k^2}{\partial t} + \sum_I \frac{1}{M_I} \nabla_I \left(A_k^2 \nabla_I S_k \right) = 0 \qquad (2.15)$$

$$\frac{\partial \rho_k}{\partial t} + \sum_I \nabla_I \mathbf{J}_{k,I} = 0 \qquad (2.16)$$

with the help of the identification of the nuclear probability density $\rho_k = |\chi_k|^2 \equiv A_k^2$, obtained directly from the definition Eq. (2.12), and with the associated current density defined as $\mathbf{J}_{k,I} = A_k^2 (\nabla_I S_k)/M_I$. This continuity equation Eq. (2.16) is independent of $\hbar$ and ensures locally the conservation of the particle probability density $|\chi_k|^2$ of the nuclei in the presence of a flux.

More important for the present purpose is a detailed discussion of the relation for the phase S_k, Eq. (2.13), of the nuclear wave function that is associated with the kth electronic state. This equation contains one term that depends explicitly on $\hbar$, a contribution that vanishes

$$\frac{\partial S_k}{\partial t} + \sum_I \frac{1}{2M_I} \left(\nabla_I S_k \right)^2 + E_k = 0 \qquad (2.17)$$

if the classical limit is taken as $\hbar \to 0$. Note that a systematic expansion in terms of $\hbar$ would, instead, lead to a hierarchy of semiclassical methods [562, 996]. The resulting equation Eq. (2.17) is now isomorphic to the equation of motion in the Hamilton–Jacobi formulation [528, 1282] of classical mechanics

$$\frac{\partial S_k}{\partial t} + H_k \left(\{ \mathbf{R}_I \}, \{ \nabla_I S_k \} \right) = 0 \qquad (2.18)$$

with the classical Hamilton function

$$H_k(\{ \mathbf{R}_I \}, \{ \mathbf{P}_I \}) = T(\{ \mathbf{P}_I \}) + V_k(\{ \mathbf{R}_I \}) \qquad (2.19)$$

for a given conserved energy $dE_k^{\text{tot}}/dt = 0$ and hence

$$\frac{\partial S_k}{\partial t} = -\left(T + E_k \right) = -E_k^{\text{tot}} = \text{const.} \qquad (2.20)$$

defined in terms of (generalized) coordinates $\{\mathbf{R}_I\}$ and their conjugate canonical momenta $\{\mathbf{P}_I\}$. With the help of the connecting transformation

$$\mathbf{P}_I \equiv \nabla_I S_k \quad \left[= M_I \frac{\mathbf{J}_{k,I}}{\rho_k} \right] \tag{2.21}$$

the Newtonian equations of motion, $\dot{\mathbf{P}}_I = -\nabla_I V_k(\{\mathbf{R}_I\})$, corresponding to the Hamilton–Jacobi form Eq. (2.17) can be read off

$$\frac{d\mathbf{P}_I}{dt} = -\nabla_I E_k \qquad \text{or}$$

$$M_I \ddot{\mathbf{R}}_I(t) = -\nabla_I V_k^{\text{BO}}(\{\mathbf{R}_I(t)\}) \tag{2.22}$$

separately for each decoupled electronic state k. Thus, the nuclei move according to *classical* mechanics in an effective potential, V_k^{BO}, which is given by the Born–Oppenheimer potential energy surface E_k obtained by solving simultaneously the *time-independent* electronic Schrödinger equation for the kth state, Eq. (2.3), at the given nuclear configuration $\{\mathbf{R}_I(t)\}$. In other words, this time-local many-body interaction potential due to the quantum electrons is a function of the set of all classical nuclear positions at time t. Since the Born–Oppenheimer total energies in a specific adiabatic electronic state yield directly the forces used in this variant of *ab initio* molecular dynamics, this particular approach is often called "Born–Oppenheimer molecular dynamics", to be discussed in more detail later in Section 2.3.

In order to present an alternative derivation, which does maintain a quantum-mechanical *time evolution* of the electrons and thus does not invoke solving the *time-independent* electronic Schrödinger equation Eq. (2.3) as before, the elegant route taken in Refs. [1516, 1517] is followed; see also Ref. [943]. To this end, the nuclear and electronic contributions to the total wave function $\Phi(\{\mathbf{r}_i\}, \{\mathbf{R}_I\}; t)$ are separated directly such that, ultimately, the classical limit can be imposed for the nuclei only. The simplest possible form is a product ansatz

$$\Phi(\{\mathbf{r}_i\}, \{\mathbf{R}_I\}; t) \approx \Psi(\{\mathbf{r}_i\}; t)\, \chi(\{\mathbf{R}_I\}; t)\, \exp\left[\frac{i}{\hbar} \int_{t_0}^{t} \tilde{E}_e(t')\, dt' \right] , \tag{2.23}$$

where the nuclear and electronic wave functions are separately normalized to unity at every instant of time, i.e. $\langle \chi; t | \chi; t \rangle = 1$ and $\langle \Psi; t | \Psi; t \rangle = 1$, respectively. In addition, a phase factor

$$\tilde{E}_e = \int \Psi^\star(\{\mathbf{r}_i\}; t)\, \chi^\star(\{\mathbf{R}_I\}; t)\, \mathcal{H}_e\, \Psi(\{\mathbf{r}_i\}; t)\, \chi(\{\mathbf{R}_I\}; t)\, d\mathbf{r} d\mathbf{R} \tag{2.24}$$

was introduced at this stage for convenience such that the final equations will look simpler; again $\int \cdots d\mathbf{r}d\mathbf{R}$ refers to the integration over all $i = 1, \ldots$ and $I = 1, \ldots$ electronic and nuclear variables $\mathbf{r} = \{\mathbf{r}_i\}$ and $\mathbf{R} = \{\mathbf{R}_I\}$, respectively. It is mentioned in passing that this approximation is called a one-determinant or single-configuration ansatz for the *total* wave function, which at the end must lead to a mean-field description of the coupled dynamics. Note in addition that this product ansatz differs, independently from the issue of phase factor, from Born's ansatz Eq. (2.5) used above in terms of *adiabatic* electronic states $\{\Psi_k\}$ even if only a single electronic state k is considered according to Eq. (2.10).

Inserting this particular separation ansatz Eq. (2.23) into Eqs. (2.1)–(2.2) yields (after multiplying from the left by $\Psi^\star$ and $\chi^\star$, integrating over nuclear and electronic coordinates, respectively, and imposing conservation $d\langle\mathcal{H}\rangle/dt \equiv 0$ of the *total* energy) the following relations

$$i\hbar\frac{\partial\Psi}{\partial t} = -\sum_i \frac{\hbar^2}{2m_\mathrm{e}}\nabla_i^2\Psi$$

$$+ \left\{\int \chi^\star(\{\mathbf{R}_I\};t)V_{\mathrm{n-e}}(\{\mathbf{r}_i\},\{\mathbf{R}_I\})\chi(\{\mathbf{R}_I\};t)\,d\mathbf{R}\right\}\Psi \quad (2.25)$$

$$i\hbar\frac{\partial\chi}{\partial t} = -\sum_I \frac{\hbar^2}{2M_I}\nabla_I^2\chi$$

$$+ \left\{\int \Psi^\star(\{\mathbf{r}_i\};t)\mathcal{H}_\mathrm{e}(\{\mathbf{r}_i\},\{\mathbf{R}_I\})\Psi(\{\mathbf{r}_i\};t)\,d\mathbf{r}\right\}\chi . \quad (2.26)$$

This set of coupled time-dependent Schrödinger equations defines the basis of the time-dependent self-consistent field (TDSCF) method introduced as early as 1930 by Dirac [344], see also Ref. [338]. Both electrons and nuclei move quantum-mechanically in *time-dependent* effective potentials, i.e. self-consistently obtained average fields, given by the expressions in the braces. These potentials are obtained from appropriate averages (defined as quantum-mechanical expectation values $\langle\ldots\rangle$) over the other class of degrees of freedom by using the nuclear and electronic wave functions, respectively. Thus, the single-determinant ansatz Eq. (2.23) produces, as already anticipated, a mean-field description of the coupled nuclear–electronic *quantum dynamics*. This is the price to pay for the simplest possible separation of electronic and nuclear variables in terms of *dynamics*.

At this stage the nuclei must again be approximated as classical point particles, however this time in the presence of electrons which do move quantum-mechanically in time according to Eq. (2.25). Invoking the same trick as before, see Eq. (2.12), but using χ as defined in Eq. (2.23) instead, one now arrives at the TDSCF equations

$$\frac{\partial S}{\partial t} + \sum_I \frac{1}{2M_I} \left(\nabla_I S\right)^2 + \int \Psi^\star \mathcal{H}_e \Psi \, d\mathbf{r} = \hbar^2 \sum_I \frac{1}{2M_I} \frac{\nabla_I^2 A}{A} \qquad (2.27)$$

$$\frac{\partial A}{\partial t} + \sum_I \frac{1}{M_I} \left(\nabla_I A\right)\left(\nabla_I S\right) + \sum_I \frac{1}{2M_I} A \left(\nabla_I^2 S\right) = 0 \qquad (2.28)$$

in terms of phase and modulus of the nuclear wave function and thus at

$$\frac{\partial S}{\partial t} + \sum_I \frac{1}{2M_I} \left(\nabla_I S\right)^2 + \int \Psi^\star \mathcal{H}_e \Psi \, d\mathbf{r} = 0 \qquad (2.29)$$

if the classical limit is taken again as $\hbar \to 0$. Correspondingly, the Newtonian equations of motion of the classical nuclei equivalent to Eq. (2.29) are given by

$$\frac{d\mathbf{P}_I}{dt} = -\nabla_I \int \Psi^\star \mathcal{H}_e \Psi \, d\mathbf{r} \qquad \text{or}$$

$$M_I \ddot{\mathbf{R}}_I(t) = -\nabla_I V_e^{\mathrm{E}}(\{\mathbf{R}_I(t)\}) \ . \qquad (2.30)$$

Here, the nuclei move according to *classical* mechanics in an effective potential, V_e^{E}, often called the Ehrenfest potential, which is given by the quantum dynamics of the electrons obtained by solving simultaneously the *time-dependent* electronic Schrödinger equation Eq. (2.25). By virtue of its definition, it is clearly seen that this time-local many-body interaction potential due to the explicit time evolution of the quantum electrons stems from averaging the electronic Hamiltonian with respect to the electronic degrees of freedom, $V_e^{\mathrm{E}} = \langle \Psi | \mathcal{H}_e | \Psi \rangle$.

However, the very TDSCF equation that describes the time evolution of the electrons, Eq. (2.25), still contains the full quantum-mechanical nuclear wave function $\chi(\{\mathbf{R}_I\}; t)$ instead of just the classical-mechanical nuclear positions $\{\mathbf{R}_I(t)\}$. In this case the classical reduction can be achieved simply by replacing the nuclear density $|\chi(\{\mathbf{R}_I\}; t)|^2$ in Eq. (2.25) in the limit $\hbar \to 0$ by a product of delta functions $\prod_I \delta(\mathbf{R}_I - \mathbf{R}_I(t))$ centered at the instantaneous positions $\{\mathbf{R}_I(t)\}$ of the classical nuclei as given by Eq. (2.30). This naive approach yields, e.g. for the position operator,

$$\int \chi^\star(\{\mathbf{R}_I\}; t) \, \mathbf{R}_I \, \chi(\{\mathbf{R}_I\}; t) \, d\mathbf{R} \xrightarrow{\hbar \to 0} \mathbf{R}_I(t) \qquad (2.31)$$

the required expectation value. This classical limit leads to a time-dependent wave equation for the electrons

$$ i\hbar \frac{\partial \Psi}{\partial t} = -\sum_i \frac{\hbar^2}{2m_e} \nabla_i^2 \Psi + V_{n-e}(\{\mathbf{r}_i\}, \{\mathbf{R}_I(t)\})\Psi $$

$$ = \mathcal{H}_e(\{\mathbf{r}_i\}, \{\mathbf{R}_I(t)\}) \ \Psi(\{\mathbf{r}_i\}, \{\mathbf{R}_I\}; t) \tag{2.32} $$

which evolves self-consistently as the classical nuclei are propagated via Eq. (2.30). Note that now $\mathcal{H}_e$ depends *parametrically* on the classical nuclear *positions* $\{\mathbf{R}_I(t)\}$ at time t through $V_{n-e}(\{\mathbf{r}_i\}, \{\mathbf{R}_I(t)\})$. This means that feedback between the classical and quantum degrees of freedom is incorporated in both directions, although in a mean-field sense only. This is at variance with the much simpler "classical path" or Mott non-SCF approach to dynamics [1516, 1517] where the quantum subsystem moves along a predefined classical trajectory.

The approach to *ab initio* molecular dynamics that relies on solving Newton's equation for the nuclei, Eq. (2.30), simultaneously with Schrödinger's equation for the electrons, Eq. (2.32), is often called "Ehrenfest molecular dynamics" in honor of Paul Ehrenfest who was the first to address the essential question[2] of how Newtonian classical dynamics of point particles can be derived from Schrödinger's time-dependent wave equation [382]. In the present case this leads to a hybrid or mixed quantum–classical approach because only the nuclei are forced to behave like classical particles, whereas the electrons are still treated as quantum objects. At variance with the Born–Oppenheimer molecular dynamics approach relying on solving a stationary Schrödinger equation, however, the electronic subsystem evolves explicitly in time according to a time-dependent Schrödinger equation, see Section 2.2 for more detail.

Although the TDSCF approach underlying Ehrenfest molecular dynamics is clearly a mean-field theory concerning the dynamical evolution, transitions between electronic states are included in this scheme at variance with the Born–Oppenheimer molecular dynamics technique. This can be made transparent by expanding the *electronic* wave function Ψ in Eq. (2.23) (as opposed to the *total* wave function Φ according to Eq. (2.5)) in a basis of

---

[2] The opening statement of P. Ehrenfest's famous 1927 paper [382] reads: "It is desirable, to be able to answer the following question as elementarily as possible: which retrospective view results when considering Newton's fundamental equations of classical mechanics from the standpoint of quantum mechanics?" Translated by the authors from "Es ist wünschenswert, die folgende Frage möglichst elementar beantworten zu können: Welcher Rückblick ergibt sich vom Standpunkt der Quantenmechanik auf die Newtonschen Grundgleichungen der klassischen Mechanik?"

electronic states Ψ_l

$$\Psi(\{\mathbf{r}_i\}, \{\mathbf{R}_I\}; t) = \sum_{l=0}^{\infty} c_l(t) \Psi_l(\{\mathbf{r}_i\}; \{\mathbf{R}_I\}) \qquad (2.33)$$

with complex time-dependent coefficients $\{c_l(t)\}$. In this case, the coefficients $|c_l(t)|^2$ satisfying $\sum_l |c_l(t)|^2 \equiv 1$ describe explicitly the time evolution of the populations (occupations) of the different states l whereas the necessary interferences between any two such states are included via the off-diagonal term, $c_k^\star c_{l\neq k}$. One possible choice for the basis functions $\{\Psi_k\}$ is the instantaneous adiabatic basis obtained from solving the time-independent electronic Schrödinger equation

$$\mathcal{H}_e(\{\mathbf{r}_i\}; \{\mathbf{R}_I\}) \Psi_k = E_k(\{\mathbf{R}_I\}) \Psi_k(\{\mathbf{r}_i\}; \{\mathbf{R}_I\}) \ , \qquad (2.34)$$

where $\{\mathbf{R}_I\}$ are the instantaneous nuclear positions at time t that are determined according to Eq. (2.30). The actual equations of motion in terms of expansion coefficients $\{c_k\}$, adiabatic energies $\{E_k\}$, and nonadiabatic couplings are presented in Section 2.2.

Here, instead, a further simplification is invoked in order to reduce Ehrenfest molecular dynamics to Born–Oppenheimer molecular dynamics. To achieve this, the electronic wave function Ψ is restricted to be the ground state adiabatic wave function Ψ_0 of $\mathcal{H}_e$ at each instant of time according to Eq. (2.34), which implies $|c_0(t)|^2 \equiv 1$ and thus a single term in the expansion Eq. (2.33). This should be a good approximation if the energy difference between Ψ_0 and the first excited state Ψ_1 is large everywhere compared to the thermal energy $k_B T$, roughly speaking. In this limit the nuclei move according to Eq. (2.30)

$$V_e^E = \int \Psi_0^\star \mathcal{H}_e \Psi_0 \, d\mathbf{r} \equiv E_0(\{\mathbf{R}_I\}) \qquad (2.35)$$

on a single adiabatic potential energy surface. This is nothing else than the ground state Born–Oppenheimer potential energy surface that is obtained by solving the time-*independent* electronic Schrödinger equation Eq. (2.34) for $k = 0$ at each nuclear configuration $\{\mathbf{R}_I\}$ generated during molecular dynamics. This leads to the identification $V_e^E \equiv E_0 = V_0^{BO}$ and thus to Eq. (2.22), i.e. in this limit the Ehrenfest potential is identical to the ground state Born–Oppenheimer (or "clamped nuclei") potential.

As a consequence of this observation, it is conceivable to fully decouple the task of generating the classical nuclear dynamics from the task of computing the quantum potential energy surface. In a first step, the global potential energy surface E_0, which depends on *all* nuclear degrees of freedom $\{\mathbf{R}_I\}$, is computed for many different nuclear configurations by solving the stationary Schrödinger equation separately for all these situations. In a second step, these data points are fitted to an analytical functional form to yield a global potential energy surface [1280], from which the gradients can be obtained analytically. In a third step, the Newtonian equations of motion are solved on this surface for many different initial conditions, producing a "swarm" of classical trajectories $\{\mathbf{R}_I(t)\}$. This is, in a nutshell, the basis of *classical trajectory calculations* on global potential energy surfaces as used very successfully to understand scattering and chemical reaction dynamics of small systems in vacuum [1284, 1514].

As already explained in the general introduction, Chapter 1, such approaches suffer severely from the "dimensionality bottleneck" as the number of active nuclear degrees of freedom increases. One traditional way out of this dilemma, making possible calculations of large systems, is to approximate the global potential energy surface

$$V_{\mathrm{e}}^{\mathrm{E}} \approx V_{\mathrm{e}}^{\mathrm{FF}}(\{\mathbf{R}_I\}) = \sum_{I=1}^{N} v_1(\mathbf{R}_I) + \sum_{I<J}^{N} v_2(\mathbf{R}_I, \mathbf{R}_J)$$

$$+ \sum_{I<J<K}^{N} v_3(\mathbf{R}_I, \mathbf{R}_J, \mathbf{R}_K) + \cdots \quad (2.36)$$

in terms of a truncated expansion of many-body contributions [25, 550, 577, 1405], which is sometimes called a "force field". At this stage, the electronic degrees of freedom are replaced approximately by a set of interaction potentials $\{v_n\}$ and are no longer included as explicit degrees of freedom when the nuclei are propagated. Thus, the mixed quantum/classical problem is reduced to purely classical mechanics, once the $\{v_n\}$ are determined. "Standard" molecular dynamics

$$M_I \ddot{\mathbf{R}}_I(t) = -\nabla_I V_{\mathrm{e}}^{\mathrm{FF}}(\{\mathbf{R}_I(t)\}) \ , \quad (2.37)$$

relies crucially on such a force field idea as opposed to *ab initio molecular dynamics*, either according to Born–Oppenheimer or Ehrenfest flavor, where the key feature is that the potential and thus the forces are obtained from solving the electronic structure problem concurrently to generating the trajectory $\{\mathbf{R}_I(t)\}$.

In force field-based molecular dynamics, typically only two-body v_2 or three-body v_3 interactions are taken into account [25, 468, 577, 1189], although very sophisticated models exist that include nonadditive many-body interactions. This amounts to a dramatic simplification and removes, in particular, the dimensionality bottleneck since the global potential surface is reconstructed from a manageable sum of additive few-body contributions. The flipside of the medal is the introduction of the drastic approximation embodied in Eq. (2.36), which basically excludes the study of chemical reactions from the realm of computer simulation.

As a result of the derivation presented above, the essential assumptions underlying standard force field-based molecular dynamics become very transparent. The electrons follow adiabatically the classical nuclear motion and can be integrated out so that the nuclei evolve on a single Born–Oppenheimer potential energy surface (typically, but not necessarily, given by the electronic ground state), which is generally approximated in terms of few-body interactions.

Actually, force field-based molecular dynamics for *many*-body systems is only made possible by somehow decomposing the global potential energy. In order to illustrate this point, consider the simulation of $N = 500$ argon atoms in the liquid phase [395] where the interactions can be described faithfully by additive two-body terms, i.e. $V_e^{FF}(\{\mathbf{R}_I\}) \approx \sum_{I<J}^{N} v_2(|\mathbf{R}_I - \mathbf{R}_J|)$. Thus, the determination of the pair potential v_2 from *ab initio* electronic structure calculations consists in computing and fitting a one-dimensional function only. The corresponding task of determining a global potential energy surface, however, amounts to doing that in about 10^{1500} dimensions, which is simply impossible. In the case of neat argon this is obviously not necessary, but this assessment changes drastically if the 500 atoms are not all identical and, for instance, are those that form a small enzyme catalyzing a particular biochemical reaction.

2.2 Ehrenfest molecular dynamics

A systematic and general way out of the dimensionality bottleneck, other than to approximate the global potential energy surface Eq. (2.36) or reduce the number of active degrees of freedom, is to take seriously the classical nuclei approximation to the TDSCF equations, Eqs. (2.30) and (2.32). This implies computing the Ehrenfest force by actually solving numerically

$$M_I \ddot{\mathbf{R}}_I(t) = -\nabla_I \int \Psi^\star \mathcal{H}_e \Psi \, d\mathbf{r}$$

$$= -\nabla_I \langle \mathcal{H}_e \rangle \tag{2.38}$$

$$ih\frac{\partial\Psi}{\partial t} = \left[-\sum_i \frac{\hbar^2}{2m_e}\nabla_i^2 + V_{n-e}(\{\mathbf{r}_i\}, \{\mathbf{R}_I(t)\})\right]\Psi$$

$$= \mathcal{H}_e\Psi \tag{2.39}$$

the coupled set of quantum/classical equations simultaneously. Thereby, the *a priori* construction of any type of potential energy surface is avoided from the outset by solving the time-*dependent* electronic Schrödinger equation "on-the-fly" as the nuclei are propagated using classical mechanics. This allows one to compute the force from $-\nabla_I\langle\mathcal{H}_e\rangle$ for each configuration $\{\mathbf{R}_I(t)\}$ generated by molecular dynamics; see Section 2.5 for the issue of using the Hellmann–Feynman forces, i.e. $-\langle\Psi|\nabla_I\mathcal{H}_e|\Psi\rangle$ instead of $-\nabla_I\langle\Psi|\mathcal{H}_e|\Psi\rangle$.

The equations of motion corresponding to Eqs. (2.38)-(2.39) can be expressed conveniently by representing the electronic wave function Ψ in terms of the instantaneous adiabatic electronic states Eq. (2.34) and the time-dependent expansion coefficients Eq. (2.33). This particular representation of Ehrenfest molecular dynamics reads [355, 1516, 1517]

$$M_I\ddot{\mathbf{R}}_I(t) = -\nabla_I\sum_k |c_k(t)|^2 E_k \tag{2.40}$$

$$= -\sum_k |c_k(t)|^2\nabla_I E_k + \sum_{k,l} c_k^\star c_l \left(E_k - E_l\right)\mathbf{d}_I^{kl} \tag{2.41}$$

$$i\hbar\dot{c}_k(t) = c_k(t)E_k - i\hbar\sum_l c_l(t)D^{kl} \ , \tag{2.42}$$

where the nonadiabatic coupling elements are given by

$$D^{kl} = \int \Psi_k^\star\frac{\partial}{\partial t}\Psi_l \, d\mathbf{r} = \sum_I \dot{\mathbf{R}}_I \int \Psi_k^\star\nabla_I\Psi_l \, d\mathbf{r} = \sum_I \dot{\mathbf{R}}_I\mathbf{d}_I^{kl} \tag{2.43}$$

with the property $\mathbf{d}_I^{kk} \equiv \mathbf{0}$ of the nonadiabatic coupling vectors for real wave functions. The Ehrenfest approach to *ab initio* molecular dynamics is thus seen to include rigorously nonadiabatic transitions between different electronic states Ψ_k and Ψ_l within the framework of classical nuclear motion and the *mean-field* (TDSCF) approximation Eq. (2.23) to the coupled problem, see e.g. Refs. [355, 1516, 1517] for reviews and Section 5.3.5 for implementations in terms of time-dependent density functional theory.

The restriction to one electronic state in the expansion Eq. (2.33), which is in most cases the ground state Ψ_0, leads to

$$M_I\ddot{\mathbf{R}}_I(t) = -\nabla_I \langle \Psi_0 | \mathcal{H}_e | \Psi_0 \rangle \qquad (2.44)$$

$$i\hbar\frac{\partial\Psi_0}{\partial t} = \mathcal{H}_e\Psi_0 \qquad (2.45)$$

as a special case of Eqs. (2.38)-(2.39); note that $\mathcal{H}_e$ is time-dependent via the nuclear coordinates $\{\mathbf{R}_I(t)\}$. A point worth mentioning here is that the propagation of the wave function is unitary by virtue of the time-dependent Schrödinger equation, i.e. the wave function preserves its norm and the set of orbitals used to build up the wave function will stay orthonormal. The numerical consequences of this observation are discussed in Section 2.6.

Ehrenfest molecular dynamics is certainly the oldest approach to "on-the-fly" molecular dynamics, and is often used to study few-body collision- and scattering-type problems [329, 330, 1001, 1514]. In other words, the Ehrenfest approach to electron dynamics has traditionally not been in widespread use for systems with many degrees of freedom, which is the typical situation in the context of condensed matter problems. More recently, however, its use in conjunction with time-dependent density functional theory to describe the electronic subsystem gained a lot of attention, see Section 5.3.5 for an outline of these methods and, for instance, Refs. [77, 369, 370, 451, 1258, 1316, 1400, 1463, 1637, 1638] for some of the implementations, but there is also progress within the realm of Hartree–Fock-based electronic structure methods [862].

2.3 Born–Oppenheimer molecular dynamics

An alternative approach to include the electronic structure in molecular dynamics simulations consists in straightforwardly solving the *static* electronic structure problem in each molecular dynamics step given the set of *fixed* nuclear positions at that instant of time. Thus, the electronic structure part is reduced to solving a time-*independent* quantum problem, e.g. by solving the time-*independent*, stationary Schrödinger equation, concurrently to propagating the nuclei according to classical mechanics. This implies that the time dependence of the electronic structure is imposed and dictated by its parametric dependence on the classical dynamics of the nuclei which it just follows. Thus, it is *not an intrinsic dynamics* as in Ehrenfest molecular dynamics. The resulting Born–Oppenheimer molecular dynamics method can be written down readily and is defined by

$$M_I\ddot{\mathbf{R}}_I(t) = -\nabla_I \min_{\Psi_0} \left\{ \langle \Psi_0 \left| \mathcal{H}_e \right| \Psi_0 \rangle \right\} \qquad (2.46)$$

$$E_0 \Psi_0 = \mathcal{H}_e \Psi_0 \qquad (2.47)$$

for the electronic ground state, which is identical to the result Eq. (2.22) derived earlier in a systematic manner from the coupled Schrödinger equation of electrons and nuclei.

Concerning the nuclear equation of motion, a profound difference with respect to Ehrenfest dynamics is that the minimum of $\langle \mathcal{H}_e \rangle$ has to be reached in each time step of a Born–Oppenheimer molecular dynamics propagation according to Eq. (2.46), for instance by diagonalizing the Hamiltonian. In Ehrenfest dynamics, on the other hand, a wave function that minimized $\langle \mathcal{H}_e \rangle$ initially will stay, in the absence of external perturbations, in its respective ground state minimum as the nuclei move according to Eq. (2.44) by virtue of the unitarity of the wave function propagation according to Eq. (2.45).

Within the framework of Born–Oppenheimer dynamics it is easily possible to apply the scheme to some specific excited electronic state Ψ_k, $k > 0$, but without considering any interferences with other states $\Psi_{l \neq k}$ nor with itself. Thus, both the nondiagonal and diagonal corrections are neglected and, hence, Born–Oppenheimer molecular dynamics should *not* be called "adiabatic molecular dynamics" as is sometimes done, since the latter method would require including the diagonal corrections $\mathcal{C}_{kk}$ as defined in Eq. (2.8) and thus solving Eq. (2.9).

For the sake of later reference, it is useful at this stage to formulate the electronic part of the Born–Oppenheimer molecular dynamics equations of motion, i.e. the stationary Schrödinger equation Eq. (2.47), for the special case of effective one-particle Hamiltonians such as Hartree–Fock theory (see Section 2.7.3 for a concise introduction to this electronic structure method). The Hartree–Fock approximation [625, 762, 985, 1423] is obtained from the variational minimum of the energy expectation value $\langle \Psi_0 \left| \mathcal{H}_e \right| \Psi_0 \rangle$ using a single Slater determinant $\Psi_0 = 1/\sqrt{N!} \det\{\phi_i\}$ to represent the exact electronic wave function subject to the constraint that the one-particle wave functions (i.e. the orbitals) ϕ_i are orthonormal, i.e. $\langle \phi_i | \phi_j \rangle = \delta_{ij}$. The corresponding constrained minimization of the total energy with respect to the orbitals

$$\min_{\{\phi_i\}} \left\{ \langle \Psi_0 \left| \mathcal{H}_e \right| \Psi_0 \rangle \right\} \Big|_{\{\langle \phi_i | \phi_j \rangle = \delta_{ij}\}} \qquad (2.48)$$

can be cast into Lagrange's formalism

$$\mathcal{L} = -\left\langle \Psi_0 \left| \mathcal{H}_e \right| \Psi_0 \right\rangle + \sum_{i,j} \Lambda_{ij} \left(\left\langle \phi_i \left| \phi_j \right\rangle - \delta_{ij} \right) \right. \tag{2.49}$$

where Λ_{ij} are the associated Lagrange multipliers that are necessary in order to impose the constraints. Unconstrained variation of this Lagrangian with respect to the orbitals

$$\frac{\delta \mathcal{L}}{\delta \phi_i^\star} \overset{!}{=} 0 \tag{2.50}$$

leads to the well-known Hartree–Fock equations

$$H_e^{\mathrm{HF}} \phi_i = \sum_j \Lambda_{ij} \phi_j \tag{2.51}$$

as discussed amply in standard textbooks [625, 762, 985, 1423]. The more familiar diagonal canonical form $H_e^{\mathrm{HF}} \phi_i = \epsilon_i \phi_i$ is obtained after a unitary transformation and H_e^{HF} denotes the effective one-particle Hamiltonian, see Section 2.7 for more details. The equations of motion resulting from the general formulas Eqs. (2.46)-(2.47) read

$$M_I \ddot{\mathbf{R}}_I(t) = -\nabla_I \min_{\{\phi_i\}} \left\{ \left\langle \Psi_0 \left| H_e^{\mathrm{HF}} \right| \Psi_0 \right\rangle \right\} \tag{2.52}$$

$$0 = -H_e^{\mathrm{HF}} \phi_i + \sum_j \Lambda_{ij} \phi_j \tag{2.53}$$

for the Hartree–Fock special case. An isomorphic set of equations is obtained readily if Kohn–Sham density functional theory [363, 397, 760, 762, 913, 1102] is used instead

$$M_I \ddot{\mathbf{R}}_I(t) = -\nabla_I \min_{\{\phi_i\}} \left\{ \left\langle \Psi_0 \left| H_e^{\mathrm{KS}} \right| \Psi_0 \right\rangle \right\} \tag{2.54}$$

$$0 = -H_e^{\mathrm{KS}} \phi_i + \sum_j \Lambda_{ij} \phi_j \ , \tag{2.55}$$

where H_e^{HF} has to be replaced by the Kohn–Sham effective one-particle Hamiltonian H_e^{KS}, see Section 2.7.2 for more details on Kohn–Sham theory. Instead of diagonalizing a one-particle Hamiltonian such as H_e^{HF} or H_e^{KS}, an alternative but equivalent approach consists in performing the constrained minimization directly according to Eq. (2.48), using nonlinear optimization techniques explicitly.

Early applications of the iterative Born–Oppenheimer molecular dynamics scheme have been performed in the framework of semiempirical approximations to the electronic structure problem [1590, 1596]. But only a few years later an *ab initio* approach was implemented within the Hartree–Fock approximation [853]. Most notably, combining the Born–Oppenheimer propagation scheme with density functional theory [1122] in conjunction with efficient orbital prediction schemes [45, 1123] and parallelization strategies [274] in the late 1980s to early 1990s was particularly successful in the realm of *ab initio* condensed matter and also cluster simulations [75, 234, 785, 786, 1123, 1307, 1419, 1559]. In addition to these efforts, Born–Oppenheimer dynamics started to become popular in more general terms in the early 1990s [423, 551, 605, 607, 620, 896], including molecular dynamics in electronically excited states [605], with the availability of more efficient quantum chemistry electronic structure codes in conjunction with sufficient computer power to solve "interesting problems", see for instance the compilation of such studies in Table 1 of an overview article [174]. More recently, a revival of these activities with greatly improved algorithms to perform Born–Oppenheimer simulations is observed [634, 795, 1203, 1552].

Undoubtedly, the breakthrough of Kohn–Sham density functional theory in the realm of chemistry - which took place around the same time in the early 1990s - also helped by improving greatly the "price/performance ratio" of all *ab initio* molecular dynamics methods, see e.g. Refs. [766, 1390, 1622]. A third, and possibly the crucial reason that established and boosted the field of *ab initio* molecular dynamics enormously [39] was the seminal contribution of the Car–Parrinello approach [222, 1216] to *ab initio* molecular dynamics. The conclusion that this particular paper was highly influential is demonstrated by the time evolution of its citation response as depicted in Fig. 1.2. At that time, this numerically highly efficient technique not only opened novel avenues to treat large-scale problems via *ab initio* molecular dynamics and catalyzed the entire field by making "interesting calculations" possible [38], but it also united the two quite distinct electronic structure communities (i.e. "Quantum Chemistry" and "Total Energy") with the computer simulation ("Computational Physics") community [39, 934]. This particularly efficient approach to performing *ab initio* simulations is presented in detail in the next section.

2.4 Car–Parrinello molecular dynamics

2.4.1 Motivation

A non-obvious approach to cut down the computational expenses of molecular dynamics, which includes the electrons as active degrees of freedom, was proposed by Roberto Car and Michele Parrinello in 1985 [222, 1216]. In retrospect it can be considered to combine the advantages of both Ehrenfest and Born–Oppenheimer molecular dynamics in an optimal way. In Ehrenfest dynamics the time scale and thus the time step to integrate Eqs. (2.44) and (2.45) simultaneously is dictated by the intrinsic dynamics of the electrons as described by the time-dependent Schrödinger equation. Since electronic motion is typically much faster than nuclear motion - being the physical basis of the adiabatic and Born–Oppenheimer approximations - the largest possible time step in Ehrenfest dynamics is that which allows us to integrate the electronic equations of motion properly. Contrary to that, there is no electron dynamics whatsoever involved in solving the Born–Oppenheimer equations of motion, Eqs. (2.46)-(2.47), because the electronic problem is treated within the time-*independent*, stationary Schrödinger equation. This implies that these equations of motion can be integrated on the time scale given by nuclear motion, which is much slower and thus allows us to use a larger molecular dynamics time step. However, this means that the electronic structure problem has to be solved self-consistently at each molecular dynamics step, whereas this is avoided in Ehrenfest dynamics due to the possibility of propagating the wave function simply by applying the Hamiltonian to an initial wave function (obtained by a single self-consistent optimization at the very beginning of such a simulation).

From an algorithmic point of view the main task achieved in ground state Ehrenfest dynamics is simply to keep the wave function automatically minimized as the nuclei are propagated, in addition to keeping the orbitals orthonormal in effective one-particle theories such as Hartree–Fock or Kohn–Sham approaches. This, however, might be achieved - in principle - by another sort of deterministic dynamics than first-order Schrödinger dynamics. In summary, the "Best of all Worlds Method" should (i) integrate the equations of motion on the (long) time scale set by the nuclear motion but nevertheless (ii) intrinsically take advantage of the smooth time evolution of the dynamically evolving electronic subsystem as much as possible. The second point allows us to circumvent explicit diagonalization or minimization to solve the electronic structure problem iteratively before the next molecular dynamics step can be made. Car–Parrinello molecular dynamics is an efficient method to satisfy requirement (ii) automatically in a numerically stable and efficient fashion, and makes an acceptable compromise concerning the length of the time step (i).

2.4.2 Car–Parrinello Lagrangian and equations of motion

The basic idea of the Car–Parrinello approach can be viewed as taking most direct advantage of the *quantum-mechanical adiabatic time scale separation* of fast electronic (quantum) and slow (classical) nuclear motion. Algorithmically, this is achieved by transforming this separation into a *classical-mechanical adiabatic energy-scale separation* in the framework of dynamical systems theory. In order to achieve this goal the two-component quantum/classical problem is mapped onto a two-component purely classical problem with two separate energy scales at the expense of losing the physical time information of the quantum subsystem dynamics. Furthermore, the central quantity, the energy of the electronic subsystem, $\langle \Psi_0 | \mathcal{H}_e | \Psi_0 \rangle$, is certainly a *function of the nuclear positions* $\{\mathbf{R}_I\}$. But at the same time it *can* be considered to be a *functional of the wave function* Ψ_0 and thus of a set of orbitals $\{\phi_i\}$ (or other functions such as two-particle geminals) used to build up this wave function (for instance in terms of a Slater determinant $\Psi_0 = 1/\sqrt{N!}\det\{\phi_i\}$ or a combination thereof). Now, in classical mechanics the force on the nuclei is obtained from the derivative of a suitable Lagrangian with respect to the nuclear positions. This suggests that a functional derivative with respect to the orbitals, which are interpreted as classical fields, might yield the correct force on the orbitals, given a suitably defined Lagrangian. In addition, possible constraints within the set of orbitals have to be imposed, such as e.g. orbital orthonormality (or generalized orthonormality conditions if overlap matrices have to be considered in nonorthogonal schemes).

In the spirit of this fundamental observation, Car and Parrinello [222] introduced the following class of Lagrangians [995]

$$
\mathcal{L}_{\mathrm{CP}} = \underbrace{\sum_I \frac{1}{2} M_I \dot{\mathbf{R}}_I^2 + \sum_i \mu \left\langle \dot\phi_i \,\middle|\, \dot\phi_i \right\rangle}_{\text{kinetic energy}} - \underbrace{\langle \Psi_0 | \mathcal{H}_e | \Psi_0 \rangle}_{\substack{\text{potential} \\ \text{energy}}} + \underbrace{constraints}_{\text{orthonormality}}
$$

(2.56)

to serve this purpose; note that sometimes a prefactor $\mu_i/2$ is used and orbital occupation numbers $f_i = 0, 1, 2$ are introduced. The corresponding Newtonian equations of motion are obtained from the associated Euler–Lagrange equations

$$
\frac{d}{dt} \frac{\partial \mathcal{L}}{\partial \dot{\mathbf{R}}_I} = \frac{\partial \mathcal{L}}{\partial \mathbf{R}_I}
$$

(2.57)

$$
\frac{d}{dt} \frac{\delta \mathcal{L}}{\delta \dot\phi_i^\star} = \frac{\delta \mathcal{L}}{\delta \phi_i^\star}
$$

(2.58)

like in classical mechanics, but here for both the nuclear positions and the orbitals. In the latter case functional derivatives have to be taken with respect to the orbitals, $\phi_i(\mathbf{r})$, which are complex scalar fields; note that $\phi_i^\star = \langle\phi_i|$ and that the constraints are holonomic [528, 1282]. Following this flow of arguments, generic Car–Parrinello equations of motion are found to be of the form

$$M_I\ddot{\mathbf{R}}_I(t) = -\frac{\partial}{\partial\mathbf{R}_I}\langle\Psi_0|\mathcal{H}_e|\Psi_0\rangle + \frac{\partial}{\partial\mathbf{R}_I}\{constraints\} \qquad (2.59)$$

$$\mu\ddot{\phi}_i(t) = -\frac{\delta}{\delta\phi_i^\star}\langle\Psi_0|\mathcal{H}_e|\Psi_0\rangle + \frac{\delta}{\delta\phi_i^\star}\{constraints\} \qquad (2.60)$$

where μ is the fictitious mass, adiabaticity or inertia parameter assigned to the orbital degrees of freedom and the units of this parameter are energy times a squared time for reasons of dimensionality.

For the very important special case of effective one-particle Hamiltonians such as those resulting from Kohn–Sham theory (see Section 2.7.2) in conjunction with position-independent constraints this Lagrangian simplifies to

$$\mathcal{L}_{\mathrm{CP}} = \sum_I \frac{1}{2}M_I\dot{\mathbf{R}}_I^2 + \sum_i \mu\left\langle\dot{\phi}_i\,\middle|\,\dot{\phi}_i\right\rangle$$

$$-\left\langle\Psi_0|H_e^{\mathrm{KS}}|\Psi_0\right\rangle + \sum_{i,j}\Lambda_{ij}\left(\langle\phi_i|\phi_j\rangle - \delta_{ij}\right) \ , \quad (2.61)$$

where the proper orbital orthonormality, $\langle\phi_i|\phi_j\rangle = \delta_{ij}$, must be imposed by Lagrange multipliers Λ_{ij}. Using Eqs. (2.57)-(2.58) generates

$$M_I\ddot{\mathbf{R}}_I(t) = -\nabla_I\left\langle\Psi_0\,\middle|\,H_e^{\mathrm{KS}}\,\middle|\,\Psi_0\right\rangle \qquad (2.62)$$

$$\mu\ddot{\phi}_i(t) = -H_e^{\mathrm{KS}}\phi_i + \sum_j\Lambda_{ij}\phi_j \ , \qquad (2.63)$$

the corresponding Car–Parrinello equations of motion.

It is important to make clear the point that all constraints inherent in a generic Lagrangian of the type Eq. (2.56) lead to associated "constraint forces" in the equations of motion by virtue of their derivatives, such as the term $\sum_j\Lambda_{ij}\phi_j$ in Eq. (2.63). In general, however, these constraints

$$constraints = constraints\ (\{\phi_i\},\{\mathbf{R}_I\}) \qquad (2.64)$$

might be not only a functional of the set of orbitals $\{\phi_i\}$, but also a function of the nuclear positions $\{\mathbf{R}_I\}$. *Both* these dependencies have to be taken into account properly in deriving the Car–Parrinello equations following from Eq. (2.56) using Eqs. (2.57)-(2.58). This aspect, which is crucial to generate a proper energy-conserving dynamical evolution, will be discussed in depth in Section 2.5. An important example where an additional dependence of the wave function constraint on nuclear positions yields a constraint contribution to the nuclear equation Eq. (2.59), due to the usage of a specific class of pseudopotentials, can be found in Ref. [816]. This issue is discussed in depth in Section 6.4, where these ultrasoft pseudopotentials are introduced.

Coming back to the Car–Parrinello equations of motion, Eqs. (2.62)-(2.63), the nuclei evolve in time at a certain (instantaneous) physical temperature $\propto \sum_I M_I \dot{\mathbf{R}}_I^2$, whereas a "fictitious temperature" $\propto \sum_i \mu \langle \dot{\phi}_i | \dot{\phi}_i \rangle$ can be associated accordingly with the electronic degrees of freedom. In this terminology, "low electronic temperature" or "cold electrons" means that the electronic subsystem is close to its instantaneous minimum energy, $\min_{\{\phi_i\}} \langle \Psi_0 | \mathcal{H}_e | \Psi_0 \rangle$, i.e. close to the exact Born–Oppenheimer surface. Thus, a ground state wave function optimized for the initial configuration of the nuclei will stay close to its ground state also during time evolution if it is kept at a sufficiently low fictitious temperature.

The remaining task is to separate in practice nuclear and electronic motion such that the fast electronic subsystem stays cold also for long times but still follows the slow nuclear motion adiabatically (or instantaneously). Simultaneously, the nuclei must nevertheless be kept at a much higher temperature. This can be achieved in nonlinear classical dynamics via decoupling of the two subsystems and (quasi-) adiabatic time evolution. This is possible if the power spectra stemming from both dynamics do not have substantial overlap of their respective vibrational density of states, so that energy transfer from the "hot nuclei" to the "cold electrons" becomes practically impossible on the relevant time scales. This amounts, in other words, to imposing and maintaining a metastability condition in a complex dynamical system for sufficiently long times. How and to what extent this is possible in practice was analyzed in detail in a pioneering technical investigation based on well-controlled model systems [1118] (see also Refs. [224, 1117, 1442] and Sections 3.2 and 3.3 in Ref. [1209]). Later, the adiabaticity issue was investigated with mathematical rigor [181], see Section 2.4.5 for a short digression, and in terms of a generalization to a second level of adiabatic decoupling [956].

2.4.3 Why does the Car–Parrinello method work?

In order to understand why Car–Parrinello methods work not only on paper, but also in practice, the dynamics generated by the Car–Parrinello Lagrangian Eq. (2.56) has been analyzed [1118], invoking a "classical dynamics perspective" of a much simplified physical system (in particular two

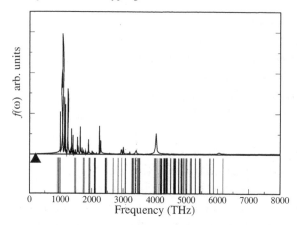

Fig. 2.1. Vibrational density of states according to Eq. (2.65) (continuous spectrum in upper part) and harmonic approximation thereof according to Eq. (2.70) (stick spectrum in lower part) of the electronic degrees of freedom compared to the highest-frequency phonon mode $\omega_{\mathrm{n}}^{\max}$ (triangle) for the model system discussed in the text.

silicon atoms on a periodic diamond lattice, local density approximation to Kohn–Sham density functional theory, norm-conserving pseudopotentials for core electrons, plane wave basis for valence orbitals, a 0.3 fs time step with $\mu = 300$ a.u., and 6.3 ps propagation time was used). A concise presentation of similar ideas can be found in Ref. [224]. Using the CPMD program package [696], the calculations from Ref. [1118] were repeated, using an equivalent parameter setting but improved sampling, in order to generate the figures used in this section to illustrate the concepts.

For this model system the vibrational density of states or power spectrum of the electronic degrees of freedom, i.e. the Fourier transform of the statistically averaged velocity autocorrelation function of the classical fields

$$f(\omega) = \int_0^\infty \cos(\omega t) \sum_i \left\langle \dot{\phi}_i; t \,\middle|\, \dot{\phi}_i; 0 \right\rangle \, dt \qquad (2.65)$$

representing the orbitals is compared to the highest-frequency phonon mode $\omega_{\mathrm{n}}^{\max}$ of the nuclear subsystem in Fig. 2.1. From this figure it is evident that, for the chosen parameters, the nuclear and electronic subsystems are dynamically separated: their power spectra do not overlap. As a result of this gap between the highest phonon frequency and the lowest orbital frequency, energy transfer from the hot nuclei to the cold electrons is expected to be prohibitively slow, a similar argument has been provided in Section 3.3 of Ref. [1209].

This is indeed the case, as can be verified in Fig. 2.2 where the conserved

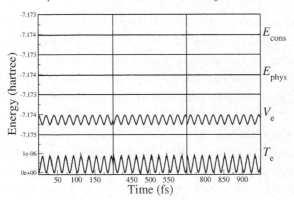

Fig. 2.2. Various energies according to Eqs. (2.66)-(2.69) for a model system propagated via Car–Parrinello molecular dynamics; for further details see text.

energy E_{cons}, physical total energy E_{phys}, electronic energy V_{e}, and fictitious kinetic energy of the electrons T_{e}

$$E_{\text{cons}} = \sum_i \mu \left\langle \dot{\phi}_i \middle| \dot{\phi}_i \right\rangle + \sum_I \frac{1}{2} M_I \dot{\mathbf{R}}_I^2 + \left\langle \Psi_0 | H_{\text{e}}^{\text{KS}} | \Psi_0 \right\rangle \qquad (2.66)$$

$$E_{\text{phys}} = \sum_I \frac{1}{2} M_I \dot{\mathbf{R}}_I^2 + \left\langle \Psi_0 | H_{\text{e}}^{\text{KS}} | \Psi_0 \right\rangle = E_{\text{cons}} - T_{\text{e}} \qquad (2.67)$$

$$V_{\text{e}} = \left\langle \Psi_0 | H_{\text{e}}^{\text{KS}} | \Psi_0 \right\rangle \qquad (2.68)$$

$$T_{\text{e}} = \sum_i \mu \left\langle \dot{\phi}_i \middle| \dot{\phi}_i \right\rangle \qquad (2.69)$$

according to the Lagrangian Eq. (2.61) and the equations of motion Eqs. (2.62)-(2.63) are shown for the same system as a function of time. First of all, there should be a conserved energy quantity according to classical dynamics since the constraints are holonomic [528, 1282]. Indeed "the Hamiltonian" or conserved energy E_{cons} is a constant of motion (with relative variations smaller than 10^{-6} and with no drift), which serves as an extremely sensitive check of the molecular dynamics algorithm. Contrary to that the electronic energy V_{e}, which is the potential energy, varies considerably as a function of time and displays a clear oscillation pattern due to the simple phonons.

Most importantly, the fictitious kinetic energy of the electrons, T_{e}, is found to perform *bound* oscillations around a *constant value*, i.e. the electrons "do not heat up" systematically in the presence of the hot nuclei; note that T_{e} is a measure for deviations from the exact Born–Oppenheimer surface.

Closer inspection shows actually two time scales of oscillations: the one visible in Fig. 2.2 (and in terms of the forces also in Fig. 2.3(a)) stems from the drag exerted by the moving nuclei on the electrons and is the mirror image of the V_e fluctuations. Superimposed on top of that (not shown, but see Fig. 2.3(b) for the underlying forces) are small-amplitude high-frequency oscillations that are intrinsic to the fictitious electron dynamics itself. Their period is only a fraction of that of the oscillations visible on the time scale of the physical time evolution. These high-frequency oscillations are actually instrumental for the stability of the Car–Parrinello dynamics, as explained below. It is important to note that already the larger variations visible in Fig. 2.2 (or Fig. 2.3(a)) are three orders of magnitude smaller than the physically meaningful oscillations of V_e. As a result, E_{phys} defined as $E_{\text{cons}} - T_e$, or equivalently as the sum of the nuclear kinetic energy and the electronic total energy (which serves as the potential energy for the nuclei), is essentially constant on the relevant energy and time scales. Thus, it behaves approximately like the strictly conserved total energy in standard molecular dynamics (with only nuclei as dynamical degrees of freedom) or in Born–Oppenheimer molecular dynamics (with fully optimized electronic degrees of freedom), and is therefore often denoted as the "physical total energy". This implies that the resulting physically significant dynamics of the nuclei yields an excellent approximation to microcanonical dynamics and, assuming ergodicity, to the microcanonical ensemble as well. The idea of adiabatic separation in molecular dynamics simulations in order to sample "fast" degrees of freedom concurrently to the propagation of "slow" degrees of freedom has been analyzed from various viewpoints in Refs. [925, 956, 1230]. In order to avoid confusion it is noted that a different explanation has been proposed in the early literature on the subject [1121] (and reiterated in a review article [1123], in particular in Sections VIII.B and VIII.C therein), which was however revised shortly after [224] along the lines of the view presented here.

Given the adiabatic separation and the stability of the propagation, another central question remains to be answered: are the forces acting on the nuclei actually the "correct" ones in Car–Parrinello molecular dynamics? As a reference we have the forces obtained from full self-consistent minimizations of the electronic energy, $\min_{\{\phi_i\}} \langle \Psi_0 | \mathcal{H}_e | \Psi_0 \rangle$, at each time step of a Car–Parrinello simulation, i.e. the Born–Oppenheimer forces corresponding to the generated configuration are obtained with extremely well-converged wave functions. The Car–Parrinello forces are indeed found to match closely the Born–Oppenheimer forces, as demonstrated in Fig. 2.3(a). In particular, the physically meaningful dynamics of the (x-component of the) force

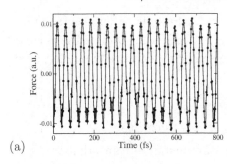

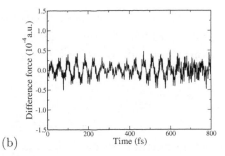

(a)

(b)

Fig. 2.3. (a) Comparison of the x-component of the force acting on one atom of a model system obtained from Car–Parrinello (solid line) and well-converged Born–Oppenheimer (dots) molecular dynamics. (b) Enlarged view of the difference between Car–Parrinello and Born–Oppenheimer forces; for further details see text.

acting on one silicon atom in the model system obtained from stable Car–Parrinello fictitious dynamics propagation of the electrons and from iterative minimizations of the electronic energy are extremely close.

The difference in the forces shown at a better resolution in panel (b) of Fig. 2.3 reveals that the gross deviations are also oscillatory but that they are four orders of magnitude smaller than the physical variations of the force resolved in Fig. 2.3(a). The latter correspond to the "large-amplitude" oscillations of T_e visible in Fig. 2.2 due to the drag of the nuclei exerted on the quasi-adiabatically following electrons. Superimposed on the gross variation in (b) are again high-frequency *bound oscillatory small-amplitude* fluctuations like those discussed already for T_e. They lead *on physically relevant time scales* (i.e. those visible in Fig. 2.3(a)) to "averaged forces" that are very close to the exact ground state Born–Oppenheimer forces. This feature is an important ingredient in the derivation of adiabatic dynamics [956, 1118]. This particular effect vanishes only in the limit of a vanishing fictitious mass $\mu \to 0$, as shown mathematically in Section 2.4.5, but can be estimated and corrected for systematically as explained in Section 2.4.9.

In conclusion, the Car–Parrinello force can be said, strictly speaking, to deviate at most instants of time from the exact Born–Oppenheimer force. However, this deviation is negligible for properly performed Car–Parrinello simulations, as explained in Section 2.4.4 due to (i) the smallness and boundedness of this difference *in conjunction with* (ii) the intrinsic averaging effect of small-amplitude high-frequency oscillations within a few molecular dynamics time steps, i.e. on the sub-femtosecond time scale which is irrelevant for *nuclear* dynamics. Clearly, there must remain small residual effects that are induced by using necessarily a strictly finite fictitious electron mass, $\mu > 0$, which will be discussed critically in Section 2.4.9 once the issues of controlling adiabaticity and thus propagating properly are explained in the next section.

2.4.4 How to control adiabaticity?

An important question is: under what circumstances can adiabatic separation be achieved, and how can it be controlled to guarantee technically correct Car–Parrinello simulations? A simple harmonic analysis [1118] of the frequency spectrum of the orbital classical fields close to the minimum defining the ground state yields

$$\omega_{ij} = \left(\frac{2(\epsilon_i - \epsilon_j)}{\mu} \right)^{1/2} , \qquad (2.70)$$

where ϵ_j and ϵ_i are the eigenvalues of occupied and unoccupied (virtual) orbitals, respectively, defined in Eq. (2.124) of the Kohn–Sham Hamiltonian that describes the model system introduced in Section 2.4.3; the case where both orbitals are occupied is also presented in Ref. [1118], see Eq. (26) therein. It can be seen from Fig. 2.1 that the harmonic approximation works faithfully as compared to the exact finite-temperature spectrum that includes anharmonic effects as well. A more general analysis of the associated small-amplitude equations of motion themselves is given in Ref. [1122] and in Section IV.A of Ref. [1123]. Since the close agreement of the harmonic approximation is in particular true for the lowest frequency $\omega_{\mathrm{e}}^{\mathrm{min}}$, the resulting handy analytic estimate for the lowest possible electronic frequency

$$\omega_{\mathrm{e}}^{\mathrm{min}} \propto \left(\frac{E_{\mathrm{gap}}}{\mu} \right)^{1/2} \qquad (2.71)$$

shows that this frequency increases like the square root of the electronic energy difference E_{gap} between the lowest unoccupied ("Lumo") and the highest occupied ("Homo") orbital. On the other hand, it increases, for a fixed energy gap, as the inverse of the square root for a decreasing fictitious mass parameter μ.

In order to guarantee adiabatic separation of the electronic and nuclear subsystems and thus stable Car–Parrinello propagation, the frequency difference $\omega_{\mathrm{e}}^{\mathrm{min}} - \omega_{\mathrm{n}}^{\mathrm{max}}$ should be large, as explained earlier. But both the highest phonon frequency $\omega_{\mathrm{n}}^{\mathrm{max}}$ and the energy gap E_{gap} are quantities that are dictated by the physics of the system. Hence, the only parameter in our hands to control adiabatic separation is the fictitious mass μ, which is therefore also called the "adiabaticity parameter". However, decreasing μ not only shifts the electronic spectrum upwards on the frequency scale as desired, but also stretches the entire frequency spectrum according to Eq. (2.70). This leads to an increase of the maximum frequency according to

$$\omega_e^{max} \propto \left(\frac{E_{cut}}{\mu}\right)^{1/2} , \qquad (2.72)$$

where E_{cut} is the largest kinetic energy in an expansion of the wave function in terms of a plane wave basis set, see Section 3.1.3.

Here, a limitation that prevents decreasing μ arbitrarily kicks in due to the maximum length of the molecular dynamics time step Δt^{max} that can be used. The time step is inversely proportional to the highest frequency in the system, which is ω_e^{max} for the present case, and thus the relation

$$\Delta t^{max} \propto \left(\frac{\mu}{E_{cut}}\right)^{1/2} \qquad (2.73)$$

limits the largest time step that is possible. It is noted in passing that similar arguments hold if localized basis sets such as Gaussians with high angular momentum basis functions and thus rapid oscillations are used (see Section 2.8.1). As a consequence, Car–Parrinello simulators have to find their way between Scylla and Charybdis, and have to make a compromise on the control parameter μ; typical values for large-gap systems are $\mu = 500\text{-}1000$ a.u., which allows for a time step of about 5-10 a.u. (0.12-0.24 fs) depending on the mass of the lightest nuclei. Using feedback algorithms it is possible to optimize μ during a particular simulation within a fixed accuracy criterion [182]. The poor man's way to keep the time step large and still increase μ in order to satisfy adiabaticity is to choose heavier nuclear masses, as customarily done with hydrogen being replaced by deuterium. That mass change depresses the largest phonon or vibrational frequency ω_n^{max} of the nuclei at the cost of renormalizing all *dynamical* quantities in the sense of classical isotope effects. Alternatively, it is possible to thermostat the motion of the electronic degrees of freedom separately from the nuclear dynamics as explained in Section 5.2.2 which also allows us to use larger time steps.

The above discussion assumed implicitly that the electronic gap is finite. What happens, however, if the electronic gap is very small or even closes $E_{gap} \to 0$ as is the case for metallic systems? According to the harmonic estimate Eq. (2.71), the fictitious mass has to be made smaller and smaller, $\mu \to 0$, in order to maintain the adiabaticity condition $\omega_e^{min} - \omega_n^{max} \gg 0$ since $\omega_n^{max} > 0$ is fixed. In the limiting case, however, all the above-given arguments break down due to the occurrence of zero-frequency electronic modes in the power spectrum, which necessarily leads to an overlap with the phonon spectrum. Following an idea of Sprik [1379] applied in a classical context it was shown that the coupling of separate Nosé–Hoover thermostats (see Section 5.2.2) to the nuclear and electronic subsystem can maintain adiabaticity by counterbalancing the energy flow from ions to electrons so that the electrons "stay cold" [146]; see Ref. [450] for a similar idea to restore adiabaticity and Refs. [144, 1030, 1510] for analyses and improvements.

Although this method has been demonstrated to work in practice [1111] even for quasi-metallic systems, this *ad hoc* cure is not entirely satisfactory from both a theoretical and a practical point of view. An alternative approach to metals based on a matrix representation in the basis of the Kohn–Sham orbitals for the one-particle statistical operator has been developed by Marzari *et al.* [961]. In this form the electron density is given by

$$n(\mathbf{r}) = \sum_{ij} f_{ij}\phi_i^\star(\mathbf{r})\phi_j(\mathbf{r}) , \qquad (2.74)$$

where the matrix f_{ij} is constrained to yield $\mathrm{Tr} f = N$ and eigenvalues bounded to $[0, 1]$. The function A to be minimized is then

$$A[T; \{\phi_i\}, \{f_{ij}\}] =$$

$$\sum_{ij} f_{ij}\langle\phi_i \mid \hat{T} + \hat{V}_{\mathrm{ext}} \mid \phi_j\rangle + E_{\mathrm{H-xc}}[n] - TS[\{f_{ij}\}] , \qquad (2.75)$$

with the terms that depend only on the charge density (Hartree and exchange-correlation) grouped together. The Fermi–Dirac entropic term $(-TS)$ is a function of the eigenvalues of the matrix f. This scheme can be cast into an extended Lagrangian [1398, 1558] and Car–Parrinello type equations of motion derived for the free parameters that include in this case also an additional set related to the occupation matrix.

Based on the above discussion it should be clear by now that for small band gap systems as well as for metals the Car–Parrinello method has to be adapted, thereby losing much of its simplicity. In particular, for strongly metallic systems the well-controlled Born–Oppenheimer iterative approach to molecular dynamics is recommended. An additional advantage in the realm of truly metallic systems is that Born–Oppenheimer dynamics is also better suited to sample many **k**-points (see Section 3.1.3), it allows more easily for fractional occupation numbers (see Section 5.3.4), and it can handle more efficiently the charge sloshing problem [1123].

2.4.5 A mathematical investigation

Up to this point the entire discussion of the stability and adiabaticity issues was based on model systems, approximate and mostly qualitative in nature. Although this approach is useful in terms of providing a feeling for the inner workings of the Car–Parrinello method, it has also been proven rigorously [181] that the deviation or the absolute error Δ_μ of the Car–Parrinello trajectory relative to the trajectory obtained on the exact Born–Oppenheimer potential energy surface is controlled by the fictitious electron mass μ:

Theorem 1 iv.): There are constants $C > 0$ and $\mu^\star > 0$ such that

$$\Delta_\mu = \left| \mathbf{R}^\mu(t) - \mathbf{R}^0(t) \right| + \left| |\psi^\mu; t\rangle - |\psi^0; t\rangle \right| \le C\mu^{1/2} \ , \ \ 0 \le t \le t^{\mathrm{max}} \quad (2.76)$$

and the fictitious kinetic energy satisfies

$$T_{\mathrm{e}} = \mu \left\langle \dot{\psi}^\mu; t \,\middle|\, \dot{\psi}^\mu; t \right\rangle \le C\mu \ , \ \ 0 \le t \le t^{\mathrm{max}} \quad\quad (2.77)$$

for all values of the parameter μ satisfying $0 < \mu \le \mu^\star$, where up to time $t^{\mathrm{max}} > 0$ there exists a unique nuclear trajectory on the exact Born–Oppenheimer surface with $\omega_{\mathrm{e}}^{\mathrm{min}} > 0$ for $0 \le t \le t^{\mathrm{max}}$, i.e. there exists "always" a finite electronic excitation gap. Here, the superscript μ or 0 indicates that the trajectory was obtained via Car–Parrinello molecular dynamics using a finite mass μ or via dynamics on the exact Born–Oppenheimer surface, respectively. The initial conditions of the Car–Parrinello trajectory are assumed to be identical to those of the Born–Oppenheimer trajectory for both the positions and momenta of the nuclei, and the initial Car–Parrinello wave function is taken to be the optimized wave function at $\mathbf{R}^0(0)$ with zero orbital velocities. Note that not only the nuclear trajectory is shown to be close to the correct one, but also the wave function is proven to stay close to the fully converged one up to time t^{max}.

Furthermore, what happens if the initial wave function at $t = 0$ does not correspond to the absolute minimum of the electronic energy $\langle \mathcal{H}_{\mathrm{e}} \rangle$ but is trapped in a higher-lying local minimum was also investigated. In this "off-equilibrium" case it is found that the propagated wave function will keep on oscillating at $t > 0$ also for $\mu \to 0$ and not even time averages converge to any of the eigenstates [181]. Note that this does not preclude performing Car–Parrinello molecular dynamics in excited states. This clearly is possible, as explained in Section 5.3, using a variety of approaches such as those based on a "minimizable" expression for the electronic energy of an excited state or using time-dependent density functional theory. However, this finding has crucial implications for electronic level-crossing situations in truly nonadiabatic simulations, see Section 5.3.3.

2.4.6 The quantum chemistry viewpoint

In order to understand Car–Parrinello molecular dynamics also from the "quantum chemistry perspective", it is useful to write these equations of motion again for the special case of the Hartree–Fock approximation and $\mathbf{R}_I$-independent constraints using the Lagrangian

$$\mathcal{L}_{\mathrm{CP}} = \sum_I \frac{1}{2} M_I \dot{\mathbf{R}}_I^2 + \sum_i \mu \left\langle \dot{\phi}_i \middle| \dot{\phi}_i \right\rangle$$

$$- \left\langle \Psi_0 \middle| H_{\mathrm{e}}^{\mathrm{HF}} \middle| \Psi_0 \right\rangle + \sum_{i,j} \Lambda_{ij} \left(\langle \phi_i | \phi_j \rangle - \delta_{ij} \right) \quad (2.78)$$

and the Euler–Lagrange equations

$$\frac{d}{dt} \frac{\partial \mathcal{L}}{\partial \dot{\mathbf{R}}_I} = \frac{\partial \mathcal{L}}{\partial \mathbf{R}_I} \quad (2.79)$$

$$\frac{d}{dt} \frac{\delta \mathcal{L}}{\delta \dot{\phi}_i^\star} = \frac{\delta \mathcal{L}}{\delta \phi_i^\star} \quad . \quad (2.80)$$

The resulting Car–Parrinello equations of motion, cf. Eqs. (2.81)-(2.82),

$$M_I \ddot{\mathbf{R}}_I(t) = -\nabla_I \left\langle \Psi_0 \middle| H_{\mathrm{e}}^{\mathrm{HF}} \middle| \Psi_0 \right\rangle \quad (2.81)$$

$$\mu \ddot{\phi}_i(t) = -H_{\mathrm{e}}^{\mathrm{HF}} \phi_i + \sum_j \Lambda_{ij} \phi_j \quad (2.82)$$

appear to be very similar to those obtained earlier for Born–Oppenheimer molecular dynamics, cf. Eqs. (2.52)-(2.53),

$$M_I \ddot{\mathbf{R}}_I(t) = -\nabla_I \min_{\{\phi_i\}} \left\{ \left\langle \Psi_0 \middle| H_{\mathrm{e}}^{\mathrm{HF}} \middle| \Psi_0 \right\rangle \right\} \quad (2.83)$$

$$0 = -H_{\mathrm{e}}^{\mathrm{HF}} \phi_i + \sum_j \Lambda_{ij} \phi_j \quad (2.84)$$

as well as to those derived for Ehrenfest molecular dynamics, cf. Eqs. (2.44)-(2.45),

$$M_I \ddot{\mathbf{R}}_I(t) = -\nabla_I \left\langle \Psi_0 \middle| H_{\mathrm{e}}^{\mathrm{HF}} \middle| \Psi_0 \right\rangle \quad (2.85)$$

$$i\hbar \frac{\partial \Psi_0}{\partial t} = H_{\mathrm{e}}^{\mathrm{HF}} \Psi_0 \quad . \quad (2.86)$$

Very loosely speaking, it could be said that the Car–Parrinello equations of motion are a "mixture" of Born–Oppenheimer and Ehrenfest molecular dynamics: they include the orthonormality constraint explictly (like in Born–Oppenheimer dynamics) but they also propagate the wave function dynamically and thus do not require explicit minimization of the total energy (like in Ehrenfest dynamics).

It is also suggestive to argue that they should become identical to the Born–Oppenheimer equations of motion if the term $|\mu\ddot{\phi}_i(t)|$ is small at any time t compared to the physically relevant forces on the right-hand side of both Eq. (2.81) and Eq. (2.82). This term being zero (or small) means that one is at (or close to) the minimum of the electronic energy $\langle\Psi_0|H_e^{\mathrm{HF}}|\Psi_0\rangle$ since time derivatives of the orbitals $\{\phi_i\}$ can be considered as variations of Ψ_0 and thus of the expectation value $\langle H_e^{\mathrm{HF}}\rangle$ itself. In other words, no forces act on the wave function if $\mu\ddot{\phi}_i \equiv 0$. But if $|\mu\ddot{\phi}_i(t)|$ is small for all i, this also implies that the associated kinetic energy $T_e = \sum_i \mu\langle\dot{\phi}_i|\dot{\phi}_i\rangle$ is small, which connects these more qualitative arguments to the preceding discussion in Sections 2.4.3 and 2.4.4. When comparing to the Ehrenfest equations of motion, it is suggested that the correct expectation value of the Hamiltonian $\langle\Psi_0|H_e^{\mathrm{HF}}|\Psi_0\rangle$, and thus the correct forces on the nuclei are obtained without explicit total energy minimization due to the additional orthonormality constraint. In conclusion, the Car–Parrinello equations are expected to produce asymptotically the correct dynamics of the nuclei and hence proper phase space trajectories in the microcanonical ensemble as demonstrated with more rigor in Section 2.4.5. The above-given heuristic qualitative analysis is put onto a quantitative basis in Section 2.4.9, where in addition analytic error estimates are derived.

Before leaving this section, it is enlightening to compare the structure of the Lagrangian Eq. (2.78) and the electronic Euler–Lagrange equation Eq. (2.80) of Car–Parrinello molecular dynamics to the analogous equations

$$\mathcal{L} = -\langle\Psi_0|\mathcal{H}_e|\Psi_0\rangle + \sum_{i,j}\Lambda_{ij}\left(\langle\phi_i|\phi_j\rangle - \delta_{ij}\right) \tag{2.87}$$

$$\frac{\delta\mathcal{L}}{\delta\phi_i^\star} = 0 \;, \tag{2.88}$$

respectively, used earlier, cf. Eqs. (2.49) and (2.50), to derive the standard Hartree–Fock approximation. The former reduce to the latter if the dynamical aspect and the associated time evolution of the wave function are neglected, that is in the (quasi-) static limit that the nuclear and electronic momenta are absent or constant. Thus, the most general Car–Parrinello theory, namely Eq. (2.56) together with Eqs. (2.57)–(2.58), can also be viewed as a systematic prescription to derive new classes of dynamical *ab initio* methods in very general terms.

2.4.7 The simulated annealing and optimization viewpoints

In the discussion given above, Car–Parrinello molecular dynamics was motivated by "combining" the attractive aspects of both Ehrenfest and Born–Oppenheimer molecular dynamics as much as possible. Looked at from another perspective, the Car–Parrinello method can also be considered as an ingenious way to perform *global* optimizations (minimizations) of nonlinear functions, here $\langle \Psi_0 | \mathcal{H}_e | \Psi_0 \rangle$, in a high-dimensional parameter space including complicated constraints. The optimization parameters are those used to represent the total wave function Ψ_0 in terms of simpler functions, for instance a set of expansion coefficients $\{c_{i\nu}\}$ of the orbitals ϕ_i in terms of Gaussians or plane waves, which are considered to be auxiliary parameters in the optimization problem. Next, combining the mimimization of the total energy expression with respect to the orbitals, i.e. $\delta \langle \Psi_0 | \mathcal{H}_e | \Psi_0 \rangle / \delta \phi_i^\star = 0$ and thus $\partial \langle \Psi_0 | \mathcal{H}_e | \Psi_0 \rangle / \partial c_{i\nu} = 0$ with relaxation of the nuclei, i.e. $\partial \langle \Psi_0 | \mathcal{H}_e | \Psi_0 \rangle / \partial \mathbf{R}_I = 0$, results in a technique where both the electronic and real space structure of matter can be determined simultaneously, as introduced in Ref. [98] (see also Ref. [97]). In conclusion, traditional "*calculations treat the relaxation of the charge density and nuclear positions consecutively, changing only one or the other at each iteration. This approach hence involves the mapping out of a section of the Born–Oppenheimer (BO) surface and the search for the energy minimum on this surface*" [98] whereas in the combined approach, "*the geometries and charge densities of general polyatomic systems can be varied simultaneously, allowing* direct *calculations of the equilibrium without intermediate knowledge of the BO surface*" [98].

Let us now discuss this very idea in the framework of molecular dynamics. Keeping the nuclei frozen for a moment, one could start the optimization procedure from a "random wave function" which certainly does not minimize the electronic energy. Thus, its fictitious kinetic energy is high, the electronic degrees of freedom are "hot". This energy, however, can be pumped out of the system by systematically cooling it to lower and lower temperatures. This can be achieved in an elegant way by adding a non-conservative damping term [1447] to the electronic Car–Parrinello equation of motion, cf. Eq. (2.60),

$$\mu\ddot{\phi}_i(t) = -\frac{\delta}{\delta\phi_i^\star}\langle\Psi_0|\mathcal{H}_e|\Psi_0\rangle + \frac{\delta}{\delta\phi_i^\star}\{constraints\} - \gamma_e\mu\dot{\phi}_i \ , \qquad (2.89)$$

where $\gamma_e > 0$ is a friction constant that governs the rate of energy dissipation via the orbital velocity-dependent last term. Alternatively, dissipation can also be enforced by simply rescaling the orbital velocities when multiplying them by a constant scaling factor < 1 after each molecular dynamics step. Note that this deterministic dynamical method is very similar in spirit to simulated annealing [746], invented in the framework of the stochastic Monte Carlo approach in the canonical ensemble. If the energy dissipation is done slowly, the wave function will find its way down to the minimum of the energy. At the end, an intricate global optimization has been performed!

If the nuclei are allowed to move according to Eq. (2.59) in the presence of an analogous friction term a combined or simultaneous optimization of both electrons and nuclei can be achieved, which amounts to a "global geometry optimization" as discussed in detail in Section 3.7.4. This special perspective is stressed in more detail in the review Ref. [485] and an implementation of such techniques within the CADPAC quantum chemistry code is described in Ref. [1620]. This particular operation mode of Car–Parrinello molecular dynamics is related to the above-sketched combined optimization technique where the goal is to optimize simultaneously both the structure of the nuclear skeleton and the electronic structure. This is achieved by considering the nuclear coordinates and the expansion coefficients of the orbitals as variation parameters on the same footing (see e.g. Refs. [98, 617, 1439] for such techniques). But Car–Parrinello molecular dynamics is much more than that because even in the case that the nuclei evolve continuously according to Newtonian dynamics at finite temperatures, it guarantees that an initially optimized wave function will stay optimized along the nuclear path so that a proper molecular dynamics trajectory is generated.

Finally, we stress that the auxiliary parameters to be optimized do not need to be electronic degrees of freedom but can represent any (pseudo-) classical field. In particular, the Car–Parrinello idea has been used for instance in polarizable fluids [1379], classical density functional theory of liquids [882–884], variational Quantum Monte Carlo [1439], and most recently to generate Maxwellian electrodynamics "on-the-fly" [4, 889, 890, 1109, 1238] where the speed of light plays the role of the adiabaticity parameter μ.

2.4.8 The extended Lagrangian viewpoint

There is still another way to look at Car–Parrinello methods to carry out dynamical electronic structure calculations, namely in the light of "extended Lagrangians" or "extended system dynamics" [31], see e.g. Refs. [25, 273, 468, 577, 1382, 1504] for introductions to extended Lagrangians. More recently even the notion of "ELMD", meaning "extended Lagrangian molecular dynamics", has been introduced to denote genuine Car–Parrinello molecular dynamics schemes [633, 634, 690, 1285]. In general, the basic idea of extended system dynamics is to couple additional, "auxiliary" degrees of freedom to the Lagrangian of interest, thereby "extending" it by increasing the dimensionality of the total phase space. These degrees of freedom are treated like classical particle coordinates, i.e. they are in general characterized by "positions", "momenta", "masses", "interactions", and a "coupling term" to the physical system's degrees of freedom such as particle positions and momenta. In order to distinguish the auxiliary from the physical dynamical variables, they are often called "fictitious degrees of freedom".

Once the extended Lagrangian is defined, the corresponding equations of motion follow from the Euler–Lagrange equations and yield a microcanonical ensemble in the extended phase space. In addition, a strictly conserved total energy can be defined which must include the proper contributions from the auxiliary degrees of freedom as well. In other words, the Hamiltonian of the physical (sub-) system fluctuates and is thus no longer (strictly) conserved and the produced ensemble is no longer the microcanonical one. Any extended system dynamics is constructed such that time averages taken in that part of the phase space associated with the physical degrees of freedom (obtained from a partial trace over the fictitious degrees of freedom) are physically meaningful. Of course, dynamics and thermodynamics of the system are affected by adding the fictitious degrees of freedom, the classic examples being temperature and pressure control by thermostats and barostats, see Section 5.2. Despite this influence due to the coupling, it is often possible to still generate an essentially unperturbed dynamical time evolution within extended systems dynamics, such as for instance in the well-known case of Nosé–Hoover thermostats at variance with velocity scaling approaches to the canonical ensemble [468]. The same is true for Car–Parrinello molecular dynamics, as will be demonstrated in detail in Section 2.4.9.

In the specific case of Car–Parrinello molecular dynamics, the basic Lagrangian that generates Newtonian dynamics of the nuclei is actually extended by classical complex scalar fields, $\{\phi_i(\mathbf{r}), \phi_i^\star(\mathbf{r})\}$, which represent the quantum wave function in terms of orbitals as introduced in Section 2.4.2.

Thus, vector products and absolute values have to be generalized to scalar products and norms of the fields, respectively and derivatives translate into functional derivatives. In addition, the "positions" of these fields $\{\phi_i\}$ actually have a physical meaning, being one-particle wave functions (orbitals), contrary to their momenta $\{\dot{\phi}_i\}$.

2.4.9 Analytic and numerical error estimates

The instantaneous Car–Parrinello forces have been shown necessarily to deviate slightly but *systematically* from the corresponding Born–Oppenheimer forces [943, 1118], as is discussed amply in Section 2.4.3 and captured graphically in Fig. 2.3(b). It has also been recognized for a long time (see for instance Ref. [142] and in particular Ref. 44 therein) that due to the small but *finite* fictitious mass that must be assigned to the electrons, the *primary effect* of $\mu > 0$ is to make the moving nuclei effectively heavier because the electronic degrees of freedom exert some excess inertia on them (i.e. they need to be accelerated whenever the nuclei accelerate), thus "dressing" the nuclei. Being largely a classical mass effect very similar to classical isotope effects, it essentially drops out in thermodynamic and structural properties, whereas the μ-renormalized nuclear masses clearly must affect truly dynamical quantities in a *systematic* way, such as for instance vibrational spectra extracted from time autocorrelation functions at finite temperatures. For such dynamical observables methods have been worked out early on (see Section V.C.2 in Ref. [142]) and used in order to correct for this *systematic* and well-controllable finite-μ error when accurate numbers are needed (see e.g. Refs. [142, 905, 942, 944] for early work). In addition it was shown a long time ago that the finite fictitious mass also affects the definition of the total momentum and its conservation law [1030, 1118], as sketched in Section 5.2.2.

However, there has been some concern in the literature (mixed with confusion due to the shortcomings of available density functionals and insufficient statistical sampling) as to the occurance, magnitude, and thus impact of finite-μ effects [53, 55, 561, 694, 838, 1301, 1440] up to the point of questioning the validity of the Car–Parrinello approach to *ab initio* molecular dynamics as such. Although the situation has been settled quickly [804, 899, 1553] in the sense of confirming, again, the viability of Car–Parrinello sampling by numerical demonstration, this aspect is discussed critically and analyzed in the following, taking several different viewpoints.

As explained in Sections 2.4.3 and 2.4.4, there is an upper limit to the magnitude of μ that can be used to generate *properly* Car–Parrinello trajectories of a desired length; recall that in the mathematical proof sketched in Section 2.4.5 there is a maximum time t^{max}, for a given choice of $\mu^\star$, up to which the generated Car–Parrinello trajectory is well-defined. Eventually, increasing the fictitious electron mass *must* destroy the adiabatic decoupling of the motion of the nuclei and the orbitals in microcanonical Car–Parrinello molecular dynamics by lowering $\omega_{\mathrm{e}}^{\mathrm{min}}$ until it hits $\omega_{\mathrm{n}}^{\mathrm{max}}$ according to Fig. 2.1. In this limit the generated Car–Parrinello dynamics is *no more adiabatic* and the electrons heat up by deviating more and more from the Born–Oppenheimer surface, i.e. the fictitious kinetic energy of the electrons T_{e} no longer oscillates about a constant but increases continuously on the relevant simulation time scale. Still, the conserved energy E^{cons} might well be constant, such that the energy pumped into the fictitious kinetic energy of the electronic degrees of freedom T_{e} must get extracted from the (physical) kinetic energy of the nuclei, thus cooling them continuously.

A simple-minded remedy to "cure" this problem would be to quench the electrons periodically so as to force them back to the Born–Oppenheimer surface (which is occasionally done in published work). This *ad hoc* approach, however, lacks any theoretical basis: the generated trajectories feature discontinuities and the computed quantities do not belong to any properly defined statistical-mechanical ensemble. On the other hand, it is possible to cope with such a situation by separately thermostatting the electrons and the nuclei, thereby keeping both subsystems fluctuating about their preset kinetic energies, see Sections 2.4.4 and 5.2.2. In particular, this assures in such fully thermostatted canonical Car–Parrinello simulations that the electrons stay close to the instantaneous Born–Oppenheimer potential energy surface such that the resulting forces oscillate about the Born–Oppenheimer forces as they should.

The situation is more complicated, but also better understood, when it comes to dynamical observables which depend on the nuclear velocities and hence on the nuclear masses and their renormalization due to the fictitious orbital dynamics [144, 1442]. In the following, these finite-μ effects are investigated in the framework of a perturbative treatment of the Car–Parrinello single-particle wave functions with respect to the corresponding Born–Oppenheimer orbitals, which allows us to work out error estimates and thus error correction schemes. In this spirit, the Car–Parrinello orbitals are decomposed as

$$\phi_i(t) = \phi_i^0(t) + \delta\phi_i(t) \qquad (2.90)$$

where $\{\phi_i^0\}$ are the fully minimized ("Born–Oppenheimer") orbitals which are defined for the instantaneous nuclear configuration $\{\mathbf{R}_I(t)\}$. Inserting this ansatz into the Car–Parrinello equations of motion Eqs. (2.62)–(2.63) and keeping only the leading-order linear terms yields

$$\mathbf{F}_{I,\alpha}^{CP}(t) = \mathbf{F}_{I,\alpha}^{BO}(t)$$

$$+ \sum_i \mu \left\{ <\ddot{\phi}_i| \frac{\partial |\phi_i^0>}{\partial \mathbf{R}_{I,\alpha}} + \frac{\partial <\phi_i^0|}{\partial \mathbf{R}_{I,\alpha}} |\ddot{\phi}_i> \right\} + \mathcal{O}\left(\delta\phi_i^2\right) \ , \quad (2.91)$$

which connects the *instantaneous* Car–Parrinello to the Born–Oppenheimer force components, $\alpha = 1, 2, 3$. In a second step, these forces have to be *averaged* over the appropriate time scale, which is intermediate between the characteristic time scales of electrons and nuclei and characterized by small-amplitude high-frequency oscillations according to Fig. 2.3(b). It can be shown that the orbital acceleration term in Eq. (2.63) vanishes, $\ddot{\phi}_i(t) \approx 0$, independently of μ *on this time scale* and that the average deviation of the force components

$$\overline{\mathbf{F}_{I,\alpha}^{CP}(t)} = \mathbf{F}_{I,\alpha}^{BO}(t) + 2 \sum_i \mu \mathrm{Re} \left\{ \sum_{J,\beta} \ddot{\mathbf{R}}_{J,\beta} \frac{\partial <\phi_i^0|}{\partial \mathbf{R}_{I,\alpha}} \frac{\partial |\phi_i^0>}{\partial \mathbf{R}_{J,\beta}} \right.$$

$$\left. + \sum_{J,K,\beta,\gamma} \dot{\mathbf{R}}_{J,\beta} \dot{\mathbf{R}}_{K,\gamma} \frac{\partial <\phi_i^0|}{\partial \mathbf{R}_{I,\alpha}} \frac{\partial^2 |\phi_i^0>}{\partial \mathbf{R}_{K,\gamma} \partial \mathbf{R}_{J,\beta}} \right\} + \mathcal{O}\left(\delta\phi_i^2\right) \quad (2.92)$$

is *linear* in the fictitious mass parameter μ so that it vanishes properly in the $\mu \to 0$ limit as it should, cf. Eq. (2.76). After this partial averaging of the underlying equations of motion, Eqs. (2.62)–(2.63), the resulting "coarse-grained" Car–Parrinello equations

$$\overline{M_I \ddot{\mathbf{R}}_I(t)} \approx \mathbf{F}_{I,\alpha}^{BO}(t) + 2 \sum_i \mu \mathrm{Re} \left\{ \sum_{J,\beta} \ddot{\mathbf{R}}_{J,\beta} \frac{\partial <\phi_i^0|}{\partial \mathbf{R}_{I,\alpha}} \frac{\partial |\phi_i^0>}{\partial \mathbf{R}_{J,\beta}} \right.$$

$$(2.93)$$

$$\left. + \sum_{J,K,\beta,\gamma} \dot{\mathbf{R}}_{J,\beta} \dot{\mathbf{R}}_{K,\gamma} \frac{\partial <\phi_i^0|}{\partial \mathbf{R}_{I,\alpha}} \frac{\partial^2 |\phi_i^0>}{\partial \mathbf{R}_{K,\gamma} \partial \mathbf{R}_{J,\beta}} \right\} + \mathcal{O}\left(\delta\phi_i^2\right)$$

$$0 \approx -H_e^{HF} \phi_i^0 + \sum_j \Lambda_{ij} \phi_j^0 + \mathcal{O}\left(\delta\phi_i^2\right) \quad (2.94)$$

where

$$\mathbf{F}_{I,\alpha}^{BO}(t) = -\nabla_I \min_{\{\phi_i\}} \left\langle \Psi_0 \left| H_e^{HF} \right| \Psi_0 \right\rangle = -\nabla_I \left\langle \Psi_0^0 \left| H_e^{HF} \right| \Psi_0^0 \right\rangle \quad (2.95)$$

can be written down approximately *on the time scale of the ionic motion.* Thus, on that relevant time scale the electronic structure self-consistency problem is solved dynamically up to second order $\mathcal{O}\left(\delta\phi_i^2\right)$, compare Eq. (2.94) to Eq. (2.53), whereas the molecular dynamics forces on the nuclei feature an excess component that is linear in the fictitious electron mass μ.

In order to proceed analytically, some assumptions have to be made. Following an early analysis [142] (in particular Section V.C.2 therein) it can be assumed that each atom retains its charge density without distortion as it moves in the field of all the other nuclei (akin to the widespread "isolated atom approximation" in X-ray scattering theory). This "rigid ion" [1442] or "infinitely dilute gas" [144] approximation allows us to obtain a handy estimate [1442] of the renormalization of the nuclear masses

$$\Delta_\mu M_I = \frac{2}{3}\frac{m_e}{\hbar^2}\sum_j \mu \left\langle \phi_j^I \left| -\frac{\hbar^2}{2m_e}\nabla_j^2 \right| \phi_j^I \right\rangle > 0 \qquad (2.96)$$

as a function of μ where ϕ_j^I are the *atomic* (valence) orbitals of atom I in vacuum; note that different definitions of μ and occupation numbers affect the numerical prefactor 2/3 and that a generalization has been derived [144] for G-dependent fictitious masses [1447]. As expected, the fictitious electron mass makes the nuclei effectively heavier, $M_I + \Delta_\mu M_I$, and the correction also vanishes in the limit $\mu \to 0$. The term in brackets is the expectation value of the (physical) quantum kinetic energy of the (valence) electrons of the Ith isolated atom, which is anyway computed in atomic electronic structure calculations when constructing pseudopotentials, see Section 3.1.5, Chapter 4, and Chapter 6. Thus, it becomes evident that systems where the electrons are strongly localized close to the nuclei will lead to more pronounced mass renormalization effects in comparison to systems with largely delocalized electrons [1442].

Most importantly, it could be shown by comparing numerical results to analytical predictions that the *major* effect of the excess force Eq. (2.92) is mostly attributable to the trivial mass renormalization Eq. (2.96) due to the nuclei dragging an essentially rigid electron cloud with them [1442]. Hence, quantities that are mass-independent in classical (statistical) mechanics, in particular thermodynamic quantities such as free energies and averaged time-independent observables such as radial distribution functions, remain largely unaffected by the finite fictitious electron mass (given that the Car–Parrinello propagation is adiabatic and continuous as required). This is different for mass-dependent quantities such as velocity or dipole time–correlation functions and the resulting power and infrared spectra, respectively, which are sensitive to mass rescaling. There are several ways to correct for these finite-μ effects. The preferred approach would be to perform simulations using different μ-values with a subsequent extrapolation, $\mu \to 0$, of the quantities of interest. Alternatively, it has been advocated [142, 144, 1442] to perform Car–Parrinello molecular dynamics simulations using for each atom its renormalized mass according to Eq. (2.96) instead of its bare mass M_I.

When it comes to extracting harmonic phonon frequencies from finite-temperature vibrational power spectra, the resulting modes can also be renormalized *a posteriori* [761, 905, 942, 944]. This is extremely simple to demonstrate for a one-dimensional harmonic oscillator with $\omega = (k/M)^{1/2}$, where k is the force constant and M the (reduced) mass, such that $\omega_M/\omega_{M'} = (M'/M)^{1/2}$ or

$$\omega_{BO} = \omega_{CP}\sqrt{1 + \frac{\Delta_\mu M}{M}} \qquad (2.97)$$

if $M' = M + \Delta_\mu M$ is the renormalized mass. A back-of-the-envelope estimate using typical quantum kinetic energies of the valence electrons and $\mu = 500$ a.u. shows that this correction amounts to frequency blue-shifts with respect to ω_{CP} of the order of about 5% of those modes that involve significant hydrogen or oxygen motion [942, 944]. These are the two atoms that are affected mostly in typical molecular applications due to their small mass and high kinetic energy, respectively; further numerical examples are provided in Ref. [144].

The generalization of this frequency renormalization to many coupled vibrational degrees of freedom is well established [1621] in terms of diagonalizing the dynamical matrix using a normal mode fit of Car–Parrinello trajectories [761, 905] and can also be used in order to compute isotope effects on harmonic vibrational frequencies [761, 905, 942, 944]. A *worst-case scenario* concerning frequency shifts induced by finite μ values has been analyzed numerically [1442] using compressed MgO at high temperature ($p \approx 900$ kbar, $T \approx 2800$ K), which is extremely ionic at these conditions with strongly localized electrons close to the oxygen cores. Using a mass of $\mu = 400$ a.u., the highest phonon frequency at about 1000 cm^{-1} is found to be red-shifted by $\approx 10\text{-}15\%$ with respect to the Born–Oppenheimer reference value, whereas this effect reduces already to $\approx 5\%$ when $\mu = 100$ a.u. is used. This extreme case should be contrasted to a covalent solid like crystalline silicon at ambient conditions [1442], where the red-shift is of the order of 1% only (with $\mu = 800$ a.u.). Overall, the finite-μ red-shift of vibrational resonances is typically in the "several percent range" and, if desired, can be corrected for *a posteriori* (by virtue of Eq. (2.96) together with Eq. (2.97)) or made arbitrarily small *a priori* (by performing simulations with small μ).

Besides systematically shifting power spectra to lower frequencies, the mass renormalization also affects the calculation of the physical temperature of the system [1442]

$$(3N - N_c) \, \frac{k_B T}{2} = \left\langle \sum_{I=1}^{N} \frac{1}{2}(M_I + \Delta_\mu M_I)\dot{\mathbf{R}}_I^2 \right\rangle \qquad (2.98)$$

from the average kinetic energy of the nuclei using the usual equipartition theorem estimator where N_c is the total number of conserved/constrained degrees of freedom [25, 468, 577, 1189]. Thus, the temperature is systematically too low if the mass renormalization effect is not taken into account but, again, this effect vanishes in the limit $\mu \to 0$ in view of Eq. (2.96). Within the isolated atom approximation it can be shown that the excess kinetic energy term of the nuclei

$$\left\langle \sum_{I=1}^{N} \frac{1}{2}\Delta_\mu M_I \dot{\mathbf{R}}_I^2 \right\rangle = \sum_i \mu \left\langle \dot{\phi}_i \,\middle|\, \dot{\phi}_i \right\rangle = T_e \qquad (2.99)$$

can be estimated from the total fictitious kinetic energy of the electrons [142, 1442], which is anyway computed during a simulation, so that the renormalization of the physical temperature can easily be corrected for.

Another possibility to investigate the effect of the electron mass on the temperature is to make use of the temperature dependence of a static system property. Within the Born–Oppenheimer approximation such properties are independent of the ionic masses, and we can therefore conclude that different values computed for different electron masses within a Car–Parrinello molecular dynamics simulation are due to the shift in temperature. The strategy to quantify the μ-effect is the following. Based on Born–Oppenheimer molecular dynamics, or experimental results, a scale (thermometer) of the property is established. In the next step the property is calculated using different fictitious electron masses within Car–Parrinello simulations and differences in temperature are normalized using the predetermined scale. Different values of the property obtained at the reference temperature can now again be translated into different temperatures and a comparison with the electron masses allows us to estimate the value for the quantities in Eq. (2.98).

In particular, accurate calculations of the oxygen-oxygen radial distribution function of liquid water will serve as the test data set and the height of the first peak, h_{OO}, is the static property of interest, assuming a linear dependence of h_{OO} on the temperature in the rather small range investigated.

Table 2.1. *Temperature dependence of the height h_{OO} of the first peak of the oxygen-oxygen radial distribution function of liquid water.*

Method	Temperature range [K]	$h_{OO}/T \times 10$	Ref.
Experiment	277-350	0.092	[658]
TIP4P-pol2	298-328	0.090	[658]
BOMD (LCAO basis set)	283-373	0.112	[1553]
BOMD (plane wave basis)	417-426	0.100	[803]

Table 2.2. *Effect of the fictitious electron mass μ on the system temperature T in Car–Parrinello molecular dynamics simulations calculated from the height h_{OO} of the first peak in the oxygen-oxygen radial distribution function of liquid water.*

μ	Temp. [K]	$h_{OO}(T)$	$h_{OO}(T_{\text{ref}})$	ΔT [K]	$\Delta T/\mu \times 100$	Ref.
0	323	3.0 ± 0.2	3.00	0	–	[804]
400	312	3.0 ± 0.2	2.91	9	2.3	[804]
800	310	2.8 ± 0.1	2.67	33	4.1	[804]

From the different values calculated in Table 2.1, an intermediate value of 0.100 is chosen for $h_{OO}/T \times 10$. Taking the consistent set of results from Kuo *et al.* [804], a shift in temperature between 2.3 and 4.1 degrees for an increase in the electron mass by 100 a.u. is found (see Table 2.2). Of course one has to consider the large error bars that are involved in such a calculation. Neither the height of the peak nor the actual temperature are known sufficiently accurately to consider the calculated temperature shifts more than rough estimates. Still, the consistency of the results points toward a real effect, and the size of the temperature shift suggests strongly that taking this effect into account would make reported literature values for h_{OO} more consistent.

2.5 What about Hellmann–Feynman forces?

An important ingredient in all dynamics methods is the efficient calculation of the forces acting on the nuclei, see Eqs. (2.44), (2.46), and (2.59). The straightforward numerical evaluation of the derivative

$$\mathbf{F}_I = -\nabla_I \langle \Psi_0 | \mathcal{H}_e | \Psi_0 \rangle \tag{2.100}$$

in terms of a finite-difference approximation of the total electronic energy is both too costly and too inaccurate for dynamical simulations. What happens if the gradients are evaluated analytically? In addition to the derivative of the Hamiltonian itself

$$\nabla_I \langle \Psi_0 | \mathcal{H}_e | \Psi_0 \rangle =$$

$$\langle \Psi_0 | \nabla_I \mathcal{H}_e | \Psi_0 \rangle + \langle \nabla_I \Psi_0 | \mathcal{H}_e | \Psi_0 \rangle + \langle \Psi_0 | \mathcal{H}_e | \nabla_I \Psi_0 \rangle \tag{2.101}$$

there are in general also contributions from variations of the wave function $\sim \nabla_I \Psi_0$. In general means here that these contributions vanish exactly

$$\mathbf{F}_I^{\mathrm{HFT}} = - \langle \Psi_0 | \nabla_I \mathcal{H}_e | \Psi_0 \rangle \tag{2.102}$$

if the wave function is an exact eigenfunction (or stationary state wave function) of the particular Hamiltonian under consideration. This is the content of the often-cited Hellmann–Feynman Theorem [416, 627, 859], which is also valid for many variational wave functions (e.g. the Hartree–Fock wave function) provided that *complete basis sets* are used. If this is not the case, which has to be assumed for any numerical calculation, the additional terms have to be evaluated explicitly.

In order to proceed, a Slater determinant $\Psi_0 = 1/\sqrt{N!} \det\{\phi_i\}$ of orbitals ϕ_i, which themselves are expanded

$$\phi_i = \sum_\nu c_{i\nu} \, f_\nu(\mathbf{r}; \{\mathbf{R}_I\}) \tag{2.103}$$

in terms of a linear combination of basis functions $\{f_\nu\}$, is used in conjunction with an effective one-particle Hamiltonian (such as e.g. in Hartree–Fock or Kohn–Sham theories). The basis functions might depend explicitly on the nuclear positions (in the case of basis functions with origin such as atom-centered orbitals), whereas the expansion coefficients always carry an implicit dependence. This means that from the outset two sorts of forces are expected

$$\nabla_I \phi_i = \sum_\nu (\nabla_I c_{i\nu}) \, f_\nu(\mathbf{r}; \{\mathbf{R}_I\}) + \sum_\nu c_{i\nu} \, (\nabla_I f_\nu(\mathbf{r}; \{\mathbf{R}_I\})) \tag{2.104}$$

in addition to the Hellmann–Feynman force Eq. (2.102).

Using such a linear expansion Eq. (2.103), the force contributions stemming from the nuclear gradients of the wave function in Eq. (2.101) can be disentangled into two terms. The first one is called "incomplete-basis-set correction" (IBS) in solid-state theory [98, 402, 1391] and corresponds to the "wave function force" [1164] or "Pulay force" in quantum chemistry [1166]. It contains the nuclear gradients of the basis functions

$$\mathbf{F}_I^{\text{IBS}} = -\sum_{i\nu\mu} c_{i\nu}^{\star} c_{i\mu} \left(\langle \nabla_I f_\nu \left| H_{\text{e}}^{\text{NSC}} - \epsilon_i \right| f_\mu \rangle + \langle f_\nu \left| H_{\text{e}}^{\text{NSC}} - \epsilon_i \right| \nabla_I f_\mu \rangle \right)$$

(2.105)

and the (in practice non-self-consistent) effective one-particle Hamiltonian [98, 1391]. The second term leads to the "non-self-consistency correction" (NSC) of the force [98, 1391]

$$\mathbf{F}_I^{\text{NSC}} = -\int d\mathbf{r} \, (\nabla_I n) \left(V^{\text{SCF}} - V^{\text{NSC}} \right)$$

(2.106)

and is governed by the difference between the self-consistent ("exact") potential or field V^{SCF} and its non-self-consistent (or approximate) counterpart V^{NSC} associated with $H_{\text{e}}^{\text{NSC}}$; $n(\mathbf{r})$ is the charge density. In summary, the total force needed in *ab initio* molecular dynamics simulations

$$\mathbf{F}_I = \mathbf{F}_I^{\text{HFT}} + \mathbf{F}_I^{\text{IBS}} + \mathbf{F}_I^{\text{NSC}}$$

(2.107)

comprises in general three qualitatively different terms; a more detailed discussion of core vs. valence states and the effect of pseudopotentials is presented in a tutorial article [402]. Assuming that self-consistency is exactly satisfied (which is *never* going to be the case in numerical calculations), the force $\mathbf{F}_I^{\text{NSC}}$ vanishes and $H_{\text{e}}^{\text{SCF}}$ has to be used in order to evaluate $\mathbf{F}_I^{\text{IBS}}$. The Pulay contribution vanishes in the limit of using a complete basis set (which is also not possible to achieve in actual calculations).

The most obvious simplification arises if the wave function is expanded in terms of originless basis functions such as plane waves, see Eq. (2.139). In this case the Pulay force vanishes exactly, which applies of course to all *ab initio* molecular dynamics schemes using that particular basis set (i.e. Ehrenfest, Born–Oppenheimer, and Car–Parrinello). However, this statement is only true for calculations where the number of plane waves is kept fixed. If the number of plane waves changes, such as in (constant-pressure) calculations with varying cell volume/shape fixing strictly the energy cutoff instead, Pulay stress contributions crop up [440, 457, 470, 530, 1547] as outlined in Section 5.2. If basis sets with origin are used instead of plane waves, Pulay forces always arise and have to be included explicitly in force calculations, see e.g. Refs. [142, 145, 869, 870, 1552] for such methods. Another interesting ramification of using plane waves (or originless basis functions in general) is noted in passing. There is no basis set superposition error

Table 2.3. *Force components in Born–Oppenheimer, Car–Parrinello and Ehrenfest* ab initio *molecular dynamics according to the decomposition Eq. (2.107).*

AIMD	Localized basis	Originless basis
BO	$\mathbf{F}_I^{\mathrm{HFT}} + \mathbf{F}_I^{\mathrm{IBS}} + \mathbf{F}_I^{\mathrm{NSC}}$	$\mathbf{F}_I^{\mathrm{HFT}} + \mathbf{F}_I^{\mathrm{NSC}}$
CP	$\mathbf{F}_I^{\mathrm{HFT}} + \mathbf{F}_I^{\mathrm{IBS}}$	$\mathbf{F}_I^{\mathrm{HFT}}$
E	$\mathbf{F}_I^{\mathrm{HFT}} + \mathbf{F}_I^{\mathrm{IBS}}$	$\mathbf{F}_I^{\mathrm{HFT}}$

(BSSE) [185] in such electronic structure calculations if the same box size and number of plane waves is used, which makes the calculation of energy differences very accurate.

A non-obvious and more delicate term in the context of *ab initio* molecular dynamics is the one stemming from non-self-consistency Eq. (2.106). This term vanishes only if the wave function Ψ_0 is an eigenfunction of the Hamiltonian *within the subspace spanned by the finite basis set used.* This demands less than the Hellmann–Feynman theorem where Ψ_0 has to be an exact eigenfunction of the Hamiltonian and a complete basis set has to be used in turn. In terms of electronic structure calculations, complete self-consistency (within a given incomplete basis set) has to be reached in order that $\mathbf{F}_I^{\mathrm{NSC}}$ vanishes. Thus, in numerical calculations the NSC term can be made arbitrarily small by optimizing the effective Hamiltonian and by determining its eigenfunctions to very high accuracy, but it can never be suppressed completely.

The crucial point is, however, that in Car–Parrinello as well as in Ehrenfest molecular dynamics it is not the minimized expectation value of the electronic Hamiltonian, i.e. $\min_{\Psi_0}\{\langle\Psi_0|\mathcal{H}_e|\Psi_0\rangle\}$, that yields the consistent forces. What is merely needed is to evaluate the expression $\langle\Psi_0|\mathcal{H}_e|\Psi_0\rangle$ with the Hamiltonian and the associated wave function available at a certain time step, compare Eq. (2.46) to Eq. (2.59) or Eq. (2.44). In other words, it is not required (concerning the present discussion of the contributions to the force!) that the expectation value of the electronic Hamiltonian is actually completely minimized for the nuclear configuration at that time step. Hence, full self-consistency is not required for this purpose in the case of Car–Parrinello (and Ehrenfest) molecular dynamics. As a consequence, the non-self-consistency correction to the force $\mathbf{F}_I^{\mathrm{NSC}}$, Eq. (2.106), is irrelevant in Car–Parrinello (and Ehrenfest) simulations, as summarized in Table 2.3.

In Born–Oppenheimer molecular dynamics, on the other hand, the expectation value of the Hamiltonian has to be minimized for each nuclear configuration before taking the gradient to obtain the consistent force! In this scheme there is (independently from the issue of Pulay forces) *always* the non-vanishing contribution of the non-self-consistency force, which is unknown by its very definition (if it were known, the problem would be solved, see Eq. (2.106)). The Born–Oppenheimer force components are compared to the other schemes in Table 2.3. It is noted in passing that there are estimation schemes available that correct *approximately* for this systematic error in Born–Oppenheimer dynamics which leads to significant time savings, see for instance Refs. [46, 786, 1167, 1203, 1552] and Section 3.7.

Heuristically, one could also argue that within Car–Parrinello dynamics the non-vanishing non-self-consistency force is kept under control, or counterbalanced, by the non-vanishing "mass times acceleration term" $\mu\ddot{\phi}_i(t) \approx 0$, which is small but not identical to zero and oscillatory. This is sufficient to keep the propagation stable, whereas $\mu\ddot{\phi}_i(t) \equiv 0$, i.e. an extremely tight minimization $\min_{\Psi_0}\{\langle\Psi_0|\mathcal{H}_e|\Psi_0\rangle\}$, is required by its very definition in order to make the Born–Oppenheimer approach stable, compare Eq. (2.82) to Eq. (2.53). Thus, from this perspective it also becomes clear that the fictitious kinetic energy of the electrons and thus their fictitious temperature is a measure for the departure from the exact Born–Oppenheimer surface during Car–Parrinello dynamics.

Finally, the present discussion shows that nowhere in these force derivations was *use made of* the Hellmann–Feynman theorem as is sometimes stated. Actually, it has been known for a long time that this theorem is quite useless for numerical electronic structure calculations, see e.g. Refs. [98, 1164, 1166] and references therein. Rather, *it turns out* that in the case of Car–Parrinello calculations using a plane wave basis the resulting relation for the force, namely Eq. (2.102), looks like the one obtained by simply invoking the Hellmann–Feynman theorem at the outset.

It is interesting to recall that the Hellmann–Feynman theorem applied to a non-eigenfunction of a Hamiltonian yields only a first-order perturbative estimate of the exact force [627, 859]. The same argument applies to *ab initio* molecular dynamics schemes in which necessary force correction terms according to Eqs. (2.105) and (2.106) are neglected without justification. Furthermore, such simulations can of course not keep the conserved energy E_{cons} defined in Eq. (2.66) strictly constant. Finally, it should be stressed that possible contributions to the force in the nuclear equation of motion as required in view of Eq. (2.59) due to *position-dependent* wave function *constraints* have to be evaluated following the same procedure. This induces similar "correction terms" to the force which have to be taken into account properly [816, 1552] in order to allow for energy-conserving molecular dynamics. As elaborated in more detail in Section 2.6, such contributions to the total force have been neglected in early implementations [604, 605] of *ab initio* molecular dynamics in the framework of Gaussian basis sets, which leads to erroneous conclusions as to the performance of Car–Parrinello vs. Born–Oppenheimer propagation [506].

2.6　Which method to choose?

Presumably the most important question for practical applications is which *ab initio* molecular dynamics method is the most efficient in terms of computer time given a specific problem. An *a priori* advantage of both the Ehrenfest and Car–Parrinello schemes over Born–Oppenheimer molecular dynamics is that no diagonalization of the Hamiltonian (or the equivalent minimization of an energy functional) is necessary, except at the very first step in order to obtain the initial wave function. The difference is, however, that the Ehrenfest time evolution according to the time-dependent Schrödinger equation Eq. (2.39) conforms to a unitary propagation [775, 776, 854]

$$\Psi(t_0 + \Delta t) = \exp\left[-i\mathcal{H}_e(t_0)\Delta t/\hbar\right]\Psi(t_0) \tag{2.108}$$

$$\Psi(t_0 + m\,\Delta t) = \exp\left[-i\mathcal{H}_e(t_0 + (m-1)\Delta t)\,\Delta t/\hbar\right]$$

$$\times \cdots$$

$$\times \exp\left[-i\mathcal{H}_e(t_0 + 2\Delta t)\,\Delta t/\hbar\right]$$

$$\times \exp\left[-i\mathcal{H}_e(t_0 + \Delta t)\,\Delta t/\hbar\right]$$

$$\times \exp\left[-i\mathcal{H}_e(t_0)\,\Delta t/\hbar\right]\Psi(t_0) \tag{2.109}$$

$$\Psi(t_0 + t^{\max}) \stackrel{\Delta t \to 0}{=} \mathsf{T}\exp\left[-\frac{i}{\hbar}\int_{t_0}^{t_0 + t^{\max}}\mathcal{H}_e(t)\,dt\right]\Psi(t_0) \tag{2.110}$$

for infinitesimally short times given by the time step $\Delta t = t^{\max}/m$; here T is the time-ordering operator and $\mathcal{H}_e(t)$ is the Hamiltonian (which is *implicitly* time-dependent via the positions $\{\mathbf{R}_I(t)\}$) evaluated at time t using e.g. split operator techniques [407]. Thus, the wave function Ψ will conserve its norm and in particular orbitals used to expand it will stay orthonormal, see e.g. Ref. [1463]. In Car–Parrinello molecular dynamics, on the contrary, the orthonormality has to be imposed "brute force" using Lagrange multipliers, which amounts to an additional orthogonalization at each molecular dynamics step. If this is not done properly, the orbitals will become nonorthogonal and thus the wave function unnormalized, see e.g. Section III.C.1 in Ref. [1123].

But this theoretical disadvantage of Car–Parrinello vs. Ehrenfest dynamics is in reality more than compensated by the possibility of using a much larger time step in order to propagate the electronic (and thus nuclear) degrees of freedom in the former scheme. In both approaches, there is the time scale inherent in the nuclear motion τ_n and the one stemming from the dynamics of the electrons τ_e. The first can be estimated by considering the highest phonon or vibrational frequency and amounts to the order of $\tau_n \sim 10^{-14}$ s (or 0.01 ps or 10 fs, assuming a maximum frequency of about 4000 cm^{-1}). This time scale depends only on the physics of the problem under consideration and yields an upper limit for the time step Δt^{max} that can be used in order to integrate the equations of motion, e.g. $\Delta t^{max} \approx \tau_n/10$.

The fastest electronic motion in Ehrenfest dynamics can be estimated within a plane wave expansion by $\omega_e^E \sim E_{cut}$, where E_{cut} is the maximum kinetic energy included in the expansion. A realistic estimate for reasonable basis sets is $\tau_e^E \sim 10^{-16}$ s, which leads to $\tau_e^E \approx \tau_n/100$. The analogous relation for Car–Parrinello dynamics reads, however, $\omega_e^{CP} \sim (E_{cut}/\mu)^{1/2}$ according to the analysis in Section 2.4, see Eq. (2.72). Thus, in addition to reducing ω_e^{CP} by introducing a finite electron mass μ, the maximum electronic frequency increases much more slowly in Car–Parrinello than in Ehrenfest molecular dynamics with increasing basis set size. An estimate for the same basis set and a typical fictitious mass yields about $\tau_e^{CP} \sim 10^{-15}$ s or $\tau_e^{CP} \approx \tau_n/10$. According to this simple estimate, the time step can be about one order of magnitude larger if Car–Parrinello second-order fictitious-time electron dynamics is used instead of Ehrenfest first-order real-time electron dynamics.

Some attempts were made to reduce the time scale and thus the time step problem inherent in Ehrenfest dynamics. In Ref. [451] the equations of motion of electrons and nuclei were integrated using two different time steps, that of the nuclei being 20 times larger than the electronic one. The powerful technology of multiple time step integration theory [1493, 1511] could also be applied in order to ameliorate the time scale disparity [1382]. A different approach, borrowed from plasma simulations, consists of decreasing the nuclear masses so that their time evolution is artificially speeded up [1463]. As a result, the *nuclear* dynamics is fictitious (in the presence of real-time electron dynamics!) and has to be rescaled to the proper mass ratio after the simulation.

Clearly, in both Ehrenfest and Car–Parrinello schemes the explicitly treated electron dynamics limits the largest time step that can be used in order to integrate simultaneously the coupled equations of motion for nuclei and electrons. This limitation does not, of course, exist in Born–Oppenheimer dynamics since there is no explicit electron dynamics so that the maximum time step is simply given by the one intrinsic to nuclear motion, i.e. $\tau_e^{\text{BO}} \approx \tau_n$. This is formally an order of magnitude advantage with respect to Car–Parrinello dynamics.

Do these back-of-the-envelope estimates have anything to do with reality? Fortunately, several state-of-the-art studies are reported in the literature for physically similar systems where all three molecular dynamics schemes have been employed. Ehrenfest simulations [451, 1316] of a dilute $K_x \cdot (KCl)_{1-x}$ melt were performed using a time step of 0.012-0.024 fs. In comparison, a time step as large as 0.4 fs could be used to produce a stable Car–Parrinello simulation of electrons in liquid ammonia [331, 332]. Since the physics of these systems has a similar nature - "unbound electrons" dissolved in liquid condensed matter (localizing as F-centers, polarons, bipolarons, etc.) - the time step difference of about a factor of 10 confirms the crude estimate given above. In a Born–Oppenheimer simulation [1350] of again $K_x \cdot (KCl)_{1-x}$ but up to a higher concentration of unbound electrons, the time step used was 0.5 fs.

The time scale advantage of Born–Oppenheimer vs. Car–Parrinello dynamics becomes more evident if the nuclear dynamics becomes fairly slow, such as in liquid sodium [787] or selenium [741] where a time step of 3 fs was used. This establishes the above-mentioned order of magnitude advantage of Born–Oppenheimer vs. Car–Parrinello dynamics in advantageous cases. However, it has to be taken into account that in simulations [741] with such a large time step dynamical information is limited to about 10 THz, which corresponds to frequencies below roughly 500 cm^{-1}. In order to resolve vibrations in molecular systems with stiff covalent bonds the time step has to be decreased to less than a femtosecond (see the estimate given above) also in Born–Oppenheimer dynamics.

The comparison of the overall performance of Car–Parrinello and Born–Oppenheimer molecular dynamics in terms of computer time is a delicate issue. For instance, it depends crucially on the choice made concerning the accuracy of the conservation of the energy E_{cons} as defined in Eq. (2.66). Thus, this issue is to some extent subject to "personal taste" as to what is considered to be a "sufficiently accurate" energy conservation. In addition, this comparison might lead to different conclusions as a function of system size: iterative optimization of a wave function - a task to be done at each time step in Born–Oppenheimer dynamics but only once in Car–Parrinello dynamics - becomes harder the larger the system is. Nevertheless, in order to shed light on this point of practical importance, two sets of simulations were performed. First, microcanonical simulations of eight silicon atoms with various parameters were investigated using Car–Parrinello and Born–Oppenheimer molecular dynamics as implemented in the CPMD simulation package [696]. This large-gap system was initially extremely well equilibrated and the runs were extended to 8 ps (and a few to 12 ps with no noticeable difference) at a temperature of about 360-370 K (with ±80 K root-mean-square fluctuations). The wave function was expanded up to $E_{cut} = 10$ Ry at the Γ-point of a simple cubic supercell and LDA was used to describe the interactions. In both cases the velocity Verlet scheme was used to integrate the equations of motion, see Eq. (3.126); it is noted in passing that the velocity Verlet algorithm [1510] allows for stable integration of the equations of motion, contrary to the statements in Ref. [1209] (see Section 3.4 and Figs. 4-5).

In Car–Parrinello molecular dynamics two different time steps were used, 5 and 10 a.u. (corresponding to about 0.12 and 0.24 fs, respectively), in conjunction with a fictitious electron mass of $\mu = 400$ a.u.; note that μ is very small for this pure silicon system and thus the time step could be increased further if desired by choosing a larger μ value for the comparison. Thus, 10 a.u. time steps also lead to perfect adiabaticity (similar to the one documented in Fig. 2.2), i.e. E_{phys} Eq. (2.67) and T_e Eq. (2.69) did not show a systematic drift relative to the energy scale set by the variations of V_e Eq. (2.68).

Within Born–Oppenheimer molecular dynamics the minimization of the energy functional was done using the highly efficient DIIS (direct inversion in the iterative subspace) scheme using 10 "history vectors", see Section 3.6. Additional savings up to a factor of two can be achieved by using an extrapolation scheme for the initial wave function (see Section 3.7). In this case, the time step was either 10 a.u. or 100 a.u., and three convergence criteria were used; note that the large time step corresponding to 2.4 fs is already beyond the limit that can be used to investigate typical *molecular* systems with feature frequencies of up to 3000 cm^{-1}. The convergence criterion is based on the largest element of the wave function gradient, which was required to be smaller than 10^{-6}, 10^{-5}, or 10^{-4} a.u.; note that the resulting energy convergence shows roughly a quadratic dependence on this criterion.

The outcome of this comparison is shown in Fig. 2.4 in terms of the time

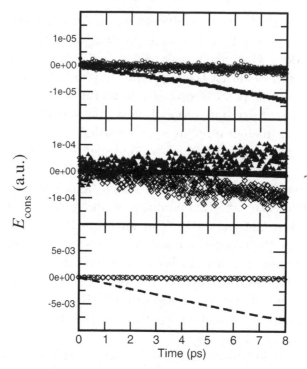

Fig. 2.4. Conserved energy E_{cons} defined in Eq. (2.66) from Car–Parrinello (CP) and Born–Oppenheimer (BO) molecular dynamics simulations of a model system for various time steps and convergence criteria using the CPMD package [696]; see text for further details and Table 2.4 for the corresponding timings. Top: solid line: CP, 5 a.u.; open circles: CP, 10 a.u.; filled squares: BO, 10 a.u., 10^{-6}. Middle: open circles: CP, 10 a.u.; filled squares: BO, 10 a.u., 10^{-6}; filled triangles: BO, 100 a.u., 10^{-6}; open diamonds: BO, 100 a.u., 10^{-5}. Bottom: open circles: CP, 10 a.u.; open diamonds: BO, 100 a.u., 10^{-5}; dashed line: BO, 100 a.u., 10^{-4}.

evolution of the conserved energy E_{cons} Eq. (2.66) on energy scales that cover more than three orders of magnitude in absolute accuracy. Within the present comparison ultimate energy stability was obtained using Car–Parrinello molecular dynamics with the shortest time step of 5 a.u., which conserves the energy of the total system to about 6×10^{-8} a.u. per picosecond, see the solid line in Fig. 2.4(top). Increasing the time step to 10 a.u. leads to an energy conservation of about 3×10^{-7} a.u./ps and much larger energy fluctuations, see open circles in Fig. 2.4(top). The computer time needed in order to generate one picosecond of Car–Parrinello trajectory decreases - to a good approximation - linearly with the increasing time step, see Table 2.4. The most stable Born–Oppenheimer run was performed with

Table 2.4. *Timings in CPU seconds and energy conservation in a.u./ps for Car–Parrinello (CP) and Born–Oppenheimer (BO) molecular dynamics simulations of a model system for 1 ps of trajectory on an IBM RS6000/Model 390 (Power2) workstation using the CPMD package [696]; see text for further details and Fig. 2.4 for corresponding energy plots.*

Method	Time step (a.u.)	Convergence (a.u.)	Conservation (a.u./ps)	Time (s)
CP	5	–	6×10^{-8}	3230
CP	7	–	1×10^{-7}	2310
CP	10	–	3×10^{-7}	1610
BO	10	10^{-6}	1×10^{-6}	16590
BO	50	10^{-6}	1×10^{-6}	4130
BO	100	10^{-6}	6×10^{-6}	2250
BO	100	10^{-5}	1×10^{-5}	1660
BO	100	10^{-4}	1×10^{-3}	1060

a time step of 10 a.u. and a convergence of 10^{-6}. This leads to an energy conservation of about 1×10^{-6} a.u./ps, see filled squares in Fig. 2.4(top).

As the maximum time step in Born–Oppenheimer dynamics is only related to the time scale associated with nuclear motion, it could be increased from 10 to 100 a.u. while keeping the convergence at the same tight limit of 10^{-6}. This worsens the energy conservation slightly (to about 6×10^{-6} a.u./ps), whereas the energy *fluctuations* increase dramatically, see filled triangles in Fig. 2.4(middle) and note the change of scale compared to Fig. 2.4(top). The overall gain is an acceleration of the Born–Oppenheimer simulation by a factor of about seven to eight, see Table 2.4. In the Born–Oppenheimer scheme, the computer time needed for a fixed amount of simulated physical time decreases only sublinearly with increasing time step, since the initial guess for the iterative minimization degrades in quality as the time step is made larger. Further savings of computer time can easily be achieved by decreasing the quality of the wave function convergence from 10^{-6} to 10^{-5} and finally to 10^{-4}, see Table 2.4. This is unfortunately tied to a significant decrease of the energy conservation from 6×10^{-6} a.u./ps at 10^{-6} (filled triangles) to about 1×10^{-3} a.u./ps at 10^{-4} (dashed line) using the same 100 a.u. time step, see Fig. 2.4(bottom) but note the change of scale compared to Fig. 2.4(middle).

In conclusion, Born–Oppenheimer molecular dynamics can be made as fast as (or even faster than) Car–Parrinello molecular dynamics (as measured by the amount of CPU time spent per picosecond) at the expense

Table 2.5. *Characteristic parameters for different Car–Parrinello (CP) and Born–Oppenheimer (BO) molecular dynamics simulations of a system of 32 water molecules; see text for details.*

Simulation	Time step [a.u.]	Convergence	Electron mass [a.u.]	Relative time
CP	3	–	300	1.74
CP	4	–	500	1.29
CP	5	–	700	1.00
BO	20	10^{-4}	–	2.08
BO	20	10^{-5}	–	2.07
BO	20	10^{-6}	–	2.98
BO	20	10^{-7}	–	3.87

of sacrificing accuracy in terms of energy conservation. In the "molecular dynamics community" there is a general consensus that this conservation law should be taken seriously being a measure of the numerical quality of the simulation. In the "quantum chemistry/total energy communities" this issue is typically of less concern. There, it is rather the quality of the convergence of the wave function or energy (as achieved in every individual molecular dynamics step) that is believed to be crucial in order to gauge the quality of a particular simulation.

Finally, attention is turned to a more challenging example: the performance and stability of Born–Oppenheimer and Car–Parrinello molecular dynamics simulations of liquid water. The test system is a well-equilibrated sample of 32 heavy water molecules in a periodic cubic supercell at a temperature close to 350 K and a density of 0.905 g/cm^3 in order to guarantee liquid-state conditions using a standard electronic structure treatment for such water simulations (BLYP density functional [84, 847], Troullier–Martins pseudopotentials [1480], and a plane wave cutoff of 70 Rydbergs). All analysis trajectories were run for two picoseconds in the microcanonical ensemble. Three trajectories with Car–Parrinello molecular dynamics using a fictitious mass of 300 a.u. together with a time step of 3 a.u. (CP300), a mass of 500 a.u. with a time step of 4 a.u. (CP500), and a mass of 700 a.u. with a time step of 5 a.u. (CP700) are analyzed. The Born–Oppenheimer simulations all use a time step of 20 a.u. and different convergence criteria for the wave function optimization. The convergence as measured by the largest element of the wave function gradient was set to 10^{-4} (BO4), 10^{-5} (BO5), 10^{-6} (BO6), and 10^{-7} (BO7), respectively. In Table 2.5 the different parameters of the simulations are collected together with the relative

Table 2.6. *Energy drifts and standard deviations (std) in Kelvin (per degree of freedom) for Car–Parrinello (CP) and Born–Oppenheimer (BO) molecular dynamics simulations of a system of 32 water molecules; see text for details.*

Simulation	Temperature [K]		Physical energy		Conserved energy	
	average	std	drift [K/ps]	std [K]	drift [K/ps]	std [K]
CP300	364	24.1	0.276	1.02	0.007	0.004
CP500	366	24.5	0.736	1.61	0.012	0.006
CP700	365	25.2	3.070	2.37	0.019	0.008
BO4	353	23.0	2.980	0.16	2.980	0.159
BO5	350	19.6	1.584	0.13	1.584	0.129
BO6	341	20.0	0.092	0.12	0.092	0.119
BO7	343	20.7	0.033	0.11	0.033	0.113

CPU time needed for a given length of the simulation. The reference is the Car–Parrinello simulation with the longest time step, CP700, which is the most efficient. For the Born–Oppenheimer simulations the DIIS method to converge the wave functions was used, in combination with a highly efficient extrapolation scheme of the initial guess (see Section 3.7). The average number of iterations needed to achieve convergence was 5, 7, 9, and 11 for convergence criteria of 10^{-4}, 10^{-5}, 10^{-6}, and 10^{-7}, respectively. The Car–Parrinello simulations turn out to be approximately a factor of two to four more efficient than the Born–Oppenheimer simulations.

The analysis of the energies along the trajectory focuses on the energy drift, being a measure of energy conservation, as well as on their standard deviations in order to quantify the energy fluctuations. The drift was calculated from the slope of a fit to a straight line and the standard deviation from the data without the drift component. In Table 2.6 the temperature of the nuclear system, as well as the drifts in the physical total energy (E_{phys} defined in Eq. (2.67), i.e. the kinetic energy of the nuclei plus the potential energy which is the Kohn–Sham total energy) and the conserved energy (E_{cons} defined in Eq. (2.66), i.e. the physical total energy plus the fictitious kinetic energy of the electrons T_e, Eq. (2.69)) are listed together with the standard deviations of these quantities. For Born–Oppenheimer dynamics the values for physical and conserved energy are of course identical, whereas for microcanonical Car–Parrinello dynamics the difference is the fictitious kinetic energy. The energy values are given in Kelvin units per degree of freedom, since thereby the possible effect on the system temperature can easily be guessed. The average temperature of the different simulations varies from about 340 K to 370 K, where the Car–Parrinello simulations show a slightly higher average temperature. The differences are most likely due to the fact that the initial conditions are not exactly the same as the systems had to be equilibrated slightly differently, i.e. the wave functions had to be equilibrated in addition to the nuclear degrees of freedom in case of the Car–Parrinello simulations. The standard deviations for the temperature in the simulations are comparable over all simulations.

As expected, the energy conservation is better for the Car–Parrinello simulations, but the difference from the *well-converged* Born–Oppenheimer runs is not large. For these runs, several tens of picoseconds of simulations could be obtained without a significant change in internal energy. The standard deviation of the conserved energy reflects the different time steps used in the Verlet algorithm (see Section 3.7.1). The fluctuations for the Born–Oppenheimer simulations are larger due to using a 20 a.u. time step, which is much longer than the corresponding values in the Car–Parrinello simulations of 3 a.u. to 5 a.u. only. Comparing the standard deviations of the physical total energy, one can see that now the values for the Car–Parrinello simulations are considerably higher. Whereas, as mentioned before, this quantity is determined only by the time step in a Born–Oppenheimer simulation, one can see the additional effect of the oscillations of the fictitious kinetic energy associated with the orbitals (discussed in Section 2.4.3) in the case of the Car–Parrinello simulations. The size of these oscillations scales with the fictitious electron mass, but it is about *two orders of magnitude* lower than the kinetic energy of the classical nuclei.

There is also a difference in the origin of the energy drift in the physical total energy. Whereas for the Born–Oppenheimer simulations the drift is related to numerical errors in the forces, caused by the incomplete optimization of the wave functions according to the discussion in Section 2.5, the corresponding drift in the Car–Parrinello simulations is caused by the flow of energy between the "cold electrons and the hot nuclei". As can be seen from the data compiled in Table 2.6 these drifts are larger the larger the fictitious electron mass. In particular, for inappropriately large μ, artifacts in the calculated averages can be observed and even a breakdown of the simulation as such can occur in the sense that the fictitious kinetic energy of the electrons, T_e, increases steadily while the conserved energy, E_{cons}, is still roughly constant. However, as mentioned before, it is not a viable solution of the problem to quench the electrons (by minimizing the energy functional) from time to time so as to bring down T_e, although this has been done in published simulations. The technically correct remedy in such cases is to use thermostats for both electronic and nuclear systems, see Section 5.2.2, which prevent this energy exchange and thus eliminate this systematic error.

Finally, it is worth commenting in this particular section on a paper entitled "A comparison of Car–Parrinello and Born–Oppenheimer generalized valence bond molecular dynamics" [506]. In this paper one (computationally expensive) term in the nuclear equations of motion is neglected [933, 1497]. It is well known that using a basis set with origin, such as Gaussians $f_\nu^G(\mathbf{r}; \{\mathbf{R}_I\})$ centered at the nuclei, see Eq. (2.138), produces various Pulay forces, see Section 2.5. In particular a linear expansion Eq. (2.103) or (2.136) based on such orbitals introduces a position dependence into the orthogonality constraint

$$\langle \phi_i | \phi_j \rangle = \sum_{\nu\mu} c_{i\nu}^\star c_{j\mu} \underbrace{\langle f_\nu^G | f_\mu^G \rangle}_{S_{\nu\mu}} = \delta_{ij} \tag{2.111}$$

that is hidden in the overlap matrix elements $S_{\nu\mu}(\{\mathbf{R}_I\})$, which are scalar products of the basis functions. According to Eq. (2.59), such a term produces a constraint force of the type

$$\sum_{ij} \Lambda_{ij} \sum_{\nu\mu} c_{i\nu}^\star c_{j\mu} \frac{\partial}{\partial \mathbf{R}_I} S_{\nu\mu}(\{\mathbf{R}_I\}) \tag{2.112}$$

in the correct Car–Parrinello equation of motion *for the nuclei* similar to the one contained in the electronic equation of motion Eq. (2.60). This term has to be included in order to yield exact Car–Parrinello trajectories and thus energy conservation, see e.g. Eq. (37) in Ref. [816] for a similar situation and Section 6.4 for the derivation of the correct energy gradients and thus proper Car–Parrinello equations of motion when using Vanderbilt's ultrasoft pseudopotentials. In the case of Born–Oppenheimer molecular dynamics, on the contrary, this term is always absent in the nuclear equation of motion, see Eq. (2.46). Thus, the particular implementation [506] underlying the comparison between Car–Parrinello and Born–Oppenheimer molecular dynamics is an approximate one from the outset concerning the Car–Parrinello part; it can be argued that this was justified in the early papers [604, 605], where the basic feasibility of both the Hartree–Fock and generalized valence bond (GVB) Car–Parrinello molecular dynamics techniques was demonstrated [603]. Most importantly, this approximation implies that the energy E_{cons} Eq. (2.66) *cannot be rigorously conserved* in this particular version of Car–Parrinello molecular dynamics. However, energy conservation of E_{cons} was used in Ref. [506] to compare the efficiency and

Table 2.7. *Comparison of Born–Oppenheimer, Car–Parrinello, and Ehrenfest* ab initio *molecular dynamics equations of motion for a single-determinant electronic structure method,* $\Psi_0 = 1/\sqrt{N!}\det\{\phi_i\}$, *with position-independent constraints.*

AIMD	Nuclei	Electronic structure
BO	$M_I \ddot{\mathbf{R}}_I(t) = -\nabla_I \min_{\{\phi_i\}} \{ \langle \Psi_0 \lvert H_e \rvert \Psi_0 \rangle \}$	$0 = -H_e \phi_i + \sum_j \Lambda_{ij} \phi_j$
CP	$M_I \ddot{\mathbf{R}}_I(t) = -\nabla_I \langle \Psi_0 \lvert H_e \rvert \Psi_0 \rangle$	$\mu \ddot{\phi}_i(t) = -H_e \phi_i + \sum_j \Lambda_{ij} \phi_j$
E	$M_I \ddot{\mathbf{R}}_I(t) = -\nabla_I \langle \Psi_0 \lvert H_e \rvert \Psi_0 \rangle$	$i\hbar \dot{\Psi}_0(t) = H_e \Psi_0$

accuracy of these two approaches for this particular implementation of *ab initio* molecular dynamics (using DIIS for the Born–Oppenheimer simulations as done in the above-given performance comparisons). Thus, the final conclusion that for "... approaches that utilize non-space-fixed bases to describe the electronic wave function, Born–Oppenheimer AIMD is the method of choice, both in terms of accuracy and speed" [506] cannot be drawn from this specific comparison for the reasons outlined above (independently of the particular basis set or electronic structure method used).

The toy system investigated here (see Fig. 2.4 and Table 2.4), i.e. eight silicon atoms in a periodic supercell, is for the purpose of comparing different approaches to *ab initio* molecular dynamics quite similar to the test system used in the assessment of Ref. [506], i.e. clusters of four or six sodium atoms (in addition, qualitatively identical results were reported in Section 4 for silicon clusters therein). Thus, it is admissible to compare the energy conservations reported in Figs. 1 and 2 of Ref. [506] to the ones depicted here in Fig. 2.4, noting that the longest simulations reported in Ref. [506] reached only 1 ps. It should be stressed that the energy conservation seen in Fig. 2.4(top) is routinely achieved in Car–Parrinello molecular dynamics simulations. Further information on the relative accuracy and efficiency of Car–Parrinello molecular dynamics compared to Born–Oppenheimer molecular dynamics for localized basis set calculations can be found in the more recent publications [633, 688–690].

Last but not least, a trend is to be observed in the new millennium that the various approaches to *ab initio* molecular dynamics – Born–Oppenheimer, Car–Parrinello, and Ehrenfest propagation – unify themselves. This is possible because it is now increasingly recognized that these methods share important similarities according to Table 2.7 (see also Table 2.3). In particular, it is now broadly acknowledged that at the heart of *any* efficient *ab initio* propagation technique must be a clever prediction scheme of the wave function, which is in practice applied to the set of orbitals as the nuclei move. Transferring ideas from Car–Parrinello fictitious dynamics, this leads to efficient extrapolation methods in order to guess a good wave function for the next molecular dynamics step, see for instance Refs. [634, 795, 862, 1167, 1203, 1552].

2.7 Electronic structure methods

2.7.1 Introduction

Up to this point, the electronic structure method to calculate the *ab initio* forces $\nabla_I \langle \Psi | \mathcal{H}_e | \Psi \rangle$ was not specified in detail. It is immediately clear that *ab initio* molecular dynamics in general, and specifically Car–Parrinello molecular dynamics, is not tied to any particular electronic structure approach, although very accurate techniques are of course prohibitively expensive. It is also evident that the strength or weakness of a particular *ab initio* molecular dynamics study is intimately connected to the strength or weakness of the underlying approximate electronic structure method chosen. This inherent limitation, i.e. mostly the particular density functional used, should not be confused with possible limitations of the algorithm employed in order to generate the dynamics as such, be it the Ehrenfest, Born–Oppenheimer, or Car–Parrinello scheme. Over the years a variety of different approaches such as density functional [75, 78, 222, 634, 787, 1122, 1123, 1286, 1611], Hartree–Fock [423, 551, 605, 607, 620, 634, 707, 853, 862, 896, 1203, 1285, 1286], generalized valence bond (GVB) [503, 504, 506, 604, 606], complete active space SCF (CASSCF) [300, 301], second-order many-body perturbation theory (MP2) [1285], full configuration interaction (FCI) [872], semiempirical [192, 229, 422, 423, 596, 1286, 1587, 1590, 1596], or other approximate [47, 371, 373, 1089, 1090, 1124, 1315] methods were combined with molecular dynamics, and this list is necessarily incomplete in view of the vast and ever-expanding literature on the subject.

The focus of the present discussion is clearly Car–Parrinello molecular dynamics in conjunction with Hohenberg–Kohn–Sham density functional theory [645, 767]. In the following, only those parts of density functional theory are presented that impact directly on *ab initio* molecular dynamics. For a deeper presentation and in particular for a discussion of the assumptions and limitations of this approach (both conceptually and in practice), the reader is referred to the existing excellent literature [363, 397, 716, 760, 762, 913, 1102, 1391]. For simplicity, the formulas are presented here for the spin-unpolarized or restricted special case only.

Following the exposition of density functional theory, the fundamentals of Hartree–Fock theory, which is the basis of a large part of quantum chemistry, are introduced for the same special case. Finally, a glimpse is given of post Hartree–Fock methods. Again, an extensive textbook literature exists for these wave function-based approaches to electronic structure calculations [625, 762, 985, 1423]. The very useful connection between the density-based and wave function-based methods goes back to Löwdin's work in the mid 1950s and is, for instance, worked out in Chapter 2.5 of Ref. [1102], where Hartree–Fock theory is formulated in density-matrix language.

2.7.2 Density functional theory

As discussed in various textbooks [363, 397, 760, 762, 913, 1102], the total ground state energy of an interacting system of electrons with classical nuclei fixed at positions $\{\mathbf{R}_I\}$ can be obtained

$$E_{\text{tot}} = \min_{\Psi_0} \left\{ \langle \Psi_0 | \mathcal{H}_e | \Psi_0 \rangle \right\} = \min_{\{\phi_i\}} E^{\text{KS}}[\{\phi_i\}] \qquad (2.113)$$

as the minimum of the Kohn–Sham energy [645, 767]

$$E^{\text{KS}}[\{\phi_i\}] = T_s[\{\phi_i\}] + \int V_{\text{ext}}(\mathbf{r}) \, n(\mathbf{r}) \, d\mathbf{r}$$

$$+ \frac{1}{2} \int V_{\text{H}}(\mathbf{r}) \, n(\mathbf{r}) \, d\mathbf{r} + E_{\text{xc}}[n] \ , \quad (2.114)$$

which is an explicit functional of the set of auxiliary functions $\{\phi_i(\mathbf{r})\}$, the Kohn–Sham orbitals, that satisfy the orthonormality relation $\langle \phi_i | \phi_j \rangle = \delta_{ij}$. This is a dramatic simplification, since the minimization with respect to all possible *many-body* wave functions $\{\Psi\}$ is replaced by a minimization with respect to a set of orthonormal one-particle functions, the Kohn–Sham orbitals $\{\phi_i\}$. The associated electronic one-body density or charge density

$$n(\mathbf{r}) = \sum_i^{\text{occ}} f_i \, | \phi_i(\mathbf{r}) |^2 \qquad (2.115)$$

is obtained from a single Slater determinant, commonly called the "Kohn–Sham determinant", built from the occupied orbitals, where $\{f_i\}$ are *integer* occupation numbers.

The first term in the Kohn–Sham functional Eq. (2.114) is the kinetic energy of a non-interacting reference system

$$T_s[\{\phi_i\}] = \sum_i^{\text{occ}} f_i \left\langle \phi_i \left| -\frac{1}{2}\nabla^2 \right| \phi_i \right\rangle \tag{2.116}$$

consisting of the same number of electrons exposed to the same external potential as in the fully interacting system. The second term comes from the fixed external potential

$$V_{\text{ext}}(\mathbf{r}) = -\sum_I \frac{Z_I}{|\mathbf{R}_I - \mathbf{r}|} + \sum_{I<J} \frac{Z_I Z_J}{|\mathbf{R}_I - \mathbf{R}_J|} \tag{2.117}$$

in which the electrons move, which comprises the Coulomb interactions between electrons and nuclei and in the definition used here also the internuclear Coulomb interactions. This term changes in the first place if core electrons are replaced by pseudopotentials, see Section 3.1.5, Chapter 4, and Chapter 6 for further details. The third term is the Hartree energy, i.e. the classical electrostatic energy of two charge clouds which stem from the electronic density, and is obtained from the Hartree potential

$$V_{\text{H}}(\mathbf{r}) = \int \frac{n(\mathbf{r}')}{|\mathbf{r} - \mathbf{r}'|}\, d\mathbf{r}'\ , \tag{2.118}$$

which in turn is related to the density via

$$\nabla^2 V_{\text{H}}(\mathbf{r}) = -4\pi n(\mathbf{r}) \tag{2.119}$$

Poisson's equation. The last contribution in the Kohn–Sham functional, the exchange–correlation energy functional $E_{\text{xc}}[n]$, is the most intricate contribution to the total electronic energy. The electronic exchange and correlation effects are lumped together and basically define this functional as the remainder between the exact energy and its Kohn–Sham decomposition in terms of the three previous contributions.

The minimum of the Kohn–Sham functional is obtained by varying the energy functional Eq. (2.114) for a fixed number of electrons with respect to the density Eq. (2.115) or with respect to the orbitals subject to the orthonormality constraint, see e.g. the discussion following Eq. (2.48) for a similar variational procedure. This leads to the Kohn–Sham equations

$$\left\{-\frac{1}{2}\nabla^2 + V_{\text{ext}}(\mathbf{r}) + V_{\text{H}}(\mathbf{r}) + \frac{\delta E_{\text{xc}}[n]}{\delta n(\mathbf{r})}\right\}\phi_i(\mathbf{r}) = \sum_j \Lambda_{ij}\phi_j(\mathbf{r}) \qquad (2.120)$$

$$\left\{-\frac{1}{2}\nabla^2 + V^{\text{KS}}(\mathbf{r})\right\}\phi_i(\mathbf{r}) = \sum_j \Lambda_{ij}\phi_j(\mathbf{r}) \qquad (2.121)$$

$$H_{\text{e}}^{\text{KS}}\phi_i(\mathbf{r}) = \sum_j \Lambda_{ij}\phi_j(\mathbf{r}) \ , \qquad (2.122)$$

which are one-electron equations involving an effective *one-particle* Hamiltonian H_{e}^{KS} with the local potential V^{KS}. Note that H_{e}^{KS} nevertheless embodies the electronic *many-body* effects by virtue of the local exchange–correlation potential

$$\frac{\delta E_{\text{xc}}[n]}{\delta n(\mathbf{r})} = V_{\text{xc}}(\mathbf{r}) \ . \qquad (2.123)$$

A unitary transformation within the space of the occupied orbitals leads to the canonical form

$$H_{\text{e}}^{\text{KS}}\phi_i = \epsilon_i\phi_i \qquad (2.124)$$

of the Kohn–Sham equations, where $\{\epsilon_i\}$ are the Kohn–Sham eigenvalues. In conventional static density functional or "band structure" calculations this set of equations has to be solved self-consistently in order to yield the density, the orbitals, and the Kohn–Sham potential for the electronic ground state [1149]. The corresponding total energy Eq. (2.114) can be written as

$$E^{\text{KS}} = \sum_i \epsilon_i - \frac{1}{2}\int V_{\text{H}}(\mathbf{r})\,n(\mathbf{r})\,d\mathbf{r} + E_{\text{xc}}[n] - \int \frac{\delta E_{\text{xc}}[n]}{\delta n(\mathbf{r})}\,n(\mathbf{r})\,d\mathbf{r} \ , \qquad (2.125)$$

where the first term, the sum over Kohn–Sham eigenvalues, is the "band-structure energy".

Thus, Eqs. (2.120)-(2.122) together with Eqs. (2.52)-(2.53) define Born–Oppenheimer molecular dynamics within Kohn–Sham density functional theory, see e.g. Refs. [45, 75, 78, 508, 786, 787, 1122, 1123, 1462, 1552, 1578, 1611] for such implementations of *ab initio* molecular dynamics. The functional derivative of the Kohn–Sham functional with respect to the orbitals, the Kohn–Sham force acting on the orbitals, can be expressed as

$$\frac{\delta E^{\text{KS}}}{\delta \phi_i^{\star}} = f_i H_{\text{e}}^{\text{KS}}\phi_i \ , \qquad (2.126)$$

which makes clear the connection to Car–Parrinello molecular dynamics, see Eq. (2.60). Thus, Eqs. (2.81)–(2.82) have to be solved with the effective one-particle Hamiltonian in the Kohn–Sham formulation Eqs. (2.120)–(2.122). In the case of Ehrenfest dynamics presented in Section 2.2, the Runge–Gross time-dependent generalization of density functional theory [42, 559, 910, 912, 1255, 1542] has to be invoked instead, see Section 5.3.5 for details on this approach.

Crucial to any application of density functional theory is the approximation of the unknown exchange and correlation functional. A discussion focused on the utilization of such functionals in the framework of *ab initio* molecular dynamics is for instance given in Ref. [1383]. Those exchange-correlation functionals that are most widely used and thus will mainly be considered in the implementation part, Section 3.3, belong to the class of the "Generalized Gradient Approximation" (GGA)

$$E_{\mathrm{xc}}^{\mathrm{GGA}}[n] = \int n(\mathbf{r}') \; \varepsilon_{\mathrm{xc}}^{\mathrm{GGA}}(n(\mathbf{r}'); \nabla n(\mathbf{r}')) \, d\mathbf{r}' \; , \qquad (2.127)$$

where the unknown functional is approximated by an integral over a function that depends only on the density and its gradient at a given point in space, see Ref. [1137] and references therein. The combined exchange–correlation function is typically split up into two additive terms ε_{x} and ε_{c} for exchange and correlation, respectively. In the simplest case it is the exchange and correlation energy density $\varepsilon_{\mathrm{xc}}^{\mathrm{LDA}}(n)$ of an interacting but homogeneous electron gas at the density given by the "local" density $n(\mathbf{r})$ at space-point $\mathbf{r}$ in the inhomogeneous system. This simple but astonishingly powerful approximation [716] is the famous "Local Density Approximation" (LDA) [767] (or "Local Spin Density" (LSD) in the spin-polarized case [1570]), and a host of different parameterizations of $\varepsilon_{\mathrm{xc}}^{\mathrm{LDA}}(n)$ exist in the literature [363, 397, 760, 762, 913, 1102]. The "Self-Interaction Correction" (SIC) [1138] as applied to LDA was assessed critically for molecules in Ref. [526] with a rather disappointing outcome, so that modified schemes [298, 1557] have been proposed for open-shell systems.

A significant improvement of the accuracy has been achieved by introducing the gradient of the density as indicated in Eq. (2.127). These GGAs (sometimes also denoted as "gradient corrected" or "semi-local" functionals) extend the applicability of density functional calculations to the realm of chemistry, see e.g. Refs. [84, 847, 1130, 1131, 1134, 1137] for a few "popular functionals" and Refs. [279, 396, 712, 719, 1361, 1395, 1397] for extensive tests on molecules, hydrogen-bonded complexes, solids, and surfaces.

According to Perdew, the GGAs are the second rung on "Jacob's Ladder from earth to heaven" [972, 1136], whereas LDA/LSD is the first one.

Another considerable advance was the successful introduction of "hybrid functionals" [85, 86] that include to some extent "Hartree–Fock" exchange according to a fixed mixing parameter or full "exact exchange" (EXX) [544] in addition to the standard GGA expression Eq. (2.127). Although such truly nonlocal and orbital-dependent exchange functionals can certainly be implemented within plane wave approaches [257, 572, 943, 1470], they are quite time-consuming in the realm of molecular dynamics simulations for the reasons outlined at the end of Section 3.3, where the algorithmic problems are presented. Still, liquid water has been investigated using several such hybrid functionals (including also full Hartree–Fock calculations) in the framework of *ab initio* molecular dynamics [1470], and static properties of solids such as lattice constants, heats of formation, and band gaps have been determined [1097].

The "Optimized Potential Method" (OPM) or "Optimized Effective Potentials" (OEP) establish another route to include "exact exchange" within density functional theory [913], see e.g. Section 13.6 in Ref. [1383] or Ref. [547] for overviews. Here, the exchange–correlation functional $E_{xc}^{OPM} = E_{xc}[\{\phi_i\}]$ depends on the individual orbitals instead of only on the density or its derivatives.

Another scheme to go beyond GGAs leads to those functionals that include higher-order powers of the gradient or the local kinetic energy density in the sense of a generalized gradient expansion beyond the first term. Promising results could be achieved by including the Laplacian or the (local) Kohn–Sham orbital kinetic energy density [167, 429, 430, 1163, 1443, 1545]. Such meta-GGAs (MGGAs) are considered to be the third rung on Jacob's ladder [1443]. The fourth rung or hyper-GGA (HGGA) [1136] uses, in addition to the ingredients of meta-GGAs, the exact exchange energy density, i.e. the local Fock exchange integral with a space-dependent mixing parameter leading to "local hybrid functionals"; note that "global hybrids" [85, 86] instead depend on the integrated Hartree–Fock exchange energy. According to this line of thinking, the fifth and final rung of Jacob's ladder, the generalized random phase approximation (GRPA), utilizes all of the Kohn–Sham orbitals, occupied as well as unoccupied [1136].

Before closing this short presentation of Kohn–Sham density functional theory, it should be stressed that the development of new parameterizations and new density functional concepts as such continues to be a very active area of research, so that currently several novel functionals are proposed per year.

2.7.3 Hartree–Fock theory

Hartree–Fock (HF) or self-consistent field (SCF) electronic structure theory can be derived by invoking the variational principle in a restricted space of wave functions as introduced in many textbooks [625, 762, 985, 1423]. In particular, the antisymmetric ground state electronic wave function is approximated by a *single* Slater determinant $\Psi_0 = 1/\sqrt{N!} \det\{\phi_i\}$ which is constructed from a set of one-particle spin orbitals $\{\phi_i\}$, the Hartree–Fock orbitals, required to be mutually orthonormal $\langle \phi_i | \phi_j \rangle = \delta_{ij}$. The corresponding variational minimum of the total electronic energy $\mathcal{H}_e$ defined in Eq. (2.2)

$$E^{\mathrm{HF}}[\{\phi_i\}] = 2 \sum_i \int \phi_i^\star(\mathbf{r}) \left[-\frac{1}{2}\nabla^2 + V_{\mathrm{ext}}(\mathbf{r}) \right] \phi_i(\mathbf{r}) \, d\mathbf{r}$$

$$+ 2 \sum_{ij} \int \int \phi_i^\star(\mathbf{r}) \phi_j^\star(\mathbf{r}') \frac{1}{|\mathbf{r} - \mathbf{r}'|} \, \phi_i(\mathbf{r}) \phi_j(\mathbf{r}') \, d\mathbf{r} \, d\mathbf{r}'$$

$$- \sum_{ij} \int \int \phi_i^\star(\mathbf{r}) \phi_j^\star(\mathbf{r}') \frac{1}{|\mathbf{r} - \mathbf{r}'|} \, \phi_j(\mathbf{r}) \phi_i(\mathbf{r}') \, d\mathbf{r} \, d\mathbf{r}' \quad (2.128)$$

yields the lowest energy and the "best" wave function within the one-determinant ansatz (in the spin-restricted form applicable for closed-shell cases); the external Coulomb potential V_{ext} was already defined in Eq. (2.117). Carrying out the constraint minimization within this ansatz (see Eq. (2.49) and the following discussion in Section 2.3 for a sketch) leads to

$$\left\{ -\frac{1}{2}\nabla^2 + V_{\mathrm{ext}}(\mathbf{r}) + 2 \sum_j \mathcal{J}_j(\mathbf{r}) - \sum_j \mathcal{K}_j(\mathbf{r}) \right\} \phi_i(\mathbf{r}) = \sum_j \Lambda_{ij} \phi_j(\mathbf{r}) \quad (2.129)$$

$$\left\{ -\frac{1}{2}\nabla^2 + V^{\mathrm{HF}}(\mathbf{r}) \right\} \phi_i(\mathbf{r}) = \sum_j \Lambda_{ij} \phi_j(\mathbf{r}) \quad (2.130)$$

$$H_e^{\mathrm{HF}} \phi_i(\mathbf{r}) = \sum_j \Lambda_{ij} \phi_j(\mathbf{r}) \quad (2.131)$$

the (spin-restricted) Hartree–Fock integro-differential equations. In analogy to the Kohn–Sham equations Eqs. (2.120)-(2.122), these are effective one-particle equations that involve an effective one-particle Hamiltonian H_e^{HF}, the Fock operator. The diagonal set of canonical orbitals

$$H_e^{\mathrm{HF}} \phi_i = \epsilon_i \phi_i \quad (2.132)$$

is obtained similarly to Eq. (2.124), where $\{\epsilon_i\}$ are the Hartree–Fock eigenvalues. The Coulomb operator

$$\mathcal{J}_j(\mathbf{r})\, \phi_i(\mathbf{r}) = \left[\int \phi_j^\star(\mathbf{r}') \frac{1}{|\mathbf{r} - \mathbf{r}'|} \phi_j(\mathbf{r}')\, d\mathbf{r}'\right]\, \phi_i(\mathbf{r}) \qquad (2.133)$$

and the exchange operator

$$\mathcal{K}_j(\mathbf{r})\, \phi_i(\mathbf{r}) = \left[\int \phi_j^\star(\mathbf{r}') \frac{1}{|\mathbf{r} - \mathbf{r}'|} \phi_i(\mathbf{r}')\, d\mathbf{r}'\right]\, \phi_j(\mathbf{r}) \qquad (2.134)$$

are most easily defined via their action on a particular orbital ϕ_i. It is found that upon acting on orbital $\phi_i(\mathbf{r})$, the exchange operator for the jth state "exchanges" $\phi_j(\mathbf{r}') \to \phi_i(\mathbf{r}')$ in the kernel as well as replaces $\phi_i(\mathbf{r}) \to \phi_j(\mathbf{r})$ in its argument, compare to the Coulomb operator. Thus, $\mathcal{K}$ is a nonlocal operator as its action on a function ϕ_i at point $\mathbf{r}$ in space requires the evaluation and thus the knowledge of that function throughout all space by virtue of $\int d\mathbf{r}'\, \phi_i(\mathbf{r}')\ldots$ the required integration. In this sense the exchange operator does not possess a simple classical interpretation like the Coulomb operator $\mathcal{J}$, which is the counterpart of the Hartree potential V_{H} in Kohn–Sham theory. The exchange operator vanishes exactly if the anti-symmetrization requirement of the wave function is relaxed, i.e. only the Coulomb contribution survives if a Hartree product is used to represent the wave function.

The force acting on the orbitals is defined

$$\frac{\delta E^{\mathrm{HF}}}{\delta \phi_i^\star} = H_{\mathrm{e}}^{\mathrm{HF}} \phi_i \qquad (2.135)$$

similarly to Eq. (2.126). At this stage, the various *ab initio* molecular dynamics schemes based on Hartree–Fock theory are defined, see Eqs. (2.52)-(2.53) for Born–Oppenheimer molecular dynamics and Eqs. (2.81)-(2.82) for Car–Parrinello molecular dynamics. In the case of Ehrenfest molecular dynamics, the time-dependent Hartree–Fock formalism [344] has to be invoked instead.

2.7.4 Post Hartree–Fock theories

Although post Hartree–Fock methods have a very unfavorable scaling of the computational cost as the number of electrons increases, a few case studies were performed with such correlated quantum chemistry techniques. For instance, *ab initio* molecular dynamics was combined with GVB [503, 504, 506, 604, 606], CASSCF [300, 301], MP2 [1285], as well as FCI [872] approaches, see also references therein. It is noted in passing that Car–Parrinello molecular dynamics can only be implemented straightforwardly if energy and wave function are "consistent". This is not the case in perturbation theories such as e.g. the widely used Møller-Plesset approach [619]: within standard MP2 the energy is correct to second order, whereas the wave function is the one given by the uncorrelated HF reference. As a result, the derivative of the MP2 energy with respect to the wave function Eq. (2.135) does not yield the correct force on the Hartree–Fock wave function in the sense of fictitious dynamics. Such problems are, of course, absent from the Born–Oppenheimer approach to sample configuration space, see e.g. Refs. [72, 708, 737] for MP2, density functional, and multi-reference CI (MRCI) *ab initio* Monte Carlo schemes or Ref. [1285] for a recent Born–Oppenheimer molecular dynamics simulation using MP2 energies and forces.

It should be kept in mind that the rapidly growing workload of post-HF calculations, although extremely powerful in principle, limits the number of explicitly treated electrons to only a few. However, the rapid development of correlated electronic structure methods that scale linearly with the number of electrons will certainly broaden the range of applicability of this class of techniques in the near future [518, 524, 1162].

2.8 Basis sets

2.8.1 Gaussians and Slater functions

Having selected a specific electronic structure method, the next choice is related to which basis set to use in order to represent the orbitals ϕ_i in terms of simple functions f_ν with well-known properties. In general, a *linear* combination of such basis functions

$$\phi_i(\mathbf{r}) = \sum_\nu c_{i\nu} f_\nu(\mathbf{r}; \{\mathbf{R}_I\}) \tag{2.136}$$

is used, which represents exactly any reasonable function in the limit of using a complete set of basis functions. In quantum chemistry, Slater-type basis functions (STOs)

$$f_\mathbf{m}^\mathrm{S}(\mathbf{r}) = N_\mathbf{m}^\mathrm{S}\, r_x^{m_x} r_y^{m_y} r_z^{m_z}\, \exp\left[-\zeta_\mathbf{m}|\mathbf{r}|\right] \tag{2.137}$$

with an exponentially decaying radial part, but in particular Gaussian-type basis functions (GTOs)

$$f_\mathbf{m}^\mathrm{G}(\mathbf{r}) = N_\mathbf{m}^\mathrm{G}\, r_x^{m_x} r_y^{m_y} r_z^{m_z}\, \exp\left[-\alpha_\mathbf{m} r^2\right] \tag{2.138}$$

as suggested by Boys, are used overwhelmingly in software packages [619]. Here, $N_{\mathbf{m}}$, $\zeta_{\mathbf{m}}$, and $\alpha_{\mathbf{m}}$ are constants that are typically kept fixed during a molecular electronic structure calculation so that only the orbital expansion coefficients $c_{i\nu}$ need to be optimized. In addition, fixed linear combinations of the above-given "primitive" basis functions can be used for a given angular momentum channel $\mathbf{m}$, which defines the "contracted" basis sets.

The Slater or Gaussian basis functions are in general centered at the positions of the nuclei ("atoms"), i.e. $\mathbf{r} \to \mathbf{r} - \mathbf{R}_I$ in Eqs. (2.137)–(2.138), which leads to the "Linear Combination of Atomic Orbitals" (LCAO) ansatz to solve differential equations algebraically. Furthermore, in particular for Gaussians the derivatives as well as the resulting matrix elements are obtained efficiently by differentiation and integration in real space. However, Pulay forces (see Section 2.5) result necessarily for such basis functions that are fixed at atoms (or bonds) if the atoms are allowed to move, either in geometry optimization or molecular dynamics schemes. This disadvantage can be circumvented by using *freely* floating Gaussians that are distributed in space [924, 1388], which form an originless basis set since it is localized but not atom-fixed.

A first generation of methods using Gaussian basis functions in the context of *ab initio* molecular dynamics was proposed roughly in the early to mid 1990s in the sense of interfacing existing electronic structure codes with a driver for molecular dynamics, see for instance Refs. [300, 301, 423, 503, 504, 506, 551, 604–607, 620, 707, 872, 896, 924]. More recently, a second generation of such approaches that take advantage more explicitly of the dynamical evolution of the electronic degrees of freedom, which is conceptually at the very root of the efficiency of the original Car–Parrinello algorithm (see Section 3.7), is developed in the framework of Car–Parrinello [633, 635, 686–690, 1203, 1204, 1285, 1286], Born–Oppenheimer [634, 1167, 1552], and Ehrenfest [862] dynamics schemes up to the point where one could say that all these approaches to *ab initio* molecular dynamics are confluent [795].

Numerical basis sets [328, 723] following the same spirit as the two analytical function-based kinds presented above are also used widely. For these basis sets the radial part of the function is typically optimized according to a prescribed recipe in an atomic calculation. These functions have an additional flexibility and can be constructed with a precise *finite* support. Especially in solid-state calculations, such basis sets can lead to extremely efficient numerical schemes with linear scaling in system size.

2.8.2 *Plane waves*

A vastly different approach has its roots in solid-state theory. Here, the periodicity of the underlying lattice produces a periodic potential and thus imposes the same periodicity on the density (implying Bloch's Theorem, Born-von Kármán periodic boundary conditions, etc., see for instance Chapter 8 in Ref. [48]). This suggests strongly using plane waves as the generic basis set in order to expand the periodic part of the orbitals (see Section 3.1.2). Plane waves are defined as

$$f_{\mathbf{G}}^{\mathrm{PW}}(\mathbf{r}) = N \exp\left[i\mathbf{Gr}\right] \ , \tag{2.139}$$

where the normalization is simply given by $N = 1/\sqrt{\Omega}$; Ω is the volume of the periodic (super-) cell. Since plane waves form a complete and orthonormal set of functions, they can be used to expand orbitals according to Eq. (2.136), where the labeling index ν is simply given by the vector $\mathbf{G}$ in reciprocal space/G-space (including only those $\mathbf{G}$-vectors that satisfy the particular periodic boundary conditions). The total electronic energy and thus its gradient is found to have a particularly simple form when expressed in plane waves [673, 1149], as discussed in Section 3.4 in quite general terms.

It is important to observe that plane waves are originless functions, i.e. they do *not* depend on the positions of the nuclei $\{\mathbf{R}_I\}$. This implies that the Pulay forces Eq. (2.105) vanish exactly even within a *finite* basis (and using a fixed number of plane waves, see the discussion related to "Pulay stress" in Section 2.5), which facilitates tremendously force and stress tensor calculations as worked out in Section 3.4. This also implies that plane waves are a very unbiased basis set in that they are "delocalized" in space and do not "favor" certain atoms or regions over others, i.e. they can be considered as an ultimately "balanced basis set" in the language of quantum chemistry. Thus, the only way to improve the quality of the basis is to increase the "energy cutoff" E_{cut}, i.e. to increase the largest $|\mathbf{G}|$-vector that is included in the finite expansion Eq. (2.136), see Section 3.1.3 and in particular Eq. (3.17) therein. This blind approach is vastly different from the traditional procedures in quantum chemistry that are needed in order to produce reliable basis sets [619, 625]. Using plane waves is also advantageous when dealing with "delocalized electrons", such as solvated electrons for instance in anionic water clusters, where the addition of diffuse functions is both necessary and tricky when using Gaussian basis sets. Another appealing feature is that derivatives in real space are simply multiplications in G-space, and both spaces can be connected efficiently via Fast Fourier Transforms (FFTs) as exploited in Section 3.1.4. Thus, one can easily evaluate operators in that space in which they are diagonal, see for instance the flow charts in Fig. 3.1 or Fig. 3.3.

According to the well-known "No Free Lunch Theorem", there cannot be only advantages connected to using plane waves. The first point is that the pseudopotential approximation as explained in Chapter 4 is connected intimately to using plane waves, why so? A plane wave basis is basically a lattice-symmetry-adapted three-dimensional Fourier decomposition of the orbitals. This means that increasingly large Fourier components are needed in order to resolve structures in real space on decreasingly small length (distance) scales. But already orbitals of first row atoms feature quite strong and rapid oscillations close to the nuclei due to the Pauli principle, which enforces a nodal structure to the wave function by imposing orthogonality of the orbitals. However, most of chemistry is ruled by the valence electrons, whereas the core electrons are essentially inert. In practice, this means that the innermost electrons can be taken out of explicit calculations. Instead, they are represented by a smooth and nodeless effective potential, the pseudopotential [280, 628, 629, 1145, 1146], see for instance Refs. [472, 1149, 1365] for reviews in the context of solid-state theory and Refs. [294, 349] for pseudopotentials in quantum chemistry (called "effective core potentials", ECPs). The basic idea is to make the resulting pseudo wave function that represents the valence electrons as smooth as possible close to the nuclear core region. This also means that properties that depend crucially on the wave function close to the core cannot be obtained straightforwardly from such calculations.

In the field of plane wave calculations the introduction of "soft" norm-conserving *ab initio* pseudopotentials as outlined in Section 4.2 has been a breakthrough, both conceptually [594] and in practice [57]. Another important contribution, especially for transition metals, was the introduction of the ultrasoft pseudopotentials by Vanderbilt [1548] (see Section 6.4) and the related projector augmented wave (PAW) method due to Blöchl [142, 145] (see Section 6.2). These approaches lead to the powerful technique of plane wave/pseudopotential electronic structure calculations in the framework of density functional theory [673, 1149]. Within this particular framework the issue of treating pseudopotentials in general terms is elaborated in more technical detail in Section 3.1.5, in Chapter 4 and in Chapter 6.

Another severe shortcoming of plane waves is the flipside of the medal of being an unbiased basis set: there is no way to shuffle more basis functions into regions in space where they are needed more than in other regions. This is particularly bad for systems with strong inhomogeneities. Such examples are all-electron calculations or the inclusion of semi-core states, a few heavy atoms in a sea of light atoms, and (semi-) finite systems such as surfaces or molecules where a large vacuum region is needed in order to allow the wave function to decay, as argued in Section 3.2.3. In the latter case quite an effort has to be expended in order to represent the large vacuum region, whereas in the former cases the cutoff parameter and thus the number of plane waves has to be increased enormously, although a better representation of the wave function would only be required in a few regions in real space. This is often referred to as the multiple length scale deficiency of plane wave calculations [223].

2.8.3 Generalized plane waves

An extremely appealing and elegant generalization of the plane wave concept [568, 569] consists of defining them in curved ξ-space

$$f_{\mathbf{G}}^{\mathrm{GPW}}(\xi) \;=\; N \det{}^{1/2} J \exp\left[i\mathbf{G}\,\mathbf{r}(\xi)\right] \tag{2.140}$$

$$\det J \;=\; \left| \frac{\partial r^i}{\partial \xi^j} \right| \;,$$

where $\det J$ is the Jacobian of the transformation from Cartesian to curvilinear coordinates $\mathbf{r} \rightarrow \xi(\mathbf{r})$ with $\xi = (\xi^1, \xi^2, \xi^3)$ and $N = 1/\sqrt{\Omega}$ as for regular plane waves. These functions are orthonormal, form a complete basis set, can be used for $\mathbf{k}$-point sampling after replacing $\mathbf{G}$ by $\mathbf{G} + \mathbf{k}$ in Eq. (2.140), are originless (but nevertheless localized) so that Pulay forces are absent, can be manipulated via efficient FFT techniques, and reduce to standard plane waves in the special case of a Euclidean space $\xi(\mathbf{r}) = \mathbf{r}$. Thus, they can be used equally well like plane waves in linear expansions of the sort Eq. (2.103) underlying most of electronic structure calculations. The Jacobian of the transformation is related to the Riemannian metric tensor

$$g_{ij} \;=\; \sum_{k=1}^{3} \frac{\partial \xi^k}{\partial r^i} \frac{\partial \xi^k}{\partial r^j}$$

$$\det J \;=\; \det{}^{-1/2} \{g_{ij}\} \tag{2.141}$$

which defines the metric of the ξ-space. The metric and thus the curvilinear coordinate system itself is considered as a variational parameter in the original fully adaptive-coordinate approach [568, 569], see also Refs. [339, 587–590, 593]. Thus, a uniform grid in curved Riemannian space is non-uniform or distorted when viewed in flat Euclidean space (where $g_{ij} = \delta_{ij}$) such that the density of grid points (or the "local" cutoff energy of the expansion in terms of **G**-vectors) is highest in regions close to the nuclei and lowest in vacuum regions, see Fig. 2 in Ref. [587].

Concerning actual calculations, this means that a lower number of generalized plane waves than standard plane waves are needed in order to achieve a given accuracy [568], see Fig. 1 in Ref. [587]. This allows even for all-electron approaches to electronic structure calculations where plane waves fail [1017, 1169, 1585, 1668]. More recently, the distortion of the metric was frozen spherically around atoms by introducing deformation functions [570, 574, 593], which leads to a concept closely connected to non-uniform atom-centered meshes in real space methods [1017, 1585, 1668] presented in Section 2.8.8. In such non-fully-adaptive (also called "locally adaptive" [593]) approaches based on using *predefined* coordinate transformations, attention has to be payed to Pulay force contributions (see Section 2.5), which have to be evaluated explicitly [570, 1017].

2.8.4 Wavelets

Similar to using generalized plane waves is the idea of exploiting the powerful multiscale properties of wavelets [3, 517]. Since this approach requires an extensive preparatory discussion of the basics (see e.g. Ref. [521] for a gentle introduction), and since it seems still quite far from being used in large-scale electronic structure calculations, the interested reader is referred to original publications [268, 312, 390, 520, 522, 1520, 1602, 1639], review articles [44, 519], and the general wavelet literature cited therein. Wavelet-based methods allow us intrinsically to exploit multiple length scales without introducing Pulay forces and can be handled efficiently by fast wavelet transforms. In addition, they are also a powerful route to linear scaling (order-N, $O(N)$) methods [518, 524, 1087], as first demonstrated in Ref. [520] with the calculation of the Hartree potential for an all-electron uranium dimer.

2.8.5 Discrete variable representations

Discrete variable representations (DVRs) have been introduced by Light and coworkers for bound state and scattering calculations, see Ref. [864] for a review. More recently, their use in the framework of *ab initio* molecular dynamics has been pioneered by Tuckerman and coworkers [848, 849, 871, 1498].

The approach is based on the use of continuous functions that satisfy the properties of position eigenfunctions on an auxiliary grid. The principle advantage of DVRs is their high localization about the auxiliary grid points.

While plane wave basis sets have the advantage of simplicity, they lack the spatial bias inherent in Gaussian basis sets. The problem of delocalization is eliminated largely with Gaussian basis sets or similar localized functions at the expense of considerably increased complexity due to the nonorthogonality of the basis functions and the introduction of a basis set superposition error. A significant advantage might be gained if the localized character of Gaussian basis sets could be achieved using simple, orthonormal basis functions that are not centered on atomic positions, thereby avoiding the complexity of Pulay forces and basis set super position errors. In this respect a DVR basis is comparable in spirit to wavelets. However, the hierarchical structure of the latter is absent in DVRs.

A one-dimensional DVR is composed of a set of N functions, $u_i(x), i = 1, \ldots, N$, and a set of N grid points, x_i, such that the basis functions satisfy

$$u_i(x_j) = \frac{\delta_{ij}}{a_i} \tag{2.142}$$

on the grid points and the complex numbers a_i define a set of quadrature weights with $w_i = |a_i|^2$. Therefore, the DVR functions behave like coordinate eigenfunctions on a particular quadrature grid and are, in this sense, the real space analogues of plane waves, which are momentum eigenfunctions. Moreover, they can be constructed in such a way that each DVR function is well localized about a point on the grid. The basis functions then satisfy orthogonality and completeness relations of the form

$$\sum_{k=1}^{N} w_k \, u_i^{\star}(x_k) \, u_j(x_k) = \delta_{ij} \tag{2.143}$$

$$\sum_{k=1}^{N} w_k \, u_k^{\star}(x_i) \, u_k(x_j) = \delta_{ij} \ . \tag{2.144}$$

DVR functions can be constructed from simpler basis functions according to the boundary conditions of the problem. A DVR appropriate for periodic boundary conditions on a one-dimensional grid of N points can, for example, be constructed from a sum of cosine functions

$$u_l(x) = \sqrt{\frac{1}{NL}} \sum_{\lambda=1}^{N} \cos[k_\lambda(x - x_l)] \ , \tag{2.145}$$

where $k_\lambda = 2\pi(\lambda - N' - 1)/L$, $\lambda, l = 1, \ldots, N$, and $N' = (N-1)/2$. The full basis set in three dimensions is then constructed from these one-dimensional

DVRs by a direct product according to

$$\varphi_{lmn}(\mathbf{r}) = u_l(x)\, v_m(y)\, q_n(z) \ . \tag{2.146}$$

The form of the terms in the energy functional follows directly from the form of the basis functions and thus the kinetic energy can be expressed analytically. Coulomb interactions can be evaluated using multigrid or multipole methods, or alternatively using a hybrid approach with an intermediate representation of the density using plane waves. Nonlocal and short-ranged parts of the local pseudopotentials are easily evaluated in real space.

The electronic charge density can be computed from the expansion of the Kohn–Sham orbitals

$$\phi_i(\mathbf{r}) = \sum_{lmn} C^i_{lmn}\varphi_{lmn}(\mathbf{r}) \ , \tag{2.147}$$

where the resulting expression

$$n(\mathbf{r}) = \sum_i f_i \left\{ \sum_{lmn} C^i_{lmn}\varphi_{lmn}(\mathbf{r}) \right\}^2 \tag{2.148}$$

simplifies considerably when the density is evaluated at the grid points. The gradient of the density has to be calculated accordingly, but the derivatives of the DVR functions can be calculated and stored on the one-dimensional grids.

An alternative method in the spirit of auxiliary basis expansion of the density is the direct representation of the density in the DVR basis

$$n(\mathbf{r}) = \sum_{lmn} N_{lmn}\varphi_{lmn}(\mathbf{r}) = \sum_{lmn} N_{lmn} u_l(x)\, v_m(y)\, q_n(z) \ , \tag{2.149}$$

which has a considerably simpler form of the gradient.

2.8.6 Augmented and mixed basis sets

Localized Gaussian basis functions on the one hand and plane waves on the other hand are certainly two extreme cases. There have been tremendous and longstanding efforts to combine such localized and originless basis functions in order to exploit their mutual strengths [880]. This has resulted in a rich variety of mixed and augmented basis sets with very specific implementation requirements. This topic will not be covered here, and the interested reader is referred to Refs. [142, 145, 869, 870, 1173, 1523, 1524, 1552] and references given therein for some recent implementations used in conjunction with *ab initio* molecular dynamics.

2.8.7 Wannier functions

An alternative to the plane wave basis set in the framework of periodic calculations in solid-state theory are Wannier functions, see for instance Section 10 in Ref. [48] for general background. These functions are obtained formally from a unitary transformation of the Bloch orbitals Eq. (3.11) and have the advantage that they can be localized exponentially under certain circumstances, thus formalizing the concept of "nearsightedness" [1162]. The maximally localized generalized Wannier functions [960] are the periodic analogues of Boys' localized orbitals defined for isolated systems. Section 7.2 is devoted to discussing the generation and properties of these particular Wannier functions. The usefulness of Wannier orbitals not only for analyzing the generated electronic structure *a posteriori* as demonstrated in Section 7.2.6, but also for performing electronic structure calculations as such has been advocated by several groups, see Refs. [18, 410, 765, 852, 960, 1464] and references given therein.

2.8.8 Real space grids

A quite different approach is to leave conventional basis set approaches altogether and resort to real space methods [83], where continuous space is replaced by a discrete space $\mathbf{r} \to r_{\mathbf{p}}$. This entails that the total energy expression and in particular the derivative operator have to be discretized in some suitable way. The high-order central-finite difference approach leads to the expression

$$
-\frac{1}{2}\nabla^2\phi_i(\mathbf{r}) \overset{h\to 0}{=} -\frac{1}{2}\left[\sum_{n_x=-N}^{N} C_{n_x}\phi_i(r_{p_x}+n_x h, r_{p_y}, r_{p_z}) \right.
$$

$$
+ \sum_{n_y=-N}^{N} C_{n_y}\phi_i(r_{p_x}, r_{p_y}+n_y h, r_{p_z})
$$

$$
\left. + \sum_{n_z=-N}^{N} C_{n_z}\phi_i(r_{p_x}, r_{p_y}, r_{p_z}+n_z h) \right] + \mathcal{O}\left(h^{2N+2}\right) \quad (2.150)
$$

for the Laplacian, which is correct up to the order h^{2N+2}. Here, h is the uniform grid spacing and $\{C_{\mathbf{n}}\}$ are known expansion coefficients that depend on the selected order of the scheme [259]. Within this approach, not only the grid spacing h but also the order are disposable parameters that can be optimized for a particular calculation. Note that the discretization points in continuous space can also be considered to constitute a sort of "finite basis set" - despite different statements in the literature - and that the "infinite basis set limit" is reached as $h \to 0$ for N fixed. A variation on the theme are Mehrstellen schemes, where the discretization of the entire differential equation and not only of the derivative operator is optimized [188].

The first real space approach devised for *ab initio* molecular dynamics was based on the lowest-order finite difference approximation in conjunction with an equally spaced cubic mesh in real space [223]. A host of other implementations of more sophisticated real space methods followed, including for instance nonuniform meshes, multigrid acceleration, different discretization techniques, and finite-element methods [20, 80, 122, 124, 259, 260, 1017, 1288, 1289, 1489–1491, 1585, 1617, 1668], see Ref. [83] for a review. Among the chief advantages of the real space methods is the idea of exploiting the "nearsightedness of electronic matter" [1162] in terms of linear scaling approaches [518, 524, 1087] that can be implemented in a natural way. In addition, the multiple length scale problem can be coped with by hierarchically adapting the grid in the framework of "multigrid methods". However, the extension to such non-uniform meshes induces the (in)famous Pulay forces (see Section 2.5) if the mesh moves as the nuclei move.

3

Implementation: using the plane wave basis set

3.1 Introduction and basic definitions

In this chapter the implementation of plane wave/pseudopotential *ab initio* molecular dynamics methods within the CPMD computer code [696] is explained. We concentrate on the basics, thus leaving advanced methods to later chapters and in addition all formulas are given for the spin-unpolarized or restricted case for simplicity. This allows us to highlight the essential features of a plane wave code, as well as the reasons for its high performance, in detail. The implementation of more sophisticated versions of the presented algorithms, as well as of the more advanced techniques in Chapter 5 is, in most cases, very similar to what is introduced here.

There are many reviews on the plane wave/pseudopotential method as such, or in connection with the Car–Parrinello algorithm. Older articles [335, 673, 1149, 1391], as well as the book by Singh [1365], concentrate on the electronic structure problem, whereas the review by Jones and Gunnarsson [716] as well as the textbooks by Martin [913] and Kohanoff [762], provide a glimpse of Car–Parrinello molecular dynamics as well. Other reviews [485, 486, 1123, 1209, 1498] introduce the plane wave/pseudopotential method with the aim of connecting it to the molecular dynamics technique.

3.1.1 Supercells and plane wave basis

The unit cell of a periodically repeated system is defined by the Bravais lattice vectors $\mathbf{a}_1$, $\mathbf{a}_2$, and $\mathbf{a}_3$ (see e.g. Ref. [48] for a general discussion). The Bravais vectors can be combined into a 3×3 matrix $\mathbf{h} = [\mathbf{a}_1, \mathbf{a}_2, \mathbf{a}_3]$ [48, 1105]. The volume Ω of the cell is calculated as the determinant of $\mathbf{h}$

$$\Omega = \det \mathbf{h} \ . \tag{3.1}$$

Furthermore, scaled coordinates $\mathbf{s}$ are introduced that are related to $\mathbf{r}$ via $\mathbf{h}$

$$\mathbf{r} = \mathbf{hs} \tag{3.2}$$

and the distances in scaled coordinates are related to distances in real coordinates by the metric tensor $\mathcal{G} = \mathbf{h}^{\mathrm{T}}\mathbf{h}$

$$(\mathbf{r}_i - \mathbf{r}_j)^2 = (\mathbf{s}_i - \mathbf{s}_j)^{\mathrm{t}}\mathcal{G}(\mathbf{s}_i - \mathbf{s}_j) \ . \tag{3.3}$$

Periodic boundary conditions can be enforced by using

$$\mathbf{r}_{\mathrm{pbc}} = \mathbf{r} - \mathbf{h}\big[\mathbf{h}^{-1}\mathbf{r}\big]_{\mathrm{NINT}} \ , \tag{3.4}$$

where $[\cdots]_{\mathrm{NINT}}$ denotes the nearest integer value so that the coordinates $\mathbf{r}_{\mathrm{pbc}}$ will always be within the box centered around the origin of the coordinate system. Reciprocal lattice vectors $\mathbf{b}_i$ are defined as

$$\mathbf{b}_i \cdot \mathbf{a}_j = 2\pi\,\delta_{ij} \tag{3.5}$$

and can also be arranged into a 3×3 matrix

$$[\mathbf{b}_1, \mathbf{b}_2, \mathbf{b}_3] = 2\pi(\mathbf{h}^{\mathrm{T}})^{-1} \ . \tag{3.6}$$

Plane waves form a complete and orthonormal basis with the above periodicity (see also the more general introduction of plane waves in Section 2.8.2)

$$f_{\mathbf{G}}^{\mathrm{PW}}(\mathbf{r}) = \frac{1}{\sqrt{\Omega}}\exp[i\mathbf{G}\cdot\mathbf{r}] = \frac{1}{\sqrt{\Omega}}\exp[2\pi\,i\,\mathbf{g}\cdot\mathbf{s}] \ , \tag{3.7}$$

with the reciprocal space vectors

$$\mathbf{G} = 2\pi(\mathbf{h}^{\mathrm{T}})^{-1}\mathbf{g} \ , \tag{3.8}$$

where $\mathbf{g} = [i, j, k]$ is a triple of integer values. Any periodic function can be expanded in this basis

$$\varphi(\mathbf{r}) = \varphi(\mathbf{r} + \mathbf{L}) = \frac{1}{\sqrt{\Omega}}\sum_{\mathbf{G}}\varphi(\mathbf{G})\exp[i\,\mathbf{G}\cdot\mathbf{r}] \ , \tag{3.9}$$

where $\varphi(\mathbf{r})$ and $\varphi(\mathbf{G})$ are related by a three-dimensional Fourier transform. The direct lattice vectors $\mathbf{L}$ connect equivalent points in different cells.

3.1.2 Plane wave expansions

The Kohn–Sham potential (see Eq. (2.121)) of a periodic system exhibits the same periodicity as the direct lattice

$$V^{\mathrm{KS}}(\mathbf{r}) = V^{\mathrm{KS}}(\mathbf{r} + \mathbf{L}) \ , \tag{3.10}$$

and the Kohn–Sham orbitals can be written in general Bloch form

$$\phi_j(\mathbf{r}, \mathbf{k}) = \exp[i\,\mathbf{k} \cdot \mathbf{r}]\ u_j(\mathbf{r}, \mathbf{k})\ , \qquad (3.11)$$

where $\mathbf{k}$, the quantum number associated with the crystal momentum, is a vector in the first Brillouin zone (see e.g. Ref. [48]). The functions $u_j(\mathbf{r}, \mathbf{k})$ have the periodicity of the direct lattice

$$u_j(\mathbf{r}, \mathbf{k}) = u_j(\mathbf{r} + \mathbf{L}, \mathbf{k})\ . \qquad (3.12)$$

The band index j runs over all states and the states have an occupation number $f_j(\mathbf{k})$ associated with them, which is a relative weight. The periodic functions $u_j(\mathbf{r}, \mathbf{k})$ are now expanded in the plane wave basis

$$u_j(\mathbf{r}, \mathbf{k}) = \frac{1}{\sqrt{\Omega}} \sum_{\mathbf{G}} c_j(\mathbf{G}, \mathbf{k}) \exp[i\mathbf{G} \cdot \mathbf{r}]\ , \qquad (3.13)$$

and the Kohn–Sham orbitals are

$$\phi_j(\mathbf{r}, \mathbf{k}) = \frac{1}{\sqrt{\Omega}} \sum_{\mathbf{G}} c_j(\mathbf{G}, \mathbf{k}) \exp[i(\mathbf{G} + \mathbf{k}) \cdot \mathbf{r}]\ , \qquad (3.14)$$

where $c_j(\mathbf{G}, \mathbf{k})$ are complex numbers. With this expansion the charge density can also be expanded into a plane wave basis

$$
\begin{aligned}
n(\mathbf{r}) &= \frac{1}{\Omega} \sum_j \int f_j(\mathbf{k}) \sum_{\mathbf{G}, \mathbf{G}'} c_j^\star(\mathbf{G}', \mathbf{k}) c_j(\mathbf{G}, \mathbf{k}) \exp[i(\mathbf{G} + \mathbf{k}) \cdot \mathbf{r}]\ d\mathbf{k} \\
&= \sum_{\mathbf{G}} n(\mathbf{G}) \exp[i\,\mathbf{G} \cdot \mathbf{r}]\ , \qquad (3.15)
\end{aligned}
$$

where the sum over $\mathbf{G}$ vectors in Eq. (3.15) extends over twice the range given by the wave function expansion. Thus, whereas for atomic orbital basis sets the number of functions needed to describe the density grows quadratically with the size of the system, there is only a linear dependence for plane waves. This is one of the main advantages of the plane wave basis.

3.1.3 Cutoffs and k-points

In actual calculations the infinite sums over $\mathbf{G}$ vectors and cells have to be truncated. Furthermore, one has to approximate the integral over the Brillouin zone by a finite sum over a set of $\mathbf{k}$-points

$$\int_{\mathrm{BZ}} d\mathbf{k} \rightarrow \sum_{\mathbf{k}} w_{\mathbf{k}}\ , \qquad (3.16)$$

where $w_{\mathbf{k}}$ are the weights of these integration points. Several schemes to choose the integration points systematically and most efficiently, including "special $\mathbf{k}$-points", are available in the literature [69, 250, 1026]; see Ref. [400] for an overview on the use of $\mathbf{k}$-points in the calculation of the electronic structure of solids. The truncation of the plane wave basis rests on the fact that the Kohn–Sham potential $V^{\mathrm{KS}}(\mathbf{G})$ converges rapidly with increasing modulus of $\mathbf{G}$. For this reason, at each $\mathbf{k}$-point, only $\mathbf{G}$ vectors with a kinetic energy lower than a given maximum cutoff

$$\frac{1}{2}\,|\mathbf{k}+\mathbf{G}|^2 \leq E_{\mathrm{cut}} \tag{3.17}$$

are included in the basis. With this choice of basis, the precision of the calculation within the approximations of density functional theory is controlled by a single parameter, namely the plane wave (energy) cutoff E_{cut}.

The number of plane waves for a given cutoff depends on the unit cell and the $\mathbf{k}$-point(s). An estimate for the size of the basis at the center of the Brillouin zone is

$$N_{\mathrm{PW}} = \frac{1}{2\pi^2}\,\Omega\,E_{\mathrm{cut}}^{3/2}\,, \tag{3.18}$$

where E_{cut} and Ω are given in atomic units, i.e. Hartree and (Bohr)^3, respectively. The basis set needed to describe the density calculated from the Kohn–Sham orbitals has a corresponding cutoff that is four times the cutoff of the orbitals. The number of plane waves needed at a given density cutoff is therefore eight times the number of plane waves needed for the orbitals.

It is a common approximation in density functional theory calculations [372, 1269] to use approximate electronic densities. Instead of using the full description, the density is expanded in an auxiliary basis. An incomplete plane wave basis can be considered as such an auxiliary basis with special properties [869, 870]. The fitting is usually done by minimizing the functional $\langle n - \tilde{n} \mid \mathcal{O} \mid n - \tilde{n} \rangle$, where n and $\tilde{n}$ are the full and approximated density, respectively, and the operator $\mathcal{O}$ provides the metric. As long as the metric operator depends only on $\mathbf{r} - \mathbf{r}'$, a simple expansion of $\tilde{n}$ in plane waves is optimal and no additional difficulties in calculation of the energy or forces appear. The only point to control is if the accuracy of the calculation is still sufficient.

Finally, sums over all unit cells in real space have to be truncated. The only term in the final energy expression with such a sum is the real space part of the Ewald sum (see Section 3.2). This term is not a major contribution to the workload in a density functional calculation, therefore its cutoff can be set rather generously.

3.1.4 Real space grid and fast Fourier transforms

A function given as a finite linear combination of plane waves can also be defined as a set of functional values on an equally spaced grid in real space. The sampling theorem (see e.g. Ref. [1161]) gives the maximal grid spacing that still allows us to hold the same information as the expansion coefficients of the plane waves. The real space sampling points $\mathbf{R}$ are defined

$$\mathbf{R} = \mathbf{h}\,\mathbf{N}\mathbf{q} \ , \tag{3.19}$$

where $\mathbf{N}$ is a diagonal matrix with the entries $1/N_s$ and $\mathbf{q}$ is a vector of integers ranging from 0 to $N_s - 1$ and $s = x, y, z$. In order to fulfill the sampling theorem N_s has to be bigger than $2\max(g_s) + 1$. In addition, N_s must be decomposable into small prime numbers (typically 2, 3, and 5) when fast Fourier transformation techniques are used (see e.g. Ref. [1161]). In practice, the smallest number N_s that fulfills the above requirements is chosen.

A periodic function can be calculated at the real space grid points $\mathbf{R}$

$$f(\mathbf{R}) = \sum_{\mathbf{G}} f(\mathbf{G}) \exp[i\,\mathbf{G} \cdot \mathbf{R}] \tag{3.20}$$

$$= \sum_{\mathbf{g}} f(\mathbf{G}) \exp\left[2\pi\, i\, \left((\mathbf{h}^\mathrm{T})^{-1}\mathbf{g}\right) \cdot (\mathbf{h}\mathbf{N}\mathbf{q})\right] \tag{3.21}$$

$$= \sum_{\mathbf{g}} f(\mathbf{G}) \exp\left[\frac{2\pi}{N_\mathrm{x}} i g_\mathrm{x} q_\mathrm{x}\right] \exp\left[\frac{2\pi}{N_\mathrm{y}} i g_\mathrm{y} q_\mathrm{y}\right] \exp\left[\frac{2\pi}{N_\mathrm{z}} i g_\mathrm{z} q_\mathrm{z}\right] . \tag{3.22}$$

The function $f(\mathbf{G})$ is zero outside the cutoff region, and the sum over the g-vectors can be extended over all indices in the cube $-\mathbf{g}_s^\mathrm{max}, \ldots, \mathbf{g}_s^\mathrm{max}$. The functions $f(\mathbf{R})$ and $f(\mathbf{G})$ are related

$$f(\mathbf{R}) \ = \ \mathrm{inv\_FT}\left[f(\mathbf{G})\right] \tag{3.23}$$

$$f(\mathbf{G}) \ = \ \mathrm{fw\_FT}\left[f(\mathbf{R})\right] \tag{3.24}$$

by three-dimensional Fourier transforms. The Fourier transforms themselves are defined by

$$[\text{inv\_FT}\,[f(\mathbf{G})]]_{uvw} = \sum_{j=0}^{N_x-1} \sum_{k=0}^{N_y-1} \sum_{l=0}^{N_z-1} f_{jkl}^{\mathbf{G}}$$

$$\times \exp\left[i\frac{2\pi}{N_x} j\,u\right] \exp\left[i\frac{2\pi}{N_y} k\,v\right] \exp\left[i\frac{2\pi}{N_z} l\,w\right] \quad (3.25)$$

and

$$[\text{fw\_FT}\,[f(\mathbf{R})]]_{jkl} = \sum_{u=0}^{N_x-1} \sum_{v=0}^{N_y-1} \sum_{w=0}^{N_z-1} f_{uvw}^{\mathbf{R}}$$

$$\times \exp\left[-i\frac{2\pi}{N_x} j\,u\right] \exp\left[-i\frac{2\pi}{N_y} k\,v\right] \exp\left[-i\frac{2\pi}{N_z} l\,w\right] \quad, \quad (3.26)$$

where the appropriate mappings of $\mathbf{q}$ and $\mathbf{g}$ to the indices

$$[u, v, w] = \mathbf{q} \tag{3.27}$$

$$[j, k, l] = \mathbf{g}_s \qquad\qquad \text{if } \mathbf{g}_s \geq 0 \tag{3.28}$$

$$[j, k, l] = N_s + \mathbf{g}_s \qquad\qquad \text{if } \mathbf{g}_s < 0 \tag{3.29}$$

have to be used. From Eqs. (3.25) and (3.26) it can be seen that the calculation of the three-dimensional Fourier transforms can be performed by a series of one-dimensional Fourier transforms. The number of transforms in each direction is $N_y N_z$, $N_x N_z$, and $N_x N_y$, respectively. Assuming that the one-dimensional transforms are performed within the fast Fourier transform framework, the number of operations per transform of length n is approximately $5n \log n$. This leads to an estimate for the number of operations for the full three-dimensional transform of $5N \log N$, where $N = N_x N_y N_z$.

3.1.5 Pseudopotentials

In order to minimize the size of the plane wave basis necessary for calculations involving many atoms, the core electrons have to be replaced by pseudopotentials. This important ingredient of any realistic calculation will be explained in full detail in Chapter 4. There the main concern will be the various classes of norm-conserving pseudopotentials, whereas generalizations will be presented in Part II on Advanced methods (see Chapter 6). At this stage, only a concise introduction to the basic concept is given and the general formulas from Chapter 4 needed in implementations are summarized.

The pseudopotential approximation in the realm of solid-state theory goes back to the work on orthogonalized plane waves [636] and core state projector methods [1146]. Empirical pseudopotentials were used in plane wave calculations [621, 1648], but new developments have increased both the efficiency and reliability of the method considerably. Pseudopotentials are required to represent the long-range interactions of the core correctly, and to produce pseudo wave function solutions that approach the full wave function outside a chosen core radius r_c. Inside this radius the pseudopotential and the wave function should be as smooth as possible, in order to allow for a small plane wave cutoff. For the pseudo wave function this requires that the nodal structure of the valence wave functions is replaced by a smooth function. In addition it is desired that a pseudopotential is transferable [432, 523], this means that one and the same pseudopotential can be used in calculations for different chemical environments, yielding results of comparable accuracy.

A first major step to achieve these conflicting goals was the introduction of "norm conservation" [1394, 1473]. Norm-conserving pseudopotentials have to be angular momentum-dependent. In their most general form they are "semi-local"

$$V^{\mathrm{PP}}(\mathbf{r}, \mathbf{r}') = \sum_{l=1}^{\infty} \sum_{m=-l}^{l} Y_{lm}^*(\omega) V_l(\mathbf{r}) \delta(\mathbf{r} - \mathbf{r}') Y_{lm}(\omega') \; , \qquad (3.30)$$

meaning that a different radial dependence, $V_l(\mathbf{r})$, is used for each angular momentum channel l that is included ($Y_{lm}(\omega)$ being the spherical harmonics and ω the Euler angles of the position vector $\mathbf{r}$); a "local" pseudopotential would only depend on the distance of the electron with respect to a given ionic core. A minimal set of requirements and a systematic construction scheme for soft, semi-local pseudopotentials were developed [57, 594]. Since then many variations of the original method have been proposed, concentrating either on an improvement in softness or transferability. Analytic representations of the core part of the potential [735, 1191, 1480, 1481] were used. Extended norm conservation [1342] was introduced to enhance transferability and new concepts to increase the softness were presented [868, 1191, 1546]. More information on pseudopotentials and their construction can be found in Chapter 4, in several review articles [472, 1149, 1365], and in modern textbooks [762, 913].

Although pseudopotentials are originally generated in a semi-local form, most applications use the fully separable form. Pseudopotentials can be transformed a posteriori to the separable form using atomic wave functions [141, 754, 1546], as explained in Section 4.3.2. Recently [525, 608], a novel type of norm-conserving pseudopotential was introduced that is separable by construction. Local and nonlocal parts of these dual-space Gaussian pseudopotentials introduced in Section 4.4 have a simple analytic form in both real and reciprocal space, and only a few parameters are needed to characterize the potential. These parameters are globally optimized in order to reproduce many properties of atoms and ensure good transferability.

A separable nonlocal pseudopotential can be put into the following general form

$$V^{\mathrm{PP}}(\mathbf{r}, \mathbf{r}') = \left(V^{\mathrm{core}}(\mathbf{r}) + \Delta V^{\mathrm{loc}}(\mathbf{r}) \right) \delta(\mathbf{r} - \mathbf{r}') + \sum_{k,l} P_k^{\star}(\mathbf{r}) w_{kl} P_l(\mathbf{r}') \; , \quad (3.31)$$

which includes all the above-mentioned types. The local part has been split into a core part ($\sim 1/r$ for $r \to \infty$) and a short-ranged local part in order to facilitate the derivation of the final energy formula. The actual form of the core potential will be defined later. The local potential $\Delta V_{\mathrm{loc}}(\mathbf{r})$ and the projectors $P_k(\mathbf{r})$ are atom-centered functions of the form

$$\varphi(\mathbf{r}) = \varphi(|\mathbf{r} - \mathbf{R}_I|) \, Y_{lm}(\theta, \phi) \; , \quad (3.32)$$

that can be expanded in plane waves

$$\varphi(\mathbf{r}) = \sum_{\mathbf{G}} \varphi(G) \exp[i\mathbf{G} \cdot \mathbf{r}] \, S_I(\mathbf{G}) \, Y_{lm}(\tilde{\theta}, \tilde{\phi}) \; , \quad (3.33)$$

where spherical polar coordinates $\mathbf{r} = (r, \theta, \phi)$ have been used and $\mathbf{G} = (G, \tilde{\theta}, \tilde{\phi})$. The matrix elements w_{kl} give the coupling strength of different projectors, and coupling between projectors of different angular momentum is not allowed. The coordinates $\mathbf{R}_I$ denote atomic positions and the structure factors S_I are defined as

$$S_I(\mathbf{G}) = \exp[-i\mathbf{G} \cdot \mathbf{R}_I] \; . \quad (3.34)$$

The functions $\varphi(G)$ are calculated from $\varphi(r)$ by a Bessel transform

$$\varphi(G) = 4\pi \, (-i)^l \int_0^\infty r^2 \, \varphi(r) \, j_l(Gr) \, dr \; , \quad (3.35)$$

where j_l are spherical Bessel functions of the first kind. The local pseudopotential and the projectors of the nonlocal part in Fourier space are given by

$$\Delta V^{\text{loc}}(\mathbf{G}) \;\; = \;\; \frac{4\pi}{\Omega} \int_0^\infty r^2 \, \Delta V^{\text{loc}}(r) j_0(Gr) \, dr \tag{3.36}$$

$$P_k(\mathbf{G}) \;\; = \;\; \frac{4\pi}{\sqrt{\Omega}} (-i)^l \int_0^\infty r^2 \, P_k(r) \, j_l(Gr) \, Y_{lm}(\tilde{\theta}, \tilde{\phi}) \, dr \;\; , \tag{3.37}$$

where lm are angular momentum quantum numbers associated with the projectors.

3.2 Electrostatic energy

3.2.1 General concepts

The electrostatic energy of a system of nuclear charges Z_I at positions $\mathbf{R}_I$ and an electronic charge distribution $n(\mathbf{r})$ consists of three parts,

$$E_{\text{ES}} = \frac{1}{2} \iint \frac{n(\mathbf{r})n(\mathbf{r}')}{|\mathbf{r} - \mathbf{r}'|} \, d\mathbf{r} \, d\mathbf{r}'$$

$$+ \sum_I \int V_I^{\text{core}}(\mathbf{r}) n(\mathbf{r}) \, d\mathbf{r} + \frac{1}{2} \sum_{I \neq J} \frac{Z_I Z_J}{|\mathbf{R}_I - \mathbf{R}_J|} \;\; , \tag{3.38}$$

the Hartree energy of the electrons, the interaction energy of the electrons with the nuclei and the internuclear interactions, see also Section 2.7.2. The Ewald method (see textbooks [25, 468] for general introductions) can be used to avoid singularities in the individual terms when the system size is infinite. In order to achieve this, a Gaussian core charge distribution associated with each nucleus

$$n_I^{\text{c}}(\mathbf{r}) = -\frac{Z_I}{\left(R_I^{\text{c}}\right)^3} \pi^{-3/2} \exp\left[-\left(\frac{\mathbf{r} - \mathbf{R}_I}{R_I^{\text{c}}} \right)^2 \right] \tag{3.39}$$

is defined. It is convenient at this point to make use of the arbitrariness in the definition of the core potential and define it to be the potential of the Gaussian charge distribution of Eq. (3.39)

$$V_I^{\text{core}}(\mathbf{r}) = \int \frac{n_I^{\text{c}}(\mathbf{r}')}{|\mathbf{r} - \mathbf{r}'|} \, d\mathbf{r}' = -\frac{Z_I}{|\mathbf{r} - \mathbf{R}_I|} \text{erf}\left[\frac{|\mathbf{r} - \mathbf{R}_I|}{R_I^{\text{c}}} \right] \;\; , \tag{3.40}$$

where erf is the error function. It is important to note that this choice is completely general as it is only necessary to correctly include the long-range $-Z_I/|\mathbf{r}|$ behavior in the potential. All deviations close to the atoms are lumped together in the short-range potential $\Delta V^{\text{loc}}(\mathbf{r})$. The interaction

energy of these Gaussian charge distributions is now added and subtracted
from the total electrostatic energy

$$E_{\mathrm{ES}} = \frac{1}{2} \int\!\!\int \frac{n(\mathbf{r})n(\mathbf{r}')}{|\mathbf{r}-\mathbf{r}'|}\, d\mathbf{r}\, d\mathbf{r}'$$

$$+ \frac{1}{2} \int\!\!\int \frac{n^{\mathrm{c}}(\mathbf{r})n^{\mathrm{c}}(\mathbf{r}')}{|\mathbf{r}-\mathbf{r}'|}\, d\mathbf{r}\, d\mathbf{r}' + \int\!\!\int \frac{n^{\mathrm{c}}(\mathbf{r})n(\mathbf{r}')}{|\mathbf{r}-\mathbf{r}'|}\, d\mathbf{r}\, d\mathbf{r}'$$

$$+ \frac{1}{2} \sum_{I \neq J} \frac{Z_I Z_J}{|\mathbf{R}_I - \mathbf{R}_J|} - \frac{1}{2} \int\!\!\int \frac{n^{\mathrm{c}}(\mathbf{r})n^{\mathrm{c}}(\mathbf{r}')}{|\mathbf{r}-\mathbf{r}'|}\, d\mathbf{r}\, d\mathbf{r}' \ , \quad (3.41)$$

where $n^{\mathrm{c}}(\mathbf{r}) = \sum_I n_I^{\mathrm{c}}(\mathbf{r})$. The first three terms can be combined to give
the electrostatic energy of a total charge distribution $n^{\mathrm{tot}}(\mathbf{r}) = n(\mathbf{r}) + n^{\mathrm{c}}(\mathbf{r})$.
The remaining terms are rewritten as a double sum over nuclei and a sum
over self-energy terms of the Gaussian charge distributions

$$E_{\mathrm{ES}} = \frac{1}{2} \int\!\!\int \frac{n^{\mathrm{tot}}(\mathbf{r})n^{\mathrm{tot}}(\mathbf{r}')}{|\mathbf{r}-\mathbf{r}'|}\, d\mathbf{r}\, d\mathbf{r}'$$

$$+ \frac{1}{2} \sum_{I \neq J} \frac{Z_I Z_J}{|\mathbf{R}_I - \mathbf{R}_J|} \mathrm{erfc}\left[\frac{|\mathbf{R}_I - \mathbf{R}_J|}{\sqrt{R_I^{\mathrm{c}\,2} + R_J^{\mathrm{c}\,2}}} \right] - \sum_I \frac{1}{\sqrt{2\pi}} \frac{Z_I^2}{R_I^{\mathrm{c}}} \ , \quad (3.42)$$

where erfc denotes the complementary error function.

3.2.2 Periodic systems

For a periodically replicated system the total energy per unit cell is derived
from the above expression by using the solution to Poisson's equation, cf.
Eq. (2.119), in Fourier space for the first term in Eq. (3.42) and making use
of the quick convergence of the second term in real space. The total charge
is expanded in plane waves with expansion coefficients (see Eq. (3.34) for
the definition of the structure factors $S_I(\mathbf{G})$)

$$n^{\mathrm{tot}}(\mathbf{G}) = n(\mathbf{G}) + \sum_I n_I^{\mathrm{c}}(\mathbf{G}) S_I(\mathbf{G}) \qquad (3.43)$$

$$= n(\mathbf{G}) - \frac{1}{\Omega} \sum_I \frac{Z_I}{\sqrt{4\pi}} \exp\left[-\frac{1}{4} G^2 R_I^{\mathrm{c}\,2} \right] S_I(\mathbf{G}) \ . \quad (3.44)$$

This leads to the electrostatic energy for a periodic system

$$E_{\mathrm{ES}} = 2\pi\,\Omega \sum_{\mathbf{G}\neq 0} \frac{|n^{\mathrm{tot}}(\mathbf{G})|^2}{G^2} + E_{\mathrm{ovrl}} - E_{\mathrm{self}} \;, \tag{3.45}$$

where

$$E_{\mathrm{ovrl}} = \sum_{I,J}{}' \sum_{\mathbf{L}} \frac{Z_I Z_J}{|\mathbf{R}_I - \mathbf{R}_J - \mathbf{L}|} \mathrm{erfc}\left[\frac{|\mathbf{R}_I - \mathbf{R}_J - \mathbf{L}|}{\sqrt{R_I^{\mathrm{c}\,2} + R_J^{\mathrm{c}\,2}}}\right] \tag{3.46}$$

and

$$E_{\mathrm{self}} = \sum_{I} \frac{1}{\sqrt{2\pi}} \frac{Z_I^2}{R_I^{\mathrm{c}}} \;. \tag{3.47}$$

Here, the sums extend over all atoms in the simulation cell, all direct lattice vectors $\mathbf{L}$, and the prime in the first sum indicates that the condition $I < J$ is imposed for the $\mathbf{L} = \mathbf{0}$ term.

3.2.3 Cluster boundary conditions

The possibility of using fast Fourier transforms in order to calculate the electrostatic energy is one of the reasons for the high performance of plane wave calculations. However, plane wave-based calculations imply periodic boundary conditions. This is appropriate for crystal calculations but very unnatural for molecular or slab calculations. For neutral systems this problem is circumvented by use of the supercell method. Namely, the molecule is periodically repeated but the distance between each molecule and its periodic images is so large that their interaction is negligible. This procedure is somewhat wasteful but can lead to satisfactory results.

Handling charged molecular systems is, however, considerably more difficult, due to the long-range Coulomb forces. A charged periodic system has infinite energy and the interaction between images cannot really be eliminated completely. In order to circumvent this problem, several solutions have been proposed. The simplest fix is to add to the system a neutralizing background charge. This is achieved trivially as the $\mathbf{G} = \mathbf{0}$ term in Eq. (3.45) is already eliminated. This leads to finite energies but does not eliminate the interaction between the images and makes the calculation of absolute energies difficult. Other solutions involve performing a set of different calculations on the system such that extrapolation to the limit of infinitely separated images is possible. It also has been shown that it is possible to estimate the correction to the total energy for the removal of the image charges [895]. Still, it does not seem easy to incorporate this scheme into the framework of molecular dynamics. Another method [123, 840] works with the separation of the long and short-range parts of the Coulomb forces. In this method the low-order multipole moments of the charge distribution are separated out and handled analytically. This method was used in the context of coupling *ab initio* and classical molecular dynamics [143] (see Section 5.5.2 for the CP-PAW/AMBER interface).

The long-range forces in Eq. (3.42) are contained in the first term. This term can be written

$$\frac{1}{2} \int\int \frac{n^{\text{tot}}(\mathbf{r}) n^{\text{tot}}(\mathbf{r}')}{|\mathbf{r} - \mathbf{r}'|} \, d\mathbf{r} \, d\mathbf{r}' = \frac{1}{2} \int V_{\text{H}}(\mathbf{r}) n^{\text{tot}}(\mathbf{r}) \, d\mathbf{r} \,, \tag{3.48}$$

where the electrostatic potential $V_{\text{H}}(\mathbf{r})$ is the solution of Poisson's equation, cf. Eq. (2.119). There are three approaches implemented in CPMD to solve Poisson's equation subject to the boundary condition $V_{\text{H}}(\mathbf{r}) \to 0$ for $\mathbf{r} \to \infty$. All of them rely on fast Fourier transforms, thus preserving the same efficient framework as for the periodic case.

The first method is due to Hockney [643] and was first applied to density functional plane wave calculations in Ref. [75]. In the following outline, for the sake of simplicity, a one-dimensional case is presented. The charge density n is assumed to be nonzero only within an interval L and sampled on N equidistant points denoted by x_p. The potential can then be written

$$V_{\text{H}}(x_p) = \frac{L}{N} \sum_{p'=-\infty}^{\infty} G(x_p - x_{p'}) n(x_{p'}) \tag{3.49}$$

$$= \frac{L}{N} \sum_{p'=0}^{N} G(x_p - x_{p'}) n(x_{p'}) \tag{3.50}$$

for $p = 0, 1, 2, \ldots, N$, where $G(x_p - x_{p'})$ is the corresponding Green's function. In Hockney's algorithm this equation is replaced by the cyclic convolution

$$\tilde{V}_{\text{H}}(x_p) = \frac{L}{N} \sum_{p'=0}^{2N+1} \tilde{G}(x_p - x_{p'}) \tilde{n}(x_{p'}) \tag{3.51}$$

where $p = 0, 1, 2, \ldots, 2N + 1$, and

$$\tilde{n}(x_p) = \begin{cases} n(x_p) & 0 \le p \le N \\ 0 & N \le p \le 2N + 1 \end{cases} \tag{3.52}$$

$$\tilde{G}(x_p) = G(x_p) \quad -(N+1) \le p \le N \tag{3.53}$$

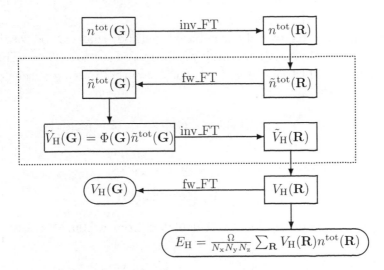

Fig. 3.1. Flow chart for the calculation of the long-range part of the electrostatic energy using the method by Hockney [643]. The part inside the dashed box is calculated most efficiently with the procedure outlined by Eastwood and Brownrigg [378].

$$\tilde{n}(x_p) = \tilde{n}(x_p + L) \tag{3.54}$$

$$\tilde{G}(x_p) = \tilde{G}(x_p + L) \ . \tag{3.55}$$

The solution $\tilde{V}_H(x_p)$ can be obtained by a series of fast Fourier transforms and has the desired property

$$\tilde{V}_H(x_p) = V_H(x_p) \quad \text{for } 0 \le p \le N \ . \tag{3.56}$$

To remove the singularity of the Green's function at $x = 0$, $G(x)$ is modified for small x and the error is corrected by using the identity

$$G(x) = \frac{1}{x}\text{erf}\left[\frac{x}{r_c}\right] + \frac{1}{x}\text{erfc}\left[\frac{x}{r_c}\right] \ , \tag{3.57}$$

where r_c is chosen such that the short-ranged part can be described accurately by a plane wave expansion with the density cutoff. In an optimized implementation Hockney's method requires a double amount of memory and two additional fast Fourier transforms on the box of double size, as illustrated graphically by the flow chart in Fig. 3.1. Hockney's method can be generalized to systems with periodicity in one (wires/polymers) and two (slabs/surfaces) dimensions. It was pointed out [378] that Hockney's method gives the exact solution to Poisson's equation for isolated systems if the boundary condition (zero density at the edges of the box) is exactly fulfilled.

A different, fully reciprocal space-based method, that can be seen as an approximation to Hockney's method, was proposed recently [929]. The final expression for the Hartree energy is also based on the splitting of the Green's function in Eq. (3.57)

$$E_{ES} = 2\pi \, \Omega \sum_{\mathbf{G}} V_{H}^{MT}(\mathbf{G}) \left(n^{tot}(\mathbf{G})\right)^{\star} + E_{ovrl} - E_{self} \; . \tag{3.58}$$

The potential function is calculated from two parts,

$$V_{H}^{MT}(\mathbf{G}) = \overline{V}_{H}(\mathbf{G}) + \tilde{V}_{H}(\mathbf{G}) \; , \tag{3.59}$$

where $\tilde{V}_{H}(\mathbf{G})$ is the analytic part, calculated from a Fourier transform of erfc

$$\tilde{V}_{H}(\mathbf{G}) = \frac{4\pi}{G^2} \left(1 - \exp\left[-\frac{G^2 r_{c}^2}{4}\right]\right) n^{tot}(\mathbf{G}) \tag{3.60}$$

and $\overline{V}_{H}(\mathbf{G})$ is calculated from a discrete Fourier transform of the Green's function on an appropriate grid. The calculation of the Green's function can be done at the beginning of the calculation and does not have to be repeated again. It is reported [929] that a cutoff of 10–20% higher than the one employed for the charge density gives converged results. The same technique can also be applied for systems that are periodic in one and two dimensions [1009].

If the boundary conditions are chosen appropriately, the discrete Fourier transforms for the calculation of $\overline{V}_{H}(\mathbf{G})$ can be performed analytically [1034] for the limiting case where $r_{c} = 0$. These boundary conditions are a sphere of radius R for an isolated cluster system. For a one-dimensional system one can choose a rod of radius R and for a two-dimensional system a slab of thickness Z. The electrostatic potentials for these systems are listed in Table 3.1, where $G_{xy} = \left[g_{x}^2 + g_{y}^2\right]^{1/2}$ and J_{n} and K_{n} are the Bessel functions of the first and second kind of integer order n.

Hockney's method requires a computational box such that the charge density is negligibly small at the edges, which is equivalent to what is needed in the supercell approach [1190]. Practical experience tells us that a minimum distance of about 3Å of all atoms from the edges of the box is sufficient for most systems. The Green's function is then applied to the charge density in a box of twice this size, and it has to be calculated only once at the beginning of the calculation. The other methods presented in this chapter require a computational box twice the size of the Hockney method, as they

Table 3.1. *Fourier space formulas for the Hartree energy of a charge distribution* $n(\mathbf{G})$, *see text for definitions.*

Dimension	Periodic	$(G^2/4\pi)V_H(\mathbf{G})$	$V_H(\mathbf{0})$
0	-	$(1 - \cos[RG])\,n(\mathbf{G})$	$2\pi R^2 n(0)$
1	z	$(1 + R\,(G_{xy}\,J_1(RG_{xy})\,K_0(Rg_z)$ $-g_z J_0(RG_{xy})\,K_1(Rg_z)))\,n(\mathbf{G})$	0
2	x, y	$(1 - (-1)^{g_z}\exp[-G_{xy}Z/2])\,n(\mathbf{G})$	0
3	x, y, z	$n(\mathbf{G})$	0

are applying the artificially periodic Green's function within the computational box. This can be equivalent to the exact Hockney method only if the box is enlarged to double the size. In plane wave calculations, computational costs grow linearly with the volume of the box. Therefore, Hockney's method will prevail over the others in terms of accuracy, speed, and memory requirements in the limit of large systems. The direct Fourier space methods have advantages through their easy implementation and if they are used with small computational boxes, i.e. for small systems in cases where high accuracy is not required. In addition, they can be of great use in calculations with classical potentials.

3.3 Exchange and correlation energy

Exchange and correlation functionals implemented in the CPMD code are of the local type with gradient corrections. This type of functional can be written as (see also Eqs. (2.127) and (2.123))

$$E_{xc} = \int \varepsilon_{xc}\,(n(\mathbf{r}), \nabla n(\mathbf{r}))\,n(\mathbf{r})\,d\mathbf{r} = \Omega \sum_{\mathbf{G}} \varepsilon_{xc}(\mathbf{G})n^\star(\mathbf{G}) \qquad (3.61)$$

with the corresponding potential

$$V_{xc}(\mathbf{r}) = \frac{\partial F_{xc}}{\partial n} - \sum_s \frac{\partial}{\partial r_s}\left[\frac{\partial F_{xc}}{\partial(\partial_s n)}\right]\ , \qquad (3.62)$$

where $F_{xc} = \varepsilon_{xc}(n, \nabla n)n$ and $\partial_s n$ is the s-component of the density gradient.

Exchange and correlation functionals have complicated analytical forms that give rise to high-frequency components in $\varepsilon_{xc}(\mathbf{G})$. Although these high-frequency components do not enter the sum in Eq. (3.61) due to the filter effect of the density, they affect the calculation of ε_{xc}. As the functionals are only local in real space, not in Fourier space, they have to be evaluated on a real space grid. The function $\varepsilon_{xc}(\mathbf{G})$ can then be calculated by a Fourier transform. Therefore, the exact calculation of E_{xc} would require a grid with infinite resolution. However, the high-frequency components are usually small and even a moderate grid gives accurate results. The use of a finite grid results in an effective redefinition of the exchange and correlation energy

$$E_{xc} = \frac{\Omega}{N_x N_y N_z} \sum_{\mathbf{R}} \varepsilon_{xc}\left(n(\mathbf{R}), \nabla n(\mathbf{R})\right) n(\mathbf{R}) = \Omega \sum_{\mathbf{G}} \tilde{\varepsilon}_{xc}(\mathbf{G}) n(\mathbf{G}) \ , \quad (3.63)$$

where $\tilde{\varepsilon}_{xc}(\mathbf{G})$ is the finite Fourier transform of $\varepsilon_{xc}\left(n(\mathbf{R}), \nabla n(\mathbf{R})\right)$. This definition of E_{xc} allows the calculation of all gradients analytically. In most applications the real space grid used in the calculation of the density and the potentials is also used for the exchange and correlation energy, but grids with higher resolution can easily be used. The density is calculated on the new grid by use of Fourier transforms and the resulting potential is transferred back to the original grid. With this procedure the different grids do not have to be commensurate.

The above redefinition has an undesired side-effect. The new exchange and correlation energy is no longer translationally invariant, because only translations by a multiple of the grid spacing do not change the total energy. This introduces a small modulation of the energy hypersurface [1615] known as "ripples". Highly accurate optimizations of structures and the calculation of harmonic frequencies can be affected by the ripples. Using a denser grid for the calculation of E_{xc} is the only solution to ameliorate these problems, as shown by the data from a representative calculation depicted in Fig. 3.2.

The calculation of a gradient-corrected functional within the plane wave framework can be conducted very efficiently using a sequence of Fourier transforms [1615]. The corresponding flow chart for such calculations is presented in Fig. 3.3. With the use of Fourier transforms the calculation of second derivatives of the charge density is avoided, leading to a numerically stable algorithm. To this end the identity

$$\frac{\partial F_{xc}}{\partial(\partial_s n)} = \frac{\partial F_{xc}}{\partial|\nabla n|} \frac{\partial_s n}{|\nabla n|} \tag{3.64}$$

is used.

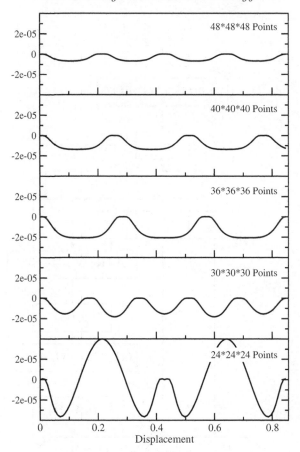

Fig. 3.2. Change in energy [a.u.] for a system of eight silicon atoms in a cubic cell displaced along the main diagonal. The calculations were done using the PBE functional [1131] and an energy cutoff of 13 Rydbergs. The grid size used in the calculation of the exchange and correlation energy is indicated in the panels.

Meta-GGA functionals [88, 166, 1051, 1052, 1443, 1545] are the next class of improved functionals and are the third rung on Perdew's "Jacob's Ladder from earth to heaven" [972, 1136], the first and second being the local density approximation (LDA) and generalized gradient approximation (GGA) functionals. Meta-GGA (MGGA) functionals depend not only on the local density and the gradient of the density but also on the kinetic energy density

$$\tau(\mathbf{r}) = \sum_i^{\text{occ}} |\nabla \phi_i(\mathbf{r})|^2 \ . \tag{3.65}$$

In a plane wave implementation, the kinetic energy density $\tau(\mathbf{r})$ can be calculated together with the electronic density. The derivatives of the Kohn–

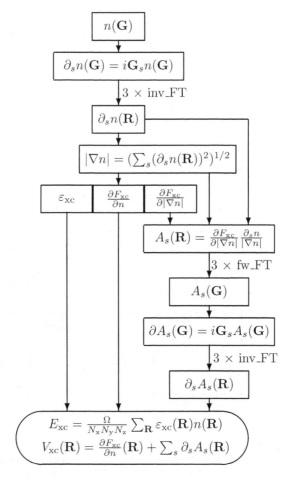

Fig. 3.3. Flow chart for the calculation of the energy and potential of a gradient-corrected exchange and correlation functional.

Sham orbitals are easily calculated in $\mathbf{G}$-space and after a Fourier transform the three functions can be squared and summed up on the real space grid. Later, $\tau(\mathbf{r})$ is needed in the calculation of the exchange and correlation energy E_{xc} and the local potential. However, it is more convenient to store the potential

$$V_\tau(\mathbf{r}) = \frac{\delta E_{xc}}{\delta \tau(\mathbf{r})} \tag{3.66}$$

alongside the local potential and to calculate the wave function forces using the formula [1052]

$$F_i^\tau(\mathbf{r}) = \nabla \cdot (V_\tau(\mathbf{r}) \, \nabla \phi_i(\mathbf{r})) \ . \tag{3.67}$$

The derivatives are calculated in $\mathbf{G}$-space and the potential $V_\tau(\mathbf{r})$ is applied on the real space mesh. This requires six additional Fourier transforms for each Kohn–Sham orbital.

This is the place to say some words on hybrid functionals, i.e. those that include "exact exchange". As mentioned in Section 2.7, this type of functional has been very popular recently, and improvements of results over GGA-type density functionals for many systems and properties have been reported. However, the calculation of the Hartree–Fock exchange causes a considerable performance problem in plane wave calculations, as can be seen from the following considerations. The Hartree–Fock exchange energy is defined as [1423]

$$E_{\text{HFx}} = \frac{1}{2} \sum_{ij} f_i f_j \int\!\!\int \frac{\rho_{ij}(\mathbf{r})\rho_{ij}(\mathbf{r}')}{|\mathbf{r} - \mathbf{r}'|} \, d\mathbf{r} d\mathbf{r}' \ , \tag{3.68}$$

where

$$\rho_{ij}(\mathbf{r}) = \phi_i(\mathbf{r})\phi_j(\mathbf{r}). \tag{3.69}$$

From this expression the wave function force is easily derived and can be calculated in Fourier space [257, 572]

$$\frac{1}{f_i}\frac{\partial E_{\text{HFx}}}{\partial c_i^\star(\mathbf{G})} = \sum_j \sum_{\mathbf{G}'} V_{\text{HFx}}^{ij}(\mathbf{G} - \mathbf{G}')c_j(\mathbf{G}') \ . \tag{3.70}$$

The force calculation is best performed in real space, whereas the potential is calculated in Fourier space. For a system with N_b electronic states and N real space grid points, a total of $5N_b^2$ three-dimensional transforms are needed, resulting in approximately $25N_b^2 N \log N$ operations needed to perform the calculation. This has to be compared to the $15N_b N \log N$ operations needed for the other Fourier transforms of the charge density and the application of the local potential and the $4N_b^2 N$ operations for the orthogonalization step. In calculations dominated by the Fourier transforms an additional factor of at least N_b is needed. If, on the other hand, orthogonalization dominates, an increase in computer time by a factor of $5 \log N$ is expected. Therefore, at least an order of magnitude more computer time is needed for calculations including exact exchange compared to ordinary density functional calculations. The exact overhead of exact exchange calculations will depend on the system studied. Slightly different performance figures can be expected from implementations using ultrasoft pseudopotentials or the projector augmented-wave (PAW) method [1096, 1097].

Ab initio molecular dynamics simulations on liquid water using hybrid functionals that include exact exhange have been performed [1470] where a special version of the CPMD code [696] was used. In particular, the simulations with hybrid functionals were performed using a two-way parallelization scheme. Besides the distribution of plane waves, orbital pairs in the exact exchange calculations were distributed over processor groups. In order to save computer time a reduced cutoff was used for the pair densities ρ_{ij}. The average CPU time per Car–Parrinello molecular dynamics step was 2.8 seconds on a computer with 576 Itanium2 (1.4 GHz) processors and a QsNet Elan4i communication system. On the same computer system but using 36 CPUs the calculation using a hybrid functional takes 31.2 seconds, whereas a GGA-type density functional requires only 0.8 seconds for a Car–Parrinello molecular dynamics step. These timings therefore verify the rough estimates made at the beginning of the section.

3.4 Total energy, gradients, and stress tensor

3.4.1 Total energy

Molecular dynamics calculations with interaction potentials derived from density functional theory require the evaluation of the total electronic energy and derivatives with respect to the parameters of the Lagrangian. In this section formulas are given in Fourier space for periodic systems. The total energy can be calculated as a sum of kinetic, external (local and nonlocal pseudopotential), exchange and correlation, and electrostatic energy (to be compared with Eq. (2.114))

$$E_{\text{tot}} = E_{\text{kin}} + E_{\text{loc}}^{\text{PP}} + E_{\text{nloc}}^{\text{PP}} + E_{\text{xc}} + E_{\text{ES}} \ . \tag{3.71}$$

The individual terms are defined by

$$E_{\text{kin}} = \sum_{\mathbf{k}} w_{\mathbf{k}} \sum_{i} \sum_{\mathbf{G}} \frac{1}{2} f_i(\mathbf{k}) \left| \mathbf{G} + \mathbf{k} \right|^2 \left| c_i(\mathbf{G}, \mathbf{k}) \right|^2 \tag{3.72}$$

$$E_{\text{loc}}^{\text{PP}} = \sum_{I} \sum_{\mathbf{G}} \Delta V_I^{\text{loc}}(\mathbf{G}) \, S_I(\mathbf{G}) n^\star(\mathbf{G}) \tag{3.73}$$

$$E_{\text{nloc}}^{\text{PP}} = \sum_{\mathbf{k}} w_{\mathbf{k}} \sum_{i} f_i(\mathbf{k}) \sum_{I} \sum_{\alpha,\beta \in I} \left(F_{I,i}^{\alpha}(\mathbf{k}) \right)^\star w_{\alpha\beta}^I F_{I,i}^{\beta}(\mathbf{k}) \tag{3.74}$$

$$E_{xc} \;=\; \Omega \sum_{\mathbf{G}} \epsilon_{xc}(\mathbf{G}) n^{\star}(\mathbf{G}) \tag{3.75}$$

$$E_{ES} \;=\; 2\pi\,\Omega \sum_{\mathbf{G}\neq 0} \frac{|n^{tot}(\mathbf{G})|^2}{G^2} + E_{ovrl} - E_{self}, \tag{3.76}$$

where $w_{\mathbf{k}}$ is defined in Eq. (3.16). The overlap between the projectors of the nonlocal pseudopotential and the Kohn–Sham orbitals has been introduced in the equation above

$$F_{I,i}^{\alpha}(\mathbf{k}) = \frac{1}{\sqrt{\Omega}} \sum_{\mathbf{G}} P_{\alpha}^{I}(\mathbf{G})\, S_I(\mathbf{G}+\mathbf{k})\, c_i^{\star}(\mathbf{G},\mathbf{k}) \; . \tag{3.77}$$

An alternative expression, using the Kohn–Sham eigenvalues $\epsilon_i(\mathbf{k})$, can also be used

$$E_{tot} = \sum_{\mathbf{k}} w_{\mathbf{k}} \sum_{i} f_i(\mathbf{k})\epsilon_i(\mathbf{k}) - \Omega \sum_{\mathbf{G}} \left(V_{xc}(\mathbf{G}) - \varepsilon_{xc}(\mathbf{G})\right) n^{\star}(\mathbf{G})$$

$$- 2\pi\,\Omega \sum_{\mathbf{G}\neq 0} \frac{|n(\mathbf{G})|^2 - |n^c(\mathbf{G})|^2}{G^2} + E_{ovrl} - E_{self} + \Delta E_{tot} \;, \tag{3.78}$$

to be compared with Eq. (2.125). The additional term ΔE_{tot} in Eq. (3.78) is required to have an expression for the energy that is quadratic in the variations of the charge density, as is true for Eq. (3.71). Without the correction term, which is zero for the exact charge density, small differences between the computed and the exact density could give rise to large errors in the total energy [258]. The correction energy can be calculated from

$$\Delta E_{tot} = -2\pi\,\Omega \sum_{\mathbf{G}\neq 0} \left(\frac{n^{in}(\mathbf{G})}{G^2} - \frac{n^{out}(\mathbf{G})}{G^2}\right) (n^{out}(\mathbf{G}))^{\star}$$

$$- \Omega \sum_{\mathbf{G}} \left(V_{xc}^{in}(\mathbf{G}) - V_{xc}^{out}(\mathbf{G})\right) (n^{out}(\mathbf{G}))^{\star}, \tag{3.79}$$

where n^{in} and n^{out} are the input and output charge densities and V_{xc}^{in} and V_{xc}^{out} the corresponding exchange and correlation potentials. This term leads to the "non-self-consistency correction" of the force as introduced in Eq. (2.106) and discussed in Section 2.5 in the framework of Car–Parrinello vs. Born–Oppenheimer dynamics.

The use of an appropriate $\mathbf{k}$-point mesh according to Eq. (3.16) is the most efficient method to calculate the total energy of a periodic system [400].

Equivalent, although not as efficient, is the usage of a supercell consisting of several replications of the unit cell in conjunction with a single integration point for the Brillouin zone. In systems where the translational symmetry is broken as in disordered systems, such as liquids, amorphous solids, or thermally excited crystals, periodic boundary conditions can still be used if combined with a supercell approach. Many systems investigated with the here-described method fall into these categories, and therefore most calculations using the Car–Parrinello molecular dynamics approach are using supercells and a single **k**-point "integration scheme". The only point calculated is the center of the Brillouin zone, **k** = **0**, the Γ-point. For the remainder of this chapter, all formulas are given for this Γ-point approximation.

3.4.2 Wave function gradient

Analytic derivatives of the total energy with respect to the parameters of the calculation are needed for stable molecular dynamics calculations. All derivatives needed are easily accessible in the plane wave/pseudopotential approach. In the following, Fourier space formulas are presented

$$
\frac{1}{f_i}\frac{\partial E_{\text{tot}}}{\partial c_i^\star(\mathbf{G})} = \frac{1}{2}\,G^2\,c_i(\mathbf{G}) + \sum_{\mathbf{G}'}\left(V^{\text{loc}}(\mathbf{G}-\mathbf{G}')\right)^\star c_i(\mathbf{G}')
$$

$$
+ \sum_I\sum_{\alpha,\beta}\left(F_{I,i}^\alpha\right)^\star w_{\alpha\beta}^I P_\beta^I(\mathbf{G})S_I(\mathbf{G})\ , \quad (3.80)
$$

where V^{loc} is the local potential

$$
V^{\text{loc}}(\mathbf{G}) = \sum_I \Delta V_I^{\text{loc}}(\mathbf{G})S_I(\mathbf{G}) + V_{\text{xc}}(\mathbf{G}) + 4\pi\,\frac{n^{\text{tot}}(\mathbf{G})}{G^2}\ . \quad (3.81)
$$

Wave function gradients are needed in both structural relaxation and optimization, as well as in the Car–Parrinello molecular dynamics approach.

3.4.3 Gradient for nuclear positions

The derivative of the total energy with respect to nuclear positions is needed for structure optimization and in molecular dynamics, that is

$$
\frac{\partial E_{\text{tot}}}{\partial R_{I,s}} = \frac{\partial E_{\text{loc}}^{\text{PP}}}{\partial R_{I,s}} + \frac{\partial E_{\text{nloc}}^{\text{PP}}}{\partial R_{I,s}} + \frac{\partial E_{\text{ES}}}{\partial R_{I,s}}\ , \quad (3.82)
$$

as the kinetic energy E_{kin} and the exchange–correlation energy E_{xc} do not depend directly on the atomic positions, the relevant parts are

$$\frac{\partial E_{\text{loc}}^{\text{PP}}}{\partial R_{I,s}} = -\Omega \sum_{\mathbf{G}} i G_s \Delta V_I^{\text{loc}}(\mathbf{G}) S_I(\mathbf{G}) n^\star(\mathbf{G}) \tag{3.83}$$

$$\frac{\partial E_{\text{nloc}}^{\text{PP}}}{\partial R_{I,s}} = \sum_i f_i \sum_{\alpha,\beta \in I} \left\{ (F_{I,i}^\alpha)^\star w_{\alpha\beta}^I \left(\frac{\partial F_{I,i}^\beta}{\partial R_{I,s}} \right) \right.$$

$$\left. + \left(\frac{\partial F_{I,i}^\alpha}{\partial R_{I,s}} \right)^\star w_{\alpha,\beta}^I F_{I,i}^\beta \right\} \tag{3.84}$$

$$\frac{\partial E_{\text{ES}}}{\partial R_{I,s}} = -\Omega \sum_{\mathbf{G} \neq 0} i G_s \frac{(n^{\text{tot}}(\mathbf{G}))^\star}{G^2} n_I^c(\mathbf{G}) S_I(\mathbf{G}) + \frac{\partial E_{\text{ovrl}}}{\partial R_{I,s}} . \tag{3.85}$$

The contribution of the projectors of the nonlocal pseudopotentials is calculated from

$$\frac{\partial F_{I,i}^\alpha}{\partial R_{I,s}} = -\frac{1}{\sqrt{\Omega}} \sum_{\mathbf{G}} i G_s P_\alpha^I(\mathbf{G}) S_I(\mathbf{G}) c_i^\star(\mathbf{G}) . \tag{3.86}$$

Finally, the real space part contribution of the Ewald sum is

$$\frac{\partial E_{\text{ovrl}}}{\partial R_{I,s}} = {\sum_{J}}' \sum_{\mathbf{L}} \left\{ \frac{Z_I Z_J}{|\mathbf{R}_I - \mathbf{R}_J - \mathbf{L}|^3} \text{erfc} \left[\frac{|\mathbf{R}_I - \mathbf{R}_J - \mathbf{L}|}{\sqrt{R_I^{c\,2} + R_J^{c\,2}}} \right] \right.$$

$$\left. + \frac{2}{\sqrt{\pi}} \frac{1}{\sqrt{R_I^{c\,2} + R_J^{c\,2}}} \frac{Z_I Z_J}{|\mathbf{R}_I - \mathbf{R}_J - \mathbf{L}|^2} \exp \left[-\frac{|\mathbf{R}_I - \mathbf{R}_J - \mathbf{L}|^2}{\sqrt{R_I^{c\,2} + R_J^{c\,2}}} \right] \right\}$$

$$\times (R_{I,s} - R_{J,s} - L_s) . \tag{3.87}$$

The self-energy E_{self} is independent of the atomic positions and does not contribute to the forces.

3.4.4 Internal stress tensor

For calculations where the supercell is changed (e.g. the combination of the Car–Parrinello method with the Parrinello–Rahman approach [118, 441] as explained in Section 5.2.3) the electronic internal stress tensor is required. The electronic part of the internal stress tensor is defined as [1055–1057]

(see also Section 5.2.3)

$$\Pi_{uv} = -\frac{1}{\Omega} \sum_s \frac{\partial E_{\text{tot}}}{\partial h_{us}} h_{sv}^{\text{T}} \ . \tag{3.88}$$

A useful identity for the derivation of the stress tensor is

$$\frac{\partial \Omega}{\partial h_{uv}} = \Omega (\mathbf{h}^{\text{T}})_{uv}^{-1} \tag{3.89}$$

in view of the relation $\Omega = \det \mathbf{h}$ introduced in Eq. (3.1). The derivatives of the total energy with respect to the components of the cell matrix $\mathbf{h}$ (see Section 3.1.1) can be performed on every part of the total energy individually, i.e.

$$\frac{\partial E_{\text{tot}}}{\partial h_{uv}} = \frac{\partial E_{\text{kin}}}{\partial h_{uv}} + \frac{\partial E_{\text{loc}}^{\text{PP}}}{\partial h_{uv}} + \frac{\partial E_{\text{nloc}}^{\text{PP}}}{\partial h_{uv}} + \frac{\partial E_{\text{xc}}}{\partial h_{uv}} + \frac{\partial E_{\text{ES}}}{\partial h_{uv}} \ . \tag{3.90}$$

Using Eq. (3.89) extensively, the derivatives can be calculated for the case of a plane wave basis in Fourier space [440] as follows

$$\frac{\partial E_{\text{kin}}}{\partial h_{uv}} = -\sum_i f_i \sum_{\mathbf{G}} \sum_s G_u G_s (\mathbf{h}^{\text{T}})_{sv}^{-1} |c_i(\mathbf{G})|^2 \tag{3.91}$$

$$\frac{\partial E_{\text{loc}}^{\text{PP}}}{\partial h_{uv}} = \Omega \sum_I \sum_{\mathbf{G}} \left(\frac{\partial \Delta V_I^{\text{loc}}(\mathbf{G})}{\partial h_{uv}} \right) S_I(\mathbf{G}) n^\star(\mathbf{G}) \tag{3.92}$$

$$\frac{\partial E_{\text{nloc}}^{\text{PP}}}{\partial h_{uv}} = \sum_i f_i \sum_I \sum_{\alpha,\beta \in I} \left\{ (F_{I,i}^\alpha)^\star w_{\alpha\beta}^I \left(\frac{\partial F_{I,i}^\beta}{\partial h_{uv}} \right) \right.$$
$$\left. + \left(\frac{\partial F_{I,i}^\alpha}{\partial h_{uv}} \right)^\star w_{\alpha,\beta}^I F_{I,i}^\beta \right\} \tag{3.93}$$

$$\frac{\partial E_{\text{xc}}}{\partial h_{uv}} = -\sum_{\mathbf{G}} n^\star(\mathbf{G}) \left[V_{\text{xc}}(\mathbf{G}) - \varepsilon_{\text{xc}}(\mathbf{G}) \right] (\mathbf{h}^{\text{T}})_{uv}^{-1}$$
$$+ \sum_s \sum_{\mathbf{G}} i G_u n^\star(\mathbf{G}) \left(\frac{\partial F_{\text{xc}}(\mathbf{G})}{\partial (\partial_s n)} \right) (\mathbf{h}^{\text{T}})_{sv}^{-1} \tag{3.94}$$

$$\frac{\partial E_{\text{ES}}}{\partial h_{uv}} = 2\pi \Omega \sum_{\mathbf{G} \neq 0} \sum_s \left\{ -\frac{|n^{\text{tot}}(\mathbf{G})|^2}{G^2} \delta_{us} + \frac{(n^{\text{tot}}(\mathbf{G}))^\star}{G^2} \right.$$
$$\left. \times \left(\frac{n^{\text{tot}}(\mathbf{G})}{G^2} + \frac{1}{2} \sum_I n_I^{\text{c}}(\mathbf{G})(R_I^{\text{c}})^2 \right) G_u G_s \right\}$$

$$\times G_u G_s (\mathbf{h}^{\mathrm{T}})_{sv}^{-1} + \frac{\partial E_{\mathrm{ovrl}}}{\partial h_{uv}} \quad . \tag{3.95}$$

Finally, the derivative of the overlap contribution to the electrostatic energy is given by

$$\frac{\partial E_{\mathrm{ovrl}}}{\partial h_{uv}} = -{\sum_{I,J}}' \sum_{\mathbf{L}} \left\{ \frac{Z_I Z_J}{|\mathbf{R}_I - \mathbf{R}_J - \mathbf{L}|^3} \mathrm{erfc} \left[\frac{|\mathbf{R}_I - \mathbf{R}_J - \mathbf{L}|}{\sqrt{R_I^{\mathrm{c}\,2} + R_J^{\mathrm{c}\,2}}} \right] \right.$$

$$\left. + \frac{2}{\sqrt{\pi}\sqrt{R_I^{\mathrm{c}\,2} + R_J^{\mathrm{c}\,2}}} \frac{Z_I Z_J}{|\mathbf{R}_I - \mathbf{R}_J - \mathbf{L}|^2} \exp \left[-\frac{|\mathbf{R}_I - \mathbf{R}_J - \mathbf{L}|^2}{\sqrt{R_I^{\mathrm{c}\,2} + R_J^{\mathrm{c}\,2}}} \right] \right\}$$

$$\times \sum_s (R_{I,u} - R_{J,u} - L_u)(R_{I,s} - R_{J,s} - L_s)(\mathbf{h}^{\mathrm{T}})_{sv}^{-1} \quad . \tag{3.96}$$

The local part of the pseudopotential $\Delta V_I^{\mathrm{loc}}(\mathbf{G})$ and the nonlocal projector functions depend on the cell matrix $\mathbf{h}$ through the volume, the Bessel transform integral, and the spherical harmonics function. Their derivatives are lengthy but easy to calculate from the definitions Eqs. (3.36) and (3.37), and read

$$\frac{\partial \Delta V_I^{\mathrm{loc}}(\mathbf{G})}{\partial h_{uv}} = -\Delta V_I^{\mathrm{loc}}(\mathbf{G})(\mathbf{h}^{\mathrm{T}})_{uv}^{-1}$$

$$+ \frac{4\pi}{\Omega} \int_0^\infty dr \, r^2 \, \Delta V^{\mathrm{loc}}(r) \left(\frac{\partial j_0(Gr)}{\partial h_{uv}} \right) Y_{lm}(\tilde{\theta}, \tilde{\phi}) \tag{3.97}$$

$$\frac{\partial F_{I,i}^\alpha}{\partial h_{uv}} = \frac{4\pi}{\sqrt{\Omega}} (-i)^l \sum_{\mathbf{G}} c_i^\star(\mathbf{G}) \, S_I(\mathbf{G})$$

$$\times \left[\left(\frac{\partial Y_{lm}(\tilde{\theta}, \tilde{\phi})}{\partial h_{uv}} - \frac{1}{2} Y_{lm}(\tilde{\theta}, \tilde{\phi})(\mathbf{h}^{\mathrm{T}})_{uv}^{-1} \right) \int_0^\infty dr \, r^2 \, P_\alpha^I(r) \, j_l(Gr) \right.$$

$$\left. + Y_{lm}(\tilde{\theta}, \tilde{\phi}) \int_0^\infty dr \, r^2 \, P_\alpha^I(r) \left(\frac{\partial j_l(Gr)}{\partial h_{uv}} \right) \right] \quad . \tag{3.98}$$

3.5 Energy and force calculations in practice

In Section 3.4 formulas for the total energy and forces were given in their Fourier space representation. Many terms are in fact calculated most easily

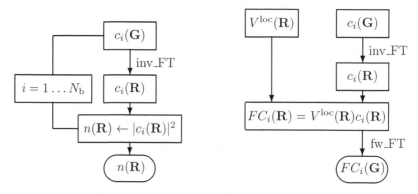

Fig. 3.4. Flow chart for the calculation of the charge density (on the left) and the force on the wave function from the local potential (on the right). The charge density calculation requires N_b (number of states) three-dimensional Fourier transforms. For the application of the local potential two Fourier transforms per state are needed. If enough memory is available the first transform can be avoided if the wave function on the real space grid is stored during the density calculation.

in this form, but some terms would require double sums over plane waves. In particular, the calculation of the charge density and the wave function gradient originating from the local potential

$$\sum_{\mathbf{G}'} \left(V^{\text{loc}}(\mathbf{G} - \mathbf{G}') \right)^{\star} c_i(\mathbf{G}') \ . \tag{3.99}$$

The expression in Eq. (3.99) is a convolution and can be calculated efficiently by a series of Fourier transforms according to the flow charts presented in Fig. 3.4. Both of these modules contain a Fourier transform of the wave functions from $\mathbf{G}$-space to the real space grid. In addition, the calculation of the wave function forces requires a back transform of the product of the local potential with the wave functions, performed on the real space grid, to Fourier space. This leads to a number of Fourier transforms that is three times the number of states in the system. If enough memory is available on the computer, the second transform of the wave functions to the grid can be avoided if the wave functions are stored in real space during the computation of the density. These modules are used further in the flow chart of the calculation of the local potential depicted in Fig. 3.5. Additional Fourier transforms are needed in this part of the calculation, however, the number of transforms does not scale with the number of electrons in the system. Additional transforms might be hidden in the module to calculate the exchange and correlation potential (see also Fig. 3.3) and the Poisson solver in cases when the Hockney method is used (see Fig. 3.1).

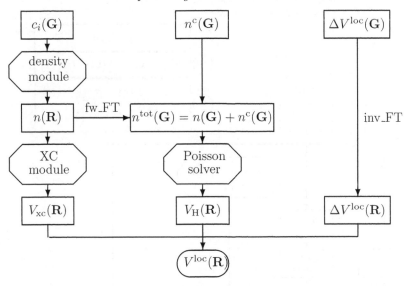

Fig. 3.5. Flow chart for the calculation of the local potential from the Kohn–Sham orbitals. This module calculates also the charge density in real and Fourier space as well as the exchange–correlation energy, Hartree energy, and local pseudopotential energy.

The calculation of the total energy, together with the local potential, is shown in Fig. 3.6. The overlap between the projectors of the nonlocal pseudopotential and the wave functions calculated in this part will be reused in the calculation of the forces on the wave functions. There are three initialization steps marked in Fig. 3.5. Step one has only to be performed at the beginning of the calculation, as the quantities $\mathbf{g}$ and E_{self} are constants. The quantities calculated in step two depend on the absolute value of the reciprocal space vectors. They have to be recalculated whenever the box matrix $\mathbf{h}$ changes. Finally, the variables in step three depend on the atomic positions and have to be calculated after each change of the nuclear positions. The flow charts of the calculation of the forces for the wave functions and the nuclei are presented in Figs. 3.7 and 3.8.

3.6 Optimizing the Kohn–Sham orbitals

Advances in the application of plane wave-based electronic structure methods are closely related to improved methods for the solution of the Kohn–Sham equations. There are now two different but equally successful approaches available. Fix-point methods based on the diagonalization of the Kohn–Sham matrix follow the more traditional ways that go back to the

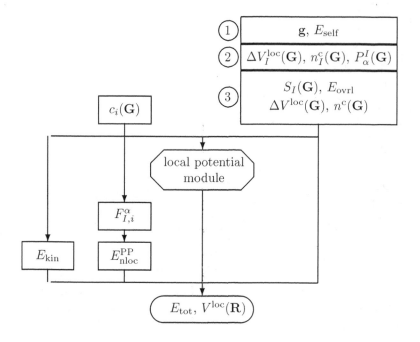

Fig. 3.6. Flow chart for the calculation of the local potential and total energy. Initialization steps are marked with numbers. Step 2 has to be repeated whenever the size of the unit cell changes. Step 3 has to be repeated whenever nuclear positions have changed.

roots of basis set methods in quantum chemistry. Direct nonlinear optimization approaches subject to a constraint were initiated by the success of the Car–Parrinello method. The following sections review some of these methods, focusing on the special problems related to the plane wave basis.

3.6.1 Initial guess

The initial guess of the Kohn–Sham orbitals is the first step to a successful calculation. One would like to introduce as much knowledge as possible into the first step of the calculation, but at the same time the procedure should be general and robust. One should also take care not to introduce artificial symmetries that may be preserved during the optimization and lead to false results. The most general initialization might be to choose the wave function coefficients from a random distribution. It makes sense to weight the random numbers by a function reflecting the relative importance of different basis functions where a good choice is a Gaussian distribution in G for the plane wave basis set. This initialization scheme clearly avoids symmetry problems

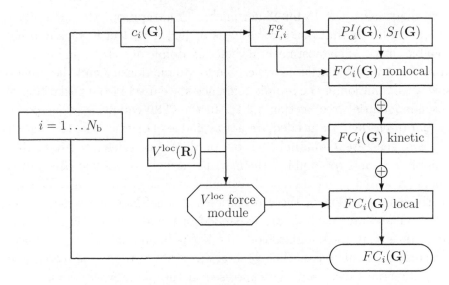

Fig. 3.7. Flow chart for the calculation of the forces on the wave functions. Notice that the calculation of the overlap terms $F^\alpha_{I,i}$ is done outside the loop over wave functions. Besides the wave functions and the local potential, the structure factors and the projectors of the nonlocal pseudopotential are input parameters to this module.

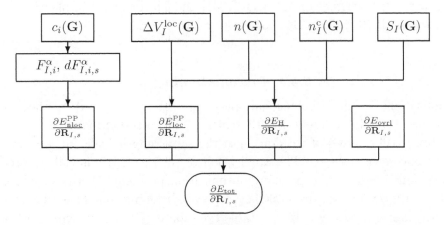

Fig. 3.8. Flow chart for the calculation of the forces on the nuclei.

but leads to energies far off the final results. Furthermore, highly tuned sophisticated optimization methods might have problems in converging. On the other hand, the random wave function method has a tendency to work in tricky cases where more intelligent methods might be misguided and thus fail.

A more educated guess is to use a superposition of atomic densities and then to diagonalize the Kohn–Sham matrix in an appropriate basis. This basis set can be the full plane wave basis or just a part of it, or any other reasonable choice. The most natural choice of atomic densities and basis sets for a plane wave calculation are the pseudo atomic density and the pseudo atomic wave function of the atomic reference state used in the generation of the pseudopotential (see Section 4.2.1). In the CPMD code this is one possibility, but often the data needed are not available. For this case, the default option is to generate a minimal basis out of Slater functions (see Eq. (2.137) in Section 2.8) and to combine them with the help of atomic occupation numbers (gathered using the usual Aufbau principle) to an atomic density. From the superposition of these densities a Kohn–Sham potential is constructed. The Slater orbitals are expanded in plane waves, and using the same routines as in the standard code the Kohn–Sham and overlap matrices are calculated in this basis. The general eigenvalue problem is solved and the eigenfunctions can easily be expressed in the plane wave basis, which are in turn used as the initial wave functions to the optimization routines. Similarly, a given plane wave representation of the total wave function can be projected onto an auxiliary set of atom-centered functions. This opens up the possibility to perform population and bond-order analyses (following for instance the schemes of Mulliken [1041] and Mayer [978]) in plane wave calculations [1270, 1271, 1311, 1312] (see Section 10.2).

3.6.2 Preconditioning

Optimizations in high-dimensional spaces are often hampered by the appearance of vastly different length scales. The introduction of a metric that brings all degrees of freedom onto the same length scale can improve convergence considerably. The application of such a metric is called "preconditioning" and is used in many optimization problems. If the variables in the optimization are decoupled, the preconditioning matrix is diagonal and becomes computationally tractable even for very large systems. Fortunately, this is to a large degree the case for a plane wave basis set. For large $\mathbf{G}$-vectors the Kohn–Sham matrix is dominated by the kinetic energy, which is diagonal in the plane wave representation. Based on this observation, efficient preconditioning schemes have been proposed [663, 786, 1447, 1462]. The preconditioner implemented in the CPMD code is based on the diagonal of the Kohn–Sham matrix $\mathbf{H}$, which is given by

$$
\begin{aligned}
K_{\mathbf{G},\mathbf{G}'} = H_{\mathbf{G},\mathbf{G}}\, \delta_{\mathbf{G},\mathbf{G}'} \qquad & \text{if } |\mathbf{G}| \geq G^c \\
K_{\mathbf{G},\mathbf{G}'} = H_{\mathbf{G}^c,\mathbf{G}^c}\, \delta_{\mathbf{G},\mathbf{G}'} \qquad & \text{if } |\mathbf{G}| \leq G^c
\end{aligned}
\qquad (3.100)
$$

where G^c is a free parameter that can be adjusted to accelerate convergence. However, it is found that the actual choice is not very critical and for practical purposes it is convenient not to fix G^c, but to use a universal constant of 0.5 Hartree for H_{G^c,G^c} that in turn defines G^c for each system.

3.6.3 Direct methods

The success of the Car–Parrinello approach stimulated interest in other methods for a direct minimization of the Kohn–Sham energy functional. These methods optimize directly the total energy using the gradient derived from the Lagrange function

$$\mathcal{L} = E^{\mathrm{KS}}(\{\phi_i\}) - \sum_{ij} \lambda_{ij}(\langle\phi_i|\phi_j\rangle - \delta_{ij}) \tag{3.101}$$

$$\frac{\delta\mathcal{L}}{\delta\phi_i^\star} = H_{\mathrm{e}}^{\mathrm{KS}}\phi_i - \sum_j \langle\phi_i|H_{\mathrm{e}}^{\mathrm{KS}}|\phi_j\rangle\phi_j \ . \tag{3.102}$$

Optimization methods differ in the way orbitals are updated. A steepest descent-based scheme

$$c_i(\mathbf{G}) \leftarrow c_i(\mathbf{G}) + \alpha K_{\mathbf{G},\mathbf{G}}^{-1} F_i(\mathbf{G}) \tag{3.103}$$

can be combined with the preconditioner and a line search option to find the optimal step size α. Nearly optimal α's can be found with an interpolation based on a quadratic polynomial. In Eq. (3.103) the symbols $F_i(\mathbf{G})$ denote the Fourier components of the wave function gradient.

Improved convergence can be achieved by replacing the steepest descent step with a search direction based on conjugate gradients [45, 508, 1174, 1462, 1578]

$$c_i(\mathbf{G}) \leftarrow c_i(\mathbf{G}) + \alpha h_i(\mathbf{G}) \ . \tag{3.104}$$

The conjugate directions are calculated from

$$h_i^{(n)}(\mathbf{G}) = \begin{cases} g_i^{(n)}(\mathbf{G}) & n = 0 \\ g_i^{(n)}(\mathbf{G}) + \gamma^{(n-1)}h_i^{n-1}(\mathbf{G}) & n = 1, 2, 3, \ldots \end{cases} \tag{3.105}$$

where

$$g_i^{(n)}(\mathbf{G}) = K_{\mathbf{G},\mathbf{G}}^{-1} F_i^{(n)}(\mathbf{G}) \tag{3.106}$$

$$\gamma^{(n)} = \frac{\sum_i \langle g_i^{(n+1)}(\mathbf{G})|g_i^{(n+1)}(\mathbf{G})\rangle}{\langle g_i^{(n)}(\mathbf{G})|g_i^{(n)}(\mathbf{G})\rangle} . \tag{3.107}$$

A very efficient implementation of this method [1462] is based on a band-by-band optimization (see Ref. [1123] for a detailed description of this approach).

The direct inversion in the iterative subspace (DIIS) method [293, 663, 1165] is a very successful extrapolation method that can be used in any kind of optimization problem. In quantum chemistry the DIIS scheme has been applied to wave function optimizations, geometry optimizations, and in post-Hartree–Fock applications. DIIS uses the information of n previous steps. Together with the position vectors $c_i^{(k)}(\mathbf{G})$, an estimate of the error vector $e_i^{(k)}(\mathbf{G})$ for each previous step k is stored. The best approximation to the final solution within the subspace spanned by the n stored vectors is obtained in a least square sense by writing

$$c_i^{(n+1)}(\mathbf{G}) = \sum_{k=1}^{n} d_k c_i^{(k)}(\mathbf{G}) \;, \tag{3.108}$$

where the d_k are subject to the restriction

$$\sum_{k=1}^{n} d_k = 1 \tag{3.109}$$

and the estimated error becomes

$$e_i^{(n+1)}(\mathbf{G}) = \sum_{k=1}^{n} d_k e_i^{(k)}(\mathbf{G}) \;. \tag{3.110}$$

The expansion coefficients d_k are calculated from a system of linear equations

$$\begin{pmatrix} b_{11} & b_{12} & \cdots & b_{1n} & 1 \\ b_{21} & b_{22} & \cdots & b_{2n} & 1 \\ \vdots & \vdots & \ddots & \vdots & \vdots \\ b_{n1} & b_{n2} & \cdots & b_{nn} & 1 \\ 1 & 1 & \cdots & 1 & 0 \end{pmatrix} \begin{pmatrix} d_1 \\ d_2 \\ \vdots \\ d_n \\ -\lambda \end{pmatrix} = \begin{pmatrix} 0 \\ 0 \\ \vdots \\ 0 \\ 1 \end{pmatrix} \tag{3.111}$$

where the b_{kl} are given by

$$b_{kl} = \sum_{i} \langle e_i^k(\mathbf{G}) | e_i^l(\mathbf{G}) \rangle \;. \tag{3.112}$$

The error vectors are not known, but can be approximated within a quadratic model

$$e_i^{(k)}(\mathbf{G}) = -K_{\mathbf{G},\mathbf{G}}^{-1} F_i^{(k)}(\mathbf{G}) \;. \tag{3.113}$$

In the same approximation, assuming $\mathbf{K}$ to be a constant, the new trial vectors are estimated to be

$$c_i(\mathbf{G}) = c_i^{(n+1)}(\mathbf{G}) + K_{\mathbf{G},\mathbf{G}}^{-1} F_i^{(n+1)}(\mathbf{G}) \ , \tag{3.114}$$

where the first derivative of the energy density functional is estimated to be

$$F_i^{(n+1)}(\mathbf{G}) = \sum_{k=1}^{n} d_k F_i^{(k)}(\mathbf{G}) \ . \tag{3.115}$$

The methods described in this section produce new trial vectors that are not orthogonal. Therefore an orthogonalization step has to be added before the new charge density is calculated

$$c_i(\mathbf{G}) \leftarrow \sum_k c_j(\mathbf{G}) X_{ji} \ . \tag{3.116}$$

There are different choices for the rotation matrix $\mathbf{X}$ that lead to orthogonal orbitals. Two of the computationally convenient choices are the Löwdin orthogonalization

$$X_{ji} = (\mathbf{S}^{-1/2})_{ji} \tag{3.117}$$

and a matrix form of the Gram-Schmidt procedure

$$X_{ji} = (\mathbf{G}^{-T})_{ji} \ , \tag{3.118}$$

where $\mathbf{S}$ is the overlap matrix and $\mathbf{G}$ is its Cholesky decomposition

$$\mathbf{S} = \mathbf{G}\mathbf{G}^T \ . \tag{3.119}$$

Recently, new methods that avoid the orthogonalization step have been introduced. One of them [1144] relies on modified functionals that can be optimized without the orthogonality constraint. These functionals, originally introduced in the context of linear scaling methods [974, 1088], have the property that their minima coincide with the original Kohn–Sham energy functional. The methods described above can be used to optimize the new functional.

Another approach [664] is to use a variable transformation from the expansion coefficients of the orbitals in plane waves to a set of non-redundant orbital rotation angles. This method was introduced in quantum chemistry [305, 362, 1465] and is used successfully in many optimization problems that involve a set of orthogonal orbitals. A generalization of the orbital rotation scheme allowed the application also for cases where the number of basis functions is orders of magnitude bigger than the number of occupied orbitals. However, no advantage is gained over the standard methods as the calculation of the gradient in the transformed variables scales the same as the orthogonalization step. In addition, there is no simple and efficient preconditioner available for the orbital rotation coordinates.

The problems associated with the orbital rotation method have been overcome by the orbital transformation method [1551]. In this method the occupied orbitals are written as a function of a set of auxiliary functions and another set of reference functions

$$\mathbf{c(x)} = \mathbf{c}_0 \cos(\mathbf{U}) + \mathbf{x}\mathbf{U}^{-1}\sin(\mathbf{U}) \tag{3.120}$$

$$\mathbf{U} = (\mathbf{x}^{\mathrm{T}}\mathbf{Sx})^{1/2} , \tag{3.121}$$

where $\mathbf{c}, \mathbf{c}_0$, and $\mathbf{x}$ are matrices with a first index running over all plane waves and the second index denoting an occupied orbital. The square matrix $\mathbf{U}$ has the dimension of the number of occupied orbitals. The overlap matrix $\mathbf{S}$ is the unit matrix if only norm-conserving pseudopotentials are used. Optimization has to be performed for all $\mathbf{x}$ with the constraint

$$\mathbf{x}^{\mathrm{T}}\mathbf{Sc}_0 = 0 . \tag{3.122}$$

This is a linear constraint that is easily handled and standard optimization methods can be applied without change. Optimization is still done using the original basis set of plane waves, and therefore the preconditioners have only to be adapted slightly.

3.6.4 Fix-point methods

Traditionally, all methods to find solutions to the Kohn–Sham equations were using matrix diagonalization methods. It quickly became clear that direct schemes can only be used for very small systems. The storage requirements of the Kohn–Sham matrix in the plane wave basis and the scaling proportional to the cube of the basis set size lead to insurmountable problems. Iterative diagonalization schemes can be adapted to the special needs of a plane wave basis, and when combined with a proper preconditioner, lead to algorithms that are comparable to the direct methods, both in memory requirements and over all scaling properties. Iterative diagonalization schemes are abundant and methods based on the Lanczos algorithm [310, 827, 1155] can be used as well as conjugate gradient techniques [209, 1462]. Very good results have been achieved by the combination of the DIIS method with the minimization of the norm of the residual vector [786, 1631].

The diagonalization methods have to be combined with an optimization method for the charge density. Methods based on mixing [7, 318], quasi-Newton algorithms [147, 194, 713], and DIIS [785, 786, 1165] are used successfully. Also, these methods use a preconditioning scheme. It was shown that the optimal preconditioning for charge density mixing is connected to the charge dielectric matrix [7, 97, 318, 641, 1550]. For a plane wave basis, the charge dielectric matrix can be approximated by expressions very close to those used for the preconditioning in the direct optimization methods. Fix-point methods have a slightly larger prefactor than most of the direct methods. Their advantage lies in the robustness and capability of treating systems with no or small band gaps.

3.7 Molecular dynamics

Numerical methods to integrate the equations of motion are an important part of every molecular dynamics program. Therefore, an extended litera-ture exists on integration techniques (see Ref. [468] and references therein). All considerations valid for the integration of equations of motion with classi-cal potentials, i.e. force fields, also apply for *ab initio* molecular dynamics if the Born–Oppenheimer dynamics approach is used. These basic techniques will not be discussed here.

A good initial guess for the Kohn–Sham optimization procedure is a cru-cial ingredient for good performance of the Born–Oppenheimer dynamics approach. An extrapolation scheme was devised [46] that makes use of the optimized wave functions from previous time steps. Other extrapolation schemes [1167, 1203, 1552] have also been used with good success. These procedures have a strong connection to the basic idea of the Car–Parrinello method, but are not essential to the method.

The rest of this section discusses the integration of the Car–Parrinello equations in their simplest form and explains the solution to the constraints equation for general geometric constraints. Finally, a special form of the equations of motion will be used for optimization purposes.

3.7.1 Car–Parrinello equations of motion

The Car–Parrinello Lagrangian and its derived equations of motion were introduced in Section 2.4. In the present section, the general expressions Eqs. (2.56), (2.59), and (2.60) are specialized to the case of a plane wave basis within Kohn–Sham density functional theory (see Section 2.7.2). This is similar to what was done in the preceding sections of this chapter with the energy expression. Specifically the functions ϕ_i, denoting the Kohn–Sham orbitals, are replaced by the expansion coefficients $c_i(\mathbf{G})$ of the plane wave basis (see Sections 3.1.1 and 3.1.2). Furthermore, it is assumed that the orthonormality constraint (see Section 2.4.2) depends only on the orbitals but not the nuclear positions (see Eqs. (2.111) and (2.112) for a more general case and Section 6.6 for the consequences in terms of the equations of motion). The equations of motion for the Car–Parrinello method in terms of the plane wave expansion coefficients are derived from the resulting extended Lagrangian

$$\mathcal{L} = \mu \sum_i \sum_{\mathbf{G}} |\dot{c}_i(\mathbf{G})|^2 + \frac{1}{2} \sum_I M_I \dot{\mathbf{R}}_I^2 - E^{\text{KS}} \left[\{\mathbf{G}\}, \{\mathbf{R}_I\} \right]$$

$$+ \sum_{ij} \Lambda_{ij} \left(\sum_{\mathbf{G}} c_i^\star(\mathbf{G}) c_j(\mathbf{G}) - \delta_{ij} \right) , \quad (3.123)$$

where μ is the fictitious electron mass, and M_I are the masses of the nuclei as usual. Because of the expansion of the Kohn–Sham orbitals in plane waves, the orthonormality constraint does not depend on the nuclear positions. For basis sets that depend on the atomic positions (such as atom-fixed Gaussian basis sets according to Eqs. (2.111) and (2.112) or for methods that introduce an atom position-dependent metric (such as ultrasoft pseudopotentials [816, 1548] and the PAW transformation [142, 790]), the integration methods have to be adapted in order to ensure a proper energy-conserving dynamics (see in particular the discussion at the end of Section 2.5). Solutions that include these cases can be found in the literature [142, 596, 665, 816] and are discussed in Part II, Advanced techniques, in Chapter 6. The Euler–Lagrange equations derived from the plane wave Lagrangian Eq. (3.123) are

$$\mu \ddot{c}_i(\mathbf{G}) = -\frac{\partial E^{\text{KS}}}{\partial c_i^\star(\mathbf{G})} + \sum_j \Lambda_{ij} c_j(\mathbf{G}) \quad (3.124)$$

$$M_I \ddot{\mathbf{R}}_I = -\frac{\partial E^{\text{KS}}}{\partial \mathbf{R}_I} . \quad (3.125)$$

The two sets of equations are coupled through the Kohn–Sham energy functional $E^{\text{KS}} \left[\{\mathbf{G}\}, \{\mathbf{R}_I\} \right]$ and special care has to be taken for the integration because of the orthonormality constraint $\sim \boldsymbol{\Lambda}$.

The integrator [1510] used in the CPMD code is based on the velocity Verlet/RATTLE algorithm [32, 1422]. The velocity Verlet algorithm requires more operations and more storage than the usual Verlet algorithm [1567], as explained for instance in Ref. [468]. However, it is much easier to incorporate temperature control via velocity scaling into the velocity Verlet algorithm. In addition, velocity Verlet allows us to change the time step trivially and is conceptually easier to handle [930, 1510]. It is defined by the following equations

$$\dot{\mathbf{R}}_I(t + \delta t) = \dot{\mathbf{R}}_I(t) + \frac{\delta t}{2M_I} \mathbf{F}_I(t)$$

$$\mathbf{R}_I(t + \delta t) = \mathbf{R}_I(t) + \delta t \, \dot{\mathbf{R}}_I(t + \delta t)$$

$$\dot{\tilde{\mathbf{c}}}_i(t + \delta t) = \dot{\mathbf{c}}_i(t) + \frac{\delta t}{2\mu} \mathbf{f}_i(t)$$

$$\tilde{\mathbf{c}}_i(t + \delta t) = \mathbf{c}_i(t) + \delta t \, \dot{\tilde{\mathbf{c}}}_i(t + \delta t)$$

$$\mathbf{c}_i(t + \delta t) = \tilde{\mathbf{c}}_i(t + \delta t) + \sum_j \mathbf{X}_{ij} \, \mathbf{c}_j(t)$$

$$\texttt{calculate} \quad \mathbf{F}_I(t + \delta t)$$

$$\texttt{calculate} \quad \mathbf{f}_i(t + \delta t)$$

$$\dot{\mathbf{R}}_I(t + \delta t) = \dot{\mathbf{R}}_I(t + \delta t) + \frac{\delta t}{2M_I} \mathbf{F}_I(t + \delta t)$$

$$\dot{\mathbf{c}}'_i(t + \delta t) = \dot{\tilde{\mathbf{c}}}_i(t + \delta t) + \frac{\delta t}{2\mu} \mathbf{f}_i(t + \delta t)$$

$$\dot{\mathbf{c}}_i(t + \delta t) = \dot{\mathbf{c}}'_i(t + \delta t) + \sum_j \mathbf{Y}_{ij} \, \mathbf{c}_j(t + \delta t) \ ,$$

where $\mathbf{R}_I(t)$ and $\mathbf{c}_i(t)$ are the atomic positions of particle I and the Kohn–Sham orbital i at time t, respectively. Here, $\mathbf{F}_I$ are the forces on atom I and $\mathbf{f}_i$ denote the forces on Kohn–Sham orbital i. The matrices $\mathbf{X}$ and $\mathbf{Y}$ are related directly to the Lagrange multipliers by

$$X_{ij} = \frac{\delta t^2}{2\mu} \Lambda_{ij}^{\mathrm{p}} \qquad (3.126)$$

$$Y_{ij} = \frac{\delta t}{2\mu} \Lambda_{ij}^{\mathrm{v}} \ . \qquad (3.127)$$

Notice that in the SHAKE/RATTLE algorithm the Lagrange multipliers needed in order to enforce the orthonormality for the positions $\mathbf{\Lambda}^{\mathrm{p}}$ and velocities $\mathbf{\Lambda}^{\mathrm{v}}$ are treated as independent variables. Denoting by $\mathbf{C}$ the matrix of wave function coefficients $c_i(\mathbf{G})$, the orthonormality constraint can be written as

$$\mathbf{C}^\dagger(t + \delta t)\mathbf{C}(t + \delta t) - \mathbf{I} = 0 \tag{3.128}$$

$$\left[\tilde{\mathbf{C}} + \mathbf{XC}\right]^\dagger \left[\tilde{\mathbf{C}} + \mathbf{XC}\right] - \mathbf{I} = 0 \tag{3.129}$$

$$\tilde{\mathbf{C}}^\dagger\tilde{\mathbf{C}} + \mathbf{X}\tilde{\mathbf{C}}^\dagger\mathbf{C} + \mathbf{C}^\dagger\tilde{\mathbf{C}}\mathbf{X}^\dagger + \mathbf{XX}^\dagger - \mathbf{I} = 0 \tag{3.130}$$

$$\mathbf{XX}^\dagger + \mathbf{XB} + \mathbf{B}^\dagger\mathbf{X}^\dagger = \mathbf{I} - \mathbf{A} , \tag{3.131}$$

where the new matrices $\mathbf{A}_{ij} = \tilde{\mathbf{c}}_i^\dagger(t + \delta t)\tilde{\mathbf{c}}_j(t + \delta t)$ and $\mathbf{B}_{ij} = \mathbf{c}_i^\dagger(t)\tilde{\mathbf{c}}_j(t + \delta t)$ have been introduced in Eq. (3.131); the unit matrix is denoted by the symbol $\mathbf{I}$. By noting that $\mathbf{A} = \mathbf{I} + \mathcal{O}(\delta t^2)$ and $\mathbf{B} = \mathbf{I} + \mathcal{O}(\delta t)$, Eq. (3.131) can be solved iteratively using

$$\mathbf{X}^{(n+1)} = \frac{1}{2}\left[\mathbf{I} - \mathbf{A} + \mathbf{X}^{(n)}\left(\mathbf{I} - \mathbf{B}\right) + \left(\mathbf{I} - \mathbf{B}\right)\mathbf{X}^{(n)} - \left(\mathbf{X}^{(n)}\right)^2\right] \tag{3.132}$$

starting from the initial guess

$$\mathbf{X}^{(0)} = \frac{1}{2}(\mathbf{I} - \mathbf{A}) . \tag{3.133}$$

In Eq. (3.132) use has been made of the fact that the matrices $\mathbf{X}$ and $\mathbf{B}$ are real and symmetric, which follows directly from their definitions. Equation (3.132) can usually be iterated to a tolerance of 10^{-6} within a few iterations.

The rotation matrix $\mathbf{Y}$ is calculated from the orthogonality condition on the orbital velocities

$$\dot{\mathbf{c}}_i^\dagger(t + \delta t)\mathbf{c}_j(t + \delta t) + \mathbf{c}_i^\dagger(t + \delta t)\dot{\mathbf{c}}_j(t + \delta t) = 0 . \tag{3.134}$$

Applying Eq. (3.134) to the trial states $\dot{\mathbf{C}}' + \mathbf{YC}$ yields a simple equation for $\mathbf{Y}$

$$\mathbf{Y} = -\frac{1}{2}(\mathbf{Q} + \mathbf{Q}^\dagger) , \tag{3.135}$$

where $\mathbf{Q}_{ij} = \mathbf{c}_i^\dagger(t + \delta t)\dot{\mathbf{c}}'_i(t + \delta t)$. The fact that $\mathbf{Y}$ can be obtained without iteration implies that the velocity constraint condition Eq. (3.134) is satisfied exactly, up to machine precision, at each time step.

3.7.2 Advanced integration

One advantage of the velocity Verlet integrator is that it can easily be combined with multiple time step algorithms [1493, 1511] and still results in reversible dynamics. The most successful implementation of such a multiple time step scheme in connection with the plane wave/pseudopotential method is the harmonic reference system idea [1122, 1511]. The high-frequency motion of the plane waves with large kinetic energy is used as a reference system for the integration. The dynamics of this reference system is harmonic and can be integrated analytically. In addition, this can be combined with the basic notion of a preconditioner already introduced in the section on optimizations (see Section 3.6.2). The introduction of a fictitious electron mass in the Car–Parrinello scheme is an algorithmic construct (see Eq. (2.60) and Section 2.4). Hence, it is allowed to generalize the idea of using a unique parameter by introducing different such masses for different "classical" degrees of freedom [1124, 1447, 1511]. In the spirit of the preconditioner Eq. (3.100) introduced in the optimization section, the new plane wave-dependent masses are chosen to be

$$\mu(\mathbf{G}) = \begin{cases} \mu & H_{\mathbf{G},\mathbf{G}} \leq \alpha \\ (\mu/\alpha) \left(\frac{1}{2} G^2 + V_{\mathbf{G},\mathbf{G}} \right) & H_{\mathbf{G},\mathbf{G}} \geq \alpha \end{cases} , \qquad (3.136)$$

where $H_{\mathbf{G},\mathbf{G}}$ and $V_{\mathbf{G},\mathbf{G}}$ denote the diagonal matrix elements of the Kohn–Sham matrix and the potential, respectively, in terms of plane waves. The reference electron mass is μ and the parameter α has been introduced before in Eq. (3.100) as $H_{\mathbf{G}^c,\mathbf{G}^c}$. Using these preconditioned masses in conjunction with the harmonic reference system, the equations of motion of the system read

$$\mu(\mathbf{G})\ddot{c}_i(\mathbf{G}) = -\lambda(\mathbf{G})c_i(\mathbf{G}) + \delta f_i(\mathbf{G}) + \sum_j \Lambda_{ij} c_j(\mathbf{G}) , \qquad (3.137)$$

where $\delta f_i(\mathbf{G})$ is the force on orbital i minus $\lambda(\mathbf{G})$. From Eq. (3.137) it is easy to see that the resulting frequencies $\omega(\mathbf{G}) = \sqrt{\lambda(\mathbf{G})/\mu(\mathbf{G})}$ are independent of $\mathbf{G}$ and that there is only one harmonic frequency equal to $\sqrt{\alpha/\mu}$. The generalized formulas for the proper integration of these equations of motion in the framework of the velocity Verlet algorithm are provided in the literature [1511].

The implications of introducing such $\mathbf{G}$-vector dependent masses can be understood by revisiting the formulas for the characteristic frequencies of the electronic system, i.e. Eqs. (2.70), (2.71), and (2.72). The masses μ are chosen such that all frequencies ω_{ij} are approximately equal, thus optimizing both adiabaticity and the molecular dynamics time step. The disadvantage of this method is that the average electron mass seen by the nuclei is drastically enhanced, leading to renormalization corrections [142] (see Section V.C.2 therein) on the nuclear masses M_I that are significantly higher than in the standard approach. In addition, the resulting corrections are no longer as simple to estimate by analytical expressions such as those provided in Section 2.4.9; of course, it is still possible to evaluate these effects numerically by performing several simulations using different reference electron masses μ.

3.7.3 Imposing geometrical constraints

Geometrical constraints are used in classical simulations to freeze fast degrees of freedom in order to allow for larger time steps. Mainly distance constraints are used, for instance to fix intramolecular covalent bonds. These type of applications of constraints are of lesser importance in *ab initio* molecular dynamics. However, in the simulation of rare events, as needed for many reactions, constraints play an important role together with the method of "Thermodynamic Integration" (TI) [468]. The "Blue Moon" or "constrained reaction coordinate dynamics" ensemble method [230, 1386] enables one to compute the potential of mean force and thus the free energy path as a function of a chosen constraint.

Free energy differences can be obtained directly from the average force of constraint and an additional geometric correction term during a molecular dynamics simulation as follows

$$\mathcal{F}(\xi_2) - \mathcal{F}(\xi_1) = \int_{\xi_1}^{\xi_2} \left\langle \frac{\partial \mathcal{H}}{\partial \xi} \right\rangle_{\xi'}^{\text{cond}} d\xi' \ , \tag{3.138}$$

where $\mathcal{F}$ is the free energy, $\xi(\mathbf{r})$ a one-dimensional reaction coordinate, $\mathcal{H}$ the Hamiltonian of the system, and $\langle \cdots \rangle_{\xi'}^{\text{cond}}$ the *conditioned average* in the constrained ensemble [1386]. By way of the Blue Moon ensemble, the statistical average is replaced by a time average over a constrained trajectory with the reaction coordinate fixed at special values, $\xi(\mathbf{R}) = \xi'$, and $\dot{\xi}(\mathbf{R}, \dot{\mathbf{R}}) = 0$. The quantity to evaluate is the mean force

$$\frac{d\mathcal{F}}{d\xi'} = \frac{\langle Z^{-1/2} [-\lambda + k_B T\ G] \rangle_{\xi'}}{\langle Z^{-1/2} \rangle_{\xi'}} \ , \tag{3.139}$$

where λ is the Lagrange multiplier used to impose the constraint, Z is a reweighting factor (the "Fixman determinant" that compensates the bias introduced by the mechanical constraint)

$$Z = \sum_I \frac{1}{M_I} \left(\frac{\partial \xi}{\partial \mathbf{R}_I} \right)^2 \ , \tag{3.140}$$

and the geometric correction factor G is given by

$$G = \frac{1}{Z^2} \sum_{I,J} \frac{1}{M_I M_J} \frac{\partial \xi}{\partial \mathbf{R}_I} \cdot \frac{\partial^2 \xi}{\partial \mathbf{R}_I \partial \mathbf{R}_J} \cdot \frac{\partial \xi}{\partial \mathbf{R}_J} \ , \tag{3.141}$$

where $\langle \cdots \rangle_{\xi'}$ is the *unconditioned average*, as obtained directly from a constrained molecular dynamics run with $\xi(\mathbf{R}) = \xi'$. The expression

$$\mathcal{F}(\xi_2) - \mathcal{F}(\xi_1) = \int_{\xi_1}^{\xi_2} \left(\frac{d\mathcal{F}}{d\xi'} \right) d\xi' \tag{3.142}$$

together with Eq. (3.139) finally defines the free energy difference and yields the free energy path along the imposed constraint. For the special case of a simple distance constraint $\xi(\mathbf{R}) = |\mathbf{R}_I - \mathbf{R}_J|$ between two nuclei I and J, the parameter Z is a constant and G = 0 (these factors are provided in Refs. [1385, 1386] for more complicated constraints).

The RATTLE algorithm allows for the calculation of the Lagrange multiplier of arbitrary constraints on geometrical variables within the velocity Verlet integrator and can be cast into the following algorithm as implemented in the CPMD code. As usual, the constraints are defined by the relations

$$\sigma_i(\{\mathbf{R}_I(t)\}) = 0 \ , \tag{3.143}$$

and the velocity Verlet algorithm can be performed with the following steps

$$\dot{\tilde{\mathbf{R}}}_I = \dot{\mathbf{R}}_I(t) + \frac{\delta t}{2M_I} \mathbf{F}_I(t)$$

$$\tilde{\mathbf{R}}_I = \mathbf{R}_I(t) + \delta t \, \dot{\tilde{\mathbf{R}}}_I$$

$$\mathbf{R}_I(t + \delta t) = \tilde{\mathbf{R}}_I + \frac{\delta t^2}{2M_I} \mathbf{g}^{\mathrm{p}}(t)$$

$$\texttt{calculate} \quad \mathbf{F}_I(t + \delta t)$$

$$\dot{\mathbf{R}}_I' = \dot{\tilde{\mathbf{R}}}_I + \frac{\delta t}{2M_I} \mathbf{F}_I(t + \delta t)$$

$$\dot{\mathbf{R}}_I(t + \delta t) = \dot{\mathbf{R}}_I' + \frac{\delta t}{2M_I} \mathbf{g}^{\mathrm{v}}(t + \delta t) \ ,$$

where the constraint forces are defined by

$$\mathbf{g}^{\mathrm{P}}(t) \;=\; -\sum_i \lambda_i^{\mathrm{p}} \frac{\partial \sigma_i(\{\mathbf{R}_I(t)\})}{\partial \mathbf{R}_I} \tag{3.144}$$

$$\mathbf{g}^{\mathrm{v}}(t) \;=\; -\sum_i \lambda_i^{\mathrm{v}} \frac{\partial \sigma_i(\{\mathbf{R}_I(t)\})}{\partial \mathbf{R}_I} \;. \tag{3.145}$$

The Lagrange multipliers have to be determined to ensure that the constraints on the positions and velocities are fulfilled exactly at the end of the time step. For the position the constraint condition is

$$\sigma_i(\{\mathbf{R}_I(t+\delta t)\}) = 0 \;, \tag{3.146}$$

which is in general a system of nonlinear equations in the set of Lagrange multipliers $\{\lambda_i^{\mathrm{p}}\}$.

These equations can be solved using a generalized Newton algorithm [1161] that can be combined with a convergence acceleration scheme based on the DIIS method [293, 1165], see also Section 3.6.3. The error vectors for a given set of Lagrange multipliers λ are calculated from

$$\mathbf{e}_i(\lambda) = -\sum_j \left(\mathbf{J}^{-1}\right)_{ij}(\lambda)\sigma_j(\lambda) \tag{3.147}$$

and the Jacobian $\mathbf{J}$ is defined by

$$J_{ij}(\lambda) \;=\; \frac{\partial \sigma_i(\lambda)}{\partial \lambda_j} \tag{3.148}$$

$$=\; \sum_{I,s} \frac{\partial \sigma_i(\lambda)}{\partial R_{I,s}(\lambda)} \frac{\partial R_{I,s}(\lambda)}{\partial \lambda_j} \tag{3.149}$$

$$=\; -\sum_{I,s} \frac{\delta t^2}{2M_I} f_{I,s}^{\mathrm{c}}(\lambda) f_{I,s}^{\mathrm{c}}(0) \;, \tag{3.150}$$

where $f_{I,s}^{\mathrm{c}}(\lambda) = \sum_i \lambda_i \partial \sigma_i / \partial R_{I,s}$. Typically only a few iterations are needed to converge the Lagrange multipliers to an accuracy of 10^{-8}.

The constraint condition for the velocities can be cast into a system of linear equations. Again, as in the case of the orthonormality constraints in the Car–Parrinello method, the Lagrange multiplier for the velocity update can be calculated exactly without making use of an iterative scheme. Defining the derivative matrix

$$\mathbf{A}_{iI} = \frac{\partial \sigma_i}{\partial \mathbf{R}_I} \;, \tag{3.151}$$

the velocity constraints are

$$\dot{\sigma}_i(t + \delta t) = 0 \qquad (3.152)$$

$$\sum_{I,s} \frac{\partial \sigma_i}{\partial R_{I,s}} \dot{R}_{I,s} = 0 \qquad (3.153)$$

$$-\sum_j \left(\sum_I \frac{\delta t^2}{2M_I} \mathbf{A}_{iI} \cdot \mathbf{A}_{jI} \right) \lambda_j^{\mathrm{v}} = \sum_I \mathbf{A}_{iI} \cdot \dot{\mathbf{R}}_I' . \qquad (3.154)$$

The only information needed to implement a new type of constraint is the formula for the value of the function and its derivative with respect to the nuclear coordinates involved in the constraint.

3.7.4 Using Car–Parrinello dynamics for optimizations

By adding a friction term, Car–Parrinello molecular dynamics can be turned into a damped second-order dynamics scheme (see also Section 2.4.7). The friction can be applied both to the nuclear degrees of freedom and the electronic coordinates. The resulting dynamics equations are a powerful method to optimize the atomic structure and the Kohn–Sham orbitals simultaneously [1123, 1447]. Harmonic reference system integration, together with **G**-vector-dependent electron masses as introduced in Section 3.7.2, are especially helpful in this context, as the generated dynamics does not have a direct physical relevance.

Introducing a friction force proportional to the constants γ_{n} and γ_{e}, the equations of motion can readily be integrated using the velocity Verlet algorithm. The friction terms translate into a simple rescaling of the velocities at the beginning and end of the time step, according to

$$\dot{\mathbf{R}}_I(t) = \gamma_{\mathrm{n}} \dot{\mathbf{R}}_I(t)$$

$$\dot{\mathbf{c}}_i(t) = \gamma_{\mathrm{e}} \dot{\mathbf{c}}_i(t)$$

velocity   Verlet update

$$\dot{\mathbf{R}}_I(t + \delta t) = \gamma_{\mathrm{n}} \dot{\mathbf{R}}_I(t + \delta t)$$

$$\dot{\mathbf{c}}_i(t + \delta t) = \gamma_{\mathrm{e}} \dot{\mathbf{c}}_i(t + \delta t) .$$

It was shown [1123, 1447] that this scheme leads to optimizations that are competitive with other methods described in Section 3.6.

Table 3.2. *Relative size of characteristic variables in two typical plane wave calculations. See text for details.*

	Silicon	Water
N_{at}	1	3
N_p	1	1
N_b	2	4
N_{PW}	53	1000
N_D	429	8000
N	1728	31250

3.8 Program organization and layout

In the practical implementation of the method, mathematical symbols have to be translated into data structures of the chosen computer language. Then mathematical formulas are set into computer code using the data structures. The layout of the data structures should be such that optimal performance for the algorithms can be achieved. The CPMD code is written in FORTRAN77 and in the following sections the most important data structures and computational kernels will be given in pseudo code form. The following variables are used to denote quantities that measure system size:

N_{at}	number of atoms
N_p	number of projectors
N_b	number of electronic bands or states
N_{PW}	number of plane waves
N_D	number of plane waves for densities and potentials
N_x, N_y, N_z	number of grid points in x, y, and z direction
$N = N_x N_y N_z$	total number of grid points.

In Table 3.2 the relative sizes of these variables are given for two systems using a representative parameter setting. The example for a silicon crystal assumes an energy cutoff of 13 Rydberg and s nonlocality for the norm-conserving pseudopotential. In the example of a water system the numbers are given per molecule. The cutoff used was 70 Rydberg and the oxygen pseudopotential has an s nonlocal part whereas the hydrogen pseudopotential is local.

3.8.1 Data structures

Important quantities in the pseudopotential plane wave method depend either not at all, linearly, or quadratically on the system size. Examples for the first kind of data are the unit cell matrix $\mathbf{h}$ and the cutoff E_{cut}. Variables with a size that grows linearly with the system are

$\mathtt{r}(3, N_{\text{at}})$	nuclear positions
$\mathtt{v}(3, N_{\text{at}})$	nuclear velocities
$\mathtt{f}(3, N_{\text{at}})$	nuclear forces
$\mathtt{g}(3, N_{\text{PW}})$	plane wave indices
$\mathtt{ipg}(3, N_{\text{PW}})$	mapping of $\mathbf{G}$-vectors (positive part)
$\mathtt{img}(3, N_{\text{PW}})$	mapping of $\mathbf{G}$-vectors (negative part)
$\mathtt{rhog}(N_{\text{PW}})$	densities $(n, n_{\text{c}}, n_{\text{tot}})$ in Fourier space
$\mathtt{vpot}(N_{\text{PW}})$	potentials $(V_{\text{loc}}, V_{\text{xc}}, V_{\text{H}})$ in Fourier space
$\mathtt{n}(N_{\text{x}}, N_{\text{y}}, N_{\text{z}})$	densities $(n, n_{\text{c}}, n_{\text{tot}})$ in real space
$\mathtt{v}(N_{\text{x}}, N_{\text{y}}, N_{\text{z}})$	potentials $(V_{\text{loc}}, V_{\text{xc}}, V_{\text{H}})$ in real space
$\mathtt{vps}(N_{\text{D}})$	local pseudopotential
$\mathtt{rpc}(N_{\text{D}})$	core charges
$\mathtt{pro}(N_{\text{PW}})$	projectors of nonlocal pseudopotential.

Pseudopotential-related quantities $\mathtt{vps}$, $\mathtt{rpc}$, and $\mathtt{pro}$ are one-dimensional in system size, but they also depend on the number of different atomic species. In the following it is assumed that only one such species is present, since it is straightforward to generalize the pseudo codes given below to more than one atomic species. For real quantities that depend on $\mathbf{G}$-vectors, only half of the values have to be stored. The other half can be recomputed when needed by using the symmetry relation

$$A(\mathbf{G}) = A^{\star}(-\mathbf{G}) \ , \tag{3.155}$$

which saves a factor of two in memory. In addition, $\mathbf{G}$-vectors are stored in a linear array instead of a three-dimensional structure since this allows one to store only nonzero variables. Because there is a spherical cutoff, another reduction by a factor of two is achieved for the memory. For the Fourier transforms the variables have to be prepared in a three-dimensional array. The mapping of the linear array to this structure is provided by the information stored in the arrays $\mathtt{ipg}$ and $\mathtt{img}$.

Most of the memory is needed for the storage of quantities that grow quadratically with system size. In order to save memory it is possible to store the structure factors only for the $\mathbf{G}$-vectors of the wave function basis or even not to store them at all. However, this requires that the missing structure factors are recomputed whenever needed. The structure factors $\mathtt{eigr}$ and the wave function-related quantities cr, cv, cf are complex numbers. Other quantities, like the local pseudopotential vps, the core charges rpc, and the projectors pro can be stored as real numbers if the factor $(-i)^l$ is excluded.

$eigr(N_D, N_{at})$ structure factors
$fnl(N_p, N_b)$ overlap of projectors and bands
$dfnl(N_p, N_b, 3)$ derivative of fnl
$smat(N_b, N_b)$ overlap matrices between bands
$cr(N_{PW}, N_b)$ bands in Fourier space
$cv(N_{PW}, N_b)$ velocity of bands in Fourier space
$cf(N_{PW}, N_b)$ forces of bands in Fourier space.

3.8.2 *Computational kernels*

Most of the calculations in a plane wave code are done in only a few, but crucial, kernel routines. These routines are discussed in the present section using a pseudo code language. Where possible, an implementation using basic linear algebra BLAS routines is given.

The first kernel is the calculation of the structure factors as defined in Eq. (3.34). The exponential function of the structure factor separates into three parts along the directions s_x, s_y, and s_z:

```
MODULE StructureFactor
FOR i=1:N_at
    s(1:3) = 2 * PI * MATMUL[htm1(1:3,1:3),r(1:3,i)]
    dp(1:3) = CMPLX[COS[s(1:3)],SIN[s(1:3)]]
    dm(1:3) = CONJG[dp(1:3)]
    e(0,1:3,i) = 1
    FOR k=1:g_max
        e(k,1:3,i) = e(k-1,1:3,i) * dp
        e(-k,1:3,i) = e(-k+1,1:3,i) * dm
    END
    FOR j=0:N_D
        eigr(j,i) = e(g(1,j),1,i) * e(g(2,j),2,i) * e(g(3,j),3,i)
    END
END
```

In the module above, htm1 is the matrix $(\mathbf{h}^t)^{-1}$.

One of the most important calculations is the inner product of two vectors in Fourier space. This kernel appears, for example, in the calculation of energies

$$e = \sum_{\mathbf{G}} A^{\star}(\mathbf{G})B(\mathbf{G}) \ . \qquad (3.156)$$

Making use of the fact that both functions are real, the sum can be restricted to half of the $\mathbf{G}$-vectors and only real operations have to be performed. This amounts to saving approximately a factor of three in the number of operations. Special care has to be taken for the zero $\mathbf{G}$-vector. It is assumed that this plane wave component is stored in the first position of the arrays:

```
MODULE DotProduct
e = A(1) * B(1)
FOR i=2: N_D
   ar = REAL(A(i))
   ai = IMAG(A(i))
   br = REAL(B(i))
   bi = IMAG(B(i))
   e = e + 2 * (ar * br + ai * bi)
END
```

This loop structure is available in the BLAS library, which is optimized on most computer architectures. To use the BLAS routines for real variables, complex numbers have to be stored as two real numbers in contiguous memory locations:

```
e = A(1) * B(1) + 2 * sdot(2 * N_D - 2,A(2),1,B(2),1)
```

The calculation of overlap matrices between sets of vectors in real space is an important task in the orthogonalization step

$$S_{ij} = \sum_{\mathbf{G}} A_i^{\star}(\mathbf{G})B_j(\mathbf{G}) \ . \qquad (3.157)$$

It can be executed efficiently by using matrix multiply routines from the BLAS library. The special case of the zero $\mathbf{G}$-vector is handled by a routine that performs a rank 1 update of the final matrix:

```
MODULE Overlap
CALL SGEMM('T','N',N_b,N_b,2*N_PW,2,&
          & ca(1,1),2*N_PW,cb(1,1),2*N_PW,0,smat,N_b)
CALL SDER(N_b,N_b,-1,ca(1,1),2*N_PW,cb(1,1),2*N_PW,smat,N_b)
```

For a symmetric overlap, additional time can be saved by using the symmetric matrix multiply routine. The overlap routines scale like $N_b^2 N_{PW}$. It is therefore very important to have an implementation of these parts that performs close to peak performance:

```
MODULE SymmetricOverlap
CALL SSYRK('U','T',N_b,2*N_PW,2,ca(1,1),2*N_PW,0,smat,N_b)
CALL SDER(N_b,N_b,-1,ca(1,1),2*N_PW,cb(1,1),2*N_PW,smat,N_b)
```

Another operation that scales as the overlap matrix calculations is the rotation of a set of wave functions in Fourier space

$$B_i(\mathbf{G}) = \sum_j A_j(\mathbf{G}) S_{ji} \; . \tag{3.158}$$

Again this kernel can be executed by using the optimized BLAS matrix multiply routines:

```
MODULE Rotation
CALL SGEMM('N','N',2*N_PW,N_b,N_b,1,ca(1,1),2*N_PW,&
        & smat,N_b,0,cb(1,1),2*N_PW)
```

The overlap calculation of the projectors of the nonlinear pseudopotential with the wave functions in Fourier space scales as $N_p N_b N_{PW}$. As the projectors are stored as real quantities, the imaginary prefactor and the structure factor have to be applied before the inner product can be calculated. The following pseudo code calculates M projectors at a time, making use of the special structure of the prefactor. Again, this allows us to do all calculations with real quantities. The code assumes that the total number of projectors is a multiple of M, but a generalization of the code to other cases is straightforward. By using batches of projectors the overlap can be calculated using matrix multiplies. The variable lp(i) holds the angular momentum of projector i:

```
MODULE FNL
FOR i=1:N_p,M
    IF (MOD(lp(i),2) == 0) THEN
        FOR j=0:M-1
            pf = -1**(lp(i+j)/2)
            FOR k=1:N_PW
                t = pro(k) * pf
```

```
            er = REAL[eigr(k,iat(i+j))]
            ei = IMAG[eigr(k,iat(i+j))]
            scr(k,j) = CMPLX[t * er,t * ei]
        END
    END
  ELSE
    FOR j=0:M-1
        pf = -1**(lp(i+j)/2+1)
        FOR k=1:N_PW
            t = pro(k) * pf
            er = REAL[eigr(k,iat(i+j))]
            ei = IMAG[eigr(k,iat(i+j))]
            scr(k,j) = CMPLX[-t * ei,t * er]
        END
    END
  END IF
  scr(1,0:M-1) = scr(1,0:M-1)/2
  CALL SGEMM('T','N',M,N_b,2*N_PW,2,&
            & scr(1,0),2*N_PW,cr(1,1),2*N_PW,0,fnl(i,1),N_p)
END
```

Fourier transform routines are assumed to work on complex data and return also arrays with complex numbers. The transform of data with the density cutoff is shown in the next two pseudo code sections assuming that a three-dimensional fast Fourier transform routine exists. This is, in fact, the case on most computers where optimized scientific libraries are available. The next two pseudo code segments show the transform of the charge density from Fourier space to real space and back:

```
MODULE INVFFT
scr(1:N_x,1:N_y,1:N_z) = 0
FOR i=1:N_D
   scr(ipg(1,i),ipg(2,i),ipg(3,i)) = rhog(i)
   scr(img(1,i),img(2,i),img(3,i)) = CONJG[rhog(i)]
END
CALL FFT3D("INV",scr)
n(1:N_x,1:N_y,1:N_z) = REAL[scr(1:N_x,1:N_y,1:N_z)]
```

```
MODULE FWFFT
scr(1:Nx,1:Ny,1:Nz) = n(1:Nx,1:Ny,1:Nz)
CALL FFT3D("FW",scr)
FOR i=1:ND
   rhog(i) = scr(ipg(1,i),ipg(2,i),ipg(3,i))
END
```

Special kernels are presented for the calculation of the density and the application of the local potential. These are the implementation of the flow charts shown in Fig. 3.4. The operation count of these routines is $N_b N \log N$ and these are the routines that take most of the computer time in most applications. Only for the largest applications possible on today's computers does the cubic scaling of the orthogonalization and the nonlocal pseudopotential become dominant. A small prefactor and the optimized implementation of the overlap are the reasons for this.

In the Fourier transforms of the wave function two properties are used in order to speed up the calculation. First, because the wave functions are real two transforms can be carried out at the same time, and second, the smaller cutoff of the wave functions can be used to avoid some parts of the transforms. The use of the sparsity in the Fourier transforms is not shown in the following modules. In an actual implementation a mask will be generated and only transforms allowed by this mask will be executed. Under optimal circumstances a gain of almost a factor of two can be achieved:

```
MODULE Density
rho(1:Nx,1:Ny,1:Nz) = 0
FOR i=1:Nb,2
   scr(1:Nx,1:Ny,1:Nz) = 0
   FOR j=1:NPW
      scr(ipg(1,i),ipg(2,i),ipg(3,i)) = c(j,i) + I * c(j,i+1)
      scr(img(1,i),img(2,i),img(3,i)) = &
         & CONJG[c(j,i) + I * c(j,i+1)]
   END
   CALL FFT3D("INV",scr)
   rho(1:Nx,1:Ny,1:Nz) = rho(1:Nx,1:Ny,1:Nz) + &
            & REAL[scr(1:Nx,1:Ny,1:Nz)]**2 + &
            & IMAG[scr(1:Nx,1:Ny,1:Nz)]**2
END
```

```
MODULE VPSI
FOR i=1:N_b,2
   scr(1:N_x,1:N_y,1:N_z) = 0
   FOR j=1:N_PW
      scr(ipg(1,i),ipg(2,i),ipg(3,i)) = c(j,i) + I * c(j,i+1)
      scr(img(1,i),img(2,i),img(3,i)) = &
            & CONJG[c(j,i) + I * c(j,i+1)]
   END
   CALL FFT3D("INV",scr)
   scr(1:N_x,1:N_y,1:N_z) = scr(1:N_x,1:N_y,1:N_z) * &
            & vpot(1:N_x,1:N_y,1:N_z)
   CALL FFT3D("FW",scr)
   FOR j=1:N_PW
      FP = scr(ipg(1,i),ipg(2,i),ipg(3,i)) &
            & + scr(img(1,i),img(2,i),img(3,i))
      FM = scr(ipg(1,i),ipg(2,i),ipg(3,i)) &
            & - scr(img(1,i),img(2,i),img(3,i))
      fc(j,i) = f(i) * CMPLX[REAL[FP],IMAG[FM]]
      fc(j,i+1) = f(i+1) * CMPLX[IMAG[FP],-REAL[FM]]
   END
END
```

4

Atoms with plane waves: accurate pseudopotentials

The replacement of chemically inactive electrons by pseudopotentials is a common method used in many kinds of electronic structure calculations. Different fields (LCAO methods, plane wave-based methods, quantum Monte Carlo methods) have developed their own flavor of such potentials. According to the general theme of this book, the focus here is on the pseudopotentials used in plane wave calculations. The norm-conserving pseudopotential approach developed in the framework of plane wave calculations provides an effective and reliable means for performing calculations on complex molecular, liquid, and solid-state systems. In this approach only the chemically active valence electrons are dealt with explicitely. The inert core electrons are eliminated within the frozen-core approximation, being considered together with the nuclei as rigid non-polarizable ion cores. All electrostatic and quantum-mechanical interactions of the valence electrons with the cores, such as the nuclear Coulomb attraction screened by the core electrons, Pauli repulsion, and exchange and correlation between core and valence electrons, are accounted for by angular momentum-dependent pseudopotentials. These should reproduce the true potential and valence orbitals outside a chosen core region but remain much weaker and smoother inside. The valence electrons are then described by smooth pseudo orbitals which play the same role as the true orbitals, but avoid the nodal structure near the nuclei that keeps the core and valence states orthogonal in an all-electron framework. The Pauli repulsion largely cancels the attractive parts of the true potential in the core region, and is built into the therefore rather weak pseudopotential. This "pseudization" of the valence wave functions, along with the removal of the core states, drastically simplifies a numerically accurate solution of the Kohn–Sham equations and Poisson's equation, and enables the use of plane waves as a basis set in electronic structure calculations.

By virtue of the norm-conservation property and when constructed carefully, as explained in the following subsections, pseudopotentials present a rather marginal approximation, and indeed allow for an adequate description of the valence electrons over the entire chemically relevant range of systems. For practical reasons, pseudopotentials should be additive and transferable. Additivity can most easily be achieved by building pseudopotentials for atoms in reference states. Transferability means that one and the same pseudopotential should be adequate for an atom in all possible chemical environments. This is especially important when a change of the environment is expected during a simulation, like in chemical reactions or for phase transitions.

4.1 Why pseudopotentials?

Pseudopotentials replace electronic degrees of freedom in the Hamiltonian by an effective potential. They lead to a reduction of the number of electrons in the system and thereby allow for faster calculation or the treatment of larger systems. An additional benefit stems from the observation that many relativistic effects are intimately connected to core electrons. Thus, these effects can be incorporated in the pseudopotentials without complicating the calculations of the final system.

Most important in the framework of the plane wave basis, pseudopotentials allow for a considerable reduction of the basis set size for the following reasons. First of all, valence states are smoother than core states already at the level of an all-electron calculation and therefore need fewer basis functions for an accurate description. Secondly, the pseudized valence wave functions are nodeless functions (in those classes of pseudopotentials that are considered here) and thus allow for an additional reduction of the basis, which is particularly gratifying when using plane waves. To illustrate this important point one can consider the 1s function of an atom approximated by an exponential function

$$\varphi_{1s}(\mathbf{r}) \sim \exp\left[-Z^\star r\right] \ , \tag{4.1}$$

where $Z^\star \approx Z$ with Z being the bare nuclear charge, and the Fourier transform of this orbital is

$$\varphi_{1s}(\mathbf{G}) \sim 16\pi \frac{Z^{5/2}}{G^2 + Z^2} \ . \tag{4.2}$$

From this formula one can estimate the relative cutoffs needed for different elements in the periodic table (see Table 4.1). According to the data

Table 4.1. *Relative plane wave cutoffs (in energy units) and resulting*
relative number of plane waves for a selection of atom types.

Atom	Nuclear charge	Cutoff	Plane waves
H	1	1	1
Li	3	4	8
C	6	9	27
Si	14	27	140
Ge	32	76	663
Sn	50	133	1534

compiled in Table 4.1, it becomes evident that the number of plane waves
needed for a certain accuracy increases with the square of the nuclear charge.
This makes calculations on all systems containing heavier elements virtually
impossible. Even calculations on lighter elements require huge resources.
Still, a number of investigations using a full potential approach with pseudopotentials have been carried out in the past. Solid LiH has been studied
by Kunc [90, 91, 799] using cutoffs ranging from 100 to 240 Rydbergs. Solid
atomic hydrogen was investigated by Natoli *et al.* [1050] at a cutoff of 150 Rydbergs. Finally, Teter [1461] calculated the equation of state of carbon in
the diamond structure using a cutoff as high as 600 Rydbergs.

4.2 Norm-conserving pseudopotentials

4.2.1 Pseudization of valence wave functions

In the following the task is to set up a collection of rules which can be
followed to generate a set of potentials with the desired properties. The
resulting potentials should be transferable between different chemical environments. This suggests that the pseudo wave functions should be identical
to the real valence wave functions in the regions of space where they overlap with other atoms. Furthermore, the pseudopotentials and pseudo wave
functions should be as smooth as possible. This is needed to keep the basis
set for the expansion of the wave functions as small as possible. A breakthrough in the construction of pseudopotentials which are both practical
and accurate came from the implementation of norm conservation in nonlocal pseudopotentials. A norm-conserving pseudopotential conserves the
normalization of the pseudo wave function inside the core region so that

the wave function outside the core resembles that of the all-electron atom as closely as possible. Norm conservation was first implemented in empirical pseudopotentials by Topp and Hopfield [1473], in *ab initio* local ionic pseudopotentials by Starkloff and Joannopoulos [1394], and stressed in the framework of creation of nonlocal pseudopotentials by Hamann, Schlüter, and Chiang (HSC) [594], which was refined later by Bachelet, Hamann, and Schlüter (BHS) [57].

Figure 4.1 shows the valence one-electron wave functions and their pseudized counterparts for a silicon atom; the recipe used to generate these particular functions is explained in Section 4.2.5. As can be seen from the graph, the two sets of functions are identical outside a given core radius. The valence orbitals were generated for the atomic ground state (s and p states) and for an excited state with the configuration [Ne]$3s^13p^{0.75}3d^{0.25}$ (d state). For s and p channels the pseudo functions differ from the valence functions considerably within the core region. Most notably of course, the pseudo orbitals are nodeless whereas the valence orbitals have nodes due to the orthogonality condition with respect to the core functions. The pseudo and valence functions for angular momentum d are almost identical.

Figure 4.2 shows the core and pseudo valence electron densities that result for a silicon atom. Note that there is only a small overlap of the two densities. This is important for a good transferability of the pseudopotentials as only first-order changes can be reproduced correctly, as will be explained later in Section 4.5. Figure 4.3 shows the pseudopotentials for s, p, and d angular momenta. These potentials are consistent with the pseudo wave functions for the corresponding reference states, and all of them converge to the limit of $-Z/r$ outside the core radius. In the following sections details of the construction of pseudopotentials will be discussed according to several widely used generation schemes.

4.2.2 Hamann-Schlüter-Chiang conditions

Norm-conserving pseudopotentials are derived from atomic reference states, calculated from the atomic Schrödinger equation

$$\left(T + V^{\text{AE}}\right)|\Psi_l\rangle = \epsilon_l|\Psi_l\rangle \ , \tag{4.3}$$

where T is the kinetic energy operator of the electrons and V^{AE} the all-electron potential derived from Kohn–Sham theory. This equation is replaced by a "valence electrons only" equation of the same form

$$\left(T + V_l^{\text{val}}\right)|\Phi_l\rangle = \tilde{\epsilon}_l|\Phi_l\rangle \tag{4.4}$$

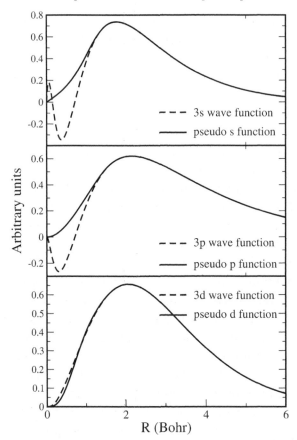

Fig. 4.1. Valence and pseudo wave functions of a silicon atom. Pseudo wave functions generated according to the Troullier–Martins scheme (see Section 4.2.5). Upper panel: ground state all-electron 3s wave function (dashed line) and pseudo s wave function (solid line). Middle panel: ground state all-electron 3p wave function (dashed line) and pseudo p wave function (solid line). Lower panel: all-electron 3d wave function (dashed line) from a $[Ne]3s^{1}3p^{0.75}3d^{0.25}$ excited state calculation and pseudo d wave function (solid line).

and Hamann, Schlüter, and Chiang [594] proposed a set of requirements that the thereby generated pseudo wave function, Ψ_l, and the pseudopotentials (part of the total atomic potential V_l^{val}, Eq. (4.4)) must satisfy.

The pseudopotential should have the following properties:

(i) Real and pseudo valence eigenvalues agree for a chosen prototype atomic configuration, i.e.

$$\epsilon_l = \tilde{\epsilon}_l \ .$$

(ii) Real and pseudo atomic wave functions agree beyond a chosen core

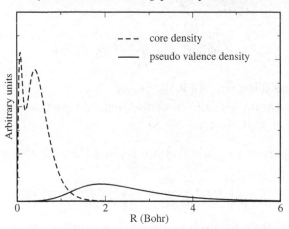

Fig. 4.2. Ground state core density (dashed line) and pseudo valence density (solid line) calculated from the pseudo wave functions shown in Fig. 4.1 of a silicon atom.

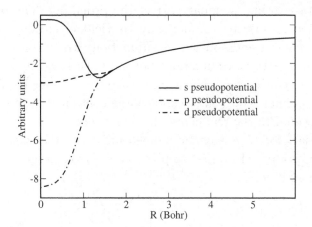

Fig. 4.3. Pseudopotentials of silicon for s, p, and d angular momentum. Pseudopotentials were calculated using the Troullier–Martins scheme (see Section 4.2.5).

radius r_c, i.e.

$$\Psi_l(r) = \Phi_l(r) \ \ \text{for} \ r \geq r_\mathrm{c} \ .$$

(iii) The integrals from 0 to R of the real and pseudo charge densities agree for $R \geq r_\mathrm{c}$ for each valence state (norm conservation), i.e.

$$\langle \Phi_l | \Phi_l \rangle_R = \langle \Psi_l | \Psi_l \rangle_R \ \ \text{for} \ R \geq r_\mathrm{c}$$

where

$$\langle \Phi_l | \Phi_l \rangle_R = \int_0^R r^2 |\Phi_l(r)|^2 dr$$

for each l and alike for $\langle \Psi_l | \Psi_l \rangle_R$.

(iv) The logarithmic derivatives of the real and pseudo wave function and their first energy derivatives agree for $r \geq r_c$.

Note that properties (iii) and (iv) are related through the identity

$$-\frac{1}{2} \left[(r\Phi)^2 \frac{d}{d\epsilon} \frac{d}{dr} \ln \Phi \right]_R = \int_0^R r^2 |\Phi|^2 dr \qquad (4.5)$$

that holds separately for each angular momentum channel l.

Furthermore, they also provide a recipe that allows us to generate pseudopotentials with the above properties [594]. The atomic potential is first multiplied by a smooth short-range cutoff function which removes the strongly attractive and singular part of the potential. The parameterized cutoff function is adjusted numerically to yield eigenvalues equal to the atomic valence levels and nodeless eigenfunctions which converge identically to the atomic valence wave functions beyond a chosen core radius. In addition, to reproduce the electrostatic and scattering properties of the real ion core with a minimal error, the pseudo charge contained in the core region is forced to be equal to the real charge in that region.

In particular, for each angular momentum l the following procedure is applied (HSC recipe). In a first step the all-electron potential is multiplied by a smoothing function

$$V_l^{(1)}(r) = V^{AE}(r) \left[1 - f_1 \left(\frac{r}{r_{c,l}} \right) \right] , \qquad (4.6)$$

where $r_{c,l}$ is the core radius, typically $\approx 0.4 - 0.6 \; r_{\max}$ with $r_{\max}$ being the outermost maximum of the real wave function. Then a function is added to the potential

$$V_l^{(2)}(r) = V_l^{(1)}(r) + c_l f_2 \left(\frac{r}{r_{c,l}} \right) , \qquad (4.7)$$

where c_l is determined such that $\tilde{\epsilon}_l = \epsilon_l$ in

$$(T + V_l^{(2)}(r)) w_l^{(2)}(r) = \tilde{\epsilon}_l w_l^{(2)}(r) \qquad (4.8)$$

is satisfied. The valence wave function is defined by

$$\Phi_l(r) = \gamma_l \left[w_l^{(2)}(r) + \delta_l r^{l+1} f_3 \left(\frac{r}{r_{c,l}} \right) \right] , \qquad (4.9)$$

where the two parameters γ_l and δ_l are chosen such that

$$\Phi_l(r) \to \Psi_l(r) \quad \text{for } r \geq r_{c,l} \tag{4.10}$$

and

$$\gamma_l^2 \int \left| w_l^{(2)}(r) + \delta_l r^{l+1} f_3\left(\frac{r}{r_{c,l}}\right) \right|^2 dr = 1 \ . \tag{4.11}$$

Finally, the Schrödinger equation has to be inverted for $\tilde{\epsilon}_l$ and $\Phi_l(r)$ to get $V_l^{\text{val}}(r)$; Hamann, Schlüter, and Chiang chose the following cutoff functions $f_1(x) = f_2(x) = f_3(x) = \exp\left[-x^4\right]$. The pseudopotential itself is defined as the unscreened potential $V_l^{\text{PP}}(r)$ calculated from

$$V_l^{\text{PP}}(r) = V_l^{\text{val}}(r) - V_{\text{H}}(n_{\text{v}}) - V_{\text{xc}}(n_{\text{v}}) \ , \tag{4.12}$$

where $V_{\text{H}}(n_{\text{v}})$ and $V_{\text{xc}}(n_{\text{v}})$ are the Hartree and exchange and correlation potentials (see Eqs. (2.118) and (2.123), respectively, in Section 2.7.2) evaluated with the pseudo valence density.

The total atomic pseudopotential in a solid-state calculation then takes the form of a sum over all angular momentum channels considered

$$V^{\text{PP}}(\mathbf{r}) = \sum_L V_L^{\text{PP}}(r) \mathbf{P}_L(\omega) \ , \tag{4.13}$$

where L denotes a combined index $\{l, m\}$, $\mathbf{P}_L(\omega)$ is the projector on the angular momentum state $\{l, m\}$, and ω are angular variables.

Clearly, these pseudopotentials are angular momentum-dependent. In other words, each angular momentum state l has its own potential that can be determined independently from the other potentials. It is therefore possible to have a different reference configuration for each angular momentum. This allows, for instance, the use of excited states or (partially) ionic states to construct the pseudopotential for l states that are not occupied in the atomic ground state in order to improve transferability properties.

4.2.3 Bachelet–Hamann–Schlüter pseudopotentials

Bachelet *et al.* [57] proposed an analytic fit to the pseudopotentials generated with the HSC recipe of the following form:

$$V^{\text{PP}}(r) = V^{\text{core}}(r) + \sum_L \Delta V_L^{\text{ion(r)}} \ , \tag{4.14}$$

$$V^{\text{core}}(r) = -\frac{Z_{\text{v}}}{r} \left[\sum_{i=1}^{2} c_i^{\text{core}} \text{erf}\left(\sqrt{\alpha_i^{\text{core}}} r\right) \right] \ , \tag{4.15}$$

$$\Delta V_L^{\mathrm{ion}}(r) = \sum_{i=1}^{3} \left(A_i + r^2 A_{i+3} \right) \exp\left[-\alpha_i r^2 \right] , \tag{4.16}$$

where the i-indexed quantities are tabulated parameters and erf is the error function. This particular analytic form was chosen in order to allow for an easy implementation into electronic structure codes, since it is important for plane wave-based codes that the Fourier transform can also be written analytically. The corresponding Bachelet–Hamann–Schlüter (BHS) pseudopotentials in Fourier space are

$$V^{\mathrm{PP}}(G) = V^{\mathrm{core}}(G) + \sum_{L} \Delta V_L^{\mathrm{ion}}(G) , \tag{4.17}$$

$$V^{\mathrm{core}}(G) = -4\pi \frac{Z_{\mathrm{v}}}{\Omega G^2} \left[\sum_{i=1}^{2} c_i^{\mathrm{core}} \exp\left[-\frac{G^2}{4\alpha_i^{\mathrm{core}}} \right] \right] , \tag{4.18}$$

$$\Delta V_L^{\mathrm{ion}}(G) = \frac{1}{\Omega} \sum_{i=1}^{3} \left[A_i \left(\frac{\pi}{\alpha_i} \right)^{\frac{3}{2}} + A_{i+3} \left(\frac{\pi}{\alpha_{i+3}} \right)^{\frac{3}{2}} (3 - G^2) \right]$$

$$\times \exp\left[-\frac{G^2}{4\alpha_i} \right] . \tag{4.19}$$

Bachelet *et al.* [57] generated pseudopotentials for almost the entire periodic table (within the local density approximation, LDA), where generalizations of the original HSC scheme [594] to include spin–orbit effects for heavy atoms were made and the cutoff functions were modified slightly to be $f_1(x) = f_2(x) = f_3(x) = \exp\left[-x^{3.5} \right]$. However, Bachelet *et al.* did not tabulate the A_i coefficients (as they are often very big numbers) but another set of numbers C_i, where

$$C_i = -\sum_{l=1}^{6} A_l Q_{il} \tag{4.20}$$

and

$$A_i = -\sum_{l=1}^{6} C_l Q_{il}^{-1} \tag{4.21}$$

with

$$Q_{il} = \begin{cases} 0 & \text{for } i > l , \\ \left[S_{il} - \sum_{k=1}^{i-1} Q_{ki}^2 \right]^{1/2} & \text{for } i = l , \\ \frac{1}{Q_{ii}} \left[S_{il} - \sum_{k=1}^{i-1} Q_{ki} Q_{kl} \right]^{1/2} & \text{for } i < l , \end{cases} \tag{4.22}$$

Table 4.2. *List of parameters cc_l that determine the optimum pseudopotential cutoff radii $r_{c,l}$ through the relation $r_{c,l} = r_{max}/cc_l$, where r_{max} is the radius of the outermost peak in the radial wave function (adapted from Bachelet et al. [57]).*

Elements	$l = 0$	$l = 1$	$l = 2$	$l = 3$
H → He	3.0	3.6	3.6	
Li	2.0	3.0	3.5	
Be → Ne	1.8	3.0	3.5	
Na	2.0	1.8	3.5	
Mg → Ar	1.8	1.45	2.2	
K → Zn	1.8	1.6	3.0	
Ga → Kr	1.8	1.7	2.0	
Rb → Cd	1.8	1.7	1.6	
In → Xe	1.6	1.7	2.0	
Cs	1.8	1.7	1.45	4.5
Ba	1.8	1.7	1.45	3.0
La → Lu	1.45	1.6	1.45	3.0
Hf → Hg	1.6	1.6	1.45	3.0
Tl → Rn	1.6	1.7	2.2	3.0
Fr	1.6	1.6	1.45	4.5
Ra → Pu	1.6	1.6	1.45	2.0

where $S_{il} = \int_0^\infty r^2 \varphi_i(r)\varphi_l(r)dr$, and

$$\varphi_k(r) = \begin{cases} \exp\left[-\alpha_k r^2\right] & \text{for } k = 1, 2, 3 \ , \\ r^2 \exp\left[-\alpha_k r^2\right] & \text{for } k = 4, 5, 6 \ . \end{cases} \tag{4.23}$$

Their list of cutoff values and atomic reference states is also useful; see Tables 4.2 and 4.3, respectively. However, it should be kept in mind that these values should only be used as starting values in the pseudopotential generation scheme.

Table 4.3. *Atomic configurations to derive the l-dependent pseudopotentials. The symbol (∗) indicates the systematic increase by one electron per increasing nuclear charge (derived from Bachelet et al. [57]).*

Elements	$l = 0$	$l = 1$	$l = 2$	$l = 3$
H	1s	$p^{0.5}$	$d^{0.5}$	
He	$1s^2$	$s^{0.8} p^{0.2}$	$s^{0.8} d^{0.2}$	
Li → Be	[He]2s∗	$p^{0.25} d^{0.25}$	as $l = 1$	

Elements	$l=0$	$l=1$	$l=2$	$l=3$
B	[He]$2s^2$ $2p$	as $l=0$	s^1 $d^{0.2}$	
C	[He]$2s^2$ $2p^2$	as $l=0$	$s^{0.75}$ p^1 $d^{0.25}$	
N $\rightarrow$ O	[He]$2s^2$ $2p^{3*}$	as $l=0$	s^1 $p^{1.75}$ $d^{0.25}$	
F	[He]$2s^2$ $2p^5$	as $l=0$	$s^{1.25}$ $p^{2.5}$ $d^{0.25}$	
Ne	[He]$2s^2$ $2p^6$	as $l=0$	s^1 $p^{2.75}$ $d^{0.25}$	
Na	[Ne]$3s$	$p^{0.25}$ $d^{0.25}$	as $l=1$	
Mg	[Ne]$3s^2$	$s^{0.5}$ $p^{0.25}$ $d^{0.25}$	as $l=1$	
Al	[Ne]$3s^2$ $3p$	as $l=0$	$s^{0.75}$ $d^{0.25}$	
Si $\rightarrow$ Ar	[Ne]$3s^2$ $3p^{2*}$	as $l=0$	s^1 $p^{0.75*}$ $d^{0.25}$	
K	[Ar]$4s$	$p^{0.25}$	$d^{0.25}$	
Ca	[Ar]$4s^2$	$s^{0.5}$ $p^{0.25}$ $d^{0.25}$	as $l=1$	
Sc $\rightarrow$ Ni	[Ar]$3d^*$ $4s^2$	$s^{0.75}$ $p^{0.25}$ d^{1*}	as $l=0$	
Cu $\rightarrow$ Zn	[Ar]$3d^{10}$ $4s^*$	$s^{0.75}$ $p^{0.25}$ d^{9*}	as $l=0$	
Ga	[Ar]$3d^{10}$ $4s^2$ $4p$	as $l=0$	$s^{0.75}$ $d^{0.25}$	
Ge $\rightarrow$ Kr	[Ar]$3d^{10}$ $4s^2$ $4p^{2*}$	as $l=0$	s^1 $p^{0.75*}$ $d^{0.25}$	
Rb	[Kr]$5s$	$p^{0.25}$	$d^{0.25}$	
Sr	[Kr]$5s^2$	$s^{0.5}$ $p^{0.25}$ $d^{0.25}$	as $l=1$	
Y $\rightarrow$ Zr	[Kr]$4d^*$ $5s^2$	$s^{0.75}$ $p^{0.25}$ d^{1*}	as $l=0$	
Nb $\rightarrow$ Mo	[Kr]$4d^{4*}$ $5s$	$s^{0.75}$ $p^{0.25}$ d^{3*}	as $l=0$	
Tc	[Kr]$4d^5$ $5s^2$	$s^{0.75}$ $p^{0.25}$ d^5	as $l=0$	
Ru $\rightarrow$ Pd	[Kr]$4d^{7*}$ $5s$	$s^{0.75}$ $p^{0.25}$ d^{6*}	as $l=0$	
Ag $\rightarrow$ Cd	[Kr]$4d^{10}$ $5s^*$	$s^{0.75}$ $p^{0.25}$ d^{9*}	as $l=0$	
In	[Kr]$4d^{10}$ $5s^2$ $5p$	as $l=0$	$s^{0.75}$ $d^{0.25}$	
Sn $\rightarrow$ Xe	[Kr]$4d^{10}$ $5s^2$ $5p^{2*}$	as $l=0$	s^1 $p^{0.75*}$ $d^{0.25}$	
Cs	[Xe]$6s$	$p^{0.25}$	$d^{0.25}$	$f^{0.25}$
Ba	[Xe]$6s^2$	$s^{0.75}$ $p^{0.25}$	$s^{0.75}$ $d^{0.25}$	$s^{0.75}$ $f^{0.25}$
La	[Xe]$4f$ $6s^2$	as $l=0$	s^2 p^6 d^1 f^1	as $l=0$
Ce	[Xe]$4f$ $5d$ $6s^2$	as $l=0$	s^2 p^6 d^1 f^2	as $l=0$
Pr $\rightarrow$ Eu	[Xe]$4f^{3*}$ $6s^2$	as $l=0$	s^2 p^6 d^1 f^{3*}	as $l=0$
Gd	[Xe]$4f^7$ $5d$ $6s^2$	as $l=0$	s^2 p^6 d^1 f^8	as $l=0$
Tb $\rightarrow$ Yb	[Xe]$4f^{9*}$ $6s^2$	as $l=0$	s^2 p^6 d^1 f^{9*}	as $l=0$
Lu	[Xe]$4f^{14}$ $5d$ $6s^2$	as $l=0$	as $l=0$	as $l=0$
Hf $\rightarrow$ Ir	[Xe]$4f^{14}$ $5d^{2*}$ $6s^2$	$s^{0.75}$ $p^{0.25}$ d^{2*}	as $l=0$	$s^{0.75}$ d^{1*} $f^{0.25}$
Pt $\rightarrow$ Au	[Xe]$4f^{14}$ $5d^{9*}$ $6s$	$s^{0.75}$ $p^{0.25}$ d^{8*}	as $l=0$	$s^{0.75}$ d^{7*} $f^{0.25}$
Hg	[Xe]$4f^{14}$ $5d^{10}$ $6s^2$	$s^{0.75}$ $p^{0.25}$ d^{10*}	as $l=0$	$s^{0.75}$ d^{9*} $f^{0.25}$
Tl	[Hg]$6p$	as $l=0$	$s^{0.75}$ $d^{0.25}$	$s^{0.75}$ $f^{0.25}$
Pb $\rightarrow$ Rn	[Hg]$6p^{2*}$	as $l=0$	s^1 $p^{0.75*}$ $d^{0.25}$	$s^{0.75}$ $p^{0.75*}$ $f^{0.2}$
Fr	[Rn]$7s$	$p^{0.25}$	$d^{0.25}$	$f^{0.25}$

Elements	$l = 0$	$l = 1$	$l = 2$	$l = 3$
Ra	[Rn]$7s^2$	$s^{0.75}\,p^{0.25}$	$s^{0.75}\,d^{0.25}$	$s^{0.75}\,f^{0.25}$
Ac	[Rn]$6d\,7s^2$	$s^{0.75}\,p^{0.25}\,d^1$	as $l = 1$	$s^{0.75}\,d^1\,f^{0.25}$
Th	[Rn]$6d^2\,7s^2$	$p^{0.25}\,d^{1.5}\,f^{0.25}$	as $l = 1$	as $l = 1$
Pa $\rightarrow$ U	[Rn]$5f^{2*}\,6d\,7s^2$	$s^1\,p^{0.5}\,d^{0.5}\,f^{1*}$	as $l = 0$	as $l = 0$
Np $\rightarrow$ Pu	[Rn]$5f^{5*}\,7s^2$	$s^{0.5}\,p^{0.5}\,d^{0.5}\,f^{3.5*}$	as $l = 0$	as $l = 0$

Using selected excited states as references for the generation of potentials for those states that are unoccupied in the ground state is just one of the possibilities. An alternative approach was introduced later by Hamann [586]. This generalized method is based on the observation that it is not necessary to use bound states as reference states in the pseudopotential generation schemes. Unbound states can be calculated at any energy by an outward integration of the radial Schrödinger equation up to a given radius R_l; Hamann proposed using $R_l \approx 2.5\ r_{c,l}$. The wave functions $\Phi_l(r)$ are then normalized within this volume

$$4\pi \int_0^{R_l} r^2\,\Phi_l(r)\,dr = 1 \ , \tag{4.24}$$

and this scheme can be used with most of the recipes in use. The pseudopotential quality does not depend on the radius R_l if properly chosen, however, the energy at which the reference states are calculated are of primary importance. They are, in fact, an additional degree of freedom that can be used in the construction procedure. The value of the highest occupied valence state is the most natural choice, but one is not restricted to that value.

For relativistic calculations the spin–orbit coupling splits up all orbitals with $l > 0$ into spin-up and spin-down orbitals with an overall angular momentum $j = l \pm 1/2$. So for each angular momentum $l > 0$, one spin-up orbital and one spin-down orbital with different wave functions and pseudopotentials exist. Bachelet $et\ al.$ [57] proposed using a weighted average and difference potential in this case. The average pseudopotential is conveniently defined as

$$V_l(r) = \frac{1}{2l + 1}[lV_{l-1/2}(r) + (l + 1)V_{l+1/2}(r)] \ , \tag{4.25}$$

weighted by the different j degeneracies of the $l \pm 1/2$ orbitals. The difference potential describes the spin–orbit coupling, and is defined as

$$\Delta V_l^{SO}(r) = \frac{2}{2l + 1}[V_{l-1/2}(r) - V_{l+1/2}(r)] \ . \tag{4.26}$$

The total pseudopotential is then given by

$$V^{PP}(r) = V^{core}(r) + \sum_L \left[\Delta V_L^{ion}(r) + \Delta V_l^{SO}(r) \, \mathbf{L} \cdot \mathbf{S} \right] , \qquad (4.27)$$

where V^{core} and ΔV_L^{ion} are now scalar relativistic quantities but with the same form as in the non-relativistic case; $\mathbf{L}$ and $\mathbf{S}$ are the atomic angular momentum and spin operators. Neglecting contributions from ΔV_l^{SO} gives an average potential that contains all scalar parts of the relativistic pseudopotential, whereas the total potential is still appropriate for Schrödinger equations yet contains relativistic effects to order α^2 (where α denotes Sommerfeld's fine-structure constant).

4.2.4 Kerker pseudopotentials

In the Kerker approach [735], pseudopotentials are constructed that satisfy the HSC conditions as well. But instead of using smooth cutoff functions (f_1, f_2, and f_3), the pseudo wave functions are constructed directly from the all-electron wave functions by replacing the all-electron wave function inside some cutoff radius by a smooth analytic function that is matched to the all-electron wave function at the cutoff radius. The four HSC properties from Section 4.2.2 then translate into a set of equations for the parameters of the particular analytic form used. After having determined the pseudo wave function the Schrödinger equation is inverted and the resulting potential unscreened. Note that the cutoff radius $r_{c,l}$ of this type of pseudopotential construction scheme is defined differently and has to be chosen considerably larger than the one used in the HSC/BHS scheme. Typically, the cutoff radius is chosen slightly smaller than R_{max}, the outermost maximum of the all-electron wave function. The analytic form proposed by Kerker is

$$\Phi_l(r) = r^{l+1} \exp\left[p(r) \right] \qquad r < r_{c,l} \qquad (4.28)$$

with l-dependent cutoff radii, $r_{c,l}$, and

$$p(r) = \alpha r^4 + \beta r^3 + \gamma r^2 + \delta . \qquad (4.29)$$

The term linear in r is missing to avoid a singularity of the potential at $r = 0$. The HSC conditions can be translated into a set of equations for the parameters $\alpha, \beta, \gamma, \delta$.

The four conditions used by Kerker in the original paper were:

(i) Real and pseudo atom have the same valence eigenvalues.
(ii) Pseudo wave functions are nodeless and identical to the real wave functions outside $r_{c,l}$.

(iii) First and second derivatives of the wave functions are matched at $r_{c,l}$.

(iv) Pseudo and real charge are identical inside $r_{c,l}$.

Conditions (i) to (iii) then read

$$\ln(P_{c,l}/r_{c,l=1}) = p(r_{c,l}) \tag{4.30}$$

$$r_{c,l}D = l + 1 + r_{c,l}\, p'(R_{c,l}) \tag{4.31}$$

$$r_{c,l}^2 V_{c,l} + (l+1)^2 - r_{c,l}^2(E_l + D^2) = r_{c,l}^2\, p''(r_{c,l}) \tag{4.32}$$

and the prime denotes differentiation with respect to r. The amplitude of the atomic radial wave function times r is denoted by $P(r)$ and $P_{c,l} = P(r_{c,l})$. Furthermore, $D = P'(r_{c,l})/P(r_{c,l})$, the value of the atomic potential at $R_{c,l}$ is given by $V_{c,l}$, and the atomic valence eigenvalue is E_l. The above equations form a set of three linear equations for α, β, and γ in terms of δ. From condition (iv) the equation

$$2\delta + \ln I - \ln A = 0 \tag{4.33}$$

is obtained, where

$$I = \int_0^{r_{c,l}} r^{2l+2} \exp\left[2\alpha r^4 + 2\beta r^3 + 2\gamma r^2\right]\, dr \tag{4.34}$$

and A is the amount of real charge contained in the core region up to $r_{c,l}$. A simple iterative scheme can now be used to solve the equations for α, β, γ, and δ. Finally, the screened pseudopotential can be calculated analytically from the four parameters:

$$V_{\text{scr}}^{\text{PP}}(r) = E_l + \lambda(2l + 2 + \lambda r^2) + 12\alpha r^2 + 6\beta r + 2\gamma \;, \tag{4.35}$$

where $\lambda = 4\alpha r^2 + 3\beta r + 2\gamma$.

4.2.5 Troullier–Martins pseudopotentials

The Kerker method was generalized by Troullier and Martins [1480] to polynomials of higher order. The rationale behind this was to use the additional parameters (the coefficients of the higher terms in the polynomial) to construct smoother pseudopotentials. The Troullier–Martins wave functions have the following form in the core region:

$$\Phi_l(r) = r^{l+1} \exp\left[p(r)\right] \qquad r < r_{c,l} \;, \tag{4.36}$$

with

$$p(r) = c_0 + c_2 r^2 + c_4 r^4 + c_6 r^6 + c_8 r^8 + c_{10} r^{10} + c_{12} r^{12} \tag{4.37}$$

and the coefficients c_n determined from

(a) Imposing norm conservation

$$2c_0 + \ln \left[\int_0^{r_{c,l}} r^{2(l+1)} \exp[2p(r) - 2c_0] \, dr \right]$$

$$= \ln \left[\int_0^{r_{c,l}} |\Psi_l(r)|^2 \, r^2 \, dr \right] . \quad (4.38)$$

(b) The fact that all-electron and pseudo wave functions are continuous for up to the fourth derivative at $r_{c,l}$

$$p(r_{c,l}) = \ln \left[\frac{P(r_{c,l})}{r_{c,l}^{l+1}} \right] , \quad (4.39)$$

$$p'(r_{c,l}) = \frac{P'(r_{c,l})}{P(r_{c,l})} - \frac{l+1}{r_{c,l}} , \quad (4.40)$$

$$p''(r_{c,l}) = 2V^{\mathrm{AE}}(r_{c,l}) - 2\epsilon_l$$
$$- \frac{2(l+1)}{r_{c,l}} p'(r_{c,l}) - [p'(r_{c,l})]^2 , \quad (4.41)$$

$$p'''(r_{c,l}) = 2\left(V^{\mathrm{AE}}\right)'(r_{c,l}) + \frac{2(l+1)}{r_{c,l}^2} p'(r_{c,l})$$
$$- \frac{2(l+1)}{r_{c,l}} p''(r_{c,l}) - 2p'(r_{c,l})p''(r_{c,l}) , \quad (4.42)$$

$$p''''(r_{c,l}) = 2\left(V^{\mathrm{AE}}\right)''(r_{c,l}) - \frac{4(l+1)}{r_{c,l}^3} p'(r_{c,l})$$
$$+ \frac{4(l+1)}{r_{c,l}^2} p''(r_{c,l}) - \frac{2(l+1)}{r_{c,l}} p'''(r_{c,l})$$
$$- 2[p''(r_{c,l})]^2 - 2p'(r_{c,l})p'''(r_{c,l}) . \quad (4.43)$$

(c) The fact that the screened pseudopotential has zero curvature at the origin

$$c_2^2 + c_4(2l + 5) = 0 . \quad (4.44)$$

In the above formulas the following symbols have been used: $P(r) = r\Psi_l(r)$ is the all-electron wave function multiplied by r, $V^{\mathrm{AE}}(r)$ the all-electron atomic screened potential, and primes denote differentiation with respect to r.

The scheme of Troullier and Martins is based on the fact that the asymptotic behavior of the pseudo wave functions depends on the matching of derivatives at the cutoff radii. In their analysis they discovered that the asymptotic behavior can be improved further by setting all odd coefficients of the polynomial to zero, which is the same as setting the odd derivatives of the potentials at the origin to zero. However, the asymptotic behavior is only helpful but not crucial in obtaining a smooth pseudopotential. Several other criteria to ensure smoothness were investigated by Troullier and Martins. One of them was the kinetic energy criterion proposed by Rappe *et al.* [1191], as discussed in the next section. They found that the best generated potentials usually exhibit a flat behavior at the origin. Therefore, the last condition to enforce zero curvature of the pseudopotential at the origin was included.

4.2.6 Kinetic energy optimized pseudopotentials

The Rappe *et al.* scheme [1191] is based on the observation that the total energy and the kinetic energy have similar convergence properties when expanded in plane waves. Therefore, the kinetic energy expansion is used as an optimization criterion in the construction of the pseudopotentials. Also, these pseudopotentials are based on an analytic representation of the pseudo wave function within $r_{c,l}$

$$\Phi_l(r) = \sum_{i=1}^{n} a_i j_l(G_i r) \qquad r < r_{c,l} \ , \tag{4.45}$$

where $j_l(G_i r)$ are spherical Bessel functions with $i - 1$ zeros at positions smaller than $r_{c,l}$. The values of G_i are fixed such that

$$\frac{j'(G_i r_{c,l})}{j(G_i r_{c,l})} = \frac{\Psi'_l(r_{c,l})}{\Psi_l(r_{c,l})} \ . \tag{4.46}$$

The conditions that are used to determine the values of a_i are:
(a) Φ_l is normalized.
(b) First and second derivatives of Φ_l are continuous at $r_{c,l}$.
(c) $\Delta E_{\text{kin}}(\{a_i\}, G_c)$ is minimal

$$\Delta E_{\text{kin}} = -\int \Phi_l^\star \nabla^2 \Phi_l \, d\mathbf{r} - \int_0^{G_c} G^2 \mid \Phi_l(G) \mid^2 \, dG \ ,$$

where ΔE_{kin} is the kinetic energy contribution above a target cutoff value G_c. The value of G_c is an additional parameter (as for example $r_{c,l}$) that has to be chosen with a reasonable value. In practice, G_c is changed until it is possible to minimize ΔE_{kin} to a small enough value.

The original scheme was analyzed by Lin *et al.* [868], and they concluded that four Bessel functions ($n = 4$) and a choice of $G_c = G_4$ (G_4 is the fourth solution of Eq. (4.46)) gives the most satisfactory results. In yet another investigation, Kresse and Hafner [788] proposed dropping the kinetic energy criterion altogether and only using three Bessel functions. Only for cutoffs with $r_{c,l}$ smaller than the outermost maximum of the wave function was it necessary to include a fourth Bessel function to be able to guarantee a nodeless pseudo wave function. The three-Bessel-function scheme also proved to be helpful in connection with unbound states.

4.3 Pseudopotentials in the plane wave basis

With the methods described in the last sections we are able to construct norm-conserving pseudopotentials for states $l = $ s, p, d, f by using reference configurations that are either the ground state of the atom or of an ion, or excited states. In principle, higher angular momentum states could also be generated, but their physical significance is questionable. In a solid-state or molecular environment, induced through the other atoms, there will be wave function components of all angular momentum character at each atom, and therefore a general procedure for all l states is needed. The general form of a pseudopotential is

$$V^{\text{PP}}(\mathbf{r}) = \sum_{l=0}^{\infty} \sum_{m=-l}^{l} V_l(r) P_{lm}(\omega) \ , \tag{4.47}$$

where $P_{lm}(\omega)$ is a projector on angular momentum functions and ω denotes angular variables. A good approximation commonly used for all pseudopotentials for angular momentum higher than a given value l_{max} is

$$V_l(r) = V^{\text{loc}}(r) \quad \text{for } l > l_{\text{max}} \ . \tag{4.48}$$

An obvious and often used choice for $V^{\text{loc}}(r)$ is the potential with the largest l-value that is occupied in the reference configuration. In this way the number of potentials with nonlocal character is reduced and computational efficiency maximized. Other choices for $V^{\text{loc}}(r)$ might be necessary for elements with unoccupied valence s or p shells or due to the problem of ghost states discussed in Section 4.3.2. Within this approximation one can rewrite

$$
\begin{aligned}
V^{\mathrm{PP}}(\mathbf{r}) &= \sum_L^{\infty} V^{\mathrm{loc}}(r) P_L(\omega) + \sum_L^{\infty} \left[V_l(r) - V^{\mathrm{loc}}(r) \right] P_L(\omega) \\
&= V^{\mathrm{loc}}(r) \sum_L^{\infty} P_L(\omega) + \sum_L^{\infty} \delta V_l(r) P_L(\omega) \\
&= V^{\mathrm{loc}}(r) + \sum_L^{l_{\max}} \delta V_l(r) P_L(\omega) ,
\end{aligned}
\tag{4.49}
$$

where the combined index $L = \{l, m\}$ has been used. The pseudopotential is now separated into two parts; the local or core pseudopotential $V^{\mathrm{loc}}(r)$ and the nonlocal pseudopotentials $\delta V_l(r) P_{lm}(\omega)$. The pseudopotentials of this type are also called semi-local, as they are local in the radial coordinate and the nonlocality is restricted to the angular part only.

The contribution of the local pseudopotential to the total energy in a Kohn–Sham calculation is of the form

$$
E_{\mathrm{loc}} = \int V^{\mathrm{loc}}(\mathbf{r}) n(\mathbf{r}) d\mathbf{r} .
\tag{4.50}
$$

It can easily be calculated together with the other local potentials. The nonlocal part needs special consideration as the operator in the plane wave basis has no simple structure, neither in real nor in reciprocal space. There are two approximations that can be used to calculate this contribution to the energy. One is based on numerical integration and the other on a projection on a local basis set, as explained in the following two sections.

4.3.1 Gauss–Hermite integration

In the first method considered the matrix elements of the nonlocal pseudopotential, which still depend on r, are given by

$$
\begin{aligned}
V^{\mathrm{nloc}}(\mathbf{G}, \mathbf{G}') &= \sum_L \frac{1}{\Omega} \int \exp\left[-i\mathbf{G} \cdot \mathbf{r}\right] \delta V_L(\mathbf{r}) \exp\left[i\mathbf{G}' \cdot \mathbf{r}\right] d\mathbf{r}
\end{aligned}
\tag{4.51}
$$

$$
= \sum_L \int_0^{\infty} \langle \mathbf{G} \mid Y_{\mathrm{L}} \rangle_\omega \, r^2 \delta V_L(r) \, \langle Y_{\mathrm{L}} \mid \mathbf{G}' \rangle_\omega \, dr ,
\tag{4.52}
$$

where $\langle \cdot \mid \cdot \rangle_\omega$ stands for an integration over the unit sphere. The integration over the radial coordinate r is replaced by a numerical quadrature

$$\int_0^\infty r^2 f(r) dr \approx \sum_i w_i f(r_i) \; , \qquad (4.53)$$

which is an approximation (since a finite number of terms is used in practice) that can be improved systematically.

The integration weights w_i and the integration points r_i are determined according to the well-known Gauss–Hermite scheme [1161]. In this approximation the nonlocal pseudopotential is

$$V^{\text{nloc}}(\mathbf{G}, \mathbf{G}') \;=\; \sum_L \frac{1}{\Omega} \sum_i w_i \, \delta V_L(r_i) \, \langle \mathbf{G} \mid Y_{\text{L}} \rangle_\omega^{r_i} \, \langle Y_{\text{L}} \mid \mathbf{G}' \rangle_\omega^{r_i} \quad (4.54)$$

$$=\; \sum_L \frac{1}{\Omega} \sum_i w_i \, \delta V_L(r_i) \, P_i^{L\star}(\mathbf{G}) \, P_i^L(\mathbf{G}') \; , \qquad (4.55)$$

where the definition for the projectors P

$$P_i^L(\mathbf{G}) = \langle Y_{\text{L}} \mid \mathbf{G} \rangle_\omega^{r_i} \qquad (4.56)$$

has been introduced. The number of projectors per atom is equal to the number of integration points (5 to 20 for low to high accuracy, respectively) multiplied by the number of angular momenta. For the case of s and p nonlocal components and 15 integration points this amounts to 60 projectors per atom. Finally, the integration of the projectors can be done analytically

$$P_i^L(\mathbf{G}) \;=\; \int_\omega Y_L^\star(\omega) \exp\left[iGr_i\right] d\omega \qquad (4.57)$$

$$=\; \int_\omega Y_L^\star(\omega) 4\pi \sum_{l=0}^\infty i^l j_l(Gr_i) \sum_{m'=-l}^l Y_{lm'}^\star(\omega) Y_{lm'}(\hat{G}) \, d\omega \quad (4.58)$$

$$=\; 4\pi i^l j_l(Gr_i) Y_L(\hat{G}) \; , \qquad (4.59)$$

where the expansion of a plane wave in spherical harmonics has been used. Here, j_l are the spherical Bessel functions and $\hat{G}$ the angular components of the Fourier vector $\mathbf{G}$.

4.3.2 Kleinman-Bylander projection

The second method considered is based on the resolution of the identity in a local, auxiliary basis set

$$\sum_{\alpha} | \chi_\alpha \rangle \langle \chi_\alpha | = 1 \; , \tag{4.60}$$

where $\{\chi_\alpha\}$ is a complete set of orthonormal functions. This identity can now be introduced into the integrals for the nonlocal part

$$V^{\text{nloc}}(\mathbf{G}, \mathbf{G}') = \sum_{L} \int_0^\infty \langle \mathbf{G} | Y_L \rangle_\omega r^2 \delta V_L(r) \langle Y_L | \mathbf{G}' \rangle_\omega \, dr$$

$$= \sum_{\alpha,\beta} \sum_{L} \int_0^\infty \langle \mathbf{G} | \chi_\alpha \rangle \langle \chi_\alpha | Y_L \rangle_\omega$$

$$\times r^2 \delta V_L(r) \langle Y_L | \chi_\beta \rangle_\omega \langle \chi_\beta | \mathbf{G}' \rangle \, dr \; , \tag{4.61}$$

and the angular integrations are easily performed using the decomposition of the basis in spherical harmonics

$$\chi_\alpha(\mathbf{r}) = \chi_\alpha^{lm}(r) Y_{lm}(\omega) \; . \tag{4.62}$$

This leads to

$$V^{\text{nloc}}(\mathbf{G}, \mathbf{G}') = \sum_{\alpha,\beta} \sum_{L} \langle \mathbf{G} | \chi_\alpha \rangle \langle \chi_\beta | \mathbf{G}' \rangle$$

$$\times \int_0^\infty r^2 \delta V_L(r) \chi_\alpha^{lm}(r) \chi_\beta^{lm}(r) \, dr \tag{4.63}$$

$$= \sum_{\alpha,\beta} \sum_{L} \langle \mathbf{G} | \chi_\alpha \rangle \, \delta V_{\alpha\beta}^l \, \langle \chi_\beta | \mathbf{G}' \rangle \; , \tag{4.64}$$

which is the nonlocal pseudopotential in fully separable form. The coupling elements of the pseudopotential

$$\delta V_{\alpha\beta}^l = \int_0^\infty r^2 \, \delta V_L(r) \chi_\alpha^{lm}(r) \, \chi_\beta^{lm}(r) \, dr \tag{4.65}$$

are independent of the plane wave basis and can be calculated for each type of pseudopotential once the expansion functions χ are specified.

The final question is how to choose the optimal set of basis functions χ, which introduces an approximation since the set has to be finite. Kleinman and Bylander (KB) [754] proposed using the eigenfunctions of the pseudo atom, i.e. the solutions to the calculations of the atomic reference state using the pseudopotential Hamiltonian. This choice of a single reference function per angular momentum guarantees nevertheless the correct result for the reference state. Now assuming that in the molecular environment only small perturbations of the wave functions close to the atoms occur, this minimal basis should still be adequate. The Kleinman-Bylander form of the projectors is

$$\sum_L \frac{\mid \chi_L \rangle \langle \delta V_L \chi_L \mid}{\langle \chi_L \mid \delta V_L \chi_L \rangle} = 1 \ , \tag{4.66}$$

where χ_L are the atomic pseudo wave functions. The plane wave matrix elements of the nonlocal pseudopotential in Kleinman-Bylander form are

$$V^{\mathrm{KB}}(\mathbf{G}, \mathbf{G}') = \frac{\langle \mathbf{G} \mid \delta V_L \chi_L \rangle \langle \delta V_L \chi_L \mid \mathbf{G}' \rangle}{\langle \chi_L \mid \delta V_L \chi_L \rangle} \ . \tag{4.67}$$

Generalizations of the Kleinman-Bylander scheme to more than one reference function were introduced by Blöchl [141] and Vanderbilt [1548] (see also Chapter 6). They make use of several reference pseudo functions $\Phi_\alpha(r)$ calculated at a set of reference energies $\{\epsilon\}$. Each of these functions is labeled with a combined index $\alpha = (l, m, \epsilon)$ and defines a function

$$\mid \chi_\alpha \rangle = -(T + V^{\mathrm{loc}} - \epsilon) \mid \Phi_\alpha \rangle \ . \tag{4.68}$$

A dual basis $\varphi_\alpha(r)$ to the functions $\Phi_\alpha(r)$

$$\langle \varphi_\alpha \mid \Phi_\beta \rangle = \delta_{\alpha\beta} \tag{4.69}$$

can be defined with

$$\varphi_\alpha(r) = \sum_\beta (\mathbf{B}^{-1})_{\alpha\beta} \chi_\beta(r) \tag{4.70}$$

and

$$B_{\alpha\beta} = \langle \Phi_\beta \mid \chi_\alpha \rangle \ . \tag{4.71}$$

The nonlocal pseudopotential in separable form is then

$$\begin{aligned} V^{\mathrm{NL}}(\mathbf{G}, \mathbf{G}') &= \sum_\alpha \langle \mathbf{G} \mid \chi_\alpha \rangle \langle \Phi_\alpha \mid \mathbf{G}' \rangle \\ &= \sum_{\alpha,\beta} \langle \mathbf{G} \mid \varphi_\alpha \rangle B_{\alpha\beta} \langle \varphi_\beta \mid \mathbf{G}' \rangle \ . \end{aligned} \tag{4.72}$$

The condition that the pseudopotential V^{NL} is Hermitian requires a generalized norm-conservation constraint on the pseudo wave functions

$$\langle \Psi_\alpha \mid \Psi_\beta \rangle - \langle \Phi_\alpha \mid \Phi_\beta \rangle = 0 \ . \tag{4.73}$$

The implementation of this constraint into the pseudopotential generation schemes might prove to be nontrivial. Kresse and Hafner [788] suggested therefore relaxing the condition to the diagonal elements only and using a symmetrized matrix

$$\tilde{B}_{\alpha\beta} = \frac{1}{2}(B_{\alpha\beta} + B_{\beta\alpha}) \tag{4.74}$$

in V^{NL}.

In transforming a semi-local to the corresponding KB-type pseudopotential one needs to make sure that the KB form does not lead to unphysical "ghost states" at energies below or near those of the physical valence states as these would undermine its transferability. Such spurious states can occur for specific (unfavorable) choices of the underlying semi-local and local pseudopotentials. They are an artefact of the KB form of the nonlocality by which the nodeless reference pseudo wave functions need to be the lowest eigenstate, unlike for the semi-local form [538, 540]. Ghost states can be avoided by using more than one reference state or by a proper choice of the local component and the cutoff radii in the basic semi-local pseudopotentials. The appearance of ghost states can be analyzed by investigating the following properties:

- Deviations of the logarithmic derivatives of the energy of the KB pseudopotential from those of the respective semi-local pseudopotential or allelectron potential.
- Comparison of the atomic bound state spectra for the semi-local and KB pseudopotentials.
- Ghost states below the valence states are identified by rigorous criteria introduced by Gonze *et al.* [538, 540].

4.4 Dual-space Gaussian pseudopotentials

Pseudopotentials in the Kleinman–Bylander form have the advantage of requiring a minimal amount of work in a plane wave calculation by still keeping most of the transferability and general accuracy of the underlying semi-local pseudopotential. However, one wonders if it would not be possible to generate pseudopotentials directly in the separable form fulfilling the Hamann–Schlüter–Chiang conditions introduced in Section 4.2.2. It was found [525, 608] that indeed it is possible to optimize a small set of parameters defining an analytical form for the local and nonlocal form of a pseudopotential that fulfills the HSC conditions and reproduces even additional properties leading to highly transferable pseudopotentials.

In the dual-space Gaussian scheme the local part of the pseudopotential is given by

$$
V^{\text{loc}}(r) = \frac{-Z_{\text{ion}}}{r} \text{erf} \left[\frac{\bar{r}}{\sqrt{2}} \right] + \exp \left[-\frac{1}{2}\bar{r}^2 \right] \left[C_1 + C_2\bar{r}^2 + C_3\bar{r}^4 + C_4\bar{r}^6 \right] \ ,
$$

$$(4.75)$$

where erf denotes the error function and $\bar{r} = r/r_{\text{loc}}$; Z_{ion} is the ionic charge of the atomic core, i.e. the total charge minus the charge of the valence electrons. The nonlocal contribution to the pseudopotential is a sum of separable terms

$$
V_l(\mathbf{r}, \mathbf{r}') = \sum_{i=1}^{3} \sum_{j=1}^{3} \sum_{m=-l}^{l} Y_{lm}(\omega) p_i^l(r) \, h_{ij}^l \, p_j^l(r) Y_{lm}^\star(\omega') \ ,
$$

$$(4.76)$$

where the projectors $p_i^l(r)$ are Gaussians of the form

$$
p_i^l(r) = \frac{\sqrt{2} r^{l+2(i-1)} \exp \left[-\frac{r^2}{2r_l^2} \right]}{r_l^{l+(4i-1)/2} \sqrt{\Gamma \left[l + \frac{4i-1}{2} \right]}} \ ,
$$

$$(4.77)$$

Γ is the gamma function, and the projectors are normalized

$$
\int_0^\infty r^2 p_i^l(r) p_i^l(r) dr = 1 \ .
$$

$$(4.78)$$

This pseudopotential also has an analytical form in Fourier space. The Fourier transform of the pseudopotential is given by

$$
V^{\text{loc}}(G) = -4\pi \frac{Z_{\text{ion}}}{G^2\Omega} \exp \left[-(Gr_{\text{loc}})^2/2 \right] + \sqrt{8\pi^3} \frac{r_{\text{loc}}^3}{\Omega} \exp \left[-(Gr_{\text{loc}})^2/2 \right]
$$
$$
\times \left\{ C_1 + C_2(3 - G^2 r_{\text{loc}}^2) + C_3(15 - 10(Gr_{\text{loc}})^2 + (Gr_{\text{loc}})^4) \right.
$$
$$
\left. + C_4(105 - 105(Gr_{\text{loc}})^2 + 21(Gr_{\text{loc}})^4 - (Gr_{\text{loc}})^6) \right\} \quad (4.79)
$$

for the local part, and

$$
V_l(\mathbf{G}, \mathbf{G}') = (-1)^l \sum_{i=1}^{3} \sum_{j=1}^{3} \sum_{m=-l}^{l} Y_{lm}(\hat{\mathbf{G}}) p_i^l(G) h_{ij}^l p_j^l(G') Y_{lm}^*(\hat{\mathbf{G}}')
$$

$$(4.80)$$

for the nonlocal part. The Fourier transform of the projectors $p_i^l(r)$ can be calculated analytically, and for the relevant cases one obtains

$$p_k^l(G) = q_k^l(Gr_l) \, \frac{\pi^{5/4} \, G^l \, \sqrt{r_l^{2l+3}}}{\sqrt{\Omega} \exp\left[\frac{1}{2}(Gr_l)^2\right]} \quad , \tag{4.81}$$

where Ω is the volume of the supercell and the functions $q_k^l(x)$ are defined by

$$q_1^0(x) = 4\sqrt{2} \, , \tag{4.82}$$

$$q_2^0(x) = 8\sqrt{\frac{2}{15}} \, (3 - x^2) \, , \tag{4.83}$$

$$q_3^0(x) = \frac{16}{3}\sqrt{\frac{2}{105}} \, (15 - 20x^2 + 4x^4) \, , \tag{4.84}$$

$$q_1^1(x) = 8\sqrt{\frac{1}{3}} \, , \tag{4.85}$$

$$q_2^1(x) = 16\sqrt{\frac{1}{105}} \, (5 - x^2) \, , \tag{4.86}$$

$$q_3^1(x) = \frac{32}{3}\sqrt{\frac{1}{1155}} \, (35 - 28x^2 + 4x^4) \, , \tag{4.87}$$

$$q_1^2(x) = 8\sqrt{\frac{2}{15}} \, , \tag{4.88}$$

$$q_2^2(x) = \frac{16}{3}\sqrt{\frac{2}{105}} \, (7 - x^2) \, , \tag{4.89}$$

$$q_3^2(x) = \frac{32}{3}\sqrt{\frac{2}{15015}} \, (63 - 36x^2 + 4x^4) \, , \tag{4.90}$$

$$q_1^3(x) = 16\sqrt{\frac{1}{105}} \, , \tag{4.91}$$

$$q_2^3(x) = \frac{32}{3}\sqrt{\frac{1}{1155}} \, (9 - x^2) \, , \tag{4.92}$$

$$q_3^3(x) = \frac{64}{45}\sqrt{\frac{1}{1001}} \, (99 - 44x^2 + 4x^4) \, . \tag{4.93}$$

Thus, in both real and Fourier space, the projectors have the form of a Gaussian multiplied by a polynomial. Due to this property, the dual-space

Gaussian pseudopotential is the optimal compromise between good convergence properties in real and Fourier space. The multiplication of the wave function with the nonlocal pseudopotential arising from an atom can be limited to a small region around the atom as the radial projectors asymptotically tend to zero outside the covalent radius of the atom. In addition, a very dense integration grid is not required, as the projector is reasonably smooth because of its good decay properties in Fourier space. The parameters of this type of pseudopotential are found by minimizing a target function. This function is built up as the sum of the differences of properties calculated from the all-electron atom and the pseudo atom. Properties included are the integrated charge and the eigenvalues of occupied and lowest unoccupied states.

4.5 Nonlinear core correction

The success of pseudopotentials in density functional calculations relies on two essential assumptions. The transferability of the core electrons to different environments and the linearization of the exchange and correlation energy. The second assumption is only valid if the frozen-core electrons and the valence state do not overlap. It leads to a useful approximation if this overlap is not substantial, as for instance shown in Fig. 4.2 for the case of the Si atom. However, if there is significant overlap between core and valence densities, the linearization will lead to reduced transferability and systematic errors. The most straightforward remedy is to include "semi-core states" in addition to the valence shell, i.e. one more inner shell is treated explicitly which is – from a chemical viewpoint – an inert core level that does not contribute to bonding. This approach leads to very hard norm-conserving pseudopotentials which call for high plane wave cutoffs, whereas suitable ultrasoft pseudopotentials (see Section 6.4) can be constructed including such semi-core states.

As an alternative solution it has been proposed to treat the nonlinear parts of the exchange and correlation energy E_{xc} explicitly [879]. This idea does not lead to an increase of the cutoff, but ameliorates the above-mentioned problems quite considerably. To achieve this, the exchange–correlation energy E_{xc} is calculated not from the bare valence density $n(\mathbf{r})$ alone, but from a modified density

$$\tilde{n}(\mathbf{r}) = n(\mathbf{r}) + \tilde{n}_{\text{core}}(\mathbf{r}) \ , \tag{4.94}$$

where $\tilde{n}_{\text{core}}(\mathbf{r})$ denotes a spherical density that is equal to the core density of

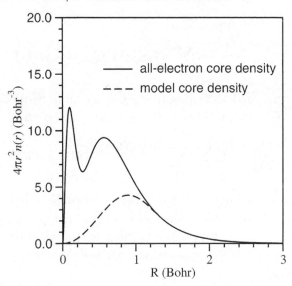

Fig. 4.4. Core density (solid line) and pseudized core density (dashed line) of a sodium atom; the two functions match at $r_0 = 1.2$ Bohr.

the atomic reference state in the region of overlap with the valence density

$$\tilde{n}_{\text{core}}(r) = n_{\text{core}}(r) \quad \text{if } r > r_0 \ . \tag{4.95}$$

Close to the nuclei a model density is chosen in order to reduce the cutoff for the plane wave expansion. Finally, the two densities and their derivatives are matched at r_0.

Popular choices for the analytical form of the pseudized core density, see Fig. 4.4, are polynomial functions as used for Kerker pseudopotentials (see Section 4.2.4) or spherical Bessel functions. This procedure leads to a modified total energy, where E_{xc} is replace by

$$E_{\text{xc}} = E_{\text{xc}}(n + \tilde{n}_{\text{core}}) \ , \tag{4.96}$$

and the corresponding potential is

$$V_{\text{xc}} = V_{\text{xc}}(n + \tilde{n}_{\text{core}}) \ . \tag{4.97}$$

The sum of all modified core densities

$$\tilde{n}_{\text{core}}(\mathbf{G}) = \sum_I \tilde{n}_{\text{core}}^I(\mathbf{G}) S_I(\mathbf{G}) \tag{4.98}$$

now depends on the nuclear positions; here, $S_I(\mathbf{G}) = \exp[-i\mathbf{G} \cdot \mathbf{R}_I]$ is again the structure factor, Eq. (3.34). Importantly, this leads to a new, additional contribution to the forces

$$\frac{\partial E_{xc}}{\partial R_{I,s}} = -\Omega \sum_{\mathbf{G}} iG_s V_{xc}^{\star}(\mathbf{G}) \tilde{n}_{core}^{I}(\mathbf{G}) S_I(\mathbf{G}) \tag{4.99}$$

which must be included according to the discussion in Section 2.5. Similarly, there is also such a Pulay contribution to the stress tensor

$$\frac{\partial E_{xc}}{\partial h_{uv}} = \Omega \sum_{I} \sum_{\mathbf{G}} \frac{\partial n_{core}^{I}(\mathbf{G})}{\partial h_{uv}} S_I(\mathbf{G}) V_{xc}(\mathbf{G}) \ . \tag{4.100}$$

The method of nonlinear core correction dramatically improves results on systems with alkali and transition metal atoms. For practical applications, one should keep in mind that the nonlinear core correction should only be applied together with pseudopotentials that were generated using the same energy expression.

4.6 Pseudopotential transferability

The radial Schrödinger equation is a second-order linear differential equation. Given the screened all-electron potential and an arbitrary energy ϵ, the solution of such an equation is uniquely defined by the value of the wave function $\Psi(r)$ and its first derivative $\Psi'(r)$ at any given point r_0. Thus, neglecting normalization, the wave function is uniquely determined by its logarithmic derivative

$$\frac{d}{dr} \ln \left[\Psi(r, \epsilon) \right]\Big|_{r=r_0} = \frac{1}{\Psi(r, \epsilon)} \frac{d\Psi(r, \epsilon)}{dr}\Big|_{r=r_0} \tag{4.101}$$

at the point r_0. If the screened all-electron potential and the pseudopotential are identical outside the radius $r_{c,l}$, then the all-electron wave function $\Psi(r)$ and the pseudo wave function $\Phi(r)$ are proportional outside $r_{c,l}$ if

$$\frac{1}{\Phi(r, \epsilon)} \frac{d\Phi(r, \epsilon)}{dr} = \frac{1}{\Psi(r, \epsilon)} \frac{d\Psi(r, \epsilon)}{dr} \tag{4.102}$$

holds.

By construction this is true for a pseudopotential obeying the Hamann–Schlüter–Chiang conditions introduced in Section 4.2.2 for the eigenvalue energies ϵ_l.

A perfect pseudopotential would fulfill this equation for all energies above the core state energies. The norm-conservation condition imposes that the above equality is closely satisfied for a region surrounding ϵ_l, which can be

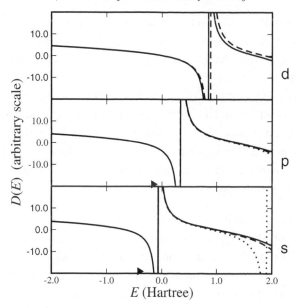

Fig. 4.5. Logarithmic derivatives of the wave functions for silicon. Pseudopotential generated according to Ref. [540], s and p angular momenta have a nonlocal potential, d channel is the local part. Upper panel: d functions; middle panel: p functions; lower panel: s functions. The derivative was taken at $r_0 = 2.7758$ a.u. All-electron functions (solid line) and pseudo wave functions from the semi-local (dashed line) and from the separable form (dotted line) are shown. Reference state energies are indicated by triangles.

traced back to the fact that the radial Schrödinger equation obeys

$$-\frac{1}{2}\frac{\partial}{\partial \epsilon}\frac{\partial}{\partial r}\ln[\Psi(r,\epsilon)]\bigg|_{\substack{\epsilon=\epsilon_l \\ r=r_{c,l}}} = \frac{1}{r_{c,l}^2\Psi^2(r_{c,l},\epsilon_l)}\int_0^{r_{c,l}} r^2\Psi^2(r,\epsilon_l)\,dr \ , \quad (4.103)$$

a version of the Friedel sum rule [594]. Comparing the logarithmic derivatives of the all-electron and the pseudopotential wave functions as a function of the energy ϵ at a radius $r_0 \geq r_{c,l}$ for the range of energies anticipated in an application (which are typically the energies of the valence and lower conduction bands in a solid) provides a quick estimate of the quality of the pseudopotential. Often r_0 is taken as the Wigner–Seitz radius, but any useful radii can be used.

However, the logarithmic derivatives are calculated assuming that the electron screening part of the potential is a constant. This is not the case when the environment of the atom changes, and therefore a logarithmic derivative comparison cannot be taken as an absolute test of the quality of an ionic pseudopotential. The pseudopotential will accurately reproduce the all-electron calculation in the reference configuration in which it was generated. In practice, it is also required to closely reproduce other all-electron calculations in different environments; in other words it should be transferable. The logarithmic derivatives provide a first test of the transferability of the screened pseudopotential. Comparing the all-electron calculations for atomic states other than the reference state provides another easy way to test the ionic pseudopotential, as shown in Fig. 4.5 for the Si atom. Doing the same comparison for typical crystals and molecules is of course a better test of the transferability of the pseudopotential, but it is also a more elaborate and costlier test.

A possible method to improve the transferability to the solid is to generate the pseudopotential using an atomic configuration that as closely as possible mimics the effect of the environment in which it will be placed. This may require the use of wave functions that are non-bound or off from eigenstates. However, this may be seen as a way to produce specialized pseudopotentials whereas typically general pseudopotentials are searched for. However, in general, ionic pseudopotentials are rather insensitive to reasonable variations in the reference atomic configuration and improvements obtained in this fashion are limited.

It was also proposed [1342] to use higher-energy derivatives of the logarithmic derivatives of the wave function in the construction of pseudopotentials, but only small improvements could be achieved. The easiest approach to increase transferability is to simply decrease the cutoff radii $r_{c,l}$ used to generate the pseudopotential and pseudo wave functions, which immediately reduces the difference between the all-electron and pseudopotential results. However, there are practical limits on how far one can decrease $r_{c,l}$; the cutoff radius has to be larger than the outermost node of the all-electron wave function to ensure a nodeless pseudo wave function. In fact, if $r_{c,l}$ is chosen too close to the node one will encounter oscillations in the pseudopotential and a steep increase in the required Fourier space expansion of the pseudo wave functions.

Since extending the energy range of correct logarithmic derivatives does not necessarily enhance – and is not a foolproof test for – transferability one has to look for other criteria. As isolated atoms are allowed to interact, states are created with eigenvalues away from the reference states and additionally, in many cases significant transfer of electrons from one atom to another occurs. Thus the changes in wave functions with eigenvalue and with occupancies are interrelated. The eigenvalue of a state is the derivative of the total energy, Eq. (2.113), with respect to the occupation of that state

$$\frac{\partial E_{\text{tot}}}{\partial f_i} = \epsilon_i \ . \tag{4.104}$$

By requiring the same eigenvalues, one has enforced the condition that small changes in the occupancies of atomic valence states yield the same energy changes in both the atom and the pseudo atom. The norm-conservation property enforces the condition that if the eigenvalues of atom and pseudo atom have the same changes, the valence wave functions for both will change in the same way at the surface of the atomic core. However, there is no guarantee that for any given perturbation the change in the eigenvalues of the pseudo atom will be correct. Therefore, it is anticipated that ensuring the correct eigenvalue changes will remove an important error of pseudopotentials. The quantity related to these changes is the hardness (see Ref. [1102] for a general discussion). The hardness was first used by Teter [1461] to construct improved pseudopotentials and later by Filippetti *et al.* [431, 432] to analyze pseudopotential transferability.

The chemical hardness matrix, or orbital hardness matrix, is defined within density functional theory as

$$H_{ij} = \frac{1}{2} \frac{\partial^2 E[n]}{\partial f_i \partial f_j} \ , \tag{4.105}$$

where $E[n]$ is the Janak functional [701] and f_i is the occupation number of the ith eigenstate, which yields

$$H_{ij} = \frac{1}{2} \frac{\partial \epsilon_i}{\partial f_j} \ , \tag{4.106}$$

together with Eq. (4.104). Thus, the hardness matrix measures the first-order change of an eigenvalue resulting from a variation of an occupation number, while allowing the total number of electrons to vary. This is exactly the quantity referred to in the last paragraph. Now, using the Hellmann–Feynman theorem one obtains

$$
\begin{aligned}
H_{ij} &= \frac{1}{2} \left\langle \Psi_i \left| \frac{\partial}{\partial f_j} [T + V_{\text{ext}} + V_{\text{H}} + V_{\text{xc}}] \right| \Psi_i \right\rangle \\
&= \frac{1}{2} \left\langle \Psi_i \left| \frac{\partial [V_{\text{H}} + V_{\text{xc}}]}{\partial n} \frac{\partial n}{\partial f_j} \right| \Psi_i \right\rangle \ .
\end{aligned}
\tag{4.107}
$$

The derivative of the electron density with respect to occupation numbers consists of a part due to the change in the screening potential with variation of the occupation number for a fixed wave function, and another part arising from the response of the wave function to the perturbation. The second part has to be calculated within linear response theory and follows the same steps as outlined in Section 7.1.2.

In chemical systems atoms may have a very anisotropic environment so that nonspherical changes of electron occupation become important. Therefore, it is natural to consider in the test for transferability not only perturbations that keep the spherical symmetry of the atom but also occupation changes that lead to nonspherical densities and screening potentials. For this purpose it is convenient to generalize occupation numbers to the concept of an occupation matrix. This concept was originally introduced by Vanderbilt and Joannopoulos [1549] to describe changes in the electron density at surfaces, but proved to be powerful also in other applications [863, 960, 1398, 1558] (see for instance Eq. (2.75) and Eq. (5.88)). Within this generalization the charge density is calculated from the Kohn–Sham orbitals $\Psi_i(r)$ and the occupation matrix f_{ij}

$$n(r) = \sum_{ij} f_{ij} \Psi_i^\star(r) \Psi_j(r) \ , \qquad (4.108)$$

where in the ground state $f_{ij} = f_i \delta_{ij}$ is valid and the standard relationship between density and wave functions in Kohn–Sham theory is recovered as well. The generalized hardness is now defined as the second derivative of the total energy with respect to any two elements of the occupation matrix

$$H_{ij,kl} = \frac{1}{2} \frac{\partial^2 E_{\text{tot}}}{\partial f_{ij} \partial f_{kl}} \ . \qquad (4.109)$$

In Table 4.4 some elements of the generalized hardness matrix for silicon are listed. The values are compared between all-electron calculations and calculations with some high-quality pseudopotentials. Excellent agreement is achieved for the ground state configuration within the local density approximation. More demanding cases are discussed in the pertinent literature [431, 432].

When making the transition from a semi-local to a pseudopotential in the Kleinman–Bylander form (see Section 4.3.2), the local part of the pseudopotential has to be chosen. In principle, one has complete freedom in choosing the local part. However, as the problem of ghost states indicates, the choice of the local part can have a dramatic influence on the robustness and transferability of the pseudopotential. Typically, the first non-occupied state for the atomic ground state is a good choice for the local part. It is therefore important that these states are also very carefully generated using either an

Table 4.4. *Generalized hardness calculated for a silicon atom in the ground state configuration. Results from all-electron and different high-quality pseudopotential calculations.*

	All-electron [432]	HGH-PP [608]	SGS-PP [540]	
		dual-space	semi-local	separable
ss,ss	0.299	0.300	0.300	0.300
ss,pp	0.270	0.270	0.270	0.270
pp,pp	0.258	0.256	0.256	0.256
sp,sp	0.026	0.036	0.036	0.036

adequate excited state or the Hamann procedure [586] for unbound states as explained in Section 4.2.3.

For very difficult cases it was proposed [431] to use a linear combination of potentials from different angular momenta as the local potential

$$V^{\mathrm{loc}} = \alpha V_{\mathrm{s}} + \beta V_{\mathrm{p}} + \gamma V_{\mathrm{d}} \ , \tag{4.110}$$

where $\alpha + \beta + \gamma = 1$ is needed for the correct long-range behavior. By optimizing the coefficients it is possible to generate separable pseudopotentials with the same transferability properties as the original semi-local potentials for a wider range of cutoffs.

4.7 Example: pseudopotentials for carbon

In the following a practical example on how to generate pseudopotentials according to the theoretical considerations from earlier sections is given. In particular, some important points that should be considered whenever a new pseudopotential is created are discussed using carbon as an example. The Troullier–Martins recipe from Section 4.2.5 will be used, and all generated pseudopotentials will be for the local density approximation with occasional comparisons to other pseudopotentials.

The reference configuration is the atomic ground state $1s^2 2s^2 2p^2$ and the potential for the unoccupied d state will be generated with the Hamann method [586]. The reference energy for the d channel is taken to be the eigenvalue of the p eigenfunction. The underlying all-electron eigenfunctions are shown in Fig. 4.6. Accordingly, the core 1s state is considerably

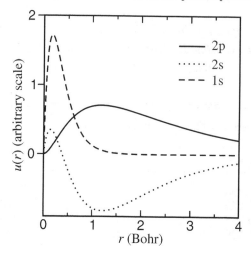

Fig. 4.6. All-electron wave functions $u(r) = r\Psi(r)$ for the carbon atom in the ground state.

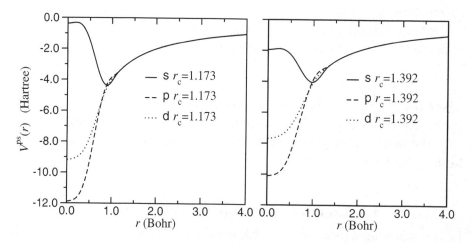

Fig. 4.7. Two sets of Troullier–Martins pseudopotentials for carbon generated by using two different cutoff radii r_c as indicated in the panels.

tighter than the valence states, so that it is not necessary to use the nonlinear core correction scheme from Section 4.5. The maxima of the functions are at 1.21 Bohr and 1.18 Bohr for 2s and 2p states, respectively. Two pseudopotentials will be generated, one with a cutoff radius close to the wave function maximum at 1.173 Bohr and another with a cutoff further out at 1.392 Bohr using the same radial cutoff numbers $r_c = r_{c,l}$ for all angular momenta l. As can be seen in Fig. 4.7, the smaller cutoff in the left panel

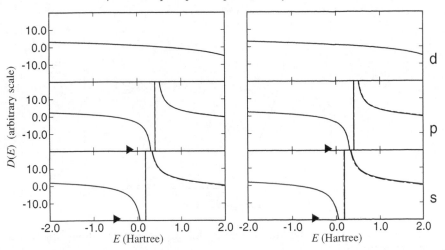

Fig. 4.8. Logarithmic derivatives (at $r_0 = 1.91$ Bohr) of the wave functions for the two sets of Troullier–Martins pseudopotentials for carbon from Fig. 4.7. Left panels: pseudopotentials with $r_c = 1.173$ Bohr. Right panels: pseudopotentials with $r_c = 1.392$ Bohr. Triangles: reference energies. Solid lines: all-electron wave functions. Dashed lines: pseudo wave functions from semi-local potentials. Dashed-dotted line: pseudo wave functions from separable form of pseudopotential.

results in deeper p and d potentials and it can be assumed that this will be reflected in a higher plane wave cutoff needed in applications.

The first test along the lines of the general discussion in Section 4.6 is to check the logarithmic derivatives of the wave functions which are depicted in Fig. 4.8 for a radius of $r_0 = 1.91$ Bohr. There seems to be no problem with either potential: the all-electron values are perfectly reproduced by both sets of potentials in the semi-local as well as in the separable forms. Based on this check, a good transferability can be expected for both pseudopotentials. Secondly, the kinetic energy criterion should be investigated in order to extract information on the plane wave energy cutoff Eq. (3.17) that is needed in actual plane wave calculations. To this end, in Fig. 4.9 the error in the kinetic energy is plotted against the plane wave cutoff. As expected from the smoothness argument based on Fig. 4.7, the pseudopotential generated with a larger r_c value converges faster. In both pseudopotentials the s wave function is considerably softer than the p function. Based on this analysis one can expect reasonably converged total energies for these two pseudopotentials at 50 Rydberg and 65 Rydberg, respectively. However, one is mostly interested in the convergence of structural parameters or energy differences of the system of interest and these quantities might converge much faster as a function of plane wave cutoff than absolute energies.

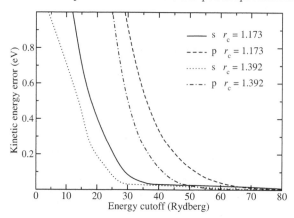

Fig. 4.9. Convergence of the kinetic energy for two sets of Troullier–Martins pseudopotentials for carbon from Fig. 4.7.

Finally, we go back to the issue of transferability. The logarithmic derivatives suffer from the problem that they are calculated from the screened pseudopotentials. In other words, they show transferability under the assumption that the electron density is not changing. A better test, still at the level of the atom, is the calculation of excited or ionic states. In Table 4.5 the eigenvalues of pseudopotential calculations for the two Troullier–Martins and a dual-space pseudopotential [608] are compared. It is seen that the states involving only s and p electrons are well reproduced by all three potentials, which also holds for excited and ionic states. On the other hand, only the dual-space pseudopotential is able to faithfully reproduce the states that involve 3d electrons. This is most likely due to the fact that the d potential has been generated at an energy of -0.199 eV and with an electronic distribution given by the ground state, whereas the states involving the 3d electrons in the table are rather far from this reference. Whether this shortcoming of the pseudopotentials will affect calculations of real systems has to be seen, as the states in question can be considered rather exotic. Independently from that, the overall quality of the dual-space Gaussian pseudopotentials (see Section 4.4) is demonstrated clearly with these examples. However, the transferability and accuracy of these pseudopotentials comes at a price, which is the rather high plane wave cutoff needed to obtain converged results.

A most critical check is to use the different pseudopotentials in calculations beyond the atom. Here, a first series of tests uses carbon in the diamond structure. All calculations were done within the unit cell (face centered cubic with two atoms using the experimental lattice constant) and the

Table 4.5. *All-electron eigenvalues (given in eV) for different states of a carbon atom. The errors of the pseudopotential calculations with respect to the all-electron energies are shown in meV. The pseudopotentials shown are the two Troullier–Martins pseudopotentials from Fig. 4.7 with cutoff radii of 1.173 Bohr (TM12) and 1.393 Bohr (TM14), respectively, and a dual-space separable potential (HGH [608]).*

State		All-electron	HGH	TM12	TM14
s^2p^2	s	−0.502	0	0	0
	p	−0.199	0	0	0
$s^{0.75}pd^{0.25}$	s	−1.530	5	71	73
	p	−1.226	0	70	69
	d	−0.345	1	500	500
s^2p	s	−0.941	0	0	0
	p	−0.629	1	1	1
s^2	s	−1.476	4	4	5
sp^3	s	−0.518	1	1	1
	p	−0.214	0	0	0
$s^{1.5}p^{1.5}d^{0.5}$	s	−0.873	1	70	71
	p	−0.562	2	70	69
	d	−0.060	1	755	756

Brillouin zone sampling used 28 special points generated with a Monkhorst-Pack [1026] grid of six points in each direction (see Section 3.1.3). Again, the energy convergence has been tested with the two Troullier–Martins pseudopotentials from Fig. 4.7 and the dual-space HGH pseudopotential [608] already used in the atom calculations in Table 4.5. The conclusions drawn from the kinetic energy criterion based on the atomic calculations are reproduced with the results in Fig. 4.10. However, the curves indicate that slightly higher cutoffs should be used for converged results with respect to the total energy. It is also immediately clear that a considerably higher cutoff is needed with the dual-space pseudopotential: only for cutoffs beyond 100 Rydberg are converged energy values achieved. All pseudopotentials give qualitatively similar results for the structural properties of diamond (see Table 4.6). Since no differences can be found for the two Troullier–Martins pseudopotentials, one can conclude that even very soft pseudopotentials are sufficient for this task.

As a second test the carbon-carbon bond length is calculated in three

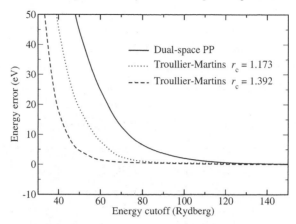

Fig. 4.10. Convergence of total energy of diamond at the experimental lattice constant (3.57Å) using different pseudopotentials as indicated; see text for further details.

Table 4.6. *Lattice constant (Å), bulk modulus (GPa), and derivatives of the bulk modulus for diamond. The pseudopotentials shown are the two Troullier–Martins pseudopotentials from Fig. 4.7 with cutoff radii of 1.173 Bohr (TM12) and 1.393 Bohr (TM14), respectively, and a dual-space separable potential (HGH [608]). Experimental values are cited from Ref. [1480].*

	Experiment	HGH	TM12	TM14
Lattice constant	3.567	3.532	3.528	3.528
Bulk modulus B_0	443	456	460	460
B_0'	4.0	3.549	3.477	3.478

simple hydrocarbon molecules (see Table 4.7). The prototypical cases span carbon single, double, and triple bonds. All calculations have been done with an energy cutoff high enough to get converged results. The dual-space pseudopotential gives bond lengths very close to the reference all-electron calculation, which is based on a calculation with a large Gaussian basis set; the maximum deviation is only 0.003Å for the triple bond. The Troullier–Martins pseudopotentials have been tested in three forms. The original semi-local form using a Gauss–Hermite integration scheme from Section 4.3.1 and the d angular momentum potential serving as local potential is denoted by

Table 4.7. *Carbon-carbon bond length (Å) in C_2H_2, C_2H_4, and C_2H_6. The all-electron values were calculated with the $6-311++G(3df,2p)$ Gaussian basis set. The pseudopotentials shown are the two Troullier–Martins pseudopotentials from Fig. 4.7 with cutoff radii of 1.173 Bohr (TM12) and 1.393 Bohr (TM14), respectively, and a dual-space separable potential (HGH [608]). Pseudopotentials in separable form using the Kleinman-Bylander scheme are denoted with the angular momentum potential used as the local potential, see text.*

	C_2H_2	C_2H_4	C_2H_6
all-electron	1.201	1.322	1.521
HGH	1.198	1.321	1.522
TM12	1.197	1.319	1.520
TM12p	1.198	1.320	1.520
TM12d	1.196	1.318	1.519
TM14	1.195	1.318	1.519
TM14p	1.198	1.321	1.523
TM14d	1.195	1.317	1.517

TM12 and TM14 for the small and large cutoff radii r_c. The Kleinman–Bylander construction from Section 4.3.2 with either s and p or only s angular momentum and the corresponding d and p potentials as local potential have also been tested (TM12p, TM12d, TM14p, TM14d). In general, the inclusion of the nonlocal p channel improves results. However, only for the soft TM14p potential should one consider the results not fully satisfactory.

Part II

Advanced techniques

5

Beyond standard *ab initio* molecular dynamics

5.1 Introduction

The discussion up to this point has revolved essentially around the "standard" *ab initio* molecular dynamics methodologies. The notion "standard" means in particular that *classical* nuclei evolve *adiabatically* in the electronic *ground state* in the *microcanonical* ensemble. In addition, it is assumed that the electronic structure of all constituents of the system is treated on an *equal footing*. This combination allows already a multitude of applications, but many circumstances exist where the underlying approximations or restrictions break down or are unsatisfactory. Among these cases are situations where:

  (i) It is necessary to keep the temperature and/or pressure constant, such as during journeys in phase diagrams or in the investigation of solid-state phase transitions.

 (ii) Pronounced free energy barriers have to be surmounted in rather short simulation times, such as during chemical reactions, conformational changes or phase transitions with high activation energies.

(iii) There is a sufficient population of excited electronic states, such as in materials with a small or vanishing electronic gap, or nonadiabatic dynamics involving a few specific excited states occurs, such as in dye molecules or chromophores after photoexcitation events.

 (iv) Light nuclei are involved in crucial steps of a process, such as in studies of proton transfer in hydrogen-bonded systems or muonium impurities in crystals.

  (v) The system is too large to be fully described quantum-mechanically, such as enzymes, where a large "biomatrix" hosts a small "hot spot" to carry out a biochemical reaction or in the field of surface chemical reactions and heterogeneous catalysis.

177

In the following sections techniques are introduced which transcend these limitations. Thus, the realm of *ab initio* molecular dynamics is considerably increased beyond the basic setup as discussed in general terms in Chapter 2 and concerning its implementation in Chapter 3. The presented "advanced simulation techniques" are selected because they cover the current state-of-the-art in the field, in addition to being available in the current version of the CPMD package [696]. However, their implementations as such are not discussed in as much detail as the basic technique.

5.2 Beyond microcanonics: thermostats, barostats, metadynamics

5.2.1 Introduction

In the framework of statistical mechanics all ensembles can be formally obtained from the microcanonical or *NVE* ensemble - where particle number, volume, and energy are the external thermodynamic control variables - by suitable Laplace transforms of its partition function; note that V (and not Ω) is used to denote the volume when it comes to labeling the various ensembles in this chapter. Thermodynamically, this corresponds to Legendre transforms of the associated thermodynamic potentials where intensive and extensive conjugate variables are interchanged. In thermodynamics, this task is achieved by a "sufficiently weak" coupling of the original system to an appropriate infinitely large bath or reservoir via a contact that establishes thermodynamic equilibrium. The same basic idea is instrumental in generating distribution functions of such ensembles by computer simulation [210, 546]. In Sections 5.2.2 and 5.2.3, two important special cases are discussed, thermostats and barostats, respectively. They are used to impose temperature instead of energy and/or pressure instead of volume, respectively, as external control parameters [25, 468, 577, 1065, 1189, 1382]. Furthermore, a method is presented in Section 5.2.3 that takes into account explicitly the pressurizing medium, such as a liquid, hosting a finite system of interest. Aspects of grand canonical simulations subject to a variable number of electrons in the system are deferred to Section 5.3.4, where electronically excited states are discussed. A technique to treat exchange of electrons between molecules and a reservoir at fixed chemical potential, i.e. redox processes, and thus a variable number of electrons in a two-state system is worked out in Ref. [1453] within the framework of *ab initio* molecular dynamics, whereas a "variational particle number approach" is proposed in Ref. [1571].

In addition to transforming the ensemble, the microcanonic formalism must be transcended in cases where the underlying free Newtonian dynamics does not sample the relevant phase space during accessible simulation times. This is particularly severe for *ab initio* molecular dynamics simulations, since the time scale is rather limited due to the concurrent electronic structure computations. There is a host of methods available that help to overcome these limitations, several of which were used in conjunction with CPMD [696] or implemented therein, such as "chemical flooding" [565, 1040], "nudged elastic band" [631, 632, 1224, 1279], "string MD" [376, 725, 1279], "action-derived MD" [11, 292, 1114–1116], "multiple steering" [231], "bias potentials" [1554], "transition path sampling" [173, 321, 496, 497], "targeted MD" [909, 1287], or various other techniques [1036, 1555, 1556]. The widely used "constrained reaction coordinate dynamics" or "Blue Moon" ensemble technique [230, 1386], which is based on applying geometric mechanical constraints, is discussed in detail in Section 3.7.3 as this method is a basis for various other techniques that build upon it. In Section 5.2.4, the powerful metadynamics approach [821], in its extended Lagrangian formulation [667], is outlined in more detail.

5.2.2 Imposing temperature: thermostats

In the limit of ergodic sampling the ensemble created by standard molecular dynamics is the microcanonical or *NVE* ensemble, where in addition the total momentum is conserved [25, 468, 577, 1189]. Thus, the temperature is not a control variable in the Newtonian approach to molecular dynamics and hence it cannot be preselected and fixed. But it is evident that also within molecular dynamics the possibility of controlling the average temperature (as obtained from the average kinetic energy of the nuclei and the energy equipartition theorem) is welcome for physical reasons. A *deterministic* algorithm to achieve temperature control in the spirit of extended system dynamics [31] by a sort of dynamical friction mechanism was devised by Nosé and Hoover [651, 1063, 1064, 1066], see e.g. Refs. [25, 468, 577, 1065, 1189, 1382] for reviews of this well-established technique. Thereby, the canonical or *NVT* ensemble is generated in the case of ergodic dynamics.

As discussed in depth in Section 2.4, the Car–Parrinello approach to *ab initio* molecular dynamics works due to a dynamical separation between the physical and fictitious temperatures of the nuclear and electronic subsystems, respectively. This separability and thus the associated metastability condition breaks down if the electronic excitation gap becomes comparable to the thermal energy or smaller, that is in particular for metallic systems. Nvertheless, in order to satisfy adiabaticity in the sense of Car and

Parrinello it was proposed to couple separate thermostats [1379] to the classical fields that stem from the electronic degrees of freedom [146, 450]; see Refs. [144, 1007, 1030, 1510] for generalizations and improvements of the original idea. Finally, the (long-term) stability of the molecular dynamics propagation can be increased due to the same mechanism, which enables one to increase the time step that still allows for adiabatic time evolution [1510]. Note that these technical reasons to include additional thermostats are by construction absent from any Born–Oppenheimer molecular dynamics scheme.

It is well known that the standard Nosé–Hoover thermostat method suffers from non-ergodicity problems for certain classes of Hamiltonians, such as the harmonic oscillator [651]. A closely related technique, the Nosé–Hoover chain thermostat [927], cures that problem and assures ergodic sampling of phase space even for the pathological harmonic oscillator. This is achieved by thermostatting the original thermostat by another thermostat, which in turn is thermostatted and so on. In addition to restoring ergodicity even with only a few thermostats in the chain, this technique is found to be much more efficient in imposing the desired temperature. Furthermore, it has been demonstrated that this method ensures establishing the correct thermal distribution for both vibrational and rotational degrees of freedom in finite molecular systems [473].

Nosé–Hoover chain thermostatted Car–Parrinello molecular dynamics was introduced in Ref. [1510]. The underlying equations of motion read

$$M_I \ddot{\mathbf{R}}_I = -\nabla_I E^{\mathrm{KS}} - M_I \dot{\xi}_1 \dot{\mathbf{R}}_I \tag{5.1}$$

$$Q_1^{\mathrm{n}} \ddot{\xi}_1 = \left[\sum_I M_I \dot{\mathbf{R}}_I^2 - g k_{\mathrm{B}} T \right] - Q_1^{\mathrm{n}} \dot{\xi}_1 \dot{\xi}_2$$

$$Q_k^{\mathrm{n}} \ddot{\xi}_k = \left[Q_{k-1}^{\mathrm{n}} \dot{\xi}_{k-1}^2 - k_{\mathrm{B}} T \right] - Q_k^{\mathrm{n}} \dot{\xi}_k \dot{\xi}_{k+1} (1 - \delta_{kK}) \quad \text{where } k = 2, \ldots, K$$

for the nuclear part and

$$\mu \ddot{\phi}_i = -H_{\mathrm{e}}^{\mathrm{KS}} \phi_i + \sum_{ij} \Lambda_{ij} \phi_j - \mu \dot{\eta}_1 \phi_i \tag{5.2}$$

$$Q_1^{\mathrm{e}} \ddot{\eta}_1 = 2 \left[\sum_i^{\mathrm{occ}} \mu \langle \dot{\phi}_i | \dot{\phi}_i \rangle - T_{\mathrm{e}}^0 \right] - Q_1^{\mathrm{e}} \dot{\eta}_1 \dot{\eta}_2$$

$$Q_l^{\mathrm{e}} \ddot{\eta}_l = \left[Q_{l-1}^{\mathrm{e}} \dot{\eta}_{l-1}^2 - \frac{1}{\beta_{\mathrm{e}}} \right] - Q_l^{\mathrm{e}} \dot{\eta}_l \dot{\eta}_{l+1} (1 - \delta_{lL}) \quad \text{where } l = 2, \ldots, L$$

for the electronic contribution. These equations are written down in density functional language (see Eq. (2.114) and Eqs. (2.120)-(2.122) for the definitions of E^{KS} and H_e^{KS}, respectively), but completely analogous expressions are operational if other electronic structure approaches are used instead, such as the Hartree–Fock method. Using separate thermostatting baths $\{\xi_k\}$ and $\{\eta_l\}$, separate chains composed of K and L coupled thermostats are attached to the nuclear and electronic equations of motion, respectively.

By inspection of Eq. (5.1) it becomes intuitively clear how the thermostat works: $\dot{\xi}_1$ can be considered as a *dynamical* friction coefficient. The resulting "dissipative dynamics" leads to non-Hamiltonian flow, but the friction term can aquire positive or negative sign according to its equation of motion. This leads to damping or acceleration of the nuclei and thus to cooling or heating if the instantaneous kinetic energy of the nuclei is higher or lower than $k_B T$, which is preset. As a result, this extended system dynamics can be shown to produce a canonical ensemble in the subspace of the nuclear coordinates and momenta. In spite of being non-Hamiltonian, Nosé–Hoover (chain) dynamics is also distinguished by conserving an energy quantity of the extended system, see Eq. (5.5).

The desired average physical temperature is given by T, and g denotes the number of dynamical degrees of freedom to which the nuclear thermostat chain is coupled (i.e. constraints imposed on the nuclei have to be subtracted). Similarly, T_e^0 is the desired fictitious kinetic energy of the electrons and $1/\beta_e$ is the associated temperature. In principle, β_e should be chosen such that $1/\beta_e = 2T_e^0/N_e$. Here, N_e is the number of dynamical degrees of freedom needed to parameterize the wave function, i.e. N_{PW} in the plane wave scheme from Section 3.8, minus the number of constraint conditions on the orbitals. It is found that this choice requires a very accurate integration of the resulting equations of motion (for instance by using a high-order Suzuki–Yoshida integrator, see Section VI.A in Ref. [1510]). However, relevant quantities are rather insensitive to the particular value, so that N_e can be replaced heuristically by N_e', which is just the number of orbitals ϕ_i used to expand the wave function [1510], i.e. N_b according to the definition in Section 3.8.

The choice of the inertia parameters or "masses" assigned to the thermostat degrees of freedom should be made such that the overlap of their power spectra and those of the thermostatted subsystems is maximal [146, 1510]. The relations

$$Q_1^n = \frac{gk_B T}{\omega_n^2} , \qquad Q_k^n = \frac{k_B T}{\omega_n^2} \qquad (5.3)$$

$$Q_1^e = \frac{2T_e^0}{\omega_e^2}, \qquad Q_l^e = \frac{1}{\beta_e \omega_e^2} \tag{5.4}$$

ensure this if ω_n is a typical phonon or vibrational frequency of the nuclear subsystem (say of the order of 2000 to 4000 cm^{-1}) and ω_e is sufficiently large compared to the maximum frequency ω_n^{max} of the nuclear power spectrum (say 10 000 cm^{-1} or larger). The integration of these equations of motion is discussed in detail in Ref. [1510] using the velocity Verlet/RATTLE algorithm.

In some instances, for example during equilibration runs, it is advantageous to go one step further and actually couple one chain of Nosé–Hoover thermostats to every individual nuclear degree of freedom, akin to what has been introduced for path integral molecular dynamics simulations [1500, 1502, 1507]. This so-called "massive thermostatting method" is found to accelerate considerably the expensive equilibration periods within *ab initio* molecular dynamics, which is useful for both Car–Parrinello and Born–Oppenheimer dynamics. In addition, it makes sure that even stiff vibrational modes, which are essentially decoupled from all other modes, get thermally populated as required by the equipartition theorem of classical statistical mechanics [984]. In the case of *ab initio* path integral molecular dynamics, on the other hand, this sampling technique turns out to be mandatory in order to establish an ergodic molecular dynamics sampling of the path integral, as explained in Section 5.4.2.

In microcanonical classical molecular dynamics simulations two quantities are conserved during a simulation, the total energy and the total momentum. The same constants of motion apply to (exact) microcanonical Born–Oppenheimer molecular dynamics because the only *dynamical* variables are the nuclear positions and momenta as in classical molecular dynamics. In microcanonical Car–Parrinello molecular dynamics the total energy of the *extended* dynamical system composed of nuclear and electronic positions and momenta, that is E_{cons} as defined in Eq. (2.66), is also conserved, see e.g. Fig. 2.2 in Section 2.4. There is also a conserved energy quantity in the case of thermostatted molecular dynamics according to Eqs. (5.1)-(5.2). Instead of Eq. (2.66), this constant of motion ("conserved energy") reads

$$E_{cons}^{NVT} = \sum_i^{occ} \mu \left\langle \dot{\phi}_i \mid \dot{\phi}_i \right\rangle + \sum_I \frac{1}{2} M_I \dot{\mathbf{R}}_I^2 + E^{KS}\left[\{\phi_i\}, \{\mathbf{R}_I\}\right]$$

$$+ \sum_{l=1}^{L} \frac{1}{2} Q_l^e \dot{\eta}_l^2 + \sum_{l=2}^{L} \frac{\eta_l}{\beta_e} + 2T_e^0 \eta_1$$

$$+ \sum_{k=1}^{K} \frac{1}{2} Q_k^n \dot{\xi}_k^2 + \sum_{k=2}^{K} k_B T \xi_k + g k_B T \xi_1 \quad (5.5)$$

for Nosé–Hoover chain thermostatted canonical Car–Parrinello molecular dynamics [1510].

In microcanonical Car–Parrinello molecular dynamics the total nuclear momentum $\mathbf{P}_n$ is no longer a constant of motion as a result of the fictitious dynamics of the wave function; this quantity as well as other symmetries and associated invariants, is discussed in Ref. [1118]. However, a generalized linear momentum which embraces the electronic degrees of freedom

$$\mathbf{P}_{CP} = \mathbf{P}_n + \mathbf{P}_e = \sum_I \mathbf{P}_I + \sum_i^{occ} \mu \left\langle \dot{\phi}_i \middle| -\frac{1}{2} \nabla_r \middle| \phi_i \right\rangle + \text{c.c.} \quad (5.6)$$

can be defined [1030, 1118]; $\mathbf{P}_I = M_I \dot{\mathbf{R}}_I$. This quantity is a constant of motion in non-thermostatted Car–Parrinello molecular dynamics due to an exact cancellation of the nuclear and electronic contributions [1030, 1118]. As a result, the nuclear momentum $\mathbf{P}_n$ fluctuates during such a run, but in practice $\mathbf{P}_n$ is conserved *on the average* as shown in Fig. 1 of Ref. [1030]. This is analogous to the behavior of the physical total energy E_{phys} Eq. (2.67), which fluctuates slightly due to the presence of the fictitious kinetic energy of the electrons T_e Eq. (2.69).

As recently outlined in detail, it is clear that the coupling of more than one thermostat to a dynamical system, such as done in Eqs. (5.1)-(5.2), destroys the conservation of momentum [1030], i.e. $\mathbf{P}_{CP}$ is no longer an invariant. In unfavorable cases, in particular in small-gap or metallic regimes where there is a substantial coupling of the nuclear and electronic subsystems, momentum can be transferred to the nuclear subsystem such that $\mathbf{P}_n$ grows in the course of a simulation. This problem can be cured by controlling the nuclear momentum (using e.g. scaling or constraint methods), so that the total nuclear momentum $\mathbf{P}_n$ remains small [1030]. As an alternative, a feedback algorithm has also been proposed [144].

5.2.3 Imposing pressure: barostats

Keeping the pressure constant is a desirable feature for many applications of molecular dynamics simulations. The concept of barostats and thus constant-pressure molecular dynamics was introduced in the framework of extended system dynamics by Hans Andersen [31], see e.g. Refs. [25, 468, 577, 1189, 1382] for introductions. This method was devised to allow for isotropic fluctuations in the volume of the supercell. A powerful extension consists in also allowing for changes of the *shape* of the supercell to occur as a result of applying external pressure [1105–1107, 1608], including the possibility of non-isotropic *external* stress [1106]. The additional fictitious degrees of freedom in the Parrinello–Rahman approach [1105–1107] are the lattice vectors of the supercell, whereas the strain tensor is the dynamical variable in the Wentzcovitch approach [1608]. These variable-cell approaches make it possible to study dynamically structural phase transitions in solids at finite temperatures. With the birth of *ab initio* molecular dynamics both approaches were combined, starting out with isotropic volume fluctuations [195] *à la* Andersen [31] and followed by Born–Oppenheimer [1609, 1613] as well as Car–Parrinello [116, 118, 440, 441] variable-cell techniques.

The basic idea to allow for changes of the cell shape consists in constructing an extended Lagrangian, where the primitive Bravais lattice vectors $\mathbf{a}_1$, $\mathbf{a}_2$, and $\mathbf{a}_3$ of the simulation cell are additional dynamical variables similar to the thermostat degree of freedom ξ, see Eq. (5.1). Using the 3×3 matrix $\mathbf{h} = [\mathbf{a}_1, \mathbf{a}_2, \mathbf{a}_3]$ (which fully defines the cell with volume Ω), the real-space position $\mathbf{R}_I$ of a particle in this original cell can be expressed as

$$\mathbf{R}_I = \mathbf{h}\mathbf{S}_I \qquad (5.7)$$

where $\mathbf{S}_I$ is a scaled coordinate with components $S_{I,u} \in [0, 1]$ that defines the position of the Ith particle in a unit cube (i.e. $\Omega_{\text{unit}} = 1$), which is the scaled cell [1105, 1106]; see Section 3.1.1 for the underlying basics. The resulting metric tensor $\mathcal{G} = \mathbf{h}^{\mathrm{T}}\mathbf{h}$ converts distances measured in scaled coordinates to distances as given by the original coordinates according to Eq. (3.3) and periodic boundary conditions are applied using Eq. (3.4).

In the case of *ab initio* molecular dynamics the orbitals have to be expressed suitably in the scaled coordinates $\mathbf{s} = \mathbf{h}^{-1}\mathbf{r}$. The normalized original orbitals $\phi_i(\mathbf{r})$ as defined in the unscaled cell $\mathbf{h}$ are transformed according to

$$\phi_i(\mathbf{r}) = \frac{1}{\sqrt{\Omega}}\, \phi_i(\mathbf{s}) \qquad (5.8)$$

satisfying

$$\int_\Omega \phi_i^\star(\mathbf{r})\phi_i(\mathbf{r})\, d\mathbf{r} = \int_{\Omega_{\text{unit}}} \phi_i^\star(\mathbf{s})\phi_i(\mathbf{s})\, d\mathbf{s} \ , \qquad (5.9)$$

so that the resulting charge density is given by

$$n(\mathbf{r}) = \frac{1}{\Omega}\, n(\mathbf{s}) \qquad (5.10)$$

in the scaled cell, i.e. the unit cube. Importantly, the scaled fields $\phi_i(\mathbf{s})$ and thus their charge density $n(\mathbf{s})$ do *not* depend on the dynamical variables associated with the cell degrees of freedom and thus can be varied independently from the cell; the original unscaled fields $\phi_i(\mathbf{r})$ do depend on the cell variables $\mathbf{h}$, via the normalization by the cell volume $\Omega = \det \mathbf{h}$ as evidenced by Eq. (5.8).

After these preliminaries a variable-cell extended Lagrangian for *ab initio* molecular dynamics can be postulated [118, 440, 441]:

$$\mathcal{L} = \sum_i \mu \left\langle \dot{\phi}_i(\mathbf{s}) \middle| \dot{\phi}_i(\mathbf{s}) \right\rangle - E^{\mathrm{KS}} \left[\{\phi_i\}, \{\mathbf{h}\mathbf{S}_I\} \right]$$

$$+ \sum_{ij} \Lambda_{ij} \left(\langle \phi_i(\mathbf{s}) | \phi_j(\mathbf{s}) \rangle - \delta_{ij} \right)$$

$$+ \sum_I \frac{1}{2} M_I \left(\dot{\mathbf{S}}_I^{\mathrm{T}} \mathcal{G} \dot{\mathbf{S}}_I \right) + \frac{1}{2} W \, \mathrm{Tr} \, \dot{\mathbf{h}}^{\mathrm{T}} \dot{\mathbf{h}} - p \, \Omega \quad , \quad (5.11)$$

with an additional nine fictitious dynamical degrees of freedom that are associated with the lattice vectors of the supercell $\mathbf{h}$. This constant-pressure Lagrangian reduces to the constant-volume Car–Parrinello Lagrangian, see e.g. Eq. (2.56) or Eq. (2.78), in the limit $\dot{\mathbf{h}} \to 0$ of a rigid cell (apart from a constant term $p \, \Omega$). Here, p defines the externally applied hydrostatic pressure, W defines the fictitious mass or inertia parameter that controls the time scale of the motion of the cell $\mathbf{h}$, and the interaction energy E^{KS} is of the form that is defined in Eq. (2.114). In particular, this Lagrangian allows for symmetry-breaking fluctuations - which might be necessary to drive a solid-state phase transformation - to take place spontaneously. The resulting equations of motion read

$$M_I \ddot{S}_{I,u} = -\sum_{v=1}^{3} \frac{\partial E^{\mathrm{KS}}}{\partial R_{I,v}} \left(\mathbf{h}^{\mathrm{T}} \right)_{vu}^{-1} - M_I \sum_{v=1}^{3} \sum_{s=1}^{3} \mathcal{G}_{uv}^{-1} \dot{\mathcal{G}}_{vs} \dot{S}_{I,s} \quad (5.12)$$

$$\mu \ddot{\phi}_i(\mathbf{s}) = -\frac{\delta E^{\mathrm{KS}}}{\delta \phi_i^\star(\mathbf{s})} + \sum_j \Lambda_{ij} \phi_j(\mathbf{s}) \quad (5.13)$$

$$W \ddot{h}_{uv} = \Omega \sum_{s=1}^{3} \left(\Pi_{us}^{\mathrm{tot}} - p \, \delta_{us} \right) \left(\mathbf{h}^{\mathrm{T}} \right)_{sv}^{-1} \quad , \quad (5.14)$$

where the total internal stress tensor

$$\Pi_{us}^{\mathrm{tot}} = \frac{1}{\Omega} \sum_I M_I \left(\dot{\mathbf{S}}_I^{\mathrm{T}} \mathcal{G} \dot{\mathbf{S}}_I \right)_{us} + \Pi_{us} \quad (5.15)$$

is the sum of the thermal contribution due to nuclear motion at finite temperature and the electronic stress tensor [1055–1057] $\mathbf{\Pi}$ which is defined according to Eq. (3.88) and the following equations in Section 3.4.

Similar to the thermostat case discussed in the previous section one can recognize a sort of frictional feedback mechanism. The average internal pressure $\langle (1/3) \, \mathrm{Tr} \, \mathbf{\Pi}^{\mathrm{tot}} \rangle$ equals the externally applied pressure p as a result of maintaining dynamically a balance between $p \, \delta$ and the instantaneous internal stress $\mathbf{\Pi}^{\mathrm{tot}}$ by virtue of the friction coefficient $\propto \dot{\mathcal{G}}$ in Eq. (5.12). Ergodic trajectories obtained from solving the associated *ab initio* equations of motion Eqs. (5.12)–(5.14) lead to a sampling according to the isobaric–isoenthalpic or *NpH* ensemble. However, the generated dynamics is fictitious, similar to the constant-temperature case discussed in the previous section. The isobaric-isothermal or *NpT* ensemble is obtained by combining barostats and thermostats, see Ref. [928] for a general formulation and Ref. [930] for reversible integration schemes.

More recently, the metadynamics approach [821], see Section 5.2.4 for an introduction, has been combined [921] with the Parrinello–Rahman technique [1105–1107]. Choosing the lattice vectors of the supercell, i.e. the matrix $\mathbf{h}$ defined in Eq. (5.7), as collective variables $\mathbf{s}$ for metadynamics, the Gibbs free energy surface is sampled efficiently in the space of cell deformations by placing Gaussians in $\mathbf{h}$-space. As a result, free energies of phase transformations involving different crystal structures can be obtained close to equilibrium pressures, i.e. without over-pressurizing and hysteresis. However, it must be stressed that this combined technique [921] is not a constant-pressure simulation method such as the Parrinello–Rahman technique [1105–1107] just outlined, but rather a method for exploring the dependence of the Gibbs free energy on the cell parameters [248, 919, 921].

An important practical issue in *ab initio* molecular dynamics simulations using the Parrinello–Rahman variable-cell technique is related to basis set superposition error (BSSE) problems caused by using a finite basis set leading to "incomplete basis set" or Pulay-type contributions to the stress tensor, as explained in Section 2.5 in general terms. Using a finite plane wave basis (together with a finite number of $\mathbf{k}$-points) in the presence of a fluctuating cell [457, 530] one has the choice of either fixing the number of plane waves or keeping the energy cutoff constant during a simulation; see Eq. (3.18) for their relation according to a rule of thumb. A constant number of plane waves implies no Pulay stress but a decreasing precision of the calculation as the volume of the supercell increases, hence leading to a systematically biased (but smooth) equation of state. The constant cutoff procedure has better convergence properties toward the infinite-basis-set limit [530]. However, it produces in general unphysical discontinuities in the total energy and thus in the equation of state at volumes where the number of plane waves changes abruptly, see e.g. Fig. 5 in Ref. [457].

Computationally, the number of plane waves has to be fixed in the framework of Car–Parrinello variable-cell molecular dynamics [118, 195, 440, 441], whereas the energy cutoff can easily be kept constant in Born–Oppenheimer approaches to variable-cell molecular dynamics [1609, 1613]. Sticking to the Car–Parrinello technique, a practical remedy [118, 440] to this problem consists in modifying the electronic kinetic energy term Eq. (3.72) in a plane wave expansion Eq. (3.71) of the Kohn–Sham functional E^{KS} Eq. (2.114)

$$E_{\text{kin}} = \sum_i \sum_{\mathbf{q}} \frac{1}{2} |\mathbf{G}|^2 |c_i(\mathbf{q})|^2 \ , \tag{5.16}$$

where the unscaled $\mathbf{G}$ and scaled $\mathbf{q} = 2\pi\mathbf{g}$ reciprocal lattice vectors are interrelated via the cell $\mathbf{h}$ according to Eq. (3.8) (thus $\mathbf{Gr} = \mathbf{qs}$) and the cutoff Eq. (3.17) is defined as $(1/2) |\mathbf{G}|^2 \leq E_{\text{cut}}$ for a fixed number of $\mathbf{q}$-vectors, see Section 3.1. The modified kinetic energy at the Γ-point of the Brillouin zone associated with the supercell reads

$$\tilde{E}_{\text{kin}} = \sum_i \sum_{\mathbf{q}} \frac{1}{2} \left| \tilde{\mathbf{G}} \left(A, \sigma, E_{\text{cut}}^{\text{eff}} \right) \right|^2 |c_i(\mathbf{q})|^2 \tag{5.17}$$

$$\left| \tilde{\mathbf{G}} \left(A, \sigma, E_{\text{cut}}^{\text{eff}} \right) \right|^2 = |\mathbf{G}|^2 + A \left\{ 1 + \text{erf} \left[\frac{\frac{1}{2}|\mathbf{G}|^2 - E_{\text{cut}}^{\text{eff}}}{\sigma} \right] \right\} \ , \tag{5.18}$$

where A, σ, and $E_{\text{cut}}^{\text{eff}}$ are positive constants and the number of scaled vectors $\mathbf{q}$, that is the number of plane waves, is kept strictly fixed.

In the limit of a vanishing smoothing ($A \to 0; \sigma \to \infty$), the constant number of plane waves result is recovered. In the limit of a sharp step function ($A \to \infty; \sigma \to 0$), all plane waves with $(1/2) |\mathbf{G}|^2 \gg E_{\text{cut}}^{\text{eff}}$ have a negligible weight in $\tilde{E}_{\text{kin}}$ and are thus effectively suppressed. This situation mimics a constant cutoff calculation at an "effective cutoff" of $\approx E_{\text{cut}}^{\text{eff}}$ within a constant number of plane waves scheme. For this trick to work, note that $E_{\text{cut}} \gg E_{\text{cut}}^{\text{eff}}$ has to be satisfied. In the case $A > 0$ the electronic stress tensor $\mathbf{\Pi}$ given by Eq. (3.88) features an additional term (due to changes in the "effective basis set" as a result of variations of the supercell), which is related to the Pulay stress [470, 1547].

Finally, the strength of the smoothing $A > 0$ should be kept as modest as possible, since the modification Eq. (5.17) of the kinetic energy leads to an increase of the highest frequency in the electronic power spectrum $\propto A$. This implies a decrease of the permissible molecular dynamics time step Δt^{max} according to Eq. (2.73). It is found that a suitably tuned set of additional parameters $(A, \sigma, E_{\text{cut}}^{\text{eff}})$ leads to an efficiently converging constant-pressure scheme, in conjunction with a fairly small number of plane waves [118, 440]. Note that the cutoff was kept strictly constant in applications of the Born–Oppenheimer implementation [1611] of variable-cell molecular dynamics [1609, 1613], but the smoothing scheme presented here could be implemented in this case as well. An efficient method to correct for the discontinuities of *static* total energy calculations performed at constant cutoff was proposed in Ref. [457]. Evidently, the best way to deal with the incomplete-basis-set problem is to increase the cutoff such that the resulting artifacts become negligible on the physically relevant energy scale.

A constant-pressure Car–Parrinello *ab initio* molecular dynamics scheme suitable for studying pressure-induced structural transformations in finite nonperiodic systems such as clusters has been introduced in Ref. [922], alternative approaches are presented in Refs. [276, 1418]. The basic idea of establishing an isotropic compression is to surround the finite cluster by an *explicit* pressurizing medium that is described by N_{L} classical point particles, such as a liquid modeled by purely repulsive soft spheres. The corresponding Lagrangian reads, cf. Eq. (5.11)

$$
\mathcal{L} = \sum_I \frac{1}{2} M_I \dot{\mathbf{R}}_I^2 + \sum_i \mu \left\langle \dot{\phi}_i(\mathbf{r}) \middle| \dot{\phi}_i(\mathbf{r}) \right\rangle - E^{\text{KS}} \left[\{\phi_i\}, \{\mathbf{R}_I\} \right]
$$

$$
+ \sum_{ij} \Lambda_{ij} \left(\langle \phi_i(\mathbf{r}) | \phi_j(\mathbf{r}) \rangle - \delta_{ij} \right) + \sum_\alpha \frac{1}{2} M_\alpha \dot{\mathbf{X}}_\alpha^2
$$

$$
- \sum_{I,\alpha} V_{\text{C–L}} \left(|\mathbf{R}_I - \mathbf{X}_\alpha| \right) - \sum_{\alpha < \beta} V_{\text{L–L}} \left(|\mathbf{X}_\alpha - \mathbf{X}_\beta| \right) \quad , \quad (5.19)
$$

where M_α is the mass of a liquid particle at position $\mathbf{X}_\alpha$ and $V_{\text{C–L}}$ as well as $V_{\text{L–L}}$ are model pair potentials that describe the interactions between the cluster and the liquid and those within the liquid. Note that the coordinates of the classical particles do not enter the electronic structure, i.e. E^{KS} does not depend on them. Thus, this approach to constant-pressure simulations can be thought of as a simple QM/MM coupling scheme in the sense of "mechanical embedding", see Section 5.5.1.

The liquid is enclosed in a box of volume Ω_{L} and subject to periodic boundary conditions, whereas cluster boundary conditions as introduced in Section 3.2.3 are used for the quantum cluster; dangling bonds are saturated by hydrogen capping atoms. How is the pressure controlled? The equation of state, $p(T)$, of purely repulsive soft spheres of the kind

$$V_{L-L}(r) = \epsilon_{L-L}\left(\frac{\sigma_{L-L}}{r}\right)^{12} \tag{5.20}$$

is known in terms of the reduced number density $\tilde{\rho}$

$$p = \frac{N_L k_B T}{\Omega_L}\, \xi(\tilde{\rho}) \tag{5.21}$$

$$\xi(\tilde{\rho}) = \frac{N_L}{\Omega_L}\, \frac{\sigma_{L-L}^3}{\sqrt{2}}\, \left(\frac{\epsilon_{L-L}}{k_B T}\right)^{1/4} \tag{5.22}$$

as a function of σ_{L-L}; the function $\xi(x)$ is available from simulations [922] and $\epsilon_{L-L} = 1$. Thus, keeping Ω_L and N_L fixed allows us to conveniently adjust the pressure for a given temperature by varying the interaction parameter ϵ_{L-L} without changing particle number and/or volume of the classical box. The isotropic pressure established thereby in the bulk liquid is transferred onto the cluster via a similar purely repulsive pairwise additive potential, $V_{C-L}(r) = \epsilon_{C-L}(\sigma_{C-L}/r)^{12}$, with suitably chosen interaction parameters. This technique has been applied to study pressure-driven transformations of silicon nanocrystals [918, 922, 923, 1022].

5.2.4 Sampling rare events and free energies: metadynamics

Studying rare events, i.e. those that occur "once in a blue moon" [230], on the time scale of typical molecular dynamics simulations, finding dynamical paths connecting reactants to products, and mapping free energy surfaces of dynamical systems is a formidable task. This is particularly true when using *ab initio* molecular dynamics to study "large" barrier-activated processes since this technique is, compared to parameterized methods, restricted to rather short trajectories in view of the burden of performing full electronic structure calculations at each propagation step. Various methods have been developed in order to address this limitation, several of which were used in conjunction with CPMD [696] or are implemented therein, such as "chemical flooding" [565, 1040], "nudged elastic band" [631, 632, 1224, 1279], "string MD" [376, 725, 1279], "action-derived MD" [11, 292, 1114–1116], "multiple steering" [231], "bias potentials" [1554], "transition path sampling" [173, 321, 496, 497], "targeted MD" [909, 1287], or various other techniques [1036, 1555, 1556]. A detailed discussion of how to implement holonomic constraints, which is the technical essence of the "Blue Moon" ensemble technique [230, 1386], but also a major ingredient of various other techniques, is given in Section 3.7.3.

An efficient and practical approach, the metadynamics technique [821], is obtained by combining a coarse graining of the underlying microscopic dynamics in conjunction with a time-dependent bias potential; various related methods exist in the literature, as outlined in Ref. [821]. In the following, the elegant and rather easy-to-use extended Lagrangian formulation [667] of metadynamics is presented, see Ref. [393] for a review. Crucial to the method is the idea of preventing the system revisiting areas in configuration space where it has been in the past. This long-term memory is slowly built up by growing an external bias potential as the system is propagated in phase space. In particular, the history-dependent bias potential is modeled by Gaussians that are dropped during propagation in the coarse-grained space spanned by a set of collective variables. Thus, starting in a deep (free) energy minimum, which is characterized by a high phase space density and thus long residence times, results in filling this basin preferentially by increasing its potential energy artificially. Once this minimum is filled the trajectory might escape the basin, by crossing a former barrier region or some "transition state", and explore another basin. This is akin to backwater filling a new reservoir in the mountains after closing a dam. The resulting dynamics is genuinely non-Markovian and thus non-Newtonian and non-conservative.

Secondly, it is crucial to introduce a small set of (scaled and dimensionless, see below) collective coordinates, $\mathbf{S} = \{S_1, \ldots, S_\alpha, \ldots, S_{N_s}\}$, which are defined in terms of the Cartesian coordinates of the nuclei, $\mathbf{R} = \{\mathbf{R}_1, \ldots, \mathbf{R}_I, \ldots, \mathbf{R}_N\}$, at time t, i.e. $\{S_\alpha(\mathbf{R}(t))\}$. Typical collective coordinates are those used in molecular dynamics subject to holonomic constraints, see Section 3.7.3, such as bond distances and (dihedral) angles or coordination numbers [1385]. The number of collective coordinates must be much smaller than the dimension of configuration space, i.e. $N_s \ll N$, thus resulting in a significant coarse graining of the underlying microscopic dynamics. Only this reduction of the dimensionality makes (numerically) possible the filling of free energy minima by placing a manageable number of Gaussians, which successively build up the time-dependent memory. On the other hand, the set of collective coordinates must be flexible enough to fully describe the process of interest, including the reactants and products in the case of chemical reactions, i.e. it should span the relevant reaction coordinate space as far as possible. Finally, this manifold must also include all relevant "slow modes", which are those that cannot be sampled within the typical time scale of the simulation technique used.

The extended Lagrangian for metadynamics as a supplement to the Car–Parrinello Lagrangian $\mathcal{L}_{CP}$, Eq. (2.56), reads

$$\mathcal{L} = \mathcal{L}_{\mathrm{CP}} + \sum_{\alpha=1}^{N_{\mathrm{s}}} \frac{1}{2} \mu_{\alpha}^{\mathrm{s}} \dot{s}_{\alpha}^{2}(t)$$

$$- \sum_{\alpha=1}^{N_{\mathrm{s}}} \frac{1}{2} k_{\alpha} \left[S_{\alpha} \left(\mathbf{R}(t) \right) - s_{\alpha}(t) \right]^{2} + V \left(t, [\mathbf{s}] \right) , \quad (5.23)$$

where the second term is the fictitious kinetic energy associated with the fictitious variables $\{\mathbf{s}_{\alpha}\}$; note that $\mathcal{L}_{\mathrm{CP}}$ could be any Lagrangian $\mathcal{L}_{0}$ suitable to describe the system of interest. These variables are the auxiliary degrees of freedom with associated fictitious masses $\{\mu_{\alpha}^{\mathrm{s}}\}$, extending the standard Lagrangian $\mathcal{L}_{\mathrm{CP}}$, which are introduced to continuously explore the space of collective coordinates $\{\mathbf{S}_{\alpha}\}$. The third term in Eq. (5.23) is a harmonic potential that restrains the value of the instantaneous collective coordinates $\mathbf{S}$ close to the corresponding dynamic auxiliary variables $\mathbf{s}$. Most important is the history- and thus time-dependent potential, the non-Markovian metapotential

$$V \left(t, [\mathbf{s}] \right) = \int_{t_0}^{t} \left| \dot{\mathbf{s}}(t') \right| W(t') \exp \left[- \frac{(\mathbf{s}(t) - \mathbf{s}(t'))^{2}}{2 \left(\Delta s^{\perp} \right)^{2}} \right]$$

$$\times \, \delta \left(\frac{\dot{\mathbf{s}}(t')}{|\dot{\mathbf{s}}(t')|} \left(\mathbf{s}(t) - \mathbf{s}(t') \right) \right) dt' , \quad (5.24)$$

which is a functional of the entire path $[\mathbf{s}]$ in the space of auxiliary variables, $\mathbf{s}(t') = \{s_1(t'), \ldots, s_{\alpha}(t'), \ldots, s_{N_{\mathrm{s}}}(t')\}$ for all $t' \in [t_0, t]$. This potential describes an N_{s}-dimensional *Gaussian tube* in $\mathbf{s}$-space whose width is governed by the parameter $\Delta s^{\perp}$ with weight given by $W(t')$; a simplified *purely Gaussian* metapotential,

$$\tilde{V} \left(t, [\mathbf{s}] \right) = \int_{t_0}^{t} W(t') \exp \left[- \frac{(\mathbf{s}(t) - \mathbf{s}(t'))^{2}}{2 \left(\Delta s^{\perp} \right)^{2}} \right] dt' , \quad (5.25)$$

is readily obtained. A discretized version of the tube potential Eq. (5.24) is derived after representing the delta-function by a sufficiently narrow Gaussian approximant

$$V \left(t, [\mathbf{s}] \right) \approx V \left(t, \{\mathbf{s}(t_i)\} \right) = \sum_{i=0}^{imax} W(t_i) \exp \left[- \frac{(\mathbf{s}(t) - \mathbf{s}(t_i))^{2}}{2 \left(\Delta s^{\perp} \right)^{2}} \right]$$

$$\times \exp \left[- \frac{[(\mathbf{s}(t_{i+1}) - \mathbf{s}(t_i)) \cdot (\mathbf{s}(t) - \mathbf{s}(t_i))]^{2}}{2 |\mathbf{s}(t_{i+1}) - \mathbf{s}(t_i)|^{4}} \right] \quad (5.26)$$

with time order $t_0 < t_1 < t_2 < \cdots < t_i < \cdots < t_{imax} < t$ and meta time step $\Delta_s t = t_{i+1} - t_i \; \forall i$. That is, the metapotential V at time t is successively constructed by superimposing a series of Gaussians centered at $\mathbf{s}(t_0)$, $\mathbf{s}(t_1)$, ..., $\mathbf{s}(t_{imax})$ in the visited region separated by discrete time intervals $\Delta_s t \gg \Delta t$. The width of the Gaussians orthogonal to the direction of motion in $\mathbf{s}$-space is given by a fixed parameter, $\Delta s^{\perp}$, whereas the width along this direction, $\Delta s^{\parallel}(t_i) = |\mathbf{s}(t_{i+1}) - \mathbf{s}(t_i)|$, is adapted to the topology of the surface at each meta time step. The prefactor that determines the magnitude or height of the Gaussians can be either fixed, $W(t_i) = \bar{W}$, or made adaptive

$$W(t_i) = \eta \sum_{\alpha} [s_{\alpha}(t_{i+1}) - s_{\alpha}(t_i)] \left\langle k_{\alpha}[S_{\alpha}(\mathbf{R}(t'')) - s_{\alpha}(t'')] \right\rangle_{t'' \in [t_i, t_{i+1}]} , \quad (5.27)$$

where $\langle \ldots \rangle_{t'' \in [t_i, t_{i+1}]}$ denotes a sliding average over the meta time step $\Delta_s t$ and $0 < \eta < 1$. Finally, the scaling of the collective coordinates $\{S_{\alpha}(\mathbf{R})\}$ is done such that the maximum amplitude of their fluctuations, $\max |S_{\alpha}(\mathbf{R}) - \langle S_{\alpha}(\mathbf{R}) \rangle|$, is similar (unity) for all α independently of their geometric nature.

The corresponding equations of motion are obtained from Eq. (5.23) together with Eq. (5.26) as usual

$$\frac{d}{dt} \frac{\partial \mathcal{L}}{\partial \dot{s}_{\alpha}} = \frac{\partial \mathcal{L}}{\partial s_{\alpha}} \tag{5.28}$$

$$\frac{d}{dt} \frac{\partial \mathcal{L}_{CP}}{\partial \dot{\mathbf{R}}_I} = \frac{\partial \mathcal{L}_{CP}}{\partial \mathbf{R}_I} \tag{5.29}$$

$$\frac{d}{dt} \frac{\delta \mathcal{L}_{CP}}{\delta \dot{\phi}_i^{\star}} = \frac{\delta \mathcal{L}_{CP}}{\delta \phi_i^{\star}} \tag{5.30}$$

from the Euler–Lagrange equations. Solving these equations generates a coarse-grained dynamics in the extended phase space of the auxiliary variables, $\{s_{\alpha}, \dot{s}_{\alpha}\}$ if they are sufficiently well adiabatically separated from all other degrees of freedom [822]. Thus, special care has to be taken in order to keep their vibrational density of states separated from that of both nuclei and electrons. It must be stressed that the trajectories generated on the manifold of the collective coordinates do not reflect the physical real-time dynamics, rather the fictitious dynamics is only used in order to efficiently explore the free energy surface in this most relevant subspace in the spirit of an extended Lagrangian scheme. Note that the original approach [821] was based on an iterative sampling strategy of the fast modes using separate molecular dynamics runs in order to sample explicitly the coarse-grained forces,

$$\frac{\partial \mathcal{F}}{\partial s_\alpha} = - \lim_{t \to \infty} \langle k_\alpha \left[S_\alpha \left(\mathbf{R}(t) \right) - s_\alpha(t) \right] \rangle \quad . \tag{5.31}$$

Closer analysis [822] of the Lagrangian formulation Eq. (5.23) uncovered that the adiabatic separation condition, i.e. sufficiently small fictitious masses $\{\mu_\alpha^s\}$ as explicitly invoked earlier [667], is not strictly required. However, it is important that the Gaussians are added sufficiently slowly in order to fill the free energy minima properly [822]. Thus, the meta time step $\Delta_s t$, i.e. the "dropping rate" of the Gaussians, must be much larger than the (*ab initio*) molecular dynamics time step, Δt, used to produce the physical fluctuations by evolving the microscopic equations of motion; typical values are $\Delta_s t = 10 - 100 \Delta t$.

In addition to exploring possible reaction pathways, the metadynamics approach also allows us to map out the free energy surface [821] in the chosen subspace spanned by $\mathbf{s}$. The latter can be obtained from the coarse-grained Boltzmann distribution function

$$\mathcal{F}(\mathbf{s}) = -k_B T \ln P(\mathbf{s}) \tag{5.32}$$

using the constrained canonical partition function

$$P(\mathbf{s}) = \frac{1}{\mathcal{Z}} \prod_\alpha \left[\delta \left(S_\alpha(\mathbf{R}) - s_\alpha \right) \right] \exp \left[-\frac{E^{KS}(\mathbf{R})}{k_B T} \right] d\mathbf{R} \quad , \tag{5.33}$$

where $\mathcal{Z}$ is the corresponding unconstrained partition function; note that $\mathcal{Z}$ and thus the proper normalization of $P(\mathbf{s})$ is not easily accessible in simulations, so that $\mathcal{F}(\mathbf{s})$ is defined only up to an additive constant. As argued in Ref. [821], and shown in Ref. [208] the free energy profile can be estimated conveniently from the accumulated metapotential Eq. (5.26)

$$\Delta \mathcal{F}(\mathbf{s}) = - \lim_{t \to \infty} V \left(t, \{ \mathbf{s}(t_i) \} \right) \tag{5.34}$$

in the limit of sufficiently long simulations. The accuracy of filling and probing free energy surfaces can be analyzed in terms of the simplified purely Gaussian metapotential Eq. (5.25) with fixed height, $W(t_i) = \bar{W}$, which allows us to derive handy error estimates such as

$$\varepsilon \propto \sqrt{\frac{\bar{W} \Delta s^\perp}{\Delta_s t}} \tag{5.35}$$

in terms of metadynamics parameters [822]; here the system-specific and temperature-dependent prefactor is omitted. Note that ε does not depend on μ_α^s and k_α, assuming that the fictitious dynamics is done properly in the coarse-grained space. A "recipe" for choosing the parameters that govern the metadynamics in the coarse-grained space, i.e. the fictitious masses μ_α^s and restraining force constants k_α, and those that control the convergence of the free energy sampling, i.e. the heights $\bar{W}$ and widths $\Delta s^\perp$ of the Gaussians in addition to the meta time step $\Delta_s t$, is provided in Ref. [392]. In particular, μ_α^s and k_α should be fixed for a specific system and set of collective coordinates in a preliminary short run in the coarse-grained space *without* dropping Gaussians, which also allows us to choose $\Delta s^\perp$ based on the resulting fluctuations of the collective variables in the reactant well. The ratio of height and dropping rate, $\bar{W}$ and $\Delta_s t$, determines the filling efficiency and ultimately increases the accuracy in the limit $\bar{W}/\Delta_s t \to 0$ of slowly added small Gaussians according to Eq. (5.35). Thus, the reconstruction of the free energy landscape can be performed in low- and high-resolution limits depending on the actual task. Finally, having mapped out the lowest free energy path and thus identified the most probable reaction path, standard methods such as umbrella sampling can be used to further refine the free energy profile along the one-dimensional reaction coordinate in terms of postprocessing [392].

The metadynamics method has already been applied to various problems in chemistry and chemical physics [269, 393, 499, 668, 1047, 1403]. In addition, it allows us to compute efficiently the energy density of states for systems described by classical statistical mechanics [1003], which might open up a new approach to simulate phase transitions.

5.3 Beyond ground states: ROKS, surface hopping, FEMD, TDDFT

5.3.1 Introduction

Extending *ab initio* molecular dynamics to a single excited state that does not couple to any other electronic state is straightforward in the framework of wave function-based methods such as Hartree–Fock [423, 551, 605, 607, 620, 707, 853, 896], generalized valence bond (GVB) [503, 504, 506, 604, 606], complete active space SCF (CASSCF) [300, 301], or full configuration interaction (FCI) [872] approaches, see Section 2.7. However, these methods are computationally quite demanding - given present-day algorithms and hardware. Impressive progress in the direction of including several excited states and nonadiabatic couplings between them have also been made in the realm of wave function-based methods, see for instance Refs. [138, 236, 277, 657, 752, 914, 916, 917, 1000, 1569, 1632] and Refs. [82, 96, 1002, 1217, 1633] for reviews.

Density functional theory offers an alternative route to solving electronic structure problems approximately and recent approaches to excited state *dynamics* within this framework look promising, see Ref. [355] for a review focusing on this very topic. Here, the following important classes of scenarios will be considered:

(i) A single excited many-electron state of well-defined symmetry is completely decoupled from all other states on the relevant time scale and thus evolves adiabatically, see Sections 5.3.2 and 5.3.6.

(ii) Two or a few such many-electron states interact explicitly via nonadiabatic couplings, see Section 5.3.3.

(iii) Many closely spaced one-electron states are populated according to a broad thermal distribution via fractional occupation numbers without specifying symmetry, see Section 5.3.4.

(iv) The time-dependent Kohn–Sham equations are propagated in real time without controlling the symmetry of the many-electron state, see Section 5.3.5.

(v) Gradients are obtained for an excited state of specified symmetry from the linear response approximation to time-dependent density functional theory, see Section 5.3.6.

The first situation applies, for instance, to large-gap molecular systems which undergo an ultrafast chemical reaction in a single excited state as a result of a vertical HOMO/LUMO or instantaneous one-particle/one-hole photoexcitation, the initial steps of *cis–trans* isomerizations and excited state proton transfers (ESPTs) being prominent examples. The second scenario, (ii), is a nontrivial generalization of the first one to cases where decay from a higher-lying state to a lower-lying state is crucial, important cases being radiationless decay mechanisms via conical intersections as encountered after *cis–trans* isomerization or excited state proton transfer reactions have taken place. Many excited states, (iii), are typical for metallic systems where states beyond the Fermi edge must be taken into account via fractional occupation numbers, or for materials at high temperatures compared to the Fermi temperature. It is already stressed at this stage that imposing the thermal Fermi–Dirac distribution via free energy functional or ensemble density functional approaches [363, 1102] leads to *incoherent* excitations without a proper time evolution of the individual state populations.

The advent of density functional theory for time-dependent systems [42, 559, 910, 912, 1255, 1542] greatly fostered the field of excited state simulations. Two major routes to *ab initio* molecular dynamics simulations relying on time-dependent density functional theory (TDDFT) exist. In the most direct implementation, (iv), the effective one-particle time-dependent

Kohn–Sham equations are propagated in real time. In this density functional theory-based Ehrenfest approach, see Section 2.2, specifying the symmetry of the excited state(s) is difficult as an effective "average state" is created according to Eq. (2.38) together with Eq. (2.39), in particular after coupling to an external electromagnetic field to model an initial laser excitation. In a complementary approach, (v), derivatives of excited state energies are obtained within the linear response approximation to time-dependent Kohn–Sham density functional theory formulated in the frequency domain, thus providing nuclear gradients for molecular dynamics. In this case, the symmetry of the excited state(s) is defined, whereas the linear response approximation requires the excited state(s) to be "sufficiently similar" to the ground state, in particular in relation to its multi-configuration character. It is noted that these excited state techniques have been implemented within the CPMD package [696].

5.3.2 *A single excited state: ROKS dynamics*

For large-gap systems with well separated electronic states it might be desirable to single out a particular state in order to allow the nuclei to move on the associated excited state potential energy surface. Approaches that rely on fractional occupation numbers, such as ensemble density functional theories - including the free energy functional discussed in the previous section - are difficult to adapt for cases where the symmetry and/or spin of the electronic state should be fixed [363]. An early approach in order to select a particular excited state was based on introducing a "quadratic restoring potential" which vanishes only at the eigenvalue of the particular state [221, 974].

A method that combines Roothaan's symmetry-adapted wave functions [1227] with Kohn–Sham density functional theory was proposed in Ref. [462] and used to simulate a photoisomerization pathway via molecular dynamics, see Refs. [554, 1077] for further analysis of the method and for refinement. Viewed from Kohn–Sham theory, this approach consists in building up the spin density of an open-shell system based on a symmetry-adapted "wave function" that is constructed from spin-restricted determinants, see also Ref. [545]. Viewed from the restricted open-shell Hartree–Fock (ROHF) theory *à la* Roothaan, it amounts essentially to replacing Hartree–Fock exchange by an approximate exchange–correlation density functional, thus the name restricted open-shell Kohn–Sham (ROKS) method. Thus, the central idea is to impose symmetry constraints on the many-electron wave function by constructing a symmetry-adapted *multi-determinantal* wave function,

$$\Psi_{SA} = \sum_\mu a_\mu \Phi_\mu \; , \tag{5.36}$$

from spin-restricted single Slater determinants, the "microstates" Φ_μ. Following Roothaan [1227], the expansion coefficients, a_μ, are chosen to be the Clebsch-Gordan coefficients for a given symmetry. The energy is then given by

$$E_{SA} = \sum_\mu c_\mu(\{a_\mu\})E_\mu \quad \text{with} \quad \sum_\mu c_\mu = 1 \; , \tag{5.37}$$

where E_μ is the total energy of the μth microstate. Within density functional theory this procedure leads to an explicitly orbital-dependent functional which was formulated for the first excited singlet state S_1 in Ref. [462]. The relation of this approach to previous theories is discussed in some detail in Ref. [462]. In particular, the success of the closely related Ziegler-Rauk-Baerends ΔSCF approach [1665], the so-called "sum method" [307, 1408, 1665], was an important stimulus.

More recently, several papers [425–428, 548, 549, 626, 1046] appeared that are similar in spirit to the method of Ref. [462]. In particular the approach of Ref. [425] and its subsequent refinements can be viewed as a generalization to arbitrary spin states of the special case (applicable to the S_1 state only) worked out in Ref. [462]. These more general methods [425–428] were derived within the framework of ensemble density functional theory, whereas a wave function perspective was the starting point in Ref. [462]. In addition to being more general, the ensemble density functional framework makes clear the point that certain symmetry restrictions apply which constrain the constants that weight the different microstates.

However, it should be acknowledged that is was demonstrated more recently that there is no Hohenberg–Kohn theorem for excited states [491]. Thus, *rigorously speaking*, no *exact* excited state functional of the excited state density exists even for a given level of excitation, whereas the excited state energy is clearly a functional of the *ground state* density by virtue of the Hohenberg–Kohn theorem [363, 760, 762, 913, 1102]. In addition, the notion of a "many-body wave function" is ill-defined in density functional methods which implies very fundamental conceptual difficulties of all approaches that mix in ideas stemming from the realm of wave function-based methods. On the other hand, there is definitively now a wealth of supporting evidence that *pragmatic approaches* to excited states within density functional theory are useful practical tools to extend *ab initio* molecular dynamics beyond the ground state.

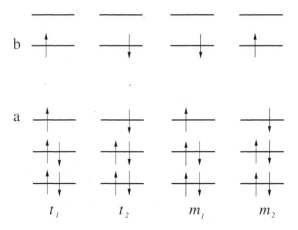

Fig. 5.1. Four possible determinants $|t_1\rangle$, $|t_2\rangle$, $|m_1\rangle$, and $|m_2\rangle$ as a result of the promotion of a single electron from the HOMO to the LUMO of a closed-shell system yielding the singly occupied orbitals ϕ_a and ϕ_b, see text for further details. Taken from Ref. [462].

   In the following, the ROKS method is outlined with the focus on performing molecular dynamics in the S_1 state. Promoting one electron from the HOMO to the LUMO ("particle-hole excitation") in a closed-shell system with $2n$ electrons assigned to n doubly occupied orbitals (that is spin-restricted orbitals that have the same spatial part for both spin-up α and spin-down β electrons) leads to four different excited wave functions or determinants, see Fig. 5.1 for a sketch. Two states $|t_1\rangle$ and $|t_2\rangle$ are energetically degenerate triplets t, whereas the two states $|m_1\rangle$ and $|m_2\rangle$ are not eigenfunctions of the total spin operator, $\hat{S}^2$, and thus degenerate "mixed states" m. Note that the m states do not correspond - as is well known - to singlet states due to significant spin contamination despite the suggestive occupation pattern in Fig. 5.1 and the fact that they are called "broken symmetry states" mainly in the literature on magnetic interactions [677, 1062, 1143].

   However, suitable Clebsch-Gordan projections of the mixed states $|m_1\rangle$ and $|m_2\rangle$ yield another triplet state $|t_3\rangle$ and the desired first excited singlet or S_1 state $|s_1\rangle$. Here, the ansatz [462] for the total energy of the S_1 state is given by

$$E_{S_1}^{\mathrm{ROKS}}\left[\{\phi_i\}\right] \;=\; 2E_m^{\mathrm{KS}}\left[\{\phi_i\}\right] - E_t^{\mathrm{KS}}\left[\{\phi_i\}\right] \;, \tag{5.38}$$

where the energies of the mixed and triplet determinants

$$E_m^{\mathrm{KS}}[\{\phi_i\}] = T_{\mathrm{s}}[\{\phi_i\}] + \int V_{\mathrm{ext}}(\mathbf{r})n(\mathbf{r})\,d\mathbf{r}$$

$$+ \frac{1}{2}\int V_{\mathrm{H}}(\mathbf{r})n(\mathbf{r})\,d\mathbf{r} + E_{\mathrm{xc}}[n_m^{\alpha}, n_m^{\beta}] \quad (5.39)$$

$$E_t^{\mathrm{KS}}[\{\phi_i\}] = T_{\mathrm{s}}[\{\phi_i\}] + \int V_{\mathrm{ext}}(\mathbf{r})n(\mathbf{r})\,d\mathbf{r}$$

$$+ \frac{1}{2}\int V_{\mathrm{H}}(\mathbf{r})n(\mathbf{r})\,d\mathbf{r} + E_{\mathrm{xc}}[n_t^{\alpha}, n_t^{\beta}] \quad (5.40)$$

are expressed in terms of (restricted) Kohn–Sham spin-density functionals constructed from the set $\{\phi_i\}$, cf. Eq. (2.114). The associated S_1 "wave function" is given by

$$|s_1[\{\phi_i\}]\rangle = \frac{1}{\sqrt{2}}\{|m_1[\{\phi_i\}]\rangle + |m_2[\{\phi_i\}]\rangle\}\ , \quad (5.41)$$

whereas the antisymmetric linear combination yields

$$|t_3[\{\phi_i\}]\rangle = \frac{1}{\sqrt{2}}\{|m_1[\{\phi_i\}]\rangle - |m_2[\{\phi_i\}]\rangle\} \quad (5.42)$$

a third triplet "wave function". Importantly, the "microstates" m_1 and m_2 are both constructed from the same set $\{\phi_i\}$ of $n+1$ spin-*restricted* orbitals. Using this particular set of orbitals the total density

$$n(\mathbf{r}) = n_m^{\alpha}(\mathbf{r}) + n_m^{\beta}(\mathbf{r}) = n_t^{\alpha}(\mathbf{r}) + n_t^{\beta}(\mathbf{r}) \quad (5.43)$$

is of course identical for both the m and t determinants, whereas their spin densities clearly differ, see Fig. 5.2. Thus, the decisive difference between the m and t functionals Eq. (5.39) and Eq. (5.40), respectively, comes exclusively from the exchange–correlation functional E_{xc}, whereas kinetic, external, and Hartree energy are identical by construction. Note that this basic philosophy can be generalized to other spin states by adapting the microstates and the corresponding coefficients suitably in Eq. (5.38) and Eq. (5.41).

Having defined a density functional for the first excited singlet state, the corresponding Kohn–Sham equations are obtained by varying the ROKS functional Eq. (5.38)

$$\frac{\delta}{\delta\phi_k^{\star}}\left\{2E_m^{\mathrm{KS}}[\{\phi_i\}] - E_t^{\mathrm{KS}}[\{\phi_i\}] - \sum_{i,j=1}^{n+1}\Lambda_{ij}(\langle\phi_i\,|\,\phi_j\rangle - \delta_{ij})\right\} = 0 \quad (5.44)$$

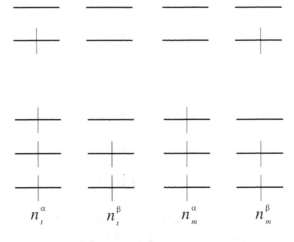

Fig. 5.2. Four patterns of spin densities n_t^α, n_t^β, n_m^α, and n_m^β corresponding to the two spin-restricted determinants $|t\rangle$ and $|m\rangle$ sketched in Fig. 5.1, see text for further details. Taken from Ref. [462].

subject to the orthonormality constraint. Following this procedure, the equation for the doubly occupied orbitals $i = 1, \ldots, n-1$ reads

$$
\left\{ -\frac{1}{2}\nabla^2 + V_{\mathrm{H}}(\mathbf{r}) + V_{\mathrm{ext}}(\mathbf{r}) \right.
$$

$$
+ V_{\mathrm{xc}}^\alpha[n_m^\alpha(\mathbf{r}), n_m^\beta(\mathbf{r})] + V_{\mathrm{xc}}^\beta[n_m^\alpha(\mathbf{r}), n_m^\beta(\mathbf{r})] - \frac{1}{2}V_{\mathrm{xc}}^\alpha[n_t^\alpha(\mathbf{r}), n_t^\beta(\mathbf{r})]
$$

$$
\left. -\frac{1}{2}V_{\mathrm{xc}}^\beta[n_t^\alpha(\mathbf{r}), n_t^\beta(\mathbf{r})] \right\} \phi_i(\mathbf{r}) = \sum_{j=1}^{n+1} \Lambda_{ij}\phi_j(\mathbf{r}) \quad (5.45)
$$

whereas

$$
\left\{ \frac{1}{2}\left[-\frac{1}{2}\nabla^2 + V_{\mathrm{H}}(\mathbf{r}) + V_{\mathrm{ext}}(\mathbf{r}) \right] + V_{\mathrm{xc}}^\alpha[n_m^\alpha(\mathbf{r}), n_m^\beta(\mathbf{r})] \right.
$$

$$
\left. -\frac{1}{2}V_{\mathrm{xc}}^\alpha[n_t^\alpha(\mathbf{r}), n_t^\beta(\mathbf{r})] \right\} \phi_a(\mathbf{r}) = \sum_{j=1}^{n+1} \Lambda_{aj}\phi_j(\mathbf{r}) \quad (5.46)
$$

and

$$
\left\{ \frac{1}{2}\left[-\frac{1}{2}\nabla^2 + V_{\mathrm{H}}(\mathbf{r}) + V_{\mathrm{ext}}(\mathbf{r}) \right] + V_{\mathrm{xc}}^\beta[n_m^\alpha(\mathbf{r}), n_m^\beta(\mathbf{r})] \right.
$$

$$-\frac{1}{2}V_{\mathrm{xc}}^{\alpha}[n_t^{\alpha}(\mathbf{r}), n_t^{\beta}(\mathbf{r})]\Big\}\phi_{\mathrm{b}}(\mathbf{r}) = \sum_{j=1}^{n+1}\Lambda_{\mathrm{b}j}\phi_j(\mathbf{r}) \quad (5.47)$$

are two *different* equations for the two singly occupied open-shell orbitals ϕ_{a} and ϕ_{b}, respectively, see Fig. 5.1. Note that these Kohn–Sham-like equations

$$H_{e,i}^{\mathrm{ROKS}}\phi_i(\mathbf{r}) = \sum_{j=1}^{n+1}\Lambda_{ij}\phi_j(\mathbf{r}) \qquad (5.48)$$

feature a Kohn–Sham-like Hamiltonian as defined in Eqs. (5.45)-(5.47), which is however orbital-dependent, hence the orbital index i, by virtue of its exchange–correlation potential $V_{\mathrm{xc}}^{\alpha}[n_m^{\alpha}, n_m^{\beta}] = \delta E_{\mathrm{xc}}[n_m^{\alpha}, n_m^{\beta}]/\delta n_m^{\alpha}$; analogous definitions hold for the β and t cases. Solving these equations self-consistently yields a set of relaxed orbitals, including the two singly occupied ones.

The set of equations Eqs. (5.45)-(5.47) could be solved by diagonalization of the corresponding ROKS Hamiltonian or alternatively by direct minimization of the associated total energy functional. The algorithm proposed in Ref. [526], which allows us to properly and efficiently minimize such orbital-dependent functionals including the orthonormality constraints, was implemented in the CPMD package [696]. Based on this minimization technique, Born–Oppenheimer molecular dynamics simulations can be performed in the first excited singlet state. Alternatively, the more efficient Car–Parrinello propagation scheme can be used once an initial set of optimized (singly and doubly occupied ROKS) orbitals with proper symmetry has been determined.

As is well known, the original approach [462] causes problems in situations where the excited state has the same spatial symmetry as the corresponding ground state, such as encountered in $\pi-\pi^*$ transitions of conjugated systems. The angle between the two singly occupied orbitals can be rotated arbitrarily if it is not fixed by symmetry constraints, thus changing the S_1 energy as a function of orbital rotation [1077]; viewed from ensemble density functional theory, these collapse phenomena have been traced back to the choice of coefficients for the symmetry adapted wave function, Eq. (5.36). In order to largely cure this problem, a generalization of the original approach was devised [554], checked [554] for various $\pi-\pi^*$ transitions, and used [1061] in an *ab initio* molecular dynamics study of photoinduced isomerizations of 1,3-butadiene and cyclohexadiene. In order to impose explicitly the generalized Brillouin theorem and thus prevent the collapse to a lower energy within the same symmetry class, an additional constraint is introduced [554]

$$H_{e,i}^{\text{ROKS}} \phi_i(\mathbf{r}) = \sum_{j=1}^{n+1} \left\{ \Lambda_{ij} \phi_j(\mathbf{r}) + \Theta_{ij} \left(\Lambda_{ji}^{\star} - \Lambda_{ij} \right) \right\} \quad (5.49)$$

via another set of real Lagrange multipliers $\{\Theta_{ij}\}$ as inspired by analogous techniques within open-shell Hartree–Fock theory, see Ref. [554] for details and references. It is found that the specific choice $\{\Theta_{ij} = 1/2\}$ yields the Goedecker–Umrigar algorithm [526] used in the original ROKS implementation [462]. This works well if the off-diagonal multipliers vanish for symmetry reasons, such as for $n - \pi^{\star}$ transitions [462]. Otherwise, unphysical solutions may exist that are lower in energy but do not fulfill the two variational conditions independently. These are characterized by a rotation between the two open-shell orbitals ϕ_a and ϕ_b, leading to singly occupied orbitals that do not obey the molecular symmetry. In addition, the energy collapses to that of the triplet state. To avoid such spurious solutions, it is sufficient to choose Θ_{ib} differently from Θ_{ja}. It is found that the choice $\Theta_{ib} = -1/2$ and $\Theta_{ja} = +1/2$ leads to good convergence [554], comparable to that of the Goedecker–Umrigar algorithm [526]. This generalization allows us to treat $\pi - \pi^{\star}$ transitions reasonably well, as checked for homologous series with conjugated double bonds using polyenes, cyanines, and protonated imines [554].

An alternative but closely related formulation of open-shell singlets was provided in Refs. [548, 549], implemented in the CPMD package, and compared to ROKS [1077]. This restricted open-shell singlet (ROSS) method [548, 549] is defined by the energy functional

$$E_{S_1}^{\text{ROSS}} [\{\phi_i\}] = T_s [\{\phi_i\}] + \int V_{\text{ext}}(\mathbf{r}) n(\mathbf{r}) \, d\mathbf{r} + \frac{1}{2} \int V_{\text{H}}(\mathbf{r}) n(\mathbf{r}) \, d\mathbf{r}$$

$$+ E_x[n_t^{\alpha}, n_t^{\beta}] + 2K[\phi_a, \phi_b] + E_c[n_m^{\alpha}, n_m^{\beta}] \ , \quad (5.50)$$

which is minimized subject to the usual orthogonality constraint, see Eq. (5.44). Thus, in both ROSS and ROKS the orbitals are obtained variationally and the efficient computation of gradients with respect to nuclear degrees of freedom, as needed for both optimization and molecular dynamics, is obtained straightforwardly. The term $K = \langle \phi_a \phi_b | \phi_a \phi_b \rangle$ is the usual exchange energy due to the electrons in the two singly occupied orbitals ϕ_a and ϕ_b from Fig. 5.1. Within Hartree–Fock theory, $2K$ is the (non-relaxed) singlet–triplet splitting, thus adding $2K$ explicitly to the lower-lying triplet energy should yield another approximation to the singlet state based on an "exact exchange" ansatz. Correlation is taken care of by adding the correlation functional E_c evaluated with the spin densities of the mixed state. The ROSS method has been implemented in the CPMD package as well, and compared to ROKS data in Ref. [1077]; the single exchange integral is calculated in Fourier space, whereas the overlap density is evaluated on the real space grid, see Section 3.3.

Much experience concerning the strengths and limitations of the ROKS approach [462] to first excited singlet states accumulated since the first benchmarks were published in 1998. There are certainly notorious cases where ROKS/ROSS fails, in particular concerning excited state ordering and when close-lying excitations are present [1077]; it is noted in passing that very accurate excited state forces are available [680] at the expense of increasing the computational effort such that *ab initio* molecular dynamics is out of reach. However, when ROKS/ROSS is applicable, the typical quality of excited state *structures* seems to be quite similar to corresponding ground state density functional theory. Thus, ROKS/ROSS structures compare in general very favorably to quantum-chemical benchmark data [353, 462, 554, 1077], and, as shown later, to optimized structures obtained from time-dependent density functional theory [1077]. Clearly, there is the known systematic underestimation of the ROKS/ROSS excitation energies in the range of about 1 eV, which appears to be an essentially constant shift of the S_0/S_1 energy gap as demonstrated for conjugated π-systems [554]; ROSS excitation energies seem to be in slightly better agreement with analogous time-dependent density functional data. It is noted in passing that the qualitative failure of ROKS highlighted in Ref. [1281] for a specific case, formaldimine $H_2C{=}NH$, could be traced back to quite unusual benchmark parameters used, see Refs. [353, 433]. Still, it must be stressed that *ab initio* molecular dynamics involving excited states requires in general very careful and extensive checking of the performance of the electronic structure treatment for every specific problem of interest!

5.3.3 A few excited states: explicit nonadiabatic dynamics

A wealth of processes involve neither only a single nor very many electronic states, but rather only "a few but important" states, typical examples being the vast classes of photo(bio)physical processes and photo(bio)chemical reactions. The development of nonadiabatic *ab initio* molecular dynamics methods is further motivated by organic photochemistry [756] and the recent experimental advances in ultrafast photochemistry [359, 1651]. Many photochemical processes take place on a sub-picosecond time scale driven by conical intersections or avoided crossings of different potential energy surfaces causing the Born–Oppenheimer approximation to break down [756, 774, 1048, 1217, 1642]. Thus, *ab initio* molecular dynamics methods that go "beyond the Born–Oppenheimer approximation" by including explicitly nonadiabatic couplings between at least two states in a fully dynamical sense are required, see Refs. [355, 937] for reviews that cover aspects of this broad topic.

To this end we refer back to Ehrenfest molecular dynamics derived in Section 2.2 as the starting point. The general equations of motion Eqs. (2.41)-(2.42) obtained in the adiabatic basis Eqs. (2.34)-(2.33) yield the expansion coefficients $c_k(t)$ whose square modulus, $|c_k(t)|^2$, can be interpreted as state occupation numbers, i.e. as probabilities of finding the system in the adiabatic state Ψ_k at time t. However, these equations were immediately simplified by completely neglecting the nonadiabatic coupling elements and vectors, D^{kl} and $\mathbf{d}_I^{kl}$, as defined in Eq. (2.43) and by restricting the propagation to the ground state Ψ_0. When is this adiabatic approximation, i.e. $|c_0(t)|^2 = 1 \, \forall \, t$, justified? The quantitative answer is embodied in the Massey parameter [965]

$$\xi = \frac{\tau_{\mathrm{n}}}{\tau_{\mathrm{e}}} = \frac{\Delta E \, L}{\hbar \dot{R}} \quad , \tag{5.51}$$

which is the ratio of the passage time of the nuclei, τ_{n}, and the characteristic time scale of electronic motion, τ_{e}, approximated in terms of the energy difference ΔE between two electronic states, a characteristic length L, and the velocity of the nuclei $\dot{R}$. Thus, nonadiabatic effects are negligible only in the limit $\xi \gg 1$, i.e. for large energy spacings and small velocities such that the replacement of Eqs. (2.41)-(2.42) by the much simpler set Eqs. (2.44)-(2.45) is reasonable.

As discussed in Section 5.3.2, efficient *adiabatic* Car–Parrinello simulations have become possible also in the first excited state using the restricted open-shell Kohn–Sham (ROKS) approach [462]. Thus, *nonadiabatic* extensions of *ab initio* molecular dynamics can be devised by coupling the ROKS S_1 excited state to the Kohn–Sham ground state S_0 as introduced in Ref. [356], see Ref. [355] for a review of the entire field of "on-the-fly" approaches to excited state dynamics. As shown in Section 5.3.2, the S_1 restricted open-shell singlet "wave function" is constructed by linearly combining the mixed determinants, m_1 and m_2 from Fig. 5.1,

$$\Psi_1 = \frac{1}{\sqrt{2}} \left\{ |\phi_1^{(1)} \bar{\phi}_1^{(1)} \phi_2^{(1)} \bar{\phi}_2^{(1)} \cdots \phi_n^{(1)} \bar{\phi}_{n+1}^{(1)} \rangle \right.$$

$$\left. + |\phi_1^{(1)} \bar{\phi}_1^{(1)} \phi_2^{(1)} \bar{\phi}_2^{(1)} \cdots \bar{\phi}_n^{(1)} \phi_{n+1}^{(1)} \rangle \right\} \ , \quad (5.52)$$

where the "ket" notation signifies Slater determinants made up of Kohn–Sham orbitals, $\phi_i^{(1)}$ (spin up) and $\bar{\phi}_i^{(1)}$ (spin down); the total number of electrons is $2n$ as in Section 5.3.2. Within the ROKS approach, see Eq. (5.44), a *single* set of orbitals $\{\phi_i^{(1)}\}$ is determined that minimizes the energy functional $E_{S_1}[\{\phi_i^{(1)}\}]$ for the first excited state. Due to this optimization the entire set of orbitals $\{\phi_i^{(1)}\}$ will, in general, differ from the set of orbitals $\{\phi_i^{(0)}\}$ that defines the ground state wave function, Ψ_0,

$$\Psi_0 = |\phi_1^{(0)} \bar{\phi}_1^{(0)} \phi_2^{(0)} \bar{\phi}_2^{(0)} \cdots \phi_l^{(0)} \bar{\phi}_l^{(0)} \rangle \ . \quad (5.53)$$

As a consequence the two state functions, Ψ_0 and Ψ_1, are nonorthogonal

$$S_{01} = S_{10} = S \ , \quad S_{kk} = 1 \ , \quad (5.54)$$

and give rise to the overlap matrix $\{S_{kl}\}$. Using these two adiabatic states in a truncated expansion, see Eq. (2.33), of the electronic wave function

$$\Psi = a_0(t)\Psi_0 \exp\left[-\frac{i}{\hbar} \int E_0 dt\right] + a_1(t)\Psi_1 \exp\left[-\frac{i}{\hbar} \int E_1 dt\right] \quad (5.55)$$

and inserting it into the time-dependent electronic wave equation, Eq. (2.32) or (2.39), yields

$$\sum_l a_l p_l (H_{kl} - E_l S_{kl}) = i\hbar \left\{ \sum_l \dot{a}_l p_l S_{kl} + \sum_l a_l p_l D^{kl} \right\} \ , \quad (5.56)$$

where the Hamiltonian matrix elements are given by

$$H_{kk} = \langle \Psi_k | H_e^{(\mathrm{RO})\mathrm{KS}} | \Psi_k \rangle = E_k \quad (5.57)$$

$$H_{01} = H_{10} = E_0 S \ ; \quad (5.58)$$

the phase factors

$$p_l = \exp\left[-\frac{i}{\hbar} \int E_l dt\right] \quad (5.59)$$

have been introduced for convenience (which amounts to switching to Dirac's "intermediate representation") and the nonadiabatic couplings D^{kl} are defined in Eq. (2.43). Solving Eq. (5.56) for the time-dependent expansion coefficients $\dot{a}_0$ and $\dot{a}_1$ leads to the set of coupled differential equations

$$\dot{a}_0 = \frac{1}{S^2 - 1} \left[i a_1 \frac{p_1}{p_0} S(E_0 - E_1) + a_1 D^{01} \frac{p_1}{p_0} - a_0 D^{10} S \right] \quad (5.60)$$

$$\dot{a}_1 = \frac{1}{S^2 - 1} \left[a_0 D^{10} \frac{p_0}{p_1} - a_1 D^{01} S - i a_1 S^2 (E_0 - E_1) \right] , \quad (5.61)$$

which can be integrated numerically using standard schemes [1161]. It is computationally attractive [597] to work with the nonadiabatic coupling *elements*, D^{kl}, instead of the nonadiabatic coupling *vectors*, $\mathbf{d}_I^{lk}$, as defined in Eq. (2.43). The orbital velocities required in the former case, $\partial_t |\Psi_l\rangle$, are explicitly available within the Car–Parrinello method, see Section 2.4.2, whereas the nuclear gradient required for the latter, $\nabla_I |\Psi_l\rangle$, must be evaluated in addition. An iterative Born–Oppenheimer-style implementation of nonadiabatic *ab initio* molecular dynamics, see Section 2.3 for the adiabatic version, is also straightforwardly possible within the outlined framework. This can be done either using a simple finite difference approach to the coupling elements or by a scheme to directly compute the nonadiabatic coupling vectors as developed in Ref. [135].

If both electronic state functions $\{\Psi_0, \Psi_1\}$ were eigenfunctions of the Kohn–Sham Hamiltonian, $|a_0|^2$ and $|a_1|^2$ would be their respective occupation numbers. However, $\{\Psi_0, \Psi_1\}$ are not proper many-body wave functions stemming from an interacting Hamiltonian (see Section 5.3.2) and a look at the normalization integral of the electronic wave function, Ψ,

$$\langle \Psi | \Psi \rangle = |a_0|^2 + |a_1|^2 + 2S \Re \left(a_0^\star a_1 \frac{p_1}{p_0} \right) = 1 \quad (5.62)$$

shows that the definition of state populations in this basis is not possible due to the nonvanishing overlap $S > 0$. Thus, an auxiliary, orthonormal basis $\{\Psi_0', \Psi_1'\}$ is introduced [356] to re-expand Ψ

$$\Psi = c_0 \Psi_0' + c_1 \Psi_1' , \quad (5.63)$$

which allows us to obtain normalized occupations numbers

$$|c_0|^2 = |a_0|^2 + S^2 |a_1|^2 + 2S \Re \left(a_0^\star a_1 \frac{p_1}{p_0} \right) \quad (5.64)$$

$$|c_1|^2 = (1 - S^2) |a_1|^2 \quad (5.65)$$

satisfying the sum rule $|c_0|^2 + |c_1|^2 = 1$ such that a density matrix $\rho_{kl}(t) = c_k^\star(t) c_l(t)$ can be defined, see Refs. [350, 355] for details. It is noted that these reorthogonalization problems, encountered when using ROKS including orbital relaxation, do not occur when all states $\{\Psi_k\}$, i.e. also the excited states, are eigenfunctions of the *same* Kohn–Sham Hamiltonian. This is the case when no orbital relaxation is performed within ROKS or when the "broken symmetry determinant" [677, 1062, 1143] is used, see Section 5.3.2, which corresponds to the full "particle-hole" excitation without any orbital relaxation and thus to the m states in Fig. 5.1. A somewhat more sophisticated approach [135] would be to make use of "Slater's transition state" determinant, which can be viewed as a simple approximation to ensemble density functional theory [363] by introducing fractional occupation numbers of exactly 1/2 for the two involved particle-hole orbitals, i.e. ϕ_a and ϕ_b in Fig. 5.1. But again, it must be stressed that in the framework of density functional theory there are no "many-body wave functions" being adiabatic eigenfunctions of the interacting Hamiltonian, as has been worked out [894] in relation to a particular TDDFT approach to nonadiabatic dynamics [289].

At this stage, at least two different avenues to obtain a *nuclear* trajectory coupled to the time evolution of the electrons accessing more than one electronic state can be followed. Clearly, nonadiabatic Ehrenfest molecular dynamics according to Eqs. (2.41)-(2.42) is a first option. However, several severe deficiencies of this approach are well documented, see for example Refs. [350, 355, 1516, 1517] and references cited therein. At the root of many of these problems is the mean-field character of Ehrenfest molecular dynamics: the force acting on the nuclei is obtained by *averaging* over all adiabatic states $\{\Psi_k\}$ used to expand the electronic wave function Ψ according to their instantaneous occupation numbers $|c_k(t)|^2$, as most directly represented by Eq. (2.40). In particular, a system that was initially prepared in a pure adiabatic state will be in a mixed state when leaving the region of strong nonadiabatic coupling. In general, the pure adiabatic character of the wave function cannot be recovered even in the asymptotic regions of configuration space. In cases where the differences in the adiabatic potential energy landscapes are pronounced, it is clear that an average potential will be unable to describe all reaction channels adequately. In particular, if one is interested in a reaction branch whose occupation number is very small, the average path is likely to diverge from the "true trajectory". Furthermore, the total wave function may contain significant contributions from adiabatic states that are energetically inaccessible in classical mechanics due to lack of sufficient energy.

A powerful alternative is Tully molecular dynamics [1101, 1515, 1518], in particular when using the ingenious "fewest switches" surface-hopping algorithm [1515] to account for nonadiabatic effects, see for instance Refs. [350,

355, 1516, 1517] for reviews and Ref. [578] for a critical comparison to computationally more demanding nonadiabatic methods. Fundamental to Tully molecular dynamics is the idea that the system is always in a particular pure state. Thus, at any moment in time, the system is propagated on some adiabatic state Ψ_k, which is selected according to its state population $|c_k|^2$ by a statistical hopping criterion. Changing adiabatic state occupations can thus result in nonadiabatic transitions between different adiabatic potential energy surfaces. At variance with the Ehrenfest approach where "the best path" is generated, an ensemble of independent trajectories must be generated and analyzed in order to extract observables. This approach to nonadiabatic *ab initio* molecular dynamics [355, 356] is obtained by integrating simultaneously the following equations:

$$M_I \ddot{\mathbf{R}}_I(t) = -\nabla_I E_k \tag{5.66}$$

$$i\hbar \dot{c}_k(t) = c_k(t) E_k - i\hbar \sum_l c_l(t) D^{kl} \tag{5.67}$$

together with Eq. (2.120) and Eqs. (5.45)-(5.47) to obtain the energies and wave functions, where the nonadiabatic coupling elements, Eq. (2.43), are defined as $D^{kl} = \int d\mathbf{r}\, \Psi_k^\star \partial_t \Psi_l$. Importantly, the forces acting on the nuclei, Eq. (5.66), are obtained from the gradient of the energy E_k of a *single* adiabatic state Ψ_k, at variance with the Ehrenfest approach where *all* states are averaged according to Eq. (2.40). The particular state populated is selected stochastically in each step of the molecular dynamics propagation using the fewest switches hopping criterion [1515]

$$P_k(t) = -\frac{\dot{\rho}_{kk}(t)\delta t}{\rho_{kk}(t)} \tag{5.68}$$

to hop from state Ψ_k to any other state in the time interval $[t, t + \delta t]$; here $\rho_{kl}(t) = c_k^\star(t) c_l(t)$ is the density matrix obtained in the auxiliary basis Eqs. (5.64)-(5.65) and δt is the molecular dynamics time step. More details of the method can be found in Refs. [350, 355, 356].

The computational effort per time step increases linearly with the number of adiabatic states involved as compared to adiabatic Car–Parrinello or Born–Oppenheimer propagation. However, the time step must be as small as the one typically used in Ehrenfest dynamics, see the discussion in Section 2.6, which is of the order of $\delta t \sim 10^{-16}$ s in order to resolve the very fast dynamics of the electrons. In a straightforward implementation this would increase the cost of nonadiabatic simulations per picosecond by another factor of 10 or 100 compared to the two corresponding adiabatic schemes;

interpolation between the time steps is used to cut down this overhead. Finally, it is required within Tully molecular dynamics to generate as many independent stochastic trajectories as possible in order to sample averages. Despite the computationally demanding character of nonadiabatic *ab initio* molecular dynamics, several case studies [351, 352, 835, 836] clearly demonstrate both its power and application range in the realm of photophysics and photochemistry, including the condensed phase.

An alternative implementation of *ab initio* Tully molecular dynamics relies on using time-dependent (Kohn–Sham) density functional theory [1255] in the frequency domain within the linear response approximation [233, 364, 559, 910, 912]. With the availability of nuclear gradients $\nabla_I E_k$ corresponding to a specific excited state k as described in Section 5.3.6, see Eq. (5.152), and the decomposition of the specific excitation into a set of orbital contributions, see Eq. (5.170), it is possible to integrate Eqs. (5.66)-(5.67) directly and to evaluate the fewest switches hopping criterion Eq. (5.68) "on-the-fly"; see Ref. [1444] for an implemenation of such a method in the CPMD package [696]. Being an alternative to using the ROKS/ROSS schemes from Section 5.3.2 in conjunction with Eqs. (5.66)-(5.67), it remains however to be explored how well singularities such as conical intersections are described when obtained from linear response of the electrons where the underlying ground state most often is of genuine multi-reference nature.

Independent of the particular choice of the electronic structure method, Tully surface-hopping approaches are computationally demanding as they require the explicit propagation of coupled time-dependent Schrödinger equations in addition to an elaborate evaluation of the hopping parameters and thus nonadiabatic couplings. A more approximate approach [1518] to dynamically couple two electronic states relies on the Landau–Zener–Stückelberg theory of nonadiabatic transitions [1048] that was introduced as early as 1932 based on a one-dimensional model. Within the *diabatic* representation and using first-order perturbation theory, the Landau–Zener nonadiabatic transition probability reads

$$P_{\mathrm{LZ}} \;=\; \exp\left[-\frac{2\pi\,|V_{01}|^2}{\hbar v |F_0(\mathbf{R}^c) - F_1(\mathbf{R}^c)|}\right]\;,\qquad (5.69)$$

where V_{01} is the constant diabatic coupling matrix element, v is the constant velocity along the reaction coordinate, and $F_k(\mathbf{R}^c)$ is the force along the reaction coordinate according to the diabatic potential energy surface of state k at the point of the avoided crossing. It is noted in passing that shortcomings of the Landau–Zener–Stückelberg theory, such as those outlined in

Chapter 3.1 of Ref. [1048], have been cured within the Zhu-Nakamura theory [1048] at the expense of losing the simplicity of the original approach.

The efficient Landau–Zener–Stückelberg theory of nonadiabatic transitions has been implemented in the CPMD package [696] in order to couple two or more electronic states of well-defined symmetry [1452]. Working within the expansion Eq. (5.55) of the electronic wave function Ψ in terms of the *adiabatic* electronic states, $\{\Psi_k\}$, the associated potential energy surfaces, $E_k(\mathbf{R}(t))$, and thus the adiabatic electronic gap, $\Delta E_{01}^{\text{adia}} = E_1(\mathbf{R}(t)) - E_0(\mathbf{R}(t))$, are available at each time t in *ab initio* molecular dynamics schemes. When it comes to the computation of the required *diabatic* Landau–Zener parameters, V_{01}, $F_0(\mathbf{R}^c)$, and $F_1(\mathbf{R}^c)$ in Eq. (5.69), further simplifications [714, 1452, 1518]

$$\tilde{P}_{\text{LZ}} \approx \exp\left[-\frac{\pi \min\left|\Delta E_{01}^{\text{adia}}\right|^2}{2\hbar \max\frac{d}{dt}\left|\Delta E_{01}^{\text{adia}}\right|}\right] \tag{5.70}$$

based on the availability of the dynamical electronic gap $\Delta E_{01}^{\text{adia}}(t)$ can be invoked. In particular, the constant diabatic coupling V_{01} has been approximated by half the minimum energy difference between the adiabatic potential energy surfaces, $V_{01} \approx |E_0(\mathbf{R}(t^c)) - E_1(\mathbf{R}(t^c))|/2$, at the avoided crossing. In addition, the diabatic force denominator at the crossing, $v|F_0(\mathbf{R}(t^c)) - F_1(\mathbf{R}(t^c))|$, has been approximated by the maximum value of the rate of change of this adiabatic electronic gap with respect to time just before entering the avoided crossing region, i.e. from the largest slope of $\Delta E_{01}^{\text{adia}}(t)$ vs. $t \lesssim t^c$ as illustrated pictorially in Fig. 5 of Ref. [714]. As a result, this simplified Landau–Zener surface-hopping scheme [1452] requires only input parameters that are obtained without any additional cost from *ab initio* molecular dynamics simulations where more than one adiabatic electronic state is propagated.

5.3.4 Many excited states: free energy functionals

The free energy functional or ensemble density functional approaches [363, 1102] to excited state molecular dynamics [13, 15, 961] are mean-field approaches similar in spirit to Ehrenfest molecular dynamics, see Section 2.2. The total wave function is first factorized into a nuclear and an electronic wave function Eq. (2.23), followed by taking the classical limit for the nuclear subsystem. Thus, classical nuclei move in the *average* field as obtained from appropriately weighting all electronic states according to Eq. (2.38) using the expansion Eq. (2.33) for the electronic wave function. A difference is that according to Ehrenfest molecular dynamics, the electrons are propagated in real time and can perform nonadiabatic transitions by virtue of direct coupling terms $\propto \mathbf{d}^{kl}$ between all states Ψ_k subject to energy conservation, see Section 2.2 and in particular Eqs. (2.41)-(2.43). The average force or Ehrenfest force is obtained by weighting the different states k according to their diagonal density matrix elements (that is $\propto |c_k(t)|^2$ in Eq. (2.41)), whereas the coherent transitions are driven by the off-diagonal contributions, $\propto c_k^\star c_l$, according to Eq. (2.41).

In the free energy approach [13, 15], the excited states are populated according to the Fermi–Dirac (finite-temperature equilibrium) distribution, which is based on the assumption that the electrons "equilibrate" more rapidly than the time scale of the nuclear motion. This means that the set of electronic states evolves at a given temperature "isothermally" (rather than adiabatically) under the inclusion of *incoherent* electronic transitions as the nuclei move. Thus, instead of computing the force acting on the nuclei from the electronic ground state energy, it is obtained from the electronic *free* energy as defined in the canonical ensemble. By allowing such electronic transitions to occur, the free energy approach transcends the usual Born–Oppenheimer approximation. However, the approximation of an instantaneous equilibration of the electronic subsystem implies that the electronic structure at a given nuclear configuration $\{\mathbf{R}_I\}$ is completely independent from previous configurations along a molecular dynamics trajectory. Due to this assumption, the notion "free energy Born–Oppenheimer approximation" was coined in Ref. [214] in a similar context. Certain non-equilibrium situations can also be modeled within the free energy approach by starting off with an initial orbital occupation pattern that does not correspond to any temperature in its thermodynamic meaning, see e.g. Refs. [1349, 1352, 1356] for such applications.

The free energy functional as defined in Ref. [15] is introduced most elegantly [12, 13] by starting the discussion for the special case of *noninteracting Fermions*

$$H_{\mathrm{s}} = -\frac{1}{2}\nabla^2 - \sum_I \frac{Z_I}{|\mathbf{R}_I - \mathbf{r}|} \tag{5.71}$$

in a *fixed* external potential due to a collection of nuclei at positions $\{\mathbf{R}_I\}$. The associated grand canonical partition function and its thermodynamic potential ("grand canonical free energy") are given by

$$\Xi_{\mathrm{s}}(\mu V T) = \det^2 \left(1 + \exp\left[-\beta\left(H_{\mathrm{s}} - \mu\right)\right]\right) \tag{5.72}$$

$$\Omega_{\rm s}(\mu V T) \;\; = \;\; -k_{\rm B}T \ln \Xi_{\rm s}(\mu V T) \;\; , \tag{5.73}$$

where μ is here the chemical potential acting on the electrons (and not the Car–Parrinello fictitious mass parameter) and the square of the determinant stems from considering the spin-unpolarized special case only. This reduces to the well-known grand canonical potential expression

$$\begin{aligned}\Omega_{\rm s}(\mu V T) \;\; &= \;\; -2k_{\rm B}T \ln \det\left(1 + \exp\left[-\beta\left(H_{\rm s} - \mu\right)\right]\right) \\[4pt] &= \;\; -2k_{\rm B}T \sum_i \ln\left(1 + \exp\left[-\beta\left(\epsilon_{\rm s}^{(i)} - \mu\right)\right]\right)\end{aligned} \tag{5.74}$$

for non-interacting spin-1/2 Fermions at temperature T, where $\{\epsilon_{\rm s}^{(i)}\}$ are the eigenvalues of a one-particle Hamiltonian such as Eq. (5.71); here the standard identity $\ln \det \mathbf{M} = \operatorname{Tr} \ln \mathbf{M}$ was invoked for positive definite $\mathbf{M}$.

According to thermodynamics, the Helmholtz free energy $\mathcal{F}(NVT)$ associated with Eq. (5.73) can be obtained from an appropriate Legendre transformation of the grand canonical free energy $\Omega(\mu V T)$

$$\mathcal{F}_{\rm s}(NVT) = \Omega_{\rm s}(\mu V T) + \mu N + \sum_{I<J} \frac{Z_I Z_J}{|\mathbf{R}_I - \mathbf{R}_J|} \tag{5.75}$$

by fixing the average number of electrons N and determining μ from the conventional thermodynamic condition

$$N = -\left(\frac{\partial \Omega}{\partial \mu}\right)_{VT} \tag{5.76}$$

using $\Omega_{\rm s}$ as defined in Eq. (5.74). In addition, the internuclear Coulomb interactions between the classical nuclei were included at this stage in Eq. (5.75). Thus, derivatives of the free energy Eq. (5.75) with respect to ionic positions $-\nabla_I \mathcal{F}_{\rm s}$ define forces on the nuclei that could be used in a (hypothetical) molecular dynamics scheme using non-interacting electrons.

The interactions between the electrons can be "switched on" by resorting to Kohn–Sham density functional theory and the concept of a non-interacting reference system. Thus, instead of using the simple one-particle Hamiltonian Eq. (5.71), the effective Kohn–Sham Hamiltonian Eq. (2.122) has to be utilized. As a result, the grand canonical free energy Eq. (5.72) can be written as

$$\Omega^{\mathrm{KS}}(\mu V T) = -2k_{\mathrm{B}}T \ln\left[\det\left(1 + \exp\left[-\beta\left(H_{\mathrm{e}}^{\mathrm{KS}} - \mu\right)\right]\right)\right] \quad (5.77)$$

$$H_{\mathrm{e}}^{\mathrm{KS}} = -\frac{1}{2}\nabla^2 - \sum_I \frac{Z_I}{|\mathbf{R}_I - \mathbf{r}|} + V_{\mathrm{H}}(\mathbf{r}) + \frac{\delta\Omega_{\mathrm{xc}}[n]}{\delta n(\mathbf{r})} \quad (5.78)$$

$$H_{\mathrm{e}}^{\mathrm{KS}}\phi_i = \epsilon_i\phi_i \ , \quad (5.79)$$

where Ω_{xc} is the exchange–correlation functional at finite temperature. By virtue of Eq. (5.74) one can immediately see that Ω^{KS} is nothing else than the "Fermi–Dirac weighted sum" of the bare Kohn–Sham eigenvalues $\{\epsilon_i\}$ at temperature T. Hence, this term is the extension to finite temperatures of the "band-structure energy" contribution (or of the "sum of orbital energies" in the analogous Hartree–Fock case [625, 985, 1423]) to the total electronic energy, see Eq. (2.125).

In order to obtain the correct total electronic free energy of the interacting electrons as defined in Eq. (2.125), the corresponding extra terms (properly generalized to finite temperatures) have to be included in Ω^{KS}. This finally allows one to write down the generalization of the Helmholtz free energy of the interacting many-electron case

$$\mathcal{F}^{\mathrm{KS}}(NVT) = \Omega^{\mathrm{KS}}(\mu V T) + \mu\int n(\mathbf{r})\,d\mathbf{r} + \sum_{I<J}\frac{Z_I Z_J}{|\mathbf{R}_I - \mathbf{R}_J|}$$

$$-\frac{1}{2}\int V_{\mathrm{H}}(\mathbf{r})\,n(\mathbf{r})\,d\mathbf{r} + \Omega_{\mathrm{xc}} - \int\frac{\delta\Omega_{\mathrm{xc}}[n]}{\delta n(\mathbf{r})}\,n(\mathbf{r})\,d\mathbf{r} \quad (5.80)$$

in the framework of a Kohn–Sham-like formulation. The corresponding one-particle density at the Γ-point is given by

$$n(\mathbf{r}) = \sum_i f_i(\beta)\,|\phi_i(\mathbf{r})|^2 \quad (5.81)$$

$$f_i(\beta) = \left(1 + \exp\left[\beta\left(\epsilon_i - \mu\right)\right]\right)^{-1} \ , \quad (5.82)$$

where the fractional occupation numbers $\{f_i\}$ are obtained from the Fermi–Dirac distribution at temperature T in terms of the Kohn–Sham eigenvalues $\{\epsilon_i\}$. Finally, *ab initio* forces can be obtained as usual from the nuclear gradient of $\mathcal{F}^{\mathrm{KS}}$, which makes molecular dynamics possible. In passing, it is noted that associating fractional occupation numbers to one-particle orbitals is also one possible route to go beyond a single-determinant ansatz for constructing the charge density [363, 1102].

By construction, the total free energy Eq. (5.80) reduces to that of the non-interacting toy model Eq. (5.75) once the electron-electron interaction is switched off. Another useful limit is the ground state limit $\beta \to \infty$, where the free energy $\mathcal{F}^{\mathrm{KS}}(NVT)$ yields the standard Kohn–Sham total energy expression E^{KS} as defined in Eq. (2.125) after invoking the appropriate limit $\Omega_{\mathrm{xc}} \to E_{\mathrm{xc}}$ as $T \to 0$. Most importantly, stability analysis [13, 15] of Eq. (5.80) shows that this functional shares the same stationary point as the exact finite-temperature functional due to Mermin [994], see e.g. the textbooks [363, 1102] for introductions to density functional formalisms at finite temperatures. This implies that the self-consistent density, which defines the stationary point of $\mathcal{F}^{\mathrm{KS}}$, is identical to the exact one. This analysis reveals furthermore that, unfortunately, this stationary point is not an extremum but a saddle point, so that no variational principle and, numerically speaking, no direct minimization algorithms can be applied. For the same reason, a Car–Parrinello fictitious dynamics approach to molecular dynamics is not a straightforward option, whereas Born–Oppenheimer dynamics based on diagonalization can be used directly. Thus, an iterative molecular dynamics scheme

$$M_I \ddot{\mathbf{R}}_I = -\nabla_I \mathcal{F}^{\mathrm{KS}}(\{\mathbf{R}(t)\}) \tag{5.83}$$

is introduced by computing the *ab initio* forces from the nuclear gradient of the free energy functional $\mathcal{F}^{\mathrm{KS}}$, taking advantage of the Hellmann–Feynman theorem.

In particular, the band-structure energy term is evaluated in the CPMD package [696] by diagonalizing the Kohn–Sham Hamiltonian after a suitable "preconditioning" [15], see Section 3.6.2. Specifically, a second-order Trotter approximation is used

$$\mathrm{Tr}\,\exp\left[-\beta H_{\mathrm{e}}^{\mathrm{KS}}\right] = \sum_i \exp\left[-\beta \epsilon_i\right] = \sum_i \rho_{ii}(\beta) \tag{5.84}$$

$$= \mathrm{Tr}\left(\left\{\exp\left[-\frac{\Delta\tau}{2}\left(-\frac{1}{2}\nabla^2\right)\right]\exp\left[-\Delta\tau V^{\mathrm{KS}}[n]\right]\right.\right.$$

$$\left.\left. \times \exp\left[-\frac{\Delta\tau}{2}\left(-\frac{1}{2}\nabla^2\right)\right]\right\} + \mathcal{O}\left(\Delta\tau^3\right)\right)^P \tag{5.85}$$

$$\approx \sum_i \{\rho_{ii}(\Delta\tau)\}^P = \sum_i \{\exp\left[-\Delta\tau \epsilon_i\right]\}^P \tag{5.86}$$

in order to compute first the diagonal elements $\rho_{ii}(\Delta\tau)$ of the "high-tempera-ture" Boltzmann operator $\rho(\Delta\tau)$; here $\Delta\tau = \beta/P$ and P is the Trotter "time slice" as introduced in Section 5.4.2. To this end, the kinetic and potential energies can be conveniently evaluated in reciprocal and real space, respectively, by using the split-operator/FFT technique [407]. The Kohn–Sham eigenvalues ϵ_i are finally obtained from the density matrix via $\epsilon_i = -(1/\Delta\tau)\ln\rho_{ii}(\Delta\tau)$. They are used in order to compute the occupation numbers $\{f_i\}$, the density $n(\mathbf{r})$, the band-structure energy Ω^{KS}, and thus the free energy Eq. (5.80).

In practice, a diagonalization/density-mixing scheme is employed in or-der to compute the self-consistent density $n(\mathbf{r})$. In a first step a suitably constructed trial input density n_{in} (see e.g. the Appendix of Ref. [1348] for such a method) is used in order to compute the potential $V^{KS}[n_{in}]$. Then the lowest-order approximant to the Boltzmann operator Eq. (5.86) is diagonal-ized using an iterative Lanczos-type method. This yields an output density n_{out} and the corresponding free energy $\mathcal{F}^{KS}[n_{out}]$. Finally, the densities are mixed and the former steps are iterated until a stationary solution n_{scf} of $\mathcal{F}^{KS}[n_{scf}]$ is achieved, see Section 3.6.4 for some details on such methods. Of course the most time-consuming part of the calculation is the iterative diagonalization. In principle this is not required, and it should be possible to compute the output density directly from the Fermi–Dirac density ma-trix even in a linear scaling scheme [518], thus circumventing the explicit calculation of the Kohn–Sham eigenstates. However, to date efforts in this direction have failed, or given methods which are too slow to be useful [12].

As a method, molecular dynamics with the free energy functional is most appropriate to use when the excitation gap is small, in cases where the gap might close during a chemical transformation, or when the temperature is extremely high. In the latter case no instabilities are encountered with this approach, which is not true for ground state *ab initio* molecular dynamics methods. The price to pay is the quite demanding iterative computation of well-converged forces. Besides allowing such applications with physically relevant excitations this method can also be combined straightforwardly with **k**-point sampling and applied to metals at "zero" temperature. In this case, the electronic "temperature" is only used as a smearing parameter of the Fermi edge by introducing fractional occupation numbers, which is known to improve greatly the convergence of these ground state electronic structure calculations [414, 471, 508, 518, 566, 786, 787, 1398, 1558, 1586, 1604, 1612].

Finite-temperature expressions for the exchange–correlation functional Ω_{xc} are available in the literature. However, for most temperatures of in-terest the corrections to the ground state expression are small and it seems justified to use one of the various well-established parameterizations of the exchange–correlation energy E_{xc} at zero temperature, see Section 2.7.

The free energy functional approach to finite electronic temperature molecular dynamics was used in order to investigate, for instance, the sound velocity of dense hydrogen at conditions on Jupiter [17], laser heating of silicon [1349, 1352] and graphite [1356], and laser-induced transformations in fullerite [488]. In all cases the electronic subsystem was highly excited, either by imposing a very high equilibrium temperature for both (classical) nuclei and electrons [17], or by creating a pronounced nonequilibrium initial population of the excited electronic states in order to model the influence of an irradiating laser pulse.

An alternative variational formulation and implementation of ensemble density functional theory directly in the framework of the Mermin–Kohn–Sham approach [363, 1102] starts by adopting, instead of Eq. (5.81), a matrix representation of the Fermi operator

$$n(\mathbf{r}) = \sum_{ij} f_{ij}(\beta)\, \phi_i^\star(\mathbf{r})\phi_j(\mathbf{r}) \tag{5.87}$$

in the basis of single-particle Kohn–Sham orbitals which yields a (fractional) occupation number matrix $\mathbf{f}(\beta)$, see Ref. [961]. In order to make this definition of occupation numbers meaningful, the constraint that the trace of the occupation number matrix yields the total number of electrons in the system, $\mathrm{Tr}\,\mathbf{f} = N$, has to be imposed in addition to requiring that its eigenvalues are the usual occupation numbers, i.e. $f_i = \mathrm{diag}\,\mathbf{f} \in [0,1]$. The free energy functional $\mathcal{A}$ to be minimized is defined as

$$\mathcal{A}[\phi,\mathbf{f}] = \sum_{ij} f_{ij} \left\langle \phi_i \left| -\frac{1}{2}\nabla^2 - \sum_I \frac{Z_I}{|\mathbf{R}_I - \mathbf{r}|} \right| \phi_j \right\rangle$$

$$+ V_{\mathrm{H}}[n] + V_{\mathrm{xc}}[n] - T\mathcal{S}[\mathbf{f}] \ , \tag{5.88}$$

where the entropy term is given by

$$\mathcal{S}[\mathbf{f}] = -k_{\mathrm{B}}\mathrm{Tr}\,\{\mathbf{f}\ln\mathbf{f} + (1-\mathbf{f})\ln(1-\mathbf{f})\} \ . \tag{5.89}$$

Now the free energy Eq. (5.88) obtained from traces of operators is covariant under both orbital and occupation number unitary transformations. This allows us to introduce a new auxiliary functional

$$\tilde{\mathcal{A}}[\phi] = \min_{\{f_{ij}\}} \mathcal{A}[\phi,\mathbf{f}] \tag{5.90}$$

that is minimized w.r.t. the occupation number matrix and thus depends only on the orbitals; it is also invariant under unitary transformations of the orbitals. This opens up the possibility of a two-step iterative minimization of $\mathcal{A}$ by decoupling the orbital and occupation number evolution. In an outer loop the orbitals get updated by searching the minimum of $\tilde{\mathcal{A}}$, whereas the occupation number matrix gets updated by minimizing the corresponding functional $\mathcal{A}$ while keeping the orbitals fixed; the orthogonality of the orbitals has to be imposed. The detailed algorithm that implements these ideas is discussed in Ref. [961].

This particular implementation of the Mermin–Kohn–Sham functional has been used to study via *ab initio* molecular dynamics the behavior of the Al(110) surface for temperatures up to 900 K [962]. Most recently, a full Car–Parrinello formulation in terms of a coupled fictitious dynamics of orbitals, occupation matrix, and unitary rotations was given as well as a hybrid scheme where the $\{f_{ij}\}$ get iteratively minimized during the fictitious time evolution of the orbitals [958].

The two methods presented so far both yield the correct ensemble density functional theory formulation for electronic systems at finite temperatures and thus require a self-consistent calculation of the occupation numbers. A similar but simplified approach to include finite electronic temperatures relies on total energy calculations where only the (fractional) occupation numbers of the Kohn–Sham single-particle states are chosen according to the Fermi–Dirac distribution for a certain temperature and chemical potential according to Eq. (5.82). Recently, such an ansatz was chosen in order to investigate the phonon response in photoexcited solid tellurium [1441].

All methods discussed so far allow for *thermal* excitations within Mermin's version of ensemble density functional theory. In addition, it is also possible to create athermal initial distributions, which is however a slightly uncontrolled approach to generate electronic excitations. The intricacies of devising a density functional formulation for a particular excited state as, for example, defined by its symmetry within the framework of an ensemble Kohn–Sham scheme are discussed in Ref. [507]. Unfortunately, this formulation relies crucially on the calculation of (nonlocal) optimized effective potentials (i.e. an orbital-dependent functional), which might hamper its use for *ab initio* molecular dynamics for computational reasons.

5.3.5 RT-TDDFT: explicit real-time propagation

The extension of Hohenberg–Kohn(–Sham) density functional theory, as summarized in Section 2.7.2, to time-dependent systems [1255] is called time-dependent density functional theory (TDDFT) or Runge–Gross theory [42,

559, 910, 912, 1255, 1542]. Contrary to the static case where the total energy can be variationally minimized in order to determine the stationary Schrödinger equation and the electronic ground state, Eq. (2.113), it is the quantum-mechanical action,

$$\mathcal{A}[\Psi] \;=\; \int_{t_i}^{t_f} \left\langle \Psi(t) \left| i\frac{\partial}{\partial t} - \mathcal{H}_e(t) \right| \Psi(t) \right\rangle dt \;, \tag{5.91}$$

that has to be stationary in the time-dependent case, $\delta\mathcal{A}/\delta\langle\Psi(t)| = 0$, in order to yield the time-dependent Schrödinger equation. Thus, the corresponding action density functional, $\mathcal{A}[n]$, must have a stationary point at the correct time-dependent density, which is obtained by solving the Euler equation, $\delta\mathcal{A}[n]/\delta n(\mathbf{r},t)$, with appropriate boundary conditions; intricacies [1541] related to the proper variational procedure and causality are reviewed in Refs. [910, 1542, 1543]. Using again the Kohn–Sham auxiliary system of non-interacting electrons subject to an external local potential, V^{KS}, the time-dependent density functional theory (TDDFT) equations

$$\left\{ -\frac{1}{2}\nabla^2 + V_{\mathrm{ext}}(\mathbf{r},t) + V_{\mathrm{H}}(\mathbf{r},t) + \frac{\delta\mathcal{A}_{\mathrm{xc}}[n]}{\delta n(\mathbf{r},t)} \right\} \phi_i(\mathbf{r},t) = i\frac{\partial}{\partial t}\phi_i(\mathbf{r},t) \tag{5.92}$$

$$\left\{ -\frac{1}{2}\nabla^2 + V^{\mathrm{KS}}(\mathbf{r},t) \right\} \phi_i(\mathbf{r},t) = i\frac{\partial}{\partial t}\phi_i(\mathbf{r},t) \tag{5.93}$$

$$H_e^{\mathrm{KS}}(t)\phi_i(\mathbf{r},t) = i\frac{\partial}{\partial t}\phi_i(\mathbf{r},t) \tag{5.94}$$

can be derived; note that the nuclei are treated as dynamical classical degrees of freedom, $\mathbf{R}_I(t)$. The time-dependent external potential might contain an additional contribution that describes the matter-electromagnetic field interaction

$$V_{\mathrm{ext}}(\mathbf{r},t) = -\sum_I \frac{Z_I}{|\mathbf{R}_I(t) - \mathbf{r}|} + \sum_{I<J} \frac{Z_I Z_J}{|\mathbf{R}_I(t) - \mathbf{R}_J(t)|}$$

$$+ f(t)\,\mathbf{r}\,\mathbf{e}\,\sin(\Omega t) \;, \tag{5.95}$$

such as for instance the coupling to an external laser beam within the electric dipole (E1) approximation; $\mathbf{e}$ and Ω are the polarization and frequency of the field and $f(t)$ is an envelope function that describes amplitude and shape of the pulse in the time domain. By construction, the density of the interacting system is obtained from the time-dependent Kohn–Sham orbitals

$$n(\mathbf{r}, t) = \sum_i^{\text{occ}} |\phi_i(\mathbf{r}, t)|^2 , \tag{5.96}$$

like in the static case from which the time-dependent Hartree potential

$$V_{\text{H}}(\mathbf{r}, t) = \int d\mathbf{r}' \, \frac{n(\mathbf{r}', t)}{|\mathbf{r} - \mathbf{r}'|} \tag{5.97}$$

is obtained straightforwardly.

The exact time-dependent exchange–correlation potential

$$\frac{\delta \mathcal{A}_{\text{xc}}[n]}{\delta n(\mathbf{r}, t)} = V_{\text{xc}}(\mathbf{r}, t) \tag{5.98}$$

has an extremely complex (and unknown) functional dependence on the density, which is nonlocal in both space and time. This implies that $V_{\text{xc}}(\mathbf{r}, t)$ at position $\mathbf{r}$ and time t can depend on the density at all other positions and all previous times. In addition, $V_{\text{xc}}(\mathbf{r}, t)$ also depends on the initial Kohn–Sham determinant and on the initial many-body wave function. A practical and widely used simplification is the adiabatic local density approximation (ALDA) [233]

$$\mathcal{A}_{\text{xc}}^{\text{ALDA}}[n] = \int_{t_i}^{t_f} \int n(\mathbf{r}', t') \, \varepsilon_{\text{xc}}^{\text{LDA}} \left(n(\mathbf{r}', t') \right) \, d\mathbf{r}' \, dt' \tag{5.99}$$

or GGA extensions thereof, which is (semi-)local in both space and time, cf. Eq. (2.127). Due to the latter property, it is expected that adiabatic functionals work best for time-dependent systems that only deviate slightly from the ground state. A severe shortcoming of ALDA and its GGA extensions when used in time-dependent calculations where electrons are pushed to regions far away from the nuclei is the incorrect asymptotic behavior of the potential. The exact exchange–correlation potential decays as $-1/r$ for neutral finite systems, while LDA and most GGAs feature a much faster exponential decay to zero. This deficiency can be cured, for instance by using asymptotically corrected GGA-type functionals or orbital-dependent exact exchange (EXX) functionals. Much more severe is the issue of taking into account the nonlocal time dependence by introducing "memory effects", which is still in its infancy; see Ref. [910] for a discussion of functionals in the context of time-dependent density functional theory.

There are two principal technical approaches to solving the time-dependent Kohn–Sham equations in real time, both of which are mean-field approaches to nonadiabatic dynamics in the spirit of Ehrenfest molecular dynamics [355].

It is possible to expand the Kohn–Sham orbitals in terms of a local basis set taking into account the nonadiabatic coupling vectors explicitly [800, 1257–1259, 1400]; an extension to include the coupling to external laser fields can also be devised [801, 1525]. Alternatively, the Kohn–Sham orbitals can be propagated directly in real time [42], which takes the nonadiabatic couplings implicitly into account provided the set of nonlinearly coupled Kohn–Sham equations, Eq. (5.92), is evolved fully self-consistently; note that the couplings can be extracted *a posteriori* from such time-dependent Kohn–Sham results [62]. To this end, the time evolution operator

$$\mathcal{U}(t_{\mathrm{f}}, t_{\mathrm{i}}) \;=\; \mathcal{T} \exp\left[-i \int_{t_{\mathrm{i}}}^{t_{\mathrm{f}}} H_{\mathrm{e}}^{\mathrm{KS}}(t')\, dt'\right] \tag{5.100}$$

must be applied to the set of occupied Kohn–Sham orbitals

$$\phi_i(\mathbf{r}, t_{\mathrm{f}}) \;=\; \mathcal{U}(t_{\mathrm{f}}, t_{\mathrm{i}})\phi_i(\mathbf{r}, t_{\mathrm{i}}) \;, \quad i = 1, 2, \ldots, \mathrm{occ} \tag{5.101}$$

where $\mathcal{T}$ ensures proper time ordering in the exponential since $H_{\mathrm{e}}^{\mathrm{KS}}(t')$ is explicitly time-dependent even in the absence of external time-varying fields. In practice, it is not possible to obtain $\phi_i(\mathbf{r}, t_{\mathrm{f}})$ directly from $\phi_i(\mathbf{r}, t_{\mathrm{i}})$ for a long propagation time interval $[t_{\mathrm{i}}, t_{\mathrm{f}}]$. Instead, it is convenient to break it up into much smaller time intervals of constant length Δt

$$\mathcal{U}(t_{\mathrm{f}}, t_{\mathrm{i}}) = \prod_{j=0}^{N-1} \mathcal{U}(t_{\mathrm{j}} + \Delta t, t_{\mathrm{j}}) \;, \tag{5.102}$$

making use of the semigroup property of $\mathcal{U}$ with $t_0 = t_{\mathrm{i}}$ and $t_N = t_{\mathrm{f}}$. This reduces the propagation to a series of subsequent small propagation steps

$$\phi_i(\mathbf{r}, t_j + \Delta t) =$$

$$\mathcal{T} \exp\left[-i \int_{t_j}^{t_j + \Delta t} H_{\mathrm{e}}^{\mathrm{KS}}(t')\, dt'\right] \phi_i(\mathbf{r}, t_j) \;, \quad i = 1, 2, \ldots, \mathrm{occ} \tag{5.103}$$

so that efficient short-time approximations to the full evolution operator can be used.

Various techniques [42, 235, 854, 1411, 1598] can be used to propagate the Kohn–Sham equations directly in real time [77, 451, 1316, 1463, 1637, 1638]. One such scheme has been implemented in the CPMD package [1451]. In particular, the set of integral equations

$$\phi_i^{(n)}(t_j + \Delta t) = \phi_i^{(0)}(t_j + \Delta t)$$

$$- i \int_{t_j}^{t_j + \Delta t} H_{\mathrm{e}}^{\mathrm{KS}}(\{\phi_j^{(n-1)}(t')\}, t')\phi_i^{(n-1)}(t') \, dt' \quad (5.104)$$

is solved iteratively at each time step Δt for all Kohn–Sham orbitals $i = 1, 2, \ldots, \mathrm{occ}$. The integral involving the Kohn–Sham Hamiltonian is solved by Chebyshev quadrature in the time domain [63, 235, 854].

Within the time-dependent Kohn–Sham approach, electronic excitations can be achieved by a small momentum kick to all orbitals [63, 1637], by coupling to external laser fields [801], or by creating a one-particle/one-hole excitation in orbital space [1451]. Common to all these approaches is the difficulty in assigning a proper symmetry to the electronic state - including the initial one - as discussed in Section 5.3.2. Furthermore, when the time-dependent Kohn–Sham equations, Eq. (5.92), are integrated together with the corresponding Newtonian equation of the nuclei, Eq. (2.38), Ehrenfest molecular dynamics is performed using average forces obtained from such a mixed electronic state, cf. Eq. (2.39). Alternatively, the linear response approximation to the time-dependent Kohn–Sham equations can be used, where gradients from electronic states of well-defined symmetry can be obtained as worked out in Section 5.3.6. In this case, however, a suitable hopping criterion – such as that of Tully or Landau–Zener – has to be used in order to couple the different electronic states as outlined in Section 5.3.3, in addition to computing the necessary nonadiabatic couplings explicitly.

5.3.6 LR-TDDFT: linear response and gradients

5.3.6.1 Time-dependent linear response method

The time-dependent Kohn–Sham equation derived according to the Runge-Gross scheme [1255] (see Refs. [42, 559, 910, 912, 1542] for reviews) from the action functional of Eq. (5.91) can be written as

$$\left[-\frac{\hbar^2}{2m_{\mathrm{e}}}\nabla^2 + V_{\mathrm{eff}}(\mathbf{r}, t) \right]\phi_n(\mathbf{r}, t) = i\hbar\frac{\partial}{\partial t}\phi_n(\mathbf{r}, t) , \quad (5.105)$$

with a local one-particle potential V_{eff} given by

$$V_{\mathrm{eff}}(\mathbf{r}, t) = V^{\mathrm{KS}}(\mathbf{r}, t) + V_{\mathrm{int}}(\mathbf{r}, t)$$

$$= \sum_I V_I(\mathbf{r} - \mathbf{R}_I) + \int \frac{n(\mathbf{r}', t)}{|\mathbf{r} - \mathbf{r}'|} \, d\mathbf{r}' + V_{\mathrm{xc}}[n](\mathbf{r}, t) + V_{\mathrm{int}}(\mathbf{r}, t)$$

$$\tag{5.106}$$

where V^{KS} is the usual Kohn–Sham potential and V_{int} is the matter-field interaction. Just as the exchange–correlation potential V_{xc} within ALDA, see Section 5.3.5 Eq. (5.99), is obtained by a functional derivative

$$V_{\mathrm{xc}}(\mathbf{r}, t) = \frac{\delta E_{\mathrm{xc}}}{\delta n(\mathbf{r}, t)} \tag{5.107}$$

so is, in principle, the matter-field interaction potential describing the perturbation

$$V_{\mathrm{int}}(\mathbf{r}, t) = \frac{\delta E_{\mathrm{int}}}{\delta n(\mathbf{r}, t)} = -L^3 \, \mathbf{E}_{\mathrm{T}} \cdot \frac{\delta \mathbf{P}}{\delta n(\mathbf{r}, t)} \tag{5.108}$$

obtained from the energy expression in the dipole approximation

$$E_{\mathrm{int}} = -L^3 \mathbf{E}_{\mathrm{T}} \cdot \mathbf{P} \tag{5.109}$$

where L^3 is the volume of the cubic supercell, $\mathbf{E}_{\mathrm{T}}$ the transversal component of the applied electric field, and $\mathbf{P}(t)$ the expectation value of the polarization per unit volume. For the dipole form of the polarization

$$\mathbf{P}(t) = -\frac{e}{L^3} \int n(\mathbf{r}, t) \mathbf{r} \, d\mathbf{r} \ , \tag{5.110}$$

evaluation of the functional derivative yields the familiar *er* potential. For periodic systems Eq. (5.110) has to be replaced by the Berry phase [1329] polarization [740, 1210], see also Section 7.2.2. The way this case has to be handled, for which the functional derivative with respect to density is not known explicitly, will be outlined in Sections 5.3.6.3 and 7.1.4.

In linear response theory the coupling to an external dynamical probe, such as V_{int} of Eq. (5.108), is treated as a harmonic perturbation of frequency ω

$$\delta V(\mathbf{r}, t) = \delta V^{+}(\mathbf{r}) \exp\left[i\omega t\right] + \delta V^{-}(\mathbf{r}) \exp\left[-i\omega t\right] \ , \tag{5.111}$$

which will induce a change in the effective potential given to first order by

$$\delta V_{\mathrm{eff}}(\mathbf{r}, t) = \delta V(\mathbf{r}, t) + \delta V_{\mathrm{SCF}}(\mathbf{r}, t) \ , \tag{5.112}$$

where

$$\delta V_{\text{SCF}}(\mathbf{r}, \pm\omega) = \int \left\{ \frac{1}{|\mathbf{r} - \mathbf{r'}|} + \left. \frac{\delta^2 E_{\text{xc}}}{\delta n(\mathbf{r})\delta n(\mathbf{r'})} \right|_{n=n^{\{0\}}} \right\} n^{\{1\}}(\mathbf{r'}, \pm\omega) \, d\mathbf{r'} \quad (5.113)$$

is the change in the self-consistent field in the frequency domain and $n^{\{0\}}$ is the (unperturbed) density in the absence of any external field. The linear density response

$$n^{\{1\}}(\mathbf{r}, \pm\omega) = 2 \sum_{i=1}^{\text{occ}} \langle \phi_i^{\{\mp\}} | \mathbf{r} \rangle \langle \mathbf{r} | \phi_i^{\{0\}} \rangle + \langle \phi_i^{\{0\}} | \mathbf{r} \rangle \langle \mathbf{r} | \phi_i^{\{\pm\}} \rangle \quad (5.114)$$

is expressed in terms of a set of linear response orbitals $\{\phi_i^{\{\pm\}}(\mathbf{r})\}$ which can be chosen to be orthogonal to the subspace of the ground state orbitals $\{\phi_i^{\{0\}}(\mathbf{r})\}$

$$\langle \phi_i^{\{\pm\}} | \phi_j^{\{0\}} \rangle = 0 . \quad (5.115)$$

This leads to the coupled perturbed Kohn–Sham equations [659]

$$\sum_{j=1}^{\text{occ}} (\epsilon_{ij} - (H^{\text{KS}} \pm \omega)\delta_{ij}) \, |\phi_j^{\{\pm\}}\rangle = Q \left(\delta V^{\{\pm\}} + \delta V_{\text{SCF}}(\pm\omega) \right) \, |\phi_i^{\{0\}}\rangle , \quad (5.116)$$

where

$$Q = 1 - \sum_{k=1}^{\text{occ}} |\phi_k^{\{0\}}\rangle\langle\phi_k^{\{0\}}| \quad (5.117)$$

is the projector on the subspace of the unperturbed unoccupied states, H^{KS} is the Kohn–Sham Hamiltonian, and ϵ_{ij} is the set of Lagrange multipliers ensuring the orthonormality of the ground state orbitals. Excitation energies correspond to the poles of the response functions [232, 233] and lead to singularity in Eq. (5.116). Thus, they are solutions to the equations

$$\sum_{j=1}^{\text{occ}} (H^{\text{KS}}\delta_{ij} - \epsilon_{ij}) \, |\phi_j^{\{\pm\}}\rangle + Q \, \delta V_{\text{SCF}}(\pm\omega) \, |\phi_i^{\{0\}}\rangle = \mp\omega \, |\phi_i^{\{\pm\}}\rangle , \quad (5.118)$$

which is the central result of linear response time-dependent density functional theory in the frequency domain [232, 233, 364, 559, 910].

An early implementation of the time-dependent linear response equations, Eq. (5.118), within the plane wave/pseudopotential framework has been carried out by Doltsinis and Sprik [357]. Following the methods established in quantum chemistry, they used an intermediate basis of canonical Kohn–Sham orbitals. In order to avoid excessive matrix sizes, the virtual space had to be truncated. Based on this general scheme, nuclear gradients and also the nonadiabatic coupling vectors have been derived more recently [354].

5.3.6.2 Tamm–Dancoff approximation

The solution of Eq. (5.118) can be cast as a non-Hermitian eigenvalue equation with eigenvalues ω^2 [233, 659]. In the Tamm–Dancoff approximation (TDA) [415, 639, 640], this can be simplified to a Hermitian problem with excitation energies as eigenvalues. By expanding the ground state Kohn–Sham orbitals $\{\phi_i^{\{0\}}\}$ and the linear response orbitals $\{\phi_i^{\{\pm\}}\}$ in the orthogonal basis set functions $\pi_p(\mathbf{r})$, assumed here to be plane waves

$$\pi_p = L^{-3/2} \exp\left[i\mathbf{G} \cdot \mathbf{r}\right] \tag{5.119}$$

with $\mathbf{G}$ being a reciprocal lattice vector of the simulation supercell,

$$\phi_i^{\{0\}}(\mathbf{r}) = \sum_{p=1}^{M} c_{pi}^{\{0\}} \pi_p(\mathbf{r}) \tag{5.120}$$

$$\phi_i^{\{\pm\}}(\mathbf{r}) = \sum_{p=1}^{M} c_{pi}^{\{\pm\}} \pi_p(\mathbf{r}) \tag{5.121}$$

and introducing new sets of coefficients

$$\begin{aligned} x_{pi} &= \left(c_{pi}^{\{+\}} + c_{pi}^{\{-\}}\right) \\ y_{pi} &= \left(c_{pi}^{\{+\}} - c_{pi}^{\{-\}}\right), \end{aligned} \tag{5.122}$$

Eq. (5.116) can be rewritten as

$$\begin{aligned} (\mathcal{A} + \mathcal{B})\mathbf{x}^\mu + \omega\mathbf{y}^\mu &= \mathbf{b}^\mu \\ (\mathcal{A} - \mathcal{B})\mathbf{y}^\mu + \omega\mathbf{x}^\mu &= \mathbf{0}, \end{aligned} \tag{5.123}$$

where the superscript μ indicates the direction of the induced polarization. The operators $\mathcal{A}$ and $\mathcal{B}$ act on a general $(M \times N)$ matrix $\mathbf{z}$ according to

$$\begin{aligned} \mathcal{A}\mathbf{z} &= (\mathbf{H}^{\text{KS}}\mathbf{z} - \mathbf{z}\epsilon) + \mathbf{QWc}^{\{0\}} \\ \mathcal{B}\mathbf{z} &= \mathbf{QWc}^{\{0\}}, \end{aligned} \tag{5.124}$$

where the Kohn–Sham Hamiltonian,

$$H_{pq}^{\text{KS}} = \langle \pi_p | H^{\text{KS}} | \pi_q \rangle, \tag{5.125}$$

Lagrange multiplier,

$$\epsilon_{ij} = \sum_{pq} (c_{pi}^{\{0\}})^{\star} H_{pq}^{KS} c_{qj}^{\{0\}} , \tag{5.126}$$

virtual state projector,

$$Q_{pq} = \delta_{pq} - \sum_{i=1}^{occ} c_{pi}^{\{0\}} (c_{qi}^{\{0\}})^{\star} , \tag{5.127}$$

and δV_{SCF},

$$W_{pq} = \langle \pi_p | \delta V_{SCF} | \pi_q \rangle \tag{5.128}$$

matrix elements are all expressed in the plane wave representation. The matrix

$$\mathbf{b}^{\mu} = 2\mathbf{Q}\mathbf{f}^{\mu}\mathbf{c}^{\{0\}} \tag{5.129}$$

includes the $(M \times M)$ perturbation matrix $\mathbf{f}$, whose elements are given by

$$f_{pq}^{\mu} = -L^3 \left\langle \pi_p \left| \frac{\delta P_{\mu}}{\delta n(\mathbf{r})} \right|_{n=n^{(0)}} \right| \pi_q \right\rangle . \tag{5.130}$$

Within the present formalism, the TDA amounts to setting the expansion coefficients $c_{pi}^{\{+\}}$ to zero in Eq. (5.121), leading to $\mathbf{x}^{\mu} = -\mathbf{y}^{\mu}$. Therefore, Eq. (5.123) is replaced by

$$(\mathcal{A} - \omega\mathbf{1})\mathbf{x}^{\mu} = \mathbf{b}^{\mu} \tag{5.131}$$

and the excitation energies are now computed from the Hermitian eigenvalue problem

$$\mathcal{A}\mathbf{x}_I = \omega_I \mathbf{x}_I . \tag{5.132}$$

Provided the eigenvectors $\mathbf{x}_I$ are normalized, the linear response can then be obtained

$$\mathbf{x}^{\mu} = \frac{\mathbf{b}^{\mu}}{\mathcal{A} - \omega\mathbf{1}} = \sum_I \frac{\mathbf{x}_I \mathbf{x}_I^T \mathbf{b}^{\mu}}{\omega_I - \omega} \tag{5.133}$$

using the spectral resolution of $\mathcal{A}$.

5.3.6.3 Dynamical polarizability and oscillator strengths in extended systems

In condensed phase molecular systems such as molecular liquids and crystals, the Clausius-Mossotti approximation used in the gas phase is not strictly valid because of the small but finite overlap of molecular charge distributions. However, if the identity of molecules can still be established, which can be considered as the very definition of a condensed molecular system, there is hope that a "modified" form of the dipole (E1) approximation might still apply. Indeed, this is the assumption underlying perturbative treatments of molecular interactions and spectroscopy (see e.g. Ref. [1405]). This approach depends crucially on an expansion in a basis of zero-order molecular orbitals of individual molecules, which has hampered for a long time the implementation in plane wave codes where such a basis is not available *a priori*. The problem has been solved in the general framework of the modern theory of polarization developed by Vanderbilt, Resta and others [740, 1210–1212, 1375]. Their approach enables one to define and compute the polarization in an extended system without any *a priori* partitioning in molecular units. The central result in this formalism is the "Berry phase" expression for the electronic polarization $\mathbf{r}$

$$P_\mu = \frac{2e}{G_\mu L^3} \operatorname{Im} \ln \det \mathbf{S} \;, \tag{5.134}$$

where μ denotes a Cartesian component $\mu = x, y, z$, assuming a cubic cell of length L, and $\mathbf{G}_\mu$ is a vector spanning the corresponding reciprocal lattice with length $G_\mu = 2\pi/L$. Here, the formula is given in the Γ-point approximation valid for large supercells and a generalization to lattices of lower symmetry is worked out in Ref. [109] and presented in Section 7.2.2. The matrix $\mathbf{S}$ is defined by the matrix elements of a phase operator

$$S_{mn} = \langle \phi_m \,|\, \exp\left[-i\,\mathbf{G}_\mu \cdot \mathbf{r}\right]|\, \phi_n \rangle \tag{5.135}$$

in terms of the set of Kohn–Sham orbitals ϕ_m (assuming a spin-restricted description). The Berry phase expression for polarization and its equivalent formulation in terms of Wannier functions introduced in Section 7.2 has initiated a revolution in electronic structure calculations in the condensed phase. The expression Eq. (5.134) was derived for ground state systems and the present application to electronic spectroscopy is based on an extension to excited states [114] achieved by replacing the Kohn–Sham orbitals in Eq. (5.135) by the time dependent Kohn–Sham orbitals $\phi_n(\mathbf{r}, t)$

$$P_\mu(t) = \frac{2e}{G_\mu L^3} \operatorname{Im} \ln \det \langle \phi_m(t) \,|\, \exp\left[-i\,\mathbf{G}_\mu \cdot \mathbf{r}\right]|\, \phi_n(t) \rangle \;. \tag{5.136}$$

The Berry phase polarization of Eq. (5.136) is not available as an explicit functional of the density. In this respect it resembles exchange–correlation functionals that contain an exact exchange contribution, such as B3LYP. The common procedure to deal with orbital-dependent density functionals in quantum chemistry calculations is to use the orbital derivative rather than the density derivative. This approach, deviating from the true spirit of the Kohn–Sham method, will also be employed here in order to deal with the polarization. In the context of a perturbation calculation, this amounts to replacing the factor $\mathbf{fc}$ in the expression Eq. (5.129) for the driving force $\mathbf{b}$ by the $M \times N$ matrix $\mathbf{d}$ defined as [1171]

$$d_{pi}^\mu = -L^3 \int \pi_p^* (\mathbf{r}) \left. \frac{\delta P_\mu [\phi]}{\delta \phi_i^* (\mathbf{r})} \right|_{\phi_i=\phi_i^{\{0\}}} d\mathbf{r} \;. \tag{5.137}$$

The key to compute this functional derivative is the expression for the first-order variation of the Berry phase

$$\delta P_\mu = -\sum_\nu \frac{2e}{L^3 |\mathbf{G}_\mu|}\, \mathrm{Im}\, \left\{ \sum_{ij}^{\mathrm{occ}} \left[\langle \delta\phi_i^{\{\nu\}} | \exp\left[-i\mathbf{G}_\mu \cdot \mathbf{r}\right] |\phi_j^{\{0\}}\rangle \right. \right.$$

$$\left. + \langle \phi_i^{\{0\}} | \exp\left[-i\mathbf{G}_\mu \cdot \mathbf{r}\right] |\phi_j^{\{\nu\}}\rangle \right] \left[\langle \phi_j^{\{0\}} | \exp\left[-i\mathbf{G}_\nu \cdot \mathbf{r}\right] |\phi_i^{\{0\}}\rangle \right]^{-1} \right\} \;, \tag{5.138}$$

where $\phi_i^{\{\nu\}} \equiv \phi_i^{\{\nu\}}(t)$ represents the linear response of $\phi_i^{\{0\}}$ to a small perturbation $E_\nu(t)$ in the Cartesian direction ν; see e.g. Ref. [1171]. This relation, Eq. (5.138), can be applied immediately to obtain the polarizability tensor $\boldsymbol{\alpha}$. In the vector notation introduced above this reads

$$\alpha_{\mu\nu}(\omega) = \frac{2}{|\mathbf{G}_\mu|}\, \mathrm{Im}\, \left[(\mathbf{x}^\nu)^T \mathbf{g}^\mu \mathbf{c} + \mathbf{c}^T \mathbf{g}^\mu \mathbf{x}^\nu \right] \left[(\mathbf{c}^T \mathbf{g}^\nu \mathbf{c})^{-1} \right]^T \;, \tag{5.139}$$

where

$$g_{pq}^\mu = \langle \pi_p | \exp\left[-i\mathbf{G}_\mu \cdot \mathbf{r}\right] |\pi_q\rangle \tag{5.140}$$

was defined. Inserting the TDA solution of Eq. (5.133) for the first-order orbitals, one can rewrite Eq. (5.139) as

$$\alpha_{\mu\nu}(\omega) = \frac{4}{|\mathbf{G}_\mu|}\, \mathrm{Im}\, \left[(\mathbf{b}^\mu)^T \mathbf{x}(\mathcal{A} - \omega\mathbf{1})^{-1}\mathbf{x}^T \mathbf{g}^\nu \mathbf{c} \right] \left[(\mathbf{c}^T \mathbf{g}^\nu \mathbf{c})^{-1} \right]^T \tag{5.141}$$

and obtain for the oscillator strength of the Ith excitation mode

$$f_I = -\frac{4}{3}\, \omega_I \sum_\mu \frac{1}{|\mathbf{G}_\mu|}\, \mathrm{Im}\, \left[(\mathbf{b}^\mu)^T \mathbf{x}_I (\mathbf{x}_I)^T \mathbf{g}^\mu \mathbf{c} \right] \left[(\mathbf{c}^T \mathbf{g}^\mu \mathbf{c})^{-1} \right]^T \tag{5.142}$$

following standard time-dependent linear response theory. For $\mathbf{b}$ one must use the generalization of Eq. (5.129) with $\mathbf{d}$ of Eq. (5.137) instead of the product $\mathbf{fc}$

$$\mathbf{b}^\mu = \mathbf{Q}\mathbf{d}^\mu = 2 \left[\mathbf{d}^\mu - \mathbf{c} \left(\mathbf{c}^T \mathbf{d}^\mu \right) \right] \;, \tag{5.143}$$

where the second equality follows as a consequence of Eq. (5.127). Substituted into Eq. (5.142), this leads after some rearrangement (note $\mathbf{c}^T\mathbf{x} = \mathbf{0}$, as the linear response orbitals are orthogonal to the ground state orbitals, Eq. (5.115)) to

$$f_I = -\frac{8}{3}\,\omega_I \sum_\mu \frac{1}{|\mathbf{G}_\mu|}\,\mathrm{Im}\,((\mathbf{d}^\mu)^T\mathbf{x}_I)(\mathbf{x}_I^T\mathbf{g}^\mu\mathbf{c})[(\mathbf{c}^T\mathbf{g}^\mu\mathbf{c})^{-1}]^T\;. \qquad (5.144)$$

The complex arithmetic in Eq. (5.144) is somewhat cumbersome. From a practical point of view its function is to ensure that the transition dipoles defined further below satisfy the Born–von Kármán periodic boundary conditions as assumed when the Γ-point approximation is enforced. With this in mind, one can simplify Eq. (5.144), while strictly adhering to periodic boundary conditions, by setting

$$\mathbf{d}^\mu \approx \frac{\mathbf{g}^\mu\mathbf{c}}{|\mathbf{G}_\mu|}\;. \qquad (5.145)$$

Introducing the periodic transition dipoles using response orbitals $\phi_i^{\{-\}}$ of the excited state in TDA

$$(\gamma_{01}^\mu)_{ij} = i\,\frac{\langle\phi_i^{\{0\}}|\exp[-i\mathbf{G}_\mu\cdot\mathbf{r}]|\phi_j^{\{-\}}\rangle}{|\mathbf{G}_\mu|}\;, \qquad (5.146)$$

$$(\gamma_{10}^\mu)_{ij} = i\,\frac{\langle\phi_i^{\{-\}}|\exp[-i\mathbf{G}_\mu\cdot\mathbf{r}]|\phi_j^{\{0\}}\rangle}{|\mathbf{G}_\mu|}\;, \qquad (5.147)$$

$$(\gamma_{00}^\mu)_{ij} = i\,\frac{\langle\phi_i^{\{0\}}|\exp[-i\mathbf{G}_\mu\cdot\mathbf{r}]|\phi_j^{\{0\}}\rangle}{|\mathbf{G}_\mu|}\;, \qquad (5.148)$$

the oscillator strength can be expressed as

$$f_I = -\frac{8}{3}\,\omega_I\sum_\mu\frac{1}{|\mathbf{G}_\mu|^2}\,\mathrm{Im}\,\gamma_{01}^\mu\gamma_{10}^\mu\,[(\gamma_{00}^\mu)^{-1}]^T\;. \qquad (5.149)$$

It is noted that, in the limit of large cell size and for localized states, $\gamma_{00}^\mu|\mathbf{G}_\mu| \longrightarrow 1i$. An expansion of the exponential function in Eq. (5.147) leads to the conventional formula for the oscillator strengths in confined systems

$$f_I = \frac{2}{3}\omega_I\sum_\mu|\mathbf{P}_{01}^\mu|^2\;, \qquad (5.150)$$

where $\boldsymbol{P}_{01}^{\mu} = \langle \Psi_0 | x_{\mu} | \Psi_1 \rangle$ is the conventional "transition dipole" for an excitation from the ground state Ψ_0 to an excited state Ψ_1 whose energies differ by $\omega_I = E_1 - E_0$.

5.3.6.4 Derivatives and LR-TDDFT molecular dynamics

Clearly, nuclear gradients of the linear response energy functional are needed in order to perform molecular dynamics simulations in a specified excited state. The calculation of analytic derivatives has a long tradition in quantum chemistry [718, 1164, 1166], and special techniques have been developed for the cases of non-variational energy expressions. The Lagrangian method [624] allows for the most compact derivation and will be applied in the following section. The derivative of the total energy of an excited state with respect to an external parameter η is given by

$$\frac{dE_{\text{tot}}[\mathbf{c}^{\{0\}}, \mathbf{x}]}{d\eta} = \frac{dE_{\text{KS}}[\mathbf{c}^{\{0\}}]}{d\eta} + \frac{dE_{\text{TDA}}[\mathbf{c}^{\{0\}}, \mathbf{x}]}{d\eta} , \qquad (5.151)$$

where E_{KS} is the Kohn–Sham ground state energy and E_{TDA} the TDA excitation energy. In particular, taking this derivative with respect to the positions of the nuclei allows one to compute the gradient

$$\nabla_I E_{\text{tot}}[\mathbf{c}^{\{0\}}, \mathbf{x}] = \nabla_I E_{\text{KS}}[\mathbf{c}^{\{0\}}] + \nabla_I E_{\text{TDA}}[\mathbf{c}^{\{0\}}, \mathbf{x}] \qquad (5.152)$$

in a specified excited state, where the first term is just the standard ground state gradient of the Kohn–Sham functional as given in Section 3.4.3. All derivatives have to be evaluated taking into account all constraints of the wave function parameters. The Lagrangian function that is variational in all wave function parameters [474, 475, 659] is given by

$$\mathcal{L}_{\text{tot}}[\mathbf{c}^{\{0\}}, \mathbf{x}, \boldsymbol{\Lambda}, \mathbf{Z}, \omega] = \mathcal{L}_{\text{KS}}[\mathbf{c}^{\{0\}}, \boldsymbol{\Lambda}] + \mathcal{L}_{\text{TDA}}[\mathbf{c}^{\{0\}}, \mathbf{x}, \omega]$$

$$+ \sum_{pi} Z_{pi} \left\{ \sum_q H_{pq}^{\text{KS}} c_{qi}^{\{0\}} - \sum_j c_{qj}^{\{0\}} \Lambda_{ji} \right\} , \qquad (5.153)$$

where $\mathcal{L}_{\text{KS}}$ is the Lagrange function of the time-independent Kohn–Sham equations

$$\mathcal{L}_{\text{KS}}[\mathbf{c}^{\{0\}}, \boldsymbol{\Lambda}] = E_{\text{KS}}[\mathbf{c}^{\{0\}}] - \sum_{ij} \Lambda_{ij} \left\{ \sum_p (c_{pi}^{\{0\}})^{\star} c_{pj}^{\{0\}} - \delta_{ij} \right\} . \qquad (5.154)$$

In analogy to Eq. (2.49), $\mathcal{L}_{\text{TDA}}$ denotes the Lagrange function of the linear response TDDFT excitation energy in the Tamm–Dancoff approximation

$$\mathcal{L}_{\text{TDA}}[\mathbf{c}^{\{0\}}, \mathbf{x}, \omega] = \mathbf{x}^{\dagger} \mathcal{A} \mathbf{x} - \omega \left(\mathbf{x}^{\dagger} \mathbf{x} - 1 \right) , \qquad (5.155)$$

and $\mathbf{Z}$ is the matrix of Lagrange multipliers associated with the stationarity condition of the Kohn–Sham orbitals. The orthogonality constraint of $\mathbf{x}$ with respect to the ground state orbitals is not associated with Lagrange multipliers, but is handled by the projector functions in $\mathcal{L}_{\text{TDA}}$.

It is assumed in the following that the ground state orbitals are optimized Kohn–Sham orbitals and that $\mathbf{x}$ is a solution to Eq. (5.132). Then the derivative of $\mathcal{L}_{\text{tot}}$ with respect to η is

$$\mathcal{L}_{\text{tot}}^{(\eta)} = \frac{\partial E_{\text{KS}}}{\partial \eta} + \frac{\partial E_{\text{TDA}}}{\partial \eta} + \sum_{pqi\sigma} Z_{pi\sigma} \frac{\partial H_{pq\sigma}^{\text{KS}}}{\partial \eta} c_{qi\sigma}^{\{0\}} . \qquad (5.156)$$

In deriving Eq. (5.156) it was assumed that the orthogonality and normalization constraints for the Kohn–Sham and response orbitals is independent from the parameter η. In the case of the plane wave basis set this is true for all of the most important types of perturbations, especially for nuclear displacements and thus gradients. Using the properties of this particular basis set, the Lagrangian Eq. (5.156) (with $\partial \mathbf{H}^{\text{KS}}/\partial \eta = \mathbf{H}^{(\eta)}$)

$$\mathcal{L}_{\text{tot}}^{(\eta)} = \sum_{pqi} (c_{pi}^{\{0\}})^{\star} H_{pq}^{(\eta)} c_{qi}^{\{0\}} + \sum_{pi} \sum_{qj} (x_{pi})^{\star} \mathcal{A}_{pi,qj}^{(\eta)} x_{qj}$$

$$+ \sum_{pqi} Z_{pi} H_{pq}^{(\eta)} c_{qi}^{\{0\}} \qquad (5.157)$$

and the derivative of the TDA energy

$$\frac{\partial E_{\text{TDA}}}{\partial \eta} = \sum_{pi} \sum_{qj} (x_{pi})^{\star} \mathcal{A}_{pi\sigma,qj}^{(\eta)} x_{qj}$$

$$= \sum_{pqi} (x_{pi})^{\star} H_{pq}^{(\eta)} x_{qi} - \sum_{pji} \sum_{uv} (x_{pi})^{\star} (c_{ui}^{\{0\}})^{\star} H_{uv}^{(\eta)} c_{vj}^{\{0\}} x_{pj} . \qquad (5.158)$$

can be simplified furthermore. Introducing the density matrices $\mathbf{P}^{(x)}$ and $\mathbf{P}^{(z)}$

$$P_{qp}^{(x)} = \sum_{i} x_{qi} (x_{pi})^{\star} - \sum_{rij} x_{rj} c_{pj}^{\{0\}} (c_{qi}^{\{0\}})^{\star} (x_{ri})^{\star} \qquad (5.159)$$

$$P_{qp}^{(z)} = \sum_{i} Z_{pi} c_{qi}^{\{0\}} \qquad (5.160)$$

and the corresponding densities $n^{(x)}$ and $n^{(z)}$, the total derivative can be written in compact form as

$$\mathcal{L}_{\text{tot}}^{(\eta)} = \sum_{pq} H_{pq}^{(\eta)} \left(P_{qp} + P_{qp}^{(x)} + P_{qp}^{(z)} \right) . \tag{5.161}$$

What still needs to be done is the calculation of the Lagrange multipliers **Z**. They can be determined from the stationarity condition of the total Lagrange function, Eq. (5.153), with respect to variations of the Kohn–Sham orbitals, i.e.

$$\frac{\partial \mathcal{L}_{\text{tot}}}{\partial \mathbf{c}^{\{0\}}} = 0 . \tag{5.162}$$

Again making use of the fact that the derivatives are taken at the point of optimized Kohn–Sham orbitals, one arrives at a system of linear equations for **Z**

$$\sum_{qj} \left(H_{pq}^{\text{KS}} \delta_{ij} - \epsilon_{ij} \delta_{pq} \right) Z_{qj}^{\star} + \sum_{qr} Q_{pr} W_{rq}[n^{(z)}] c_{qi}^{\{0\}} = u_{pi} . \tag{5.163}$$

This equation is known as the "Handy-Schaefer Z vector" equation [599] and has the same form as the coupled perturbed Kohn–Sham equations from static density functional perturbation theory, see Eq. (5.116). It only differs from these equations by its right-hand side **u** to be derived below. From Eq. (5.163) it also becomes clear that **Z** fulfills the same orthogonality constraint as a linear response orbital

$$\sum_{p} Z_{pi} c_{pj}^{\{0\}} = 0 , \tag{5.164}$$

and therefore has the correct number of degrees of freedom. The vector u is calculated from

$$u_{pi} = \sum_{rk} \sum_{qj} (x_{rk})^{\star} \frac{\partial \mathcal{A}_{rk,qj}}{\partial (c_{pi}^{\{0\}})^{\star}} x_{qj} . \tag{5.165}$$

Special care has to be taken to include the projections on the virtual states correctly. The final result is

$$u_{pi} = \sum_{rq} Q_{pr} \left\{ W_{rq}[n^{(x)}] c_{qi}^{\{0\}} + W_{rq}[n^{\{1\}}] x_{qi} \right.$$

$$\left. + x_{rj} \sum_{s} \left((c_{qj}^{\{0\}})^{\star} W_{qs}[n^{\{1\}}] c_{si}^{\{0\}} \right) + W_{rq}^{(2)}[n^{\{1\}}] c_{qi}^{\{0\}} \right\} , \tag{5.166}$$

where $\mathbf{W}^{(2)}$ is the matrix representation of the potential from the third functional derivative of the exchange–correlation energy

$$W_{pq}^{(2)}(\delta n) = \int \phi_p^{\star}(\mathbf{r})\phi_q(\mathbf{r})\,d\mathbf{r} \times$$

$$\int \left.\frac{\delta^3 E_{\mathrm{xc}}}{\delta n(\mathbf{r})\delta n(\mathbf{r}')\delta n(\mathbf{r}'')}\right|_{n^{\{0\}}} n^{\{1\}}(\mathbf{r}')n^{\{1\}}(\mathbf{r}'')\,d\mathbf{r}'d\mathbf{r}'' \ . \quad (5.167)$$

Knowing the derivatives of a certain excited state with respect to the nuclear coordinates, Eq. (5.152), allows one to perform adiabatic *ab initio* molecular dynamics simulations in the linear response TDDFT formalism as exemplified with some applications collected in Section 9.13 and also in Section 9.14. In order to access the nonadiabatic regime it can be combined with surface-hopping schemes along the lines discussed in Section 5.3.3 in conjunction with nonadiabatic coupling vectors [135, 354]. In this case, however, the usefulness of the LR-TDDFT approach [364] to excited states has to be tested carefully close to important regions of configuration space such as avoided crossings or conical intersections.

5.3.6.5 *Analysis of electronic excitations*

Once the eigenvectors $\mathbf{x}_I$ of Eq. (5.132) are known, they can be used to compute the relative contribution of Kohn–Sham single-particle excitations to a given electronic transition I and thus to assign its character (see Section 4.5 in Ref. [232]). To this end a set of L virtual orbitals $\{\psi_i\}$, in canonical form, needs to be obtained from the ground state Kohn–Sham potential which can be expressed in a plane wave expansion Eq. (5.119) as

$$|\psi_i\rangle = \sum_{p=1}^{M} v_{pi}\pi_p \ . \quad (5.168)$$

The set of virtual orbitals is supposed to be large enough

$$|\phi^{\{-\}}\rangle = \sum_{j=1}^{L} w_{i\to j}\psi_j \quad (5.169)$$

to allow for a complete expansion of the linear response orbitals associated with the transition I. The "weight" of the transition from the occupied Kohn–Sham orbital i to the virtual orbital j for the excitation I, $w_{i\to j}$, is given by

$$w_{i\to j} = \langle\psi_j|\phi^{\{-\}}\rangle = \sum_{p=1}^{M} v_{pj}^{\star}x_{pi} \ . \quad (5.170)$$

An important additional piece of information obtained in any *ab initio* molecular dynamics simulation is the modulation of the ionic motion by thermal fluctuations. The excitation frequency ω_I and the oscillator strength f_I of a transition I follow the adiabatic motion of the molecules, producing a time series of excitation frequencies and oscillator strengths, $\omega_I(t)$ and $f_I(t)$, respectively. Thus, sharp single transition lines I get broadened thermally into bands with spectral profiles $F_I(\omega)$, which can be expressed as

$$F_I(\omega) = \frac{1}{t_{\mathrm{f}} - t_{\mathrm{i}}} \int_{t_{\mathrm{i}}}^{t_{\mathrm{f}}} f_I(t)\, \delta(\omega_I(t) - \omega)\, dt = \frac{1}{N_R} \sum_{R=1}^{N_R} f_I^R \delta(\omega_I^R - \omega) , \quad (5.171)$$

where $t_{\mathrm{f}} - t_{\mathrm{i}}$ is the duration of the molecular dynamics simulation. The second identity gives the discretized estimator used to compute $F_I(\omega)$ and has the form of an average over a sample of N_R nuclear configurations $\{\mathbf{R}(t)\}$. As for any such statistical average, it must be guaranteed that the configurations are generated based on a proper thermodynamic ensemble and that the sample yields a sufficient numerical representation of this ensemble using, for instance, the methods from Sections 5.2.2 and 5.2.3 to generate isothermal or isobaric ensembles. The line shape $F_I(\omega)$ must be distinguished from the normalized Excitation Energy Density of States (EEDOS) defined as

$$\mathrm{EEDOS}_I(\omega) = \frac{1}{t_{\mathrm{f}} - t_{\mathrm{i}}} \int_{t_{\mathrm{i}}}^{t_{\mathrm{f}}} \delta(\omega_I(t) - \omega)\, dt = \frac{1}{N_R} \sum_{R=1}^{N_R} \delta(\omega_I^R - \omega) , \quad (5.172)$$

describing the distribution of frequencies ω_I without the weights of the intensities. The overall intensity of a band

$$\bar{f}_I = \frac{1}{t_{\mathrm{f}} - t_{\mathrm{i}}} \int_{t_{\mathrm{i}}}^{t_{\mathrm{f}}} f_I(t)\, dt = \frac{1}{N_R} \sum_{R=1}^{N_R} f_I^R \quad (5.173)$$

can be quantified by the *time-averaged* oscillator strength of a mode I. The more general case where additional quantum-mechanical fluctuations such as zero-point motion or tunneling are included in the framework of *ab initio* path integrals is presented in Section 5.4.4.

5.4 Beyond classical nuclei: path integrals and quantum corrections

5.4.1 Introduction

Up to this point the nuclei were always approximated as classical point particles, which is customarily done in standard molecular dynamics. There are, however, many situations where quantum effects, such as zero-point vibrations and tunneling, play an important or even crucial role and cannot be neglected if the simulation aims at being realistic - which is the generic goal of *ab initio* simulations. The *ab initio* path integral technique [946] and its extension to quasiclassical time evolution [956], which is implemented in the CPMD package [696], is able to cope with such situations at finite temperatures. The central idea of this class of methods is to quantize the nuclei using Feynman's path integrals and, at the same time, to include the electronic degrees of freedom akin to *ab initio* molecular dynamics - that is "on-the-fly". The main ingredients and approximations underlying the *ab initio* path integral approach [931, 937, 946, 948, 956, 1507] are:

- The adiabatic separation of electrons and nuclei where the electrons are kept in their ground state without any coupling to electronically excited states (Born–Oppenheimer or "clamped nuclei" approximation).
- Using an approximate electronic structure approach such as density functional theory in order to calculate the interactions efficiently "on-the-fly".
- Approximating the continuous path integral for the nuclei by a finite discretization (Trotter factorization) and neglecting the indistinguishability of identical nuclei (Boltzmann statistics).
- Using finite supercells with periodic boundary conditions and finite sampling times (finite size and finite time effects) as usual.

Thus, quantum zero-point motion and tunneling effects, as well as thermal fluctuations, are included at some preset temperature without further simplifications consisting, for example, in quasiclassical or quasiharmonic approximations, restricting the Hilbert space, or artificially reducing the dimensionality of the problem. Together, this makes the technique unique in being able to cope with quantum effects in complex molecular condensed phase environments, such as hydrogen-bonded liquids or solids [939].

An extension of the static *ab initio* path integral method derived in Section 5.4.2 to quasiclassical dynamics is the centroid molecular dynamics approximation, its "adiabatic sampling" version being presented in Section 5.4.3. Furthermore, this particular path integral approach has been extended to simulate quantum-broadening effects on photoabsorption cross-sections at finite temperature and thus optical spectra by combining it with time-dependent Kohn–Sham density functional theory. As also shown in Section 5.4.4, this formalism allows in addition the proper derivation of the classical formula where only thermal fluctuations are taken into account. The "quantum correction factor" approach is explained in Section 5.4.5 in view of its widespread use in standard *ab initio* molecular dynamics simulations. There, it is for instance used in order to improve infrared spectra as obtained from the Fourier transform of classical time autocorrelation functions. The chapter closes in Section 5.4.6 with a short overview of techniques that can be used in order to cope with quantum nuclei and electrons at the same time.

5.4.2 Ab initio *path integrals: statics*

The *ab initio* path integral technique relies crucially on using the Feynman–Kac formulation [417, 418, 724] of quantum-statistical mechanics in terms of finite temperature path (functional) integrals, which are used to describe the nuclear degrees of freedom in this case. For a general introduction into this field the nonfamiliar reader is referred to standard textbooks [419, 420, 753], whereas the numerical evaluation of path integrals by statistical sampling in computer simulations [73, 454, 455, 717, 1031] is explained for instance in Refs. [125, 246, 251, 253, 509, 945, 1290, 1499, 1502] with greatly varying focuses.

There are at least two distinct ways to derive the *ab initio* path integral method [946]. The one followed in the original derivation [948] relies on using the path integral formulation [419, 420, 753] for the *coupled system of electrons and nuclei* in the real space continuum limit in conjunction with employing the Feynman–Vernon influence functional approach [420, 1605] to decouple the electronic subsystem from the set of nuclei. Later, in an alternative derivation [251, 1013], the path integral has been discretized from the outset and the electrons and nuclei have been decoupled at this stage. This is done in terms of a mixed "real space/wave function" complete basis set, thus introducing already at this stage Schrödinger's formulation of the electronic subsystem by invoking the Born–Oppenheimer approximation. Although concise and elegant, the former approach requires familiarity with the concept of influence functionals and the statistical–mechanical interpretation of path integrals [419]. Here, it is more instructive to derive *ab initio* path integrals more explicitly by starting with Trotter-discretizing the partition function and evaluating the matrix elements in a mixed basis set that is tailored to the adiabatic approximation. This didactic derivation, which shows transparently where the adiabatic and subsequently the Born–Oppenheimer approximations simplify the path integration, is a generalization of earlier approaches [237, 251, 1013].

The present derivation of the expressions for *ab initio* path integrals is based on assuming the non-relativistic standard many-body Hamiltonian introduced earlier in Eq. (2.2) in terms of the nuclear $\mathbf{R} = \{\mathbf{R}_I\}$ and electronic $\mathbf{r} = \{\mathbf{r}_i\}$ coordinates. The corresponding (exact) partition function of the quantum-statistical canonical ensemble reads

$$\mathcal{Z} = \text{Tr} \exp\left[-\beta\mathcal{H}\right] \tag{5.174}$$

$$= \int \sum_k \rho\left(\mathbf{R}, \Psi_k; \mathbf{R}, \Psi_k; \beta\right) d\mathbf{R} \tag{5.175}$$

$$= \int \cdots \int \sum_k \langle \mathbf{R}, \Psi_k(\mathbf{R})| \tag{}$$

$$\times \exp\left[-\beta\left\{-\sum_{I=1}^{N}\frac{\hbar^2\nabla_I^2}{2M_I} + \mathcal{H}_\text{e}\right\}\right] |\Psi_k(\mathbf{R}), \mathbf{R}\rangle \prod_{I=1}^{N} d\mathbf{R}_I , \tag{5.176}$$

where the trace of the density matrix (or thermal propagator) at inverse temperature $\beta = 1/k_\text{B}T$, $\rho\left(\mathbf{R}'', \Psi_{k''}; \mathbf{R}', \Psi_{k'}; \beta\right)$, can be evaluated in a mixed basis combining the position representation for the nuclear degrees of freedom with the energy representation for the electrons [237, 251, 1013]. In particular, a product basis, $|\Psi_k(\mathbf{R}), \mathbf{R}\rangle = |\Psi_k(\mathbf{R})\rangle|\mathbf{R}\rangle$, involving the complete and orthonormal adiabatic basis set

$$\mathcal{H}_\text{e}(\mathbf{R})|\Psi_k(\mathbf{R})\rangle = E_k(\mathbf{R})|\Psi_k(\mathbf{R})\rangle \tag{5.177}$$

$$\text{with} \quad \langle\Psi_{k'}(\mathbf{R})|\Psi_k(\mathbf{R})\rangle = \delta_{k'k} \text{ at each } \mathbf{R} \tag{5.178}$$

of the electronic subsystem, which depends on the nuclear positions as parameters, can be used for the electronic degrees of freedom, cf. Eq. (2.34). In the completeness relation

$$\int \sum_k |\Psi_k(\mathbf{R}), \mathbf{R}\rangle\langle\mathbf{R}, \Psi_k(\mathbf{R})| \, d\mathbf{R} = 1 \tag{5.179}$$

the integration over the nuclear positions has to be extended over the full available space and the summation over the electronic states k must also include all continuum states if necessary.

Since the nuclear kinetic energy operator and $\mathcal{H}_\text{e}$ in the Boltzmann factor do not commute as the electronic Hamiltonian depends also on the nuclear degrees of freedom, the lowest-order Trotter factorization [753]

$$\exp\left[-\beta\left\{-\sum_I\frac{\hbar^2\nabla_I^2}{2M_I} + \mathcal{H}_\text{e}\right\}\right] =$$

$$\lim_{P\to\infty}\left(\exp\left[\frac{\beta}{P}\sum_I\frac{\hbar^2\nabla_I^2}{2M_I}\right]\exp\left[-\frac{\beta}{P}\mathcal{H}_\text{e}\right]\right)^P \tag{5.180}$$

can be invoked in order to decouple the electronic and nuclear contributions to the propagator. This allows one, together with inserting as usual $P-1$ times the completeness relation Eq. (5.179), to *exactly* rewrite the partition function as a nested product of high-temperature density matrices

$$
\mathcal{Z} = \lim_{P\to\infty} \int \int \cdots \int \sum_{k^{(1)}} \sum_{k^{(2)}} \cdots \sum_{k^{(P)}}
$$

$$
\times \rho\left(\mathbf{R}^{(1)}, \Psi_{k^{(1)}}; \mathbf{R}^{(P)}, \Psi_{k^{(P)}}; \beta/P\right)
$$

$$
\times \rho\left(\mathbf{R}^{(P)}, \Psi_{k^{(P)}}; \mathbf{R}^{(P-1)}, \Psi_{k^{(P-1)}}; \beta/P\right)
$$

$$
\cdots
$$

$$
\times \rho\left(\mathbf{R}^{(2)}, \Psi_{k^{(2)}}; \mathbf{R}^{(1)}, \Psi_{k^{(1)}}; \beta/P\right) \, d\mathbf{R}^{(1)} \, d\mathbf{R}^{(2)} \cdots d\mathbf{R}^{(P)} \qquad (5.181)
$$

$$
= \lim_{P\to\infty} \prod_{s=1}^{P} \left[\int \cdots \int \right.
$$

$$
\times \sum_{k^{(s)}} \rho\left(\mathbf{R}^{(s+1)}, \Psi_{k^{(s+1)}}; \mathbf{R}^{(s)}, \Psi_{k^{(s)}}; \beta/P\right) \left] \prod_{s=1}^{P} \prod_{I=1}^{N} d\mathbf{R}_I^{(s)} \qquad (5.182)
$$

$$
= \lim_{P\to\infty} \prod_{s=1}^{P} \left[\int \cdots \int \sum_{k^{(s)}} \langle \mathbf{R}^{(s+1)}, \Psi_{k^{(s+1)}}(\mathbf{R}^{(s+1)})| \exp\left[\frac{\beta}{P} \sum_{I} \frac{\hbar^2 \nabla_I^2}{2M_I} \right] \right.
$$

$$
\times \exp\left[-\frac{\beta}{P}\mathcal{H}_e \right] |\Psi_{k^{(s)}}(\mathbf{R}^{(s)}), \mathbf{R}^{(s)}\rangle \left] \prod_{s=1}^{P} \prod_{I=1}^{N} d\mathbf{R}_I^{(s)} \qquad (5.183)
$$

where the trace condition imposes periodic (or cyclic) boundary conditions, $\mathbf{R}^{(P+1)} = \mathbf{R}^1$ and $\Psi_{k^{(P+1)}} = \Psi_{k^{(1)}}$, on the Trotter discretization parameter $s = 1, \ldots, P$. The latter parameter P is sometimes called imaginary or Euclidean "time" (note that $\hbar\beta$ has the dimension of time), and the temperature of the discretized density matrix in Eq. (5.176) is P times higher than in the full propagator in Eq. (5.175).

One remaining, essential task is the evaluation of the high-temperature density matrix elements, $\rho(s+1; s; \beta/P)$, as defined in Eqs. (5.183) together with Eq. (5.182), see Ref. [237]. By virtue of the eigenvalue equation Eq. (5.177) and the adiabatic product basis set, $\{|\Psi_{k^{(s)}}(\mathbf{R}^{(s)}), \mathbf{R}^{(s)}\rangle =$

$|\Psi_{k^{(s)}}(\mathbf{R}^{(s)})\rangle|\mathbf{R}^{(s)}\rangle\}$, the effect of the electronic Hamiltonian

$$\rho(s+1;s;\beta/P) = \langle\mathbf{R}^{(s+1)}|\langle\Psi_{k^{(s+1)}}(\mathbf{R}^{(s+1)})|\exp\left[\frac{\beta}{P}\sum_I\frac{\hbar^2\nabla_I^2}{2M_I}\right]$$

$$\times |\Psi_{k^{(s)}}(\mathbf{R}^{(s)})\rangle|\mathbf{R}^{(s)}\rangle\exp\left[-\frac{\beta}{P}E_{k^{(s)}}(\mathbf{R}^{(s)})\right] \quad (5.184)$$

is readily expressed by the adiabatic energy in electronic state $\Psi_{k^{(s)}}$ evaluated with the corresponding nuclear configuration $\mathbf{R}^{(s)}$ at time slice s, $E_{k^{(s)}}(\mathbf{R}^{(s)})$, which can be pulled out of the integral as usual. However, the matrix element of the nuclear kinetic energy operator cannot be evaluated exactly since the adiabatic electronic basis depends on the nuclear positions at each time slice, $\Psi_{k^{(s)}}(\mathbf{R}^{(s)})$, cf. Eq. (5.177). Assuming, however, that the variations in the nuclear degrees of freedom are small, the exponential can be expanded in a Taylor series

$$\exp\left[\frac{\beta}{P}\sum_I\frac{\hbar^2\nabla_I^2}{2M_I}\right] =$$

$$1 + \frac{\beta}{P}\sum_I\frac{\hbar^2\nabla_I^2}{2M_I} + \left(\frac{\beta}{P}\sum_I\frac{\hbar^2\nabla_I^2}{2M_I}\right)^2 + \mathcal{O}\left[(\beta/P)^3\right] , \quad (5.185)$$

which leads to

$$\rho(s+1;s;\beta/P) = \langle\mathbf{R}^{(s+1)}|\mathbf{R}^{(s)}\rangle\langle\Psi_{k^{(s+1)}}(\mathbf{R}^{(s+1)})|\Psi_{k^{(s)}}(\mathbf{R}^{(s)})\rangle$$

$$\times \exp\left[-\frac{\beta}{P}E_{k^{(s)}}(\mathbf{R}^{(s)})\right] + \frac{\beta}{P}\sum_I\frac{\hbar^2}{2M_I}\left\{\right.$$

$$\langle\mathbf{R}^{(s+1)}|\nabla_I^2|\mathbf{R}^{(s)}\rangle\langle\Psi_{k^{(s+1)}}(\mathbf{R}^{(s+1)})|\Psi_{k^{(s)}}(\mathbf{R}^{(s)})\rangle$$

$$+ \langle\mathbf{R}^{(s+1)}|\mathbf{R}^{(s)}\rangle\langle\Psi_{k^{(s+1)}}(\mathbf{R}^{(s+1)})|\nabla_I^2|\Psi_{k^{(s)}}(\mathbf{R}^{(s)})\rangle$$

$$+ \langle\mathbf{R}^{(s+1)}|\nabla_I|\mathbf{R}^{(s)}\rangle\langle\Psi_{k^{(s+1)}}(\mathbf{R}^{(s+1)})|\nabla_I|\Psi_{k^{(s)}}(\mathbf{R}^{(s)})\rangle$$

$$+ \mathcal{O}\left[(\beta/P)^2\right] \left.\right\}\exp\left[-\frac{\beta}{P}E_{k^{(s)}}(\mathbf{R}^{(s)})\right] ; \quad (5.186)$$

note that for simplicity this Taylor expansion is written down explicitly to first order only and thus $\mathcal{O}\left[(\beta/P)^2\right]$, whereas the second-order term is included in Eq. (5.185) as well. At this stage, the *adiabatic approximation* is applied, which amounts to assuming that the variation of the electronic wave functions as a result of changes in the nuclear coordinates is small such that nuclear gradients and all higher-order derivatives of electronic wave functions can ultimately be neglected, $\nabla_I^n | \Psi_{k^{(s)}}(\mathbf{R}^{(s)})\rangle = 0$ with $n = 1, 2, \ldots$. Neglecting these derivative terms in Eq. (5.186) yields the following zeroth order approximation:

$$\rho(s+1; s; \beta/P) \overset{\text{adia}}{\approx} \Bigg\{ \langle \mathbf{R}^{(s+1)} | \mathbf{R}^{(s)} \rangle \langle \Psi_{k^{(s+1)}}(\mathbf{R}^{(s+1)}) | \Psi_{k^{(s)}}(\mathbf{R}^{(s)}) \rangle$$

$$+ \langle \mathbf{R}^{(s+1)} | \frac{\beta}{P} \sum_I \frac{\hbar^2 \nabla_I^2}{2M_I} | \mathbf{R}^{(s)} \rangle \langle \Psi_{k^{(s+1)}}(\mathbf{R}^{(s+1)}) | \Psi_{k^{(s)}}(\mathbf{R}^{(s)}) \rangle$$

$$+ \langle \mathbf{R}^{(s+1)} | \left(\frac{\beta}{P} \sum_I \frac{\hbar^2 \nabla_I^2}{2M_I} \right)^2 | \mathbf{R}^{(s)} \rangle \langle \Psi_{k^{(s+1)}}(\mathbf{R}^{(s+1)}) | \Psi_{k^{(s)}}(\mathbf{R}^{(s)}) \rangle$$

$$+ \mathcal{O}\left[(\beta/P)^3 \right] \Bigg\} \exp\left[-\frac{\beta}{P} E_{k^{(s)}}(\mathbf{R}^{(s)}) \right] \tag{5.187}$$

$$= \langle \mathbf{R}^{(s+1)} | \exp\left[\frac{\beta}{P} \sum_I \frac{\hbar^2 \nabla_I^2}{2M_I} \right] | \mathbf{R}^{(s)} \rangle$$

$$\times \langle \Psi_{k^{(s+1)}}(\mathbf{R}^{(s+1)}) | \Psi_{k^{(s)}}(\mathbf{R}^{(s)}) \rangle \exp\left[-\frac{\beta}{P} E_{k^{(s)}}(\mathbf{R}^{(s)}) \right] ,$$
$$\tag{5.188}$$

where all remaining expansion terms in Eq. (5.187) have been re-summed exactly in Eq. (5.188), i.e. up to *infinite* order in terms of the original Taylor expansion given in Eq. (5.185), which yields again the original exponential term of the nuclear kinetic energy operator. As a result, the expectation value of the exponential of the kinetic energy operator can now be evaluated without any reference to electronic states.

Similarly, the projection of the electronic wave function at imaginary time s onto that at $s + 1$ can be evaluated in the spirit of the adiabatic approximation by expanding the wave function at time s, $\Psi_k^{(s)}(\mathbf{R}^{(s)})$, around the nuclear positions $\mathbf{R}^{(s+1)}$ at time $s + 1$

$$\langle \Psi_{k^{(s+1)}}(\mathbf{R}^{(s+1)})|\Psi_{k^{(s)}}(\mathbf{R}^{(s)})\rangle = \langle \Psi_{k^{(s+1)}}(\mathbf{R}^{(s+1)})| \Big\{ |\Psi_{k^{(s)}}(\mathbf{R}^{(s+1)})\rangle$$

$$+ \sum_I \nabla_I |\Psi_{k^{(s)}}(\mathbf{R}^{(s+1)})\rangle (\mathbf{R}^{(s+1)} - \mathbf{R}^{(s)})$$

$$+ \mathcal{O}\left[(\mathbf{R}^{(s+1)} - \mathbf{R}^{(s)})^2\right] \Big\} \qquad (5.189)$$

$$\overset{\text{adia}}{\approx} \langle \Psi_{k^{(s+1)}}(\mathbf{R}^{(s+1)})|\Psi_{k^{(s)}}(\mathbf{R}^{(s+1)})\rangle$$

$$= \delta_{k^{(s+1)}k^{(s)}} \ , \qquad (5.190)$$

which yields the orthogonality relation, Eq. (5.178), within the adiabatic approximation, i.e. $\nabla_I^n |\Psi_{k^{(s)}}(\mathbf{R}^{(s)})\rangle = 0$ with $n = 1, 2, \ldots$. As a result, the high-temperature density matrix in Eq. (5.182) simplifies to

$$\rho(s+1; s; \beta/P) \overset{\text{adia}}{\approx} \langle \mathbf{R}^{(s+1)}| \exp\left[\frac{\beta}{P} \sum_I \frac{\hbar^2 \nabla_I^2}{2M_I}\right] |\mathbf{R}^{(s)}\rangle$$

$$\times \delta_{k^{(s+1)}k^{(s)}} \exp\left[-\frac{\beta}{P} E_{k^{(s)}}(\mathbf{R}^{(s)})\right] \qquad (5.191)$$

$$= \delta_{k^{(s+1)}k^{(s)}} A_P^N \exp\left[-\beta\Big\{ \sum_{I=1}^N \frac{1}{2} M_I \omega_P^2 \left(\mathbf{R}_I^{(s)} - \mathbf{R}_I^{(s+1)}\right)^2\right.$$

$$\left. + \frac{1}{P} E_{k^{(s)}}(\mathbf{R}^{(s)}) \Big\}\right] \qquad (5.192)$$

$$\text{with} \quad A_P = \left(\frac{M_I P}{2\pi \hbar^2 \beta}\right)^{3/2} \qquad (5.193)$$

$$\text{and} \quad \omega_P^2 = \frac{P}{\hbar^2 \beta^2} \ , \qquad (5.194)$$

where the exponential of the nuclear kinetic energy operator in Eq. (5.191) can now be evaluated [419, 420, 753] as usual, analytically in momentum space. The result of these Gaussian integrations is the harmonic interactions that couple neighboring imaginary time slices.

The final expression, Eq. (5.192), can now be used to assemble the complete

Trotter product, Eq. (5.182),

$$
\mathcal{Z}_{\text{adia}} = \lim_{P \to \infty} \prod_{s=1}^{P} \left[\int \cdots \int \sum_{k(s)} \delta_{k(s+1)\,k(s)} A_P^N \right.
$$

$$
\times \exp \left[-\beta \left\{ \sum_{I=1}^{N} \frac{1}{2} M_I \omega_P^2 \left(\mathbf{R}_I^{(s)} - \mathbf{R}_I^{(s+1)} \right)^2 + \frac{1}{P} E_{k(s)} (\mathbf{R}^{(s)}) \right\} \right]
$$

$$
\times \prod_{s=1}^{P} \prod_{I=1}^{N} d\mathbf{R}_I^{(s)} \tag{5.195}
$$

$$
= \sum_{k} \lim_{P \to \infty} \prod_{s=1}^{P} \left[\prod_{I=1}^{N} \int \right]
$$

$$
\times \exp \left[-\beta \sum_{s=1}^{P} \left\{ \sum_{I=1}^{N} \frac{1}{2} M_I \omega_P^2 \left(\mathbf{R}_I^{(s)} - \mathbf{R}_I^{(s+1)} \right)^2 + \frac{1}{P} E_k (\mathbf{R}^{(s)}) \right\} \right]
$$

$$
\times \prod_{s=1}^{P} \prod_{I=1}^{N} A_P \, d\mathbf{R}_I^{(s)} \ , \tag{5.196}
$$

which yields the canonical partition function, Eq. (5.174), expressed as a path integral in the *adiabatic approximation*. Note that as a result of the product of Kronecker deltas in Eq. (5.195), each of them connecting two adiabatic states $\Psi_k^{(s)}$ and $\Psi_k^{(s+1)}$ at neighboring imaginary times s and $s+1$ under periodic boundary conditions $s = P+1 = 1$, only one sum, $\sum_{k(1)} = \sum_k$, survives. Physically, this remaining sum in Eq. (5.196) allows *all* adiabatic electronic states, $\{\Psi_k\}$, to be populated according to Boltzmann's thermal distribution by sampling the nuclear fluctuations with a weight given by the associated energy eigenvalue, $\propto \exp[-\beta E_k / P]$. Thus, the contribution of excited states $k > 0$ to the partition function and thus to observables is suppressed exponentially at a given temperature as a function of the energy gap with respect to the ground state, $\propto \exp[-\beta(E_k - E_0)/P]$, separately for *each* nuclear configuration $\{\mathbf{R}_I^{(s)}\}$ sampled in the path integration.

A further approximation is obtained if $\beta(E_k - E_0)$ is large for all nuclear configurations, i.e. $(E_k - E_0) \gg k_B T$, which amounts to assuming "electronic ground state dominance" and thus invoking the *Born–Oppenheimer approximation* on top of the adiabatic approximation. Hence, the sum $\sum_k$

collapses to taking into account only a single electronic state

$$
\mathcal{Z}_{\mathrm{BO}} = \lim_{P \to \infty} \prod_{s=1}^{P} \left[\prod_{I=1}^{N} \int \right]
$$

$$
\times \exp\left[-\beta \sum_{s=1}^{P} \left\{ \sum_{I=1}^{N} \frac{1}{2} M_I \omega_P^2 \left(\mathbf{R}_I^{(s)} - \mathbf{R}_I^{(s+1)} \right)^2 + \frac{1}{P} E_0(\mathbf{R}^{(s)}) \right\} \right]
$$

$$
\times \prod_{s=1}^{P} \prod_{I=1}^{N} A_P \, d\mathbf{R}_I^{(s)} \tag{5.197}
$$

in the Born–Oppenheimer approximation, which is chosen to be the ground electronic state Ψ_0 in this case. This is the basic working equation underlying the *ab initio* path integral technique [946, 948] where $E_0(\mathbf{R}^{(s)})$ is computed explicitly from electronic structure theory, i.e. by solving Eq. (5.177) for each nuclear configuration, $\{\mathbf{R}_I^{(s)}\}$, at each imaginary time slice s.

Alternatively, the final result, Eq. (5.197), can be obtained more readily using the Feynman–Vernon path integral influence functional formalism [420, 1605]. This approach allows one to carry out the partial trace over the electronic degrees of freedom separately for *frozen* paths of the nuclei as invoked in the original derivation [948]. Thus, the canonical partition function, Eq. (5.174), with the Hamiltonian defined in Eq. (2.2) is written directly as a continuum path integral,

$$
\mathcal{Z} = \oint' \oint' \exp\left[-\frac{1}{\hbar} \int_0^{\hbar\beta} d\tau \, \mathcal{L}_{\mathrm{E}} \left(\{\dot{\mathbf{R}}_I(\tau)\}, \{\mathbf{R}_I(\tau)\}; \{\dot{\mathbf{r}}_i(\tau)\}, \{\mathbf{r}_i(\tau)\} \right) \right] \mathcal{D}\mathbf{r} \, \mathcal{D}\mathbf{R}
$$

$$
\tag{5.198}
$$

including both nuclei *and* electrons on the same footing, where

$$
\mathcal{L}_{\mathrm{E}} = T(\dot{\mathbf{R}}) + V(\mathbf{R}) + T(\dot{\mathbf{r}}) + V(\mathbf{r}) + V(\mathbf{R}, \mathbf{r})
$$

$$
= \sum_I \frac{1}{2} M_I \left(\frac{d\mathbf{R}_I}{d\tau} \right)^2 + \sum_{I<J} \frac{e^2 Z_I Z_J}{|\mathbf{R}_I - \mathbf{R}_J|}
$$

$$
+ \sum_i \frac{1}{2} m_e \left(\frac{d\mathbf{r}_i}{d\tau} \right)^2 + \sum_{i<j} \frac{e^2}{|\mathbf{r}_i - \mathbf{r}_j|} - \sum_{I,i} \frac{e^2 Z_I}{|\mathbf{R}_I - \mathbf{r}_i|} \tag{5.199}
$$

denotes the *Euclidean* Lagrangian [419, 420, 753] known from classical mechanics. The primes in Eq. (5.198) indicate that the proper sums over all permutations corresponding to Bose–Einstein and/or Fermi–Dirac statistics have to be included. It is important to note that in Eqs. (5.198) and (5.199) the positions $\mathbf{R}$ and $\mathbf{r}$ are not operators but simply classical *functions* of the imaginary time $\tau \in [0, \hbar\beta]$, which parameterizes fluctuations around the classical path. This implies that the dots denote here derivatives with respect to *imaginary* time τ as defined in Eq. (5.199). According to Eq. (5.198), exact quantum mechanics at finite temperature $T = 1/k_B\beta$ is recovered if all closed paths $[\mathbf{R}; \mathbf{r}]$ of "length" $\hbar\beta$ are summed up and weighted with the exponential of the Euclidean action measured in units of $\hbar$. The partial trace over the electronic subsystem can be written down formally and exactly,

$$\mathcal{Z} = \oint' \exp\left[-\frac{1}{\hbar} \int_0^{\hbar\beta} T(\dot{\mathbf{R}}) \, d\tau\right] \mathcal{Z}_e\,[\mathbf{R}] \; \mathcal{D}\mathbf{R} \qquad (5.200)$$

with the aid of an influence functional [420, 1605],

$$\mathcal{Z}_e[\mathbf{R}] = \oint' \exp\left[-\frac{1}{\hbar} \int_0^{\hbar\beta} d\tau \, (T(\dot{\mathbf{r}}) + V(\mathbf{r}) + V(\mathbf{R}, \mathbf{r}) + V(\mathbf{R})) \, d\tau\right] \mathcal{D}\mathbf{r}$$
$$(5.201)$$

due to the electrons. Note that $\mathcal{Z}_e[\mathbf{R}]$ is a complicated and unknown functional for a given nuclear path configuration $[\{\mathbf{R}_I\}]$. As a consequence, the interactions between the nuclei become highly nonlocal in imaginary time due to memory effects.

In the standard Born–Oppenheimer approximation [179, 180, 771, 811], see Section 2.1, the nuclei are frozen in some configuration ("clamped nuclei") and the complete electronic problem is solved for this single static configuration. In addition to the nondiagonal correction terms that are already neglected in the adiabatic approximation, the diagonal terms are now neglected as well. Thus, the potential for the nuclear motion is simply defined as the bare electronic eigenvalues obtained from a series of fixed nuclear configurations.

In the statistical-mechanical formulation of the problem Eqs. (5.200)-(5.201), the Born–Oppenheimer approximation amounts to a "quenched average": at imaginary time τ the nuclei are frozen at a particular configuration $\mathbf{R}(\tau)$ and the electrons explore their configuration space subject only to that single configuration. This implies that the electronic degrees of freedom at different imaginary times τ and τ' become completely decoupled. Thus, the electronic influence functional $\mathcal{Z}_e[\mathbf{R}]$ has to be local in τ and becomes particularly simple; a discussion of adiabatic corrections in the path integral formulation can be found in Ref. [214]. For each τ the influence functional $\mathcal{Z}_e[\mathbf{R}]$ is given by the partition function of the electronic subsystem evaluated for the respective nuclear configuration $\mathbf{R}(\tau)$. Assuming that the temperature is small compared to the gap in the electronic spectrum, only the electronic ground state with energy $E_0\left(\mathbf{R}(\tau)\right)$ defined earlier, Eq. (5.177), is populated. This electronic ground state dominance reduces the influence *functional* to a *function*, which yields the following simple expression:

$$\mathcal{Z}_e[\mathbf{R}] \overset{\mathrm{BO}}{\approx} \mathcal{Z}_e\left(\mathbf{R}(\tau)\right) = \exp\left[-\frac{1}{\hbar}\int_0^{\hbar\beta} E_0(\mathbf{R}(\tau))\,d\tau\right]\,, \tag{5.202}$$

and thus the final result

$$\mathcal{Z}_{\mathrm{BO}} = \oint \exp\left[-\frac{1}{\hbar}\int_0^{\hbar\beta}\left(T(\dot{\mathbf{R}}) + E_0(\mathbf{R})\right)d\tau\right]\mathcal{D}\mathbf{R} \tag{5.203}$$

together with Eq. (5.200). Here, nuclear exchange is neglected by assuming that the nuclei are distinguishable so that they can be treated within Maxwell-Boltzmann quantum statistics, which corresponds to the Hartree approximation for the *nuclear* density matrix. Note, however, that the correct Fermi–Dirac statistics for the electrons is included in the definition of the electronic ground state potential $E_0(\mathbf{R})$, since it is the lowest eigenvalue according to the underlying Schrödinger equation, Eq. (5.177).

The partition function Eq. (5.203) together with the Coulomb Hamiltonian Eq. (2.2) leads, after applying the lowest-order Trotter factorization [753], to the following discretized expression:

$$\mathcal{Z}_{\mathrm{BO}} = \lim_{P\to\infty}\prod_{s=1}^{P}\left[\prod_{I=1}^{N}\int\right]$$

$$\times \exp\left[-\beta\sum_{s=1}^{P}\left\{\sum_{I=1}^{N}\frac{1}{2}M_I\omega_P^2\left(\mathbf{R}_I^{(s)} - \mathbf{R}_I^{(s+1)}\right)^2 + \frac{1}{P}E_0(\mathbf{R}^{(s)})\right\}\right]$$

$$\times \prod_{s=1}^{P}\prod_{I=1}^{N} A_P\,d\mathbf{R}_I^{(s)} \tag{5.204}$$

for the *ab initio* path integral with A_P and ω_P^2 as defined earlier in Eq. (5.193) and Eq. (5.194), respectively. Note that this Born–Oppenheimer partition function is identical to Eq. (5.197), which has been obtained from a very different derivation. As a result of the discretization, the continuous parameter $\tau \in [0, \hbar\beta]$ has been replaced by P Trotter time slices $s = 1, \ldots, P$ of

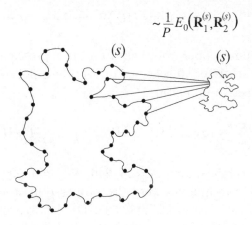

$$\sim \frac{1}{P}E_0\!\left(\mathbf{R}_1^{(s)},\mathbf{R}_2^{(s)}\right)$$

Fig. 5.3. Sketch of a diatomic molecule consisting of a light (left, large spread of the path) and a heavy (right, compact path) nucleus in the path integral representation ("ring polymer isomorphism"). The two nuclei $(\mathbf{R}_1^{(1)},\mathbf{R}_1^{(2)},\ldots,\mathbf{R}_1^{(P)})$ and $(\mathbf{R}_2^{(1)},\mathbf{R}_2^{(2)},\ldots,\mathbf{R}_2^{(P)})$ interact at each time slice s via $E_0(\mathbf{R}_1^{(s)},\mathbf{R}_2^{(s)})/P$ obtained from electronic structure calculations in the ground state. The interaction within each ring polymer $I=1,2$ is given by harmonic springs $M_I\omega_P^2(\mathbf{R}_I^{(s)}-\mathbf{R}_I^{(s+1)})^2/2$ between nearest neighbors only and is subject to periodic boundary conditions, $\mathbf{R}_I^{(P+1)}=\mathbf{R}_I^{(1)}$.

"duration" $\Delta\tau=\hbar\beta/P$. The discretized paths

$$\left\{\{\mathbf{R}_I\}^{(s)}\right\}=\left(\{\mathbf{R}_I\}^{(1)};\ldots;\{\mathbf{R}_I\}^{(P)}\right)$$

$$=\left(\mathbf{R}_1^{(1)},\ldots,\mathbf{R}_N^{(1)};\ldots;\mathbf{R}_1^{(P)},\ldots,\mathbf{R}_N^{(P)}\right) \tag{5.205}$$

have to be closed due to the trace condition $\mathbf{R}_I(0)\equiv\mathbf{R}_I(\hbar\beta)$, i.e. they are periodic in imaginary time τ which implies $\mathbf{R}_I^{(P+1)}=\mathbf{R}_I^{(1)}$. Note that Eq. (5.204) is an *exact* reformulation of Eq. (5.203) (and not an approximation thereof) in the limit of an infinitely fine discretization $P\to\infty$ of the paths.

The effective classical partition function Eq. (5.204) with a fixed discretization P is isomorphic to that for N polymers each comprised of P monomers [246, 253, 254, 420, 509]. Each quantum degree of freedom is found to be represented by a ring polymer or necklace as sketched in Fig. 5.3. The intrapolymeric interactions stem from the kinetic energy $T(\dot{\mathbf{R}})$ and consist of harmonic nearest-neighbor couplings $\propto\omega_P^2$ along the closed chain. The interpolymeric interaction is given by the scaled potential $E_0^{(s)}/P$ which is only evaluated for configurations $\{\mathbf{R}_I\}^{(s)}$ at the *same* imaginary time slice s.

In order to evaluate operators based on an expression like Eq. (5.204), most numerical path integral schemes utilize Metropolis Monte Carlo sampling with the effective potential

$$V_{\text{eff}} = \sum_{s=1}^{P} \left\{ \sum_{I=1}^{N} \frac{1}{2} M_I \omega_P^2 \left(\mathbf{R}_I^{(s)} - \mathbf{R}_I^{(s+1)} \right)^2 + \frac{1}{P} E_0(\mathbf{R}_I^{(s)}) \right\} \quad (5.206)$$

of the isomorphic classical system [125, 246, 251, 253, 509, 945, 1290, 1502]. Molecular dynamics techniques were also proposed in order to sample paths in configuration space, see Refs. [211, 585, 1108, 1157, 1176] for pioneering work and Refs. [1499, 1502] for authoritative reviews. Based on the effective potential Eq. (5.206), an extended Lagrangian

$$\mathcal{L}_{\text{PIMD}} = \sum_{s=1}^{P} \left\{ \sum_{I=1}^{N} \left(\frac{1}{2M_I'} \left(\mathbf{P}_I^{(s)} \right)^2 - \frac{1}{2} M_I \omega_P^2 \left(\mathbf{R}_I^{(s)} - \mathbf{R}_I^{(s+1)} \right)^2 \right) \right.$$

$$\left. - \frac{1}{P} E_0(\mathbf{R}_I^{(s)}) \right\} \quad (5.207)$$

can be introduced straightforwardly in terms of the "primitive coordinates" $\mathbf{R}_I^{(s)}$ by formally adding $N \times P$ fictitious momenta $\mathbf{P}_I^{(s)}(t')$ and corresponding fictitious masses M_I', which are unphysical auxiliary parameters associated with each particle I, whereas the physical masses of the nuclei are determined by M_I. At this stage it is important to be aware that the time dependence of positions and momenta, and thus the time evolution in the *extended phase space* as generated by Eq. (5.207), has no physical meaning. The sole use of "time" t' is to parameterize the deterministic dynamical exploration of *configuration space*; an approximate time evolution in the framework of a quasiclassical interpretation of the generated dynamics will be presented in Section 5.4.3. The trajectories of the positions in configuration space can, however, be analyzed similarly to the ones obtained from the stochastic dynamics that underlies the Monte Carlo method.

The crucial ingredient in *ab initio* [931, 946, 948, 1507] as opposed to standard [246, 251, 253, 509, 945, 1290, 1502] path integral simulations consists in computing the interactions E_0 "on-the-fly", like in *ab initio* molecular dynamics. In analogy to this case, both the Car–Parrinello and Born–Oppenheimer approaches from Sections 2.4 and 2.3, respectively, can be combined with any electronic structure method. The first implementation [946] was based on the Car–Parrinello/density functional combination from Section 2.4, which leads to the following extended Lagrangian:

$$\mathcal{L}_{\mathrm{AIPI}} = \frac{1}{P} \sum_{s=1}^{P} \left\{ \sum_i \mu \left\langle \dot{\phi}_i^{(s)} \middle| \dot{\phi}_i^{(s)} \right\rangle - E^{\mathrm{KS}} \left[\{\phi_i\}^{(s)}, \{\mathbf{R}_I\}^{(s)} \right] \right.$$

$$\left. + \sum_{ij} \Lambda_{ij}^{(s)} \left(\left\langle \phi_i^{(s)} \middle| \phi_j^{(s)} \right\rangle - \delta_{ij} \right) \right\}$$

$$+ \sum_{s=1}^{P} \left\{ \sum_{I=1}^{N} \frac{1}{2} M_I' \left(\dot{\mathbf{R}}_I^{(s)} \right)^2 - \sum_{I=1}^{N} \frac{1}{2} M_I \omega_P^2 \left(\mathbf{R}_I^{(s)} - \mathbf{R}_I^{(s+1)} \right)^2 \right\} , \quad (5.208)$$

where the interaction energy $E^{\mathrm{KS}}[\{\phi_i\}^{(s)}, \{\mathbf{R}_I\}^{(s)}]$ at time slice s is defined in Eq. (2.114); note that here and in the following the dots denote derivatives with respect to propagation time t' and that $E_0^{\mathrm{KS}} = \min E^{\mathrm{KS}}$. The standard Car–Parrinello Lagrangian, see e.g. Eq. (2.56) or Eq. (2.78), is recovered in the limit $P = 1$, which corresponds to classical nuclei. Mixed classical/quantum systems can easily be treated within the path integral representation of quantum mechanics by representing an arbitrary subset of the nuclei in Eq. (5.208) with only one imaginary time slice. It is mentioned in passing that this is an elegant "QM/MM" coupling scheme, which in practice, unfortunately, can only be used to separate nuclear degrees of freedom into quantum and classical ones [1506].

This simplest formulation of *ab initio* path integrals, however, is insufficient for the following reason: ergodicity of the trajectories and adiabaticity in the sense of Car–Parrinello simulations are not guaranteed. It has been known since the very first molecular dynamics computer experiments that quasiharmonic systems (such as coupled stiff harmonic oscillators subject to weak anharmonicities, i.e. the famous "Fermi-Pasta-Ulam chains" [409]) can easily lead to nonergodic behavior in the sampling of phase space [308, 453]. Similarly, "microcanonical" path integral molecular dynamics simulations might lead to an insufficient exploration of configuration space depending on the parameter settings [585]. The severity of this nonergodicity problem is governed by the stiffness of the harmonic intrachain coupling $\propto \omega_P^2$ and the anharmonicity of the overall potential surface $\propto E^{\mathrm{KS}}/P$, which establishes the coupling of the modes. For a better and better discretization P the harmonic energy term dominates according to $\sim P$, whereas the coupling that governs mode mixing decreases like $\sim 1/P$. This problem can be cured by attaching Nosé–Hoover chain thermostats [927] *individually* to all path integral degrees of freedom [1500, 1507], which is the "massive thermostatting method" introduced in Section 5.2.2; see Refs. [1499, 1502] for reviews.

The second issue is related to the separation of the power spectra associated with nuclear and electronic subsystems during Car–Parrinello *ab initio* molecular dynamics, which is instrumental for maintaining adiabaticity as explained in Section 2.4 for the standard approach. In *ab initio* molecular dynamics with classical nuclei the highest phonon or vibrational frequency $\omega_n^{\max}$ is dictated by the physics of the system, see e.g. Fig. 2.1. This means in particular that an upper limit is given by stiff intramolecular vibrations which do not exceed $\omega_n^{\max} \leq 5000$ cm^{-1} or 150 THz. In *ab initio* path integral simulations, on the contrary, $\omega_n^{\max}$ is given by ω_P, which actually diverges with increasing discretization as $\sim \sqrt{P}$. The simplest remedy would be to compensate this artifact by decreasing the fictitious electron mass μ until the power spectra are again separated for a fixed value of P and thus ω_P. This, however, would lead to a prohibitively small time step because $\Delta t^{\max} \propto \sqrt{\mu}$. This dilemma has been solved by thermostatting the electronic degrees of freedom as well [946, 948, 1507], as proposed earlier for systems with small or vanishing band gaps such as metals [146, 1510]; see Section 5.2 for a more detailed discussion in the context of metals. The equations of motion are not given here, but deferred instead to Section 5.4.3 where they are presented for another set of dynamical variables, normal modes.

Finally, it is known that diagonalizing the harmonic spring interaction in Eq. (5.208) leads to more efficient propagators in terms of normal modes or "staging coordinates" [1500, 1507]. The normal mode transformation and the resulting Nosé–Hoover chain thermostatted equations of motion will be outlined in the following section, see in particular Eqs. (5.223)–(5.229). In addition to keeping the average temperature fixed, it is also possible to generate path trajectories in the isobaric–isothermal NpT ensemble [926, 1502]. Instead of using Car–Parrinello fictitious dynamics in order to evaluate the interaction energy in Eq. (5.207), which is implemented in the CPMD package [696], it is evident that the Born–Oppenheimer approach from Section 2.3 or the free energy functional from Section 5.3.4 can also be used. This route eliminates the Car–Parrinello adiabaticity problem and was taken up, for instance, in Refs. [74, 264, 747, 1014, 1015, 1492, 1579, 1582].

A final observation concerning parallel supercomputers might be useful, see also Section 8.1. It is evident from the Lagrangian Eq. (5.208) and the resulting equations of motion (e.g. Eqs. (5.223)–(5.229)) that most of the numerical workload comes from calculating the *ab initio* forces on the nuclei. Given a fixed path configuration Eq. (5.205), the P underlying electronic structure problems are independent from each other and can be solved without communication on P nodes of a distributed memory machine. Communication is only necessary to distribute the final result, essentially the forces, to a special node that computes the quantum kinetic contribution to the energy and integrates finally the equations of motion; this approach assumes that the P different wave functions can all be kept in memory during the simulation. It is even conceivable to distribute this task in a loosely connected farm environment over many workstations/PCs or on clustered supercomputers such that "meta-computing" is accessible. Thus, the algorithm is "embarrassingly parallel" provided that the memory per node is sufficient to solve the complete Kohn–Sham problem at a given time slice. If this is not the case, or in order to speed up *ab initio* path integral simulations even further if lots of nodes are available, the individual electronic structure calculations themselves can be parallelized by introducing another hierarchical parallelization level as outlined in Chapter 8. Finally, if memory is a bottleneck, the Born–Oppenheimer sampling approach to *ab initio* path integrals allows us to keep just a single copy of the wave function in memory, which serves only as the initial guess for all P iterative wave function optimizations that can be performed one after the other.

5.4.3 Ab initio *path centroids: dynamics*

Initially, the molecular dynamics approach [1499, 1502] to path integral simulations [211, 585, 1108, 1157, 1176] was invented merely as a convenient algorithmic trick in order to sample paths in configuration space in a continuous Newtonian dynamics fashion, instead of using the Monte Carlo approach that relies on stochastic Markovian processes and thus on discrete dynamics. This paradigm changed with the introduction of the "centroid molecular dynamics" (CMD) technique [216], see Refs. [217–219, 1180–1182, 1184, 1576] for background information, and more recently the "ring polymer molecular dynamics" (RPMD) method [186, 290, 291, 1006].

In a nutshell, it is found within the centroid molecular dynamics approach [216] that the time evolution of the centers

$$\mathbf{R}_I^c(t) = \frac{1}{P} \sum_{s'=1}^{P} \mathbf{R}_I^{(s')}(t) \tag{5.209}$$

of the closed Feynman paths which represent the quantum nuclei, the "path centroids", contains quasiclassical information about the true quantum dynamics. In particular, it has been shown that the *approximate* centroid correlation function $\langle \mathbf{R}^c(0)\mathbf{R}^c(t) \rangle_{\mathrm{CMD}}$ obtained within the centroid molecular dynamics approach [216]

$$C_{\mathrm{Kubo}}(t) = \left\langle \frac{1}{\beta} \int_0^\beta \hat{\mathbf{R}}(0)\hat{\mathbf{R}}(t + i\hbar\lambda)\, d\lambda \right\rangle \qquad (5.210)$$

$$\equiv \langle \mathbf{R}^c(0)\mathbf{R}^c(t) \rangle \qquad (5.211)$$

$$\approx \langle \mathbf{R}^c(0)\mathbf{R}^c(t) \rangle_{\mathrm{CMD}} = C_{\mathrm{CMD}}(t) \qquad (5.212)$$

is a short-time approximation to the Kubo-transformed quantum correlation function [794] with the usual definition of Heisenberg operators $\hat{\mathbf{O}}(t) = \exp[i\hat{\mathcal{H}}t/\hbar]\hat{\mathbf{O}}\exp[-i\hat{\mathcal{H}}t/\hbar]$. Kubo's correlation function, $C_{\mathrm{Kubo}}(t)$, is identical to the *exact* centroid correlation function, $\langle \mathbf{R}^c(0)\mathbf{R}^c(t)\rangle$, for operators that are linear functions of position or momentum operators [702, 1180, 1182]. The centroid molecular dynamics approach can be shown to be exact for harmonic potentials and to have the correct classical limit. The path centroids move in an effective potential which is generated by all the other modes of the paths at the given temperature. This effective potential thus includes the effects of quantum fluctuations on the (quasiclassical) time evolution of the centroid degrees of freedom. Roughly speaking, the trajectory of the path centroids can be regarded as a classical trajectory of the system, which is approximately "renormalized" due to quantum effects. Furthermore, it has been shown that the leading error of the short-time limit of position and velocity autocorrelation functions obtained with centroid molecular dynamics is $\mathcal{O}(t^6)$ and $\mathcal{O}(t^4)$, respectively, whereas it is $\mathcal{O}(t^8)$ and $\mathcal{O}(t^6)$ in ring polymer molecular dynamics [186]. In summary, approximate quasiclassical time-correlation functions such as Eq. (5.212) and thus spectral information obtained from its Fourier transform can be extracted from appropriately performed path integral molecular dynamics simulations.

The original centroid molecular dynamics technique [216–219, 1576] relies on the use of model potentials as standard time-independent path integral simulations. This limitation was overcome independently in Refs. [956, 1119] by combining *ab initio* path integrals with centroid molecular dynamics. The resulting technique, *ab initio* centroid molecular dynamics, can be considered as a quasiclassical generalization of standard *ab initio* molecular dynamics. At the same time, it preserves the virtues of the *ab initio* path integral technique [931, 946, 948, 1507] to generate exact time-independent quantum equilibrium averages.

Here, the adiabatic formulation [215, 219, 925] of *ab initio* centroid molecular dynamics [956] is discussed. In close analogy to *ab initio* molecular dynamics with classical nuclei, the effective centroid potential is also generated "on-the-fly" as the centroids are propagated. This is achieved by singling out the centroid coordinates in terms of a normal mode transformation [275] and accelerating the dynamics of all non-centroid modes artificially by assigning appropriate fictitious masses. At the same time, the fictitious electron dynamics à la Car–Parrinello is kept in order to calculate efficiently the *ab initio* forces on *all* modes from the electronic structure. This makes it necessary to maintain two levels of adiabaticity in the course of simulations, which has been analyzed theoretically in Section 2.1 of Ref. [956].

The partition function Eq. (5.204), formulated in the "primitive" path variables $\{\mathbf{R}_I\}^{(s)}$, is first transformed [1502, 1507] to a representation in terms of the normal modes $\{\mathbf{u}_I\}^{(s)}$, which diagonalize the harmonic nearest-neighbor harmonic coupling [275]. The transformation follows from the Fourier expansion of a given cyclic path

$$\mathbf{R}_I^{(s)} = \sum_{s'=1}^{P} \mathbf{a}_I^{(s')} \exp\left[2\pi i(s-1)(s'-1)/P\right] \ , \qquad (5.213)$$

where the coefficients $\{\mathbf{a}_I\}^{(s)}$ are complex numbers. The normal mode variables $\{\mathbf{u}_I\}^{(s)}$ are then given in terms of the expansion coefficients according to

$$\mathbf{u}_I^{(1)} = \mathbf{a}_I^{(1)}$$

$$\mathbf{u}_I^{(P)} = \mathbf{a}_I^{((P+2)/2)}$$

$$\mathbf{u}_I^{(2s-2)} = \text{Re}\,(\mathbf{a}_I^{(s)})$$

$$\mathbf{u}_I^{(2s-1)} = \text{Im}\,(\mathbf{a}_I^{(s)}) \ . \qquad (5.214)$$

Associated with the normal mode transformation is a set of normal mode frequencies $\{\lambda\}^{(s)}$ given by

$$\lambda^{(2s-1)} = \lambda^{(2s-2)} = 2P\left[1 - \cos\left(\frac{2\pi(s-1)}{P}\right)\right] \ , \qquad (5.215)$$

with $\lambda^{(1)} = 0$ and $\lambda^{(P)} = 4P$. Equation (5.213) is equivalent to direct diagonalization of the matrix

$$\mathbf{A}_{ss'} = 2\delta_{ss'} - \delta_{s,s'-1} - \delta_{s,s'+1} \qquad (5.216)$$

with the path periodicity condition $\mathbf{A}_{s0} = \mathbf{A}_{sP}$ and $\mathbf{A}_{s,P+1} = \mathbf{A}_{s1}$ and subsequent use of the unitary transformation matrix $\mathbf{U}$ to transform from the "primitive" variables $\{\mathbf{R}_I\}^{(s)}$ to the normal mode variables $\{\mathbf{u}_I\}^{(s)}$,

$$\mathbf{R}_I^{(s)} = \sqrt{P} \sum_{s'=1}^{P} \mathbf{U}_{ss'}^{\dagger} \mathbf{u}_I^{(s')}$$

$$\mathbf{u}_I^{(s)} = \frac{1}{\sqrt{P}} \sum_{s'=1}^{P} \mathbf{U}_{ss'} \mathbf{R}_I^{(s')} \ . \tag{5.217}$$

The eigenvalues of $\mathbf{A}$ when multiplied by P are precisely the normal mode frequencies $\{\lambda\}^{(s)}$. Since the transformation is unitary, its Jacobian is unity. Finally, it is convenient to define a set of normal mode masses

$$M_I^{(s)} = \lambda^{(s)} M_I \tag{5.218}$$

that vary along the imaginary time axis $s = 1, \ldots, P$, where $\lambda^{(1)} = 0$ for the centroid mode $\mathbf{u}_I^{(1)}$.

Based on these transformations the Lagrangian corresponding to the *ab initio* path integral expressed in normal modes is obtained [1507]:

$$\mathcal{L}_{\text{AIPI}} = \frac{1}{P} \sum_{s=1}^{P} \left\{ \sum_i \mu \left\langle \dot{\phi}_i^{(s)} \middle| \dot{\phi}_i^{(s)} \right\rangle - E^{\text{KS}} \left[\{\phi_i\}^{(s)}, \left\{ \mathbf{R}_I \left(\mathbf{u}_I^{(1)}, \ldots, \mathbf{u}_I^{(P)} \right) \right\}^{(s)} \right] \right.$$

$$\left. + \sum_{ij} \Lambda_{ij}^{(s)} \left(\left\langle \phi_i^{(s)} \middle| \phi_j^{(s)} \right\rangle - \delta_{ij} \right) \right\}$$

$$+ \sum_{s=1}^{P} \left\{ \sum_{I=1}^{N} \frac{1}{2} M_I'^{(s)} \left(\dot{\mathbf{u}}_I^{(s)} \right)^2 - \sum_{I=1}^{N} \frac{1}{2} M_I^{(s)} \omega_P^2 \left(\mathbf{u}_I^{(s)} \right)^2 \right\} \ , \tag{5.219}$$

where the masses $M_I'^{(s)}$ will be defined later, see Eq. (5.230). As indicated, the electronic energy E^{KS} is always evaluated in practice in terms of the "primitive" path variables $\{\mathbf{R}_I\}^{(s)}$ in Cartesian space. The necessary transformation to switch back and forth between "primitive" and normal mode variables is easily performed, as given by the relations Eq. (5.217).

The chief advantage of the normal mode representation Eq. (5.217) for the present purpose is that the lowest-order normal mode $\mathbf{u}_I^{(1)}$

$$\mathbf{u}_I^{(1)} = \mathbf{R}_I^{\text{c}} = \frac{1}{P} \sum_{s'=1}^{P} \mathbf{R}_I^{(s')} \tag{5.220}$$

turns out to be identical to the centroid $\mathbf{R}_I^{\text{c}}$ of the path that represents the

Ith nucleus. The centroid force can also be obtained from the matrix $\mathbf{U}$ according to [1507]

$$\frac{\partial E}{\partial \mathbf{u}_I^{(1)}} = \frac{1}{P} \sum_{s'=1}^{P} \frac{\partial E^{(s')}}{\partial \mathbf{R}_I^{(s')}} , \qquad (5.221)$$

since $\mathbf{U}_{1s} = \mathbf{U}_{s1}^{\dagger} = 1/\sqrt{P}$ and the remaining normal mode forces are given by

$$\frac{\partial E}{\partial \mathbf{u}_I^{(s)}} = \frac{1}{\sqrt{P}} \sum_{s'=1}^{P} \mathbf{U}_{ss'} \frac{\partial E^{(s')}}{\partial \mathbf{R}_I^{(s')}} \qquad \text{for } s = 2, \ldots, P \qquad (5.222)$$

in terms of the "primitive" forces $-\partial E^{(s)}/\partial \mathbf{R}_I^{(s)}$. Here, E on the left-hand side with no superscript (s) refers to the average electronic energy $E = (1/P) \sum_{s=1}^{P} E^{(s)}$ from which the forces have to be derived. Thus, the force Eq. (5.221) acting on each *centroid* variable $\mathbf{u}_I^{(1)}$, $I = 1, \ldots, N$, is exactly the force averaged over imaginary time $s = 1, \ldots, P$, i.e. the *centroid force* on the Ith nucleus as already given in Eq. (2.21) of Ref. [1507]. This is the desired relation which allows in centroid molecular dynamics, the centroid forces to be obtained simply as the average force which acts on the lowest-order normal mode Eq. (5.220). The non-centroid normal modes $\mathbf{u}_I^{(s)}$, $s = 2, 3, \ldots, P$ of the paths establish the effective potential in which the centroid moves.

At this stage the equations of motion for adiabatic *ab initio* centroid molecular dynamics [956] can be obtained from the Euler–Lagrange equations. These equations of motion read

$$M_I^{'(1)} \ddot{\mathbf{u}}_I^{(1)} = -\frac{1}{P} \sum_{s=1}^{P} \frac{\partial E\left[\{\phi_i\}^{(s)}, \{\mathbf{R}_I\}^{(s)}\right]}{\partial \mathbf{R}_I^{(s)}} \qquad (5.223)$$

$$M_I^{'(s)} \ddot{u}_{I,\alpha}^{(s)} = -\frac{\partial}{\partial u_{I,\alpha}^{(s)}} \frac{1}{P} \sum_{s'=1}^{P} E\left[\{\phi_i\}^{(s')}, \left\{\mathbf{R}_I\left(u_I^{(1)}, \ldots, u_I^{(P)}\right)\right\}^{(s')}\right]$$

$$- M_I^{(s)} \omega_P^2 u_{I,\alpha}^{(s)} - M_I^{'(s)} \dot{\xi}_{I,\alpha,1}^{(s)} \dot{u}_{I,\alpha}^{(s)} , \qquad s = 2, \ldots, P \quad (5.224)$$

$$\mu \ddot{\phi}_i^{(s)} = -\frac{\delta E\left[\{\phi_i\}^{(s)}, \{\mathbf{R}_I\}^{(s)}\right]}{\delta \phi_i^{\star(s)}}$$

$$+ \sum_j \Lambda_{ij}^{(s)} \phi_j^{(s)} - \mu \dot{\eta}_1^{(s)} \dot{\phi}_i^{(s)} \ , \quad s = 1, \ldots, P \quad (5.225)$$

where $u_{I,\alpha}^{(s)}$ denotes the Cartesian components of a given normal mode vector $\mathbf{u}_I^{(s)} = (u_{I,1}^{(s)}, u_{I,2}^{(s)}, u_{I,3}^{(s)})$. In the present scheme, independent Nosé–Hoover chain thermostats [927] of length K are coupled to all non-centroid mode degrees of freedom $s = 2, \ldots, P$

$$Q^{\mathrm{n}} \ddot{\xi}_{I,\alpha,1}^{(s)} = \left[M_I^{\prime(s)} \left(\dot{u}_{I,\alpha}^{(s)}\right)^2 - k_{\mathrm{B}}T\right] - Q^{\mathrm{n}} \dot{\xi}_{I,\alpha,1}^{(s)} \dot{\xi}_{I,\alpha,2}^{(s)} \quad (5.226)$$

$$Q^{\mathrm{n}} \ddot{\xi}_{I,\alpha,k}^{(s)} = \left[Q^{\mathrm{n}} \left(\dot{\xi}_{I,\alpha,k-1}^{(s)}\right)^2 - k_{\mathrm{B}}T\right]$$

$$- Q^{\mathrm{n}} \dot{\xi}_{I,\alpha,k}^{(s)} \dot{\xi}_{I,\alpha,k+1}^{(s)} (1 - \delta_{kK}) \ , \quad k = 2, \ldots, K \quad (5.227)$$

and all orbitals at a given imaginary time slice s are thermostatted by one such thermostat chain of length L

$$Q_1^{\mathrm{e}} \ddot{\eta}_1^{(s)} = 2\left[\sum_i \mu \left\langle \dot{\phi}_i^{(s)} \middle| \dot{\phi}_i^{(s)} \right\rangle - T_{\mathrm{e}}^0\right] - Q_1^{\mathrm{e}} \dot{\eta}_1^{(s)} \dot{\eta}_2^{(s)} \quad (5.228)$$

$$Q_l^{\mathrm{e}} \ddot{\eta}_l^{(s)} = \left[Q_{l-1}^{\mathrm{e}} \left(\dot{\eta}_{l-1}^{(s)}\right)^2 - \frac{1}{\beta_{\mathrm{e}}}\right]$$

$$- Q_l^{\mathrm{e}} \dot{\eta}_l^{(s)} \dot{\eta}_{l+1}^{(s)} (1 - \delta_{lL}) \ , \quad l = 2, \ldots, L \ ; \quad (5.229)$$

note that for standard *ab initio* path integral runs as discussed in the previous section, the centroid mode should be thermostatted as well. The desired fictitious kinetic energy of the electronic subsystem T_{e}^0 can be determined based on a short equivalent classical Car–Parrinello run with $P = 1$ and using again the relation $1/\beta_{\mathrm{e}} = 2T_{\mathrm{e}}^0/6N_{\mathrm{e}}'$ where N_{e}' is the number of orbitals (see Section 5.2.2). The mass parameters $\{Q_l^{\mathrm{e}}\}$ associated with the orbital thermostats are the same as those defined in Eq. (5.4), whereas the single mass parameter Q^{n} for the nuclei is determined by the harmonic interaction and is given by $Q^{\mathrm{n}} = k_{\mathrm{B}}T/\omega_P^2 = \beta/P$. The characteristic thermostat frequency of the electronic degrees of freedom ω_e should again lie above the frequency spectrum associated with the fictitious nuclear dynamics. This is the method that is implemented in the CPMD package [696].

An important issue for adiabatic *ab initio* centroid molecular dynamics [956] is how to establish the time scale separation of the non-centroid modes compared to the centroid modes. This is guaranteed if the *fictitious* normal mode masses $M_I'^{(s)}$ are taken to be

$$M_I'^{(1)} = M_I$$

$$M_I'^{(s)} = \gamma \, M_I^{(s)} \ , \quad s = 2, \ldots, P \ , \tag{5.230}$$

where M_I is the *physical* nuclear mass, $M_I^{(s)}$ are the normal mode masses Eq. (5.218), and γ is the "centroid adiabaticity parameter"; note that this corrects a misprint of the definition of $M_I'^{(s)}$ for $s \geq 2$ in Ref. [956]. By choosing $0 < \gamma \ll 1$, the required time scale separation between the centroid and non-centroid modes can be controlled such that the motion of the non-centroid modes is artificially accelerated, see Section 3 in Ref. [956] for a systematic study of the γ-dependence. Thus, the centroids with associated physical nuclear masses move quasiclassically in real time in the centroid effective potential, whereas the fast dynamics of all other nuclear modes $s > 1$ is fictitious and serves only to generate the centroid effective potential "on-the-fly". In this sense γ (or rather γM_I) is similar to μ, the *electronic* adiabaticity parameter in Car–Parrinello molecular dynamics.

5.4.4 Ab initio *path integrals: spectroscopy*

Computing optical spectra, in particular total photoabsorption cross-sections, is an important tool in order to investigate both the structure and dynamics of clusters and molecules. In recent years it has become increasingly evident that it is not sufficient to compute only "stick spectra" (subject to *a posteriori* uniform broadening) for a few selected *static* configurations leaving out *fluctuations*. Initially, small metal clusters have been investigated within the framework of purely thermal, i.e. *classical* fluctuations, and time-dependent density functional theory to treat electronic excitations, see for instance Refs. [583, 783, 1035, 1091, 1092] and the methods for the analysis of electronic excitations in Section 5.3.6. More recently, a general first principles approach [320] to the calculation of photoabsorption cross-sections at *finite* temperatures, which includes *quantum* fluctuations of the nuclear skeleton, has been introduced. Technically, the formulation relies on a combination of imaginary time *ab initio* path integrals as outlined in Section 5.4.2 in order to sample the nuclear quantum-thermal motion "on-the-fly" and on time-dependent Kohn–Sham density functional theory in the linear response approximation, see Section 5.3.6, for computing the electronic excitations. Thus, the impact of initial state fluctuation effects on optical spectra due to zero-point motion, tunneling, and thermal excitations of the nuclei is taken into account.

The starting point is the standard expression [1606] for the total single-photon absorption cross-section $\sigma(\omega)$ from a set of initial rovibronic states $|\Phi_i\rangle$ into final states $|\Phi_f\rangle$

$$\sigma(\omega) = \left(\frac{\pi e^2 \omega}{3\hbar\epsilon_0 c}\right) \sum_{f,i} \rho_i \, |\langle\Phi_f|\hat{\mathcal{R}}|\Phi_i\rangle|^2 \, \delta(\omega - \omega_{f,i}) \ , \tag{5.231}$$

where $\omega_{f,i} = [E_f - E_i]/\hbar$ with E_f and E_i denoting the energies of $|\Phi_f\rangle$ and $|\Phi_i\rangle$, respectively (note that i and f are multi-indices). The total dipole moment operator, $\hat{\mathbf{M}} = e\hat{\mathcal{R}}$ where $\hat{\mathcal{R}} = \hat{\mathbf{r}} + \hat{\mathbf{R}}$, acts on the set of both electronic and nuclear coordinates, respectively. The initial states, $\{|\Phi_i\rangle\}$, are assumed to be populated according to the Boltzmann distribution $\rho_i = \exp[-\beta E_i]/\mathcal{Z}$. Within the Born–Oppenheimer approximation, i.e. using only one term of the expansion Eq. (2.5), the exact rovibronic wave function is factorized according to $\Phi_i(\mathbf{r}, \mathbf{R}) \approx \Psi_n(\mathbf{r}; \mathbf{R})\chi_{nk}(\mathbf{R})$, where n and k label the electronic and associated rovibrational nuclear states, respectively. In the electronic wave functions $\Psi_n(\mathbf{r}; \mathbf{R})$ the coordinate $\mathbf{R}$ denotes the clamped nuclear positions in the sense of fixed parameters. For *electronic* transitions $n \to m$ which are considered here, the transition dipole can be simplified, $\langle\Phi_f|\hat{\mathcal{R}}|\Phi_i\rangle \approx \langle\Psi_m(\mathbf{R})|\hat{\mathbf{r}}|\Psi_n(\mathbf{R})\rangle \, \langle\chi_{m\ell}|\chi_{nk}\rangle$, since $\langle\Psi_m(\mathbf{R})|\Psi_n(\mathbf{R})\rangle \equiv 0$ for $n \neq m$ according to Eq. (5.178).

When fluctuations of the nuclei are important it is mandatory to take into account the $\mathbf{R}$-dependence of the transition dipole moment. Using the Born–Oppenheimer approximation and concentrating on excitations out of the electronic ground state $n = 0$, Eq. (5.231) turns into

$$\sigma(\omega) = \left(\frac{\pi e^2 \omega}{3\hbar\epsilon_0 c}\right) \sum_{m\ell,k} \rho_{0k} \int\int \langle\Psi_0(\mathbf{R}')|\hat{\mathbf{r}}|\Psi_m(\mathbf{R}')\rangle \, \chi_{0k}^\star(\mathbf{R}')\chi_{m\ell}(\mathbf{R}')$$

$$\times \chi_{m\ell}^\star(\mathbf{R})\chi_{0k}(\mathbf{R})\langle\Psi_m(\mathbf{R})|\hat{\mathbf{r}}|\Psi_0(\mathbf{R})\rangle \, \delta(\omega - \omega_{m\ell,0k}) \, d\mathbf{R} \, d\mathbf{R}' \ , \tag{5.232}$$

where the transition dipole has to be evaluated as a function of all nuclear coordinates. Here, $\rho_{0k} = \exp[-\beta E_{0k}]/\mathcal{Z}_0$ is the corresponding initial canonical distribution function for the *nuclear* degrees of freedom assuming that the system is initially in the electronic ground state. Following Ref. [842], we approximate at each nuclear configuration $\mathbf{R}$ the energy $\hbar\omega_{m\ell,0k}$ by the difference between the excited and the ground state potential energy surfaces $V_m(\mathbf{R})$ and $V_0(\mathbf{R})$, respectively, i.e. by the difference in the electronic energies of $\Psi_m(\mathbf{r};\mathbf{R})$ and $\Psi_0(\mathbf{r};\mathbf{R})$. Thus, $\omega_{m\ell,0k} \approx \omega_{m,0}(\mathbf{R}) \equiv [V_m(\mathbf{R}) - V_0(\mathbf{R})]/\hbar$. Using $\sum_\ell \chi_{m\ell}(\mathbf{R}')\chi^\star_{m\ell}(\mathbf{R}) = \delta(\mathbf{R}' - \mathbf{R})$, Eq. (5.232) readily simplifies to

$$\sigma(\omega) = \left(\frac{\pi e^2 \omega}{3\hbar\epsilon_0 c}\right) \sum_{m,k} \rho_{0k} \int |\langle\Psi_m(\mathbf{R})|\hat{\mathbf{r}}|\Psi_0(\mathbf{R})\rangle|^2$$

$$\times \chi^\star_{0k}(\mathbf{R})\chi_{0k}(\mathbf{R})\, \delta(\omega - \omega_{m,0}(\mathbf{R}))\, d\mathbf{R} \ . \quad (5.233)$$

The range of validity of Eq. (5.233), and in particular the consistent classical limit of $\sigma(\omega)$ is discussed controversially in the literature [783, 1092].

In order to clarify its validity, an alternative derivation [842] of Eq. (5.233) is considered. Employing the Fourier representation of the delta function, the Born–Oppenheimer approximation, and the Hamiltonian operators $\hat{H}_0(\mathbf{R}) = \hat{T} + \hat{V}_0(\mathbf{R})$ and $\hat{H}_m(\mathbf{R}) = \hat{T} + \hat{V}_m(\mathbf{R})$ of the nuclear (rovibrational) wave functions $\chi_{0k}(\mathbf{R})$ and $\chi_{m\ell}(\mathbf{R})$, respectively,

$$\sigma(\omega) = \left(\frac{\pi e^2 \omega}{3\hbar\epsilon_0 c}\right) \sum_{m\ell,k} \rho_{0k} \int \int \left(\frac{1}{2\pi}\right) \int_{-\infty}^{\infty} \exp[-i\omega t]\chi^\star_{0k}(\mathbf{R}')$$

$$\times \langle\Psi_0(\mathbf{R}')|\hat{\mathbf{r}}|\Psi_m(\mathbf{R}')\rangle\, \chi_{m\ell}(\mathbf{R}')\ \langle\Psi_m(\mathbf{R})|\hat{\mathbf{r}}|\Psi_0(\mathbf{R})\rangle\chi^\star_{m\ell}(\mathbf{R})$$

$$\times \exp[i\hat{H}_m(\mathbf{R})t/\hbar]\, \exp[-i\hat{H}_0(\mathbf{R})t/\hbar]\, \chi_{0k}(\mathbf{R})\, dt\, d\mathbf{R}\, d\mathbf{R}' \quad (5.234)$$

$$\approx \left(\frac{\pi e^2 \omega}{3\hbar\epsilon_0 c}\right) \sum_{m\ell,k} \rho_{0k} \int \int \left(\frac{1}{2\pi}\right) \int_{-\infty}^{\infty} \exp[-i\omega t]\chi^\star_{0k}(\mathbf{R}')$$

$$\times \langle\Psi_0(\mathbf{R}')|\hat{\mathbf{r}}|\Psi_m(\mathbf{R}')\rangle\, \chi_{m\ell}(\mathbf{R}')\ \langle\Psi_m(\mathbf{R})|\hat{\mathbf{r}}|\Psi_0(\mathbf{R})\rangle\chi^\star_{m\ell}(\mathbf{R})$$

$$\times \exp[i\,(V_m(\mathbf{R}) - V_0(\mathbf{R}))\,t/\hbar]\, \chi_{0k}(\mathbf{R})\, dt\, d\mathbf{R}\, d\mathbf{R}' \quad (5.235)$$

is obtained from Eq. (5.232). Carrying out the time integration in Eq. (5.235) to recover the delta function and employing the definition of $\omega_{m,0}(\mathbf{R})$ again leads to Eq. (5.233). In the derivation of Eq. (5.233), in the step from Eq. (5.234) to Eq. (5.235), besides the Born–Oppenheimer approximation an *additional* approximation was invoked, namely the commutators between the operator $\hat{T}$ of the kinetic energy and the potentials $V_0(\mathbf{R})$ and $V_m(\mathbf{R})$ were neglected. This approximation, and thus Eq. (5.233), improves systematically the larger the mass of the nuclei and/or the smaller the curvature of the involved potential energy surfaces. This is often the case for "soft mode systems", such as for example many fluxional molecules or metal clusters and therefore justifies the use of Eq. (5.233) to compute their photoabsorption spectra.

Exploiting the fact that ω can be replaced by $\omega_{m,0}(\mathbf{R})$ due to the delta function, the operator $\hat{\mathcal{O}}(\mathbf{R})$

$$\hat{\mathcal{O}}(\mathbf{R}) = \left(\frac{\pi e^2 \omega}{3\hbar\epsilon_0 c}\right)$$

$$\times \sum_m \omega_{m,0}(\mathbf{R})\,|\langle\Psi_m(\mathbf{R})|\hat{\mathbf{r}}|\Psi_0(\mathbf{R})\rangle|^2\;\delta(\omega - \omega_{m,0}(\mathbf{R}))$$

$$= \left(\frac{2\pi^2 e^2}{m_e 4\pi\epsilon_0 c}\right)\sum_m f_{m,0}(\mathbf{R})\,\delta(\omega - \omega_{m,0}(\mathbf{R})) \tag{5.236}$$

is defined in the position representation, where $f_{m,0}(\mathbf{R})$ is the usual oscillator strength [1606] at the instantaneous nuclear configuration $\mathbf{R}$. These oscillator strengths f can be computed for a given excitation mode within the framework of linear response time-dependent density functional theory according to Eq. (5.149) derived in Section 5.3.6. Using this notation, Eq. (5.233) can be expressed as a trace in the space of rovibrational nuclear wave functions of the initial electronic state

$$\sigma(\omega) = \sum_k \langle\chi_{0k}|\hat{\rho}_0\hat{\mathcal{O}}|\chi_{0k}\rangle = \mathrm{Tr}\,\hat{\rho}_0\,\hat{\mathcal{O}} \tag{5.237}$$

$$= \int \langle\mathbf{R}'|\hat{\rho}_0|\mathbf{R}'\rangle\mathcal{O}(\mathbf{R}')\,d\mathbf{R}'\ , \tag{5.238}$$

which in turn can be re-expressed in the position basis of the nuclei due to the invariance of the trace, where $\hat{\mathcal{O}}$ is diagonal in view of Eq. (5.236); $\hat{\rho}_0$ is the nuclear density operator associated with the Boltzmann factor ρ_{0k} introduced above.

This formulation of a thermal operator expectation value is now directly amenable to a path integral representation of the photoabsorption cross-section in coordinate space

$$\sigma(\omega) = \frac{1}{\mathcal{Z}_0}\int\int_{\mathbf{R}(0)=\mathbf{R}'}^{\mathbf{R}(\hbar\beta)=\mathbf{R}'}\exp\left[-\frac{1}{\hbar}S_E[\mathbf{R}]\right]\mathcal{O}(\mathbf{R}')\,\mathcal{D}\mathbf{R}\,d\mathbf{R}'\ , \tag{5.239}$$

where $\mathcal{S}_{\mathrm{E}}$

$$\mathcal{S}_{\mathrm{E}}[\mathbf{R}] = \int_0^{\hbar\beta} \left\{ \frac{1}{2} M \left(\frac{d\mathbf{R}}{d\tau} \right)^2 + E_0 \left(\mathbf{R}(\tau) \right) \right\} d\tau \qquad (5.240)$$

is the classical Euclidean action. In order to use this formulation for actual computations, the continuum expression must be discretized [753] as in Section 5.4.2

$$\sigma(\omega) = \lim_{P \to \infty} \frac{1}{\mathcal{Z}_0^{[P]}} \prod_{s=1}^{P} \left[\prod_{I=1}^{N} \int \right]$$

$$\times \exp\left[-\beta \sum_{s=1}^{P} \left\{ \sum_{I=1}^{N} \frac{1}{2} M_I \omega_P^2 \left(\mathbf{R}_I^{(s)} - \mathbf{R}_I^{(s+1)} \right)^2 \right. \right.$$

$$\left. \left. + \frac{1}{P} E_0(\mathbf{R}^{(s)}) \right\} \right] \frac{1}{P} \sum_{s'=1}^{P} \mathcal{O}(\mathbf{R}^{(s')}) \prod_{s=1}^{P} \prod_{I=1}^{N} A_P \, d\mathbf{R}_I^{(s)} , \qquad (5.241)$$

with $(\mathbf{R}_1^{(s)}, \ldots, \mathbf{R}_I^{(s)}, \ldots, \mathbf{R}_N^{(s)}) = \mathbf{R}^{(s)}$ being the Cartesian positions of N nuclei with masses M_I at imaginary time slice $s = 1, \ldots, P$; here $\mathcal{Z}_0^{[P]}$ denotes the partition function for finite Trotter number P and A_P, as well as ω_P^2, are defined in Eqs. (5.193) and (5.194), respectively. In the limit $P \to \infty$ the evaluation of the nuclear quantum effects in the initial electronic ground state, by virtue of Eq. (5.241), is exactly equivalent to the corresponding wave function representation Eq. (5.233) and in a sense finite P corresponds to a truncation of the basis set when evaluating the sum $\sum_k$ over nuclear rovibrational states in the Schrödinger representation, Eq. (5.233).

In the classical or high-temperature limit, which is easily obtained in discretized path integral expressions using $P \to 1$ thus neglecting all nuclear quantum effects, this quantum formulation reduces without further assumptions or approximations to

$$\sigma_{\mathrm{CL}}(\omega) = \frac{1}{\mathcal{Z}_0^{\mathrm{CL}}} \prod_{I=1}^{N} \left[\left(\frac{M_I}{2\pi\hbar^2\beta} \right)^{3/2} \int \right]$$

$$\times \exp\left[-\beta E_0 \left(\{\mathbf{R}_I\} \right) \right] \mathcal{O}(\{\mathbf{R}_I\}) \, d\mathbf{R}_1 \cdots d\mathbf{R}_N , \qquad (5.242)$$

which has been used to compute optical spectra of metal clusters [583, 1035, 1091]; $\mathcal{Z}_0^{\mathrm{CL}} = \prod_{I=1}^{N} [(M_I/2\pi\hbar^2\beta)^{3/2} \int] \exp[-\beta E_0(\{\mathbf{R}_I\})] d\mathbf{R}_I \cdots d\mathbf{R}_N$ is the associated classical partition function. This transparent derivation of

Eq. (5.242) yields the classical limit of Eq. (5.231) subject to the approximations leading to Eq. (5.235) and thus to Eq. (5.233).

In practice, the evaluation of the classical approximation to optical spectra, Eq. (5.242), is straightforward. Canonical *ab initio* molecular dynamics simulations are performed at the desired temperature and the oscillator strengths, $f_{m,0}(\{\mathbf{R}_I\})$ introduced in Eq. (5.236), are evaluated at the instantaneous nuclear configurations $\{\mathbf{R}_I\}$ via time-dependent density functional linear response theory described in Section 5.3.6 (see in particular Eq. (5.149) and Eq. (5.171) therein). In the quantum limit, Eq. (5.241), *ab initio* path integral molecular dynamics simulations must be performed as introduced in Section 5.4.2 and $f_{m,0}(\{\mathbf{R}_I\}^{(s)})$ must be evaluated separately at each time slice $\{\mathbf{R}_I\}^{(s)}$, $s = 1, \ldots, P$. Depending on temperature and system of interest, the influence of quantum fluctuations might be pronounced, as demonstrated in Ref. [320] for a small lithium cluster.

5.4.5 *Quantum corrections of classical susceptibilities: infrared spectra*

The path integral-based techniques introduced earlier in this chapter are very powerful but computationally quite demanding, typically a factor of 10 to 100 in comparison to *ab initio* molecular dynamics using classical nuclei. Within the framework of the latter, it is also straightforward to compute autocorrelation functions of classical observables, which upon Fourier-transforming yield spectra, susceptibilities, and response functions. Clearly, the classical limit of nuclear motion introduces an approximation, which in the case of spectra or susceptibilities is particularly severe since the fundamental "detailed balance" condition, which imposes a certain asymmetry, is violated. In the following, the concept of the "quantum correction factors", which cure this qualitative shortcoming approximately, is introduced.

A very popular approach to compute spectral cross-sections or absorption coefficients is to *exactly* transform them from the Schrödinger representation of quantum mechanics, such as for instance Eq. (5.231) in Section 5.4.4, to the Heisenberg representation. This amounts to getting rid of operator matrix elements using stationary wave functions at the expense of introducing quantum time-correlation functions of the associated time-dependent Heisenberg operators. The starting point for the following discussion is the well-known formula [984] for the infrared absorption *coefficient* per unit length in the dipole (E1) approximation, which is given in the Schrödinger representation by

$$\alpha(\omega) = \left[\frac{4\pi^2 \omega}{3V \hbar c\, n(\omega)} \right] (1 - \exp\left[-\beta\hbar\omega\right])$$

$$\times \sum_{\ell k} \rho_{0k} |\langle \chi_{0\ell} | \hat{\mathbf{M}} | \chi_{0k} \rangle|^2 \delta(\omega - \omega_{f,i}) \quad (5.243)$$

for isotropic systems in the limit of weak external fields in the electronic ground state within the Born–Oppenheimer approximation. Here, $\chi_{0\ell}$ / χ_{0k} are the final/initial nuclear wave functions in the electronic ground state, V is the volume of the sample with index of refraction $n(\omega)$ at temperature $T = 1/k_B\beta$, and ρ_{0k} is the canonical probability of the system being in the kth nuclear initial state, $\chi_{0k}(\mathbf{R})$, given that it is in the electronic ground state, $\Psi_0(\mathbf{r}; \mathbf{R})$. Note that $n \approx 1$ for gas phase spectroscopy, that the absorption *cross-section* is defined without the normalization by V, and that the additional prefactor $1/3$ stems from isotropically averaging over the polarization directions of the radiation field. The above-used matrix element of the total dipole moment operator is defined as

$$\langle \chi_{0\ell} | \hat{\mathbf{M}} | \chi_{0k} \rangle = \int \chi_{0\ell}^\star(\mathbf{R}) \left[\mathbf{M}^{\mathrm{el}}(\mathbf{R}) + \mathbf{M}^{\mathrm{nuc}} \right] \chi_{0k}(\mathbf{R})\, d\mathbf{R} \quad (5.244)$$

$$\text{with} \quad \mathbf{M}^{\mathrm{el}}(\mathbf{R}) = \int \Psi_0^\star(\mathbf{r}; \mathbf{R}) \left(\sum_i e\mathbf{r}_i \right) \Psi_0(\mathbf{r}; \mathbf{R})\, d\mathbf{r} \quad (5.245)$$

$$\text{and} \quad \mathbf{M}^{\mathrm{nuc}}(\mathbf{R}) = \sum_I q_I \mathbf{R}_I , \quad (5.246)$$

where $\mathbf{M}^{\mathrm{nuc}}(\mathbf{R}_1, \ldots, \mathbf{R}_N)$ is the contribution of the set of nuclei (with charges q_I at positions $\mathbf{R}_I$) to the total dipole moment (operator) and $\mathbf{M}^{\mathrm{el}}(\mathbf{R})$ is the *electronic* contribution in the electronic ground state at the same nuclear configuration $\mathbf{R}$. Since the position operator is ill-defined within *periodic* boundary conditions (see Section 7.2.2), the computation of the electronic contribution, $\mathbf{M}^{\mathrm{el}}$ according to Eq. (5.245), to the total dipole moment is tricky and has been solved only in the mid-1990s in the framework of the modern theory of polarization as explained in Section 7.2, see also Section 5.3.6.

In the Heisenberg representation, this expression for α can be rewritten in terms of a quantum-mechanical dipole autocorrelation function [542, 984]

$$\alpha(\omega) = \left[\frac{4\pi^2\omega}{3V\hbar cn(\omega)}\right](1 - \exp[-\beta\hbar\omega])I(\omega) \qquad (5.247)$$

$$= \left[\frac{4\pi^2\omega}{3V\hbar cn(\omega)}\right](1 - \exp[-\beta\hbar\omega])$$

$$\times \frac{1}{2\pi}\int_{-\infty}^{\infty}\exp[-i\omega t]\langle\hat{\mathbf{M}}(0)\hat{\mathbf{M}}(t)\rangle\,dt \qquad (5.248)$$

$$= \left[\frac{4\pi^2\omega}{3V\hbar cn(\omega)}\right]\frac{V\hbar}{\pi}\chi''(\omega) \ , \qquad (5.249)$$

where $I(\omega)$ is the (isotropic) lineshape function or spectral density and $\chi''(\omega)$ is the imaginary part of static (isotropic) susceptibility which is connected via $\epsilon(\omega) = \epsilon' - i\epsilon'' = 1 + 4\pi\chi(\omega)$ to the complex dielectric function; $\langle\hat{O}\rangle = \mathrm{Tr}\,(\hat{\rho}_0\,\hat{O})$ denotes the canonical thermal average where $\hat{\rho}_0 = \exp[-\beta\hat{\mathcal{H}}_n]/\mathcal{Z}_0$ is the density operator for the nuclear degrees of freedom in the electronic ground state, see also Eqs. (5.237)-(5.238). Note that the standard expression [225] for (isotropic) condensed matter systems in terms of the imaginary part of the trace of the susceptibility tensor χ

$$\alpha(\omega) = \left(\frac{4V\pi^2\omega}{3\hbar cn(\omega)}\right)2\tanh\left[\frac{1}{2}\beta\hbar\omega\right]$$

$$\times \frac{1}{2\pi}\int_{-\infty}^{\infty}\exp[-i\omega t]\left\langle\left\{\hat{\mathbf{P}}(0),\hat{\mathbf{P}}(t)\right\}\right\rangle\,dt \qquad (5.250)$$

$$= \left(\frac{4\pi\omega}{cn(\omega)}\right)\mathrm{Im}\,\frac{1}{3}\sum_{i=1}^{3}\chi_{ii}(\omega) \ , \qquad (5.251)$$

is obtained from Eq. (5.248) once the bulk electronic polarization $\hat{\mathbf{P}} = \hat{\mathbf{M}}/V$ and Kubo's symmetrized product $\{\hat{\mathbf{O}}(0), \hat{\mathbf{O}}(t)\} = (\hat{\mathbf{O}}(0)\hat{\mathbf{O}}(t) + \hat{\mathbf{O}}(t)\hat{\mathbf{O}}(0))/2$ are introduced.

An advantage of the Heisenberg representation in comparison to wave function-based formulations is its formal similarity to classical mechanics, which suggests simply replacing the dipole moment Heisenberg operators, $\hat{\mathbf{M}}(t)$ in Eq. (5.248), by the classical dipole moment, $\mathbf{M}(t)$. However, the relation resulting from this dramatic simplification

$$\alpha_{\mathrm{CL}}(\omega) = \left[\frac{4\pi^2\omega}{3V\hbar cn(\omega)}\right](1 - \exp[-\beta\hbar\omega])I_{\mathrm{CL}}(\omega) \qquad (5.252)$$

$$= \left[\frac{4\pi^2 \omega}{3V \hbar c n(\omega)} \right] (1 - \exp[-\beta\hbar\omega])$$

$$\times \frac{1}{2\pi} \int_{-\infty}^{\infty} \exp[-i\omega t] \langle \mathbf{M}(0)\mathbf{M}(t) \rangle \, dt \; , \tag{5.253}$$

in terms of the classical dipole autocorrelation function, $C_{\mathrm{CL}}(t) = \langle \mathbf{M}(0)\mathbf{M}(t) \rangle$, can be shown to not satisfy the important detailed balance condition,

$$I(\omega) = \exp[\beta\hbar\omega]I(-\omega) \; . \tag{5.254}$$

In the classical limit, the detailed balance condition evidently goes over to the trivial relation $\lim_{T\to\infty} \exp[\beta\hbar\omega] = 1$, which implies that $I_{\mathrm{CL}}(\omega)$ will be an *even* function of time as opposed to the correct asymmetry embodied in Eq. (5.254).

In the sense of an intermediate step or quasiclassical approximation, the idea of quantum correction factors was introduced a long time ago, see for instance Ref. [1183] for literature on this subject. The absorption coefficient is decomposed

$$\alpha_{\mathrm{QC}}(\omega) = \left[\frac{4\pi^2 \omega}{3V \hbar c n(\omega)} \right] (1 - \exp[-\beta\hbar\omega]) \; Q_{\mathrm{QC}}(\omega) \; I_{\mathrm{CL}}(\omega) \tag{5.255}$$

into an invariant prefactor [...], the factor $(1 - \exp[-\beta\hbar\omega])$ that defines together with the lineshape function the susceptibility according to Eq. (5.249), and a frequency- and temperature-dependent function $Q_{\mathrm{QC}}(\omega)$. The latter is often called "quantum correction factor", and is a correction to the lineshape function or to the susceptibility as defined in Eqs. (5.247), (5.248), and (5.249). There are various such quantum correction factors Q_{QC} discussed in the literature [1183]:

- Standard Approximation (SA)

$$Q_{\mathrm{SA}} = \frac{2}{1 + \exp[-\beta\hbar\omega]} \; ; \tag{5.256}$$

- Harmonic Correction (HC)

$$Q_{\mathrm{HC}} = \frac{\beta\hbar\omega}{1 - \exp[-\beta\hbar\omega]} \; ; \tag{5.257}$$

- Schofield Correction (SC)

$$Q_{\text{SC}} = \exp\left[+\beta\hbar\omega/2\right] \tag{5.258}$$

$$= \frac{1}{I_{\text{CL}}(\omega)}\frac{1}{2\pi}\int_{-\infty}^{\infty}\exp\left[-i\omega t\right]C_{\text{CL}}\left(t+i\beta\hbar/2\right)\,dt\; ; \tag{5.259}$$

- Egelstaff Correction (EC)

$$Q_{\text{EC}} = \frac{1}{I_{\text{CL}}(\omega)}\frac{1}{2\pi}\int_{-\infty}^{\infty}\exp\left[-i\omega t\right]C_{\text{CL}}\left(\sqrt{t(t+i\beta\hbar)}\right)\,dt \tag{5.260}$$

$$= \frac{\exp\left[+\beta\hbar\omega/2\right]}{I_{\text{CL}}(\omega)}$$

$$\times \frac{1}{2\pi}\int_{-\infty}^{\infty}\exp\left[-i\omega t\right]C_{\text{CL}}\left(\sqrt{t^2+(\beta\hbar/2)^2}\right)\,dt\;. \tag{5.261}$$

Notably, each of the above quantum corrections is *constructed* such that the quantum-corrected lineshape function $I_{\text{QC}}(\omega) = Q_{\text{QC}}(\omega)\,I_{\text{CL}}(\omega)$ satisfies the principle of detailed balance Eq. (5.254), which amounts to a "de-symmetrization" in view of the classical limit $\lim_{T\to\infty}\exp[\beta\hbar\omega] = 1$. However, their frequency dependencies turn out to be vastly different: the Schofield correction, for instance, has a tendency to suppress intensities at low frequencies, whereas the standard approximation has the opposite effect of strongly enhancing intensities in the limit of high frequencies. Based on an entirely theoretical analysis of the quantum and classical versions of the fluctuation–dissipation theorem, it has been argued that the harmonic correction, $Q_{\text{HC}}(\omega)$, should perform best for infrared (vibrational) spectroscopy [1183]. Indeed, it affects the low- and high-frequency regimes in a more balanced way than for instance the Schofield correction or the standard approximation. Independent simulation studies [671, 841, 1291, 1292] seem to support this conclusion based on numerical evidence.

Clearly, using a multiplicative quantum correction factor, preferentially the harmonic correction Eq. (5.257), in front of Fourier-transformed classical autocorrelation functions, Eq. (5.255), is a very efficient way to improve results obtained by *ab initio* molecular dynamics simulations. However, based on the same theoretical analysis of the fluctuation-dissipation theorem, it has also been argued [1183] that the only sound way to further improve the deficiencies introduced by classical mechanics is to improve the correlation function itself, $C_{\text{CL}}(t) \to C_{\text{QC}}(t)$, and not the quantum correction factor $Q_{\text{QC}}(\omega)$.

A quasiclassical improvement of classical autocorrelation functions can be obtained within the centroid molecular dynamics approach outlined in Section 5.4.3. In particular, $C_{CMD}(t)$ should be a reasonable approximation to $C_{Kubo}(t)$ according to Eqs. (5.210)–(5.212) for the dipole operator. Hence, the infrared absorption coefficient reads

$$\alpha_{CMD}(\omega) = \left[\frac{4\pi^2\omega}{3V\hbar c n(\omega)}\right]\beta\hbar\omega$$

$$\times \frac{1}{2\pi}\int_{-\infty}^{\infty}\exp\left[-i\omega t\right]\langle\mathbf{M}^{(c)}(0)\mathbf{M}^{(c)}(t)\rangle_{CMD}\,dt\ ,\quad (5.262)$$

where $\mathbf{M}^{(c)}(t)$ denotes the centroid of the total dipole moment at time t and $\langle\cdots\rangle_{CMD}$ is the average in the ensemble generated by centroid molecular dynamics. Indeed, it has been shown [1183] numerically for various one-dimensional tunable model potentials that this approximation outperforms all quantum correction factors tested, including the harmonic correction $Q_{HC}(\omega)$; still, the deviation of α_{CMD} from the exact result α can be pronounced depending on the model potential used. However, computing the autocorrelation function of $\mathbf{M}^{(c)}(t)$ amounts to performing full *ab initio* path integral molecular dynamics simulations from which the dipole centroids are obtained, which increases the computational effort considerably as compared to autocorrelating the classical observable $\mathbf{M}(t)$.

5.4.6 *Related* ab initio *quantum approaches*

It is evident from the outset that both the Car–Parrinello and Born–Oppenheimer approaches to *ab initio* molecular dynamics can be used in order to compute the forces on the quantum nuclei from "on-the-fly" electronic structure calculations. The Born–Oppenheimer variant has been utilized in a variety of investigations ranging from clusters to molecular solids [74, 264, 747, 1014, 1015, 1492, 1579, 1582]. A similar scheme using Hartree–Fock interactions has been proposed and applied to small molecules in the gas phase [1082, 1334, 1335], as well as its MP2 extension [612, 1083, 1333, 1429–1432]. Closely related to the *ab initio* path integral approach as discussed in previous chapters is a method that is based on Monte Carlo sampling of the path integral [1599]. It is similar in spirit to the Born–Oppenheimer implementation of *ab initio* path integrals as long as only time-averaged static observables are calculated. A semiempirical ("CNDO" and "INDO") version of Born–Oppenheimer *ab initio* path integral simulations was also devised [1535] and applied to study muonated organic molecules [1534, 1535]. A centroid molecular dynamics extension has also been proposed [1081].

A non-self-consistent approach to *ab initio* path integral calculations was advocated and used in a series of publications devoted to study the interplay of nuclear quantum effects and electronic structure in unsaturated hydrocarbons such as benzene or ethylene [170, 172, 877, 1179, 1185–1187, 1294, 1295]. According to this philosophy, an ensemble of nuclear path configurations, Eq. (5.205), is first generated at finite temperature with the aid of a parameterized model potential (or using a tight-binding Hamiltonian [1179]). In a second, independent step electronic structure calculations (using Pariser–Parr–Pople, Hubbard, or Hartree–Fock Hamiltonians) are performed for this fixed ensemble of discretized quantum paths. The crucial difference compared to the self-consistent approaches presented here is that the creation of the thermal ensemble and the subsequent analysis of its electronic properties is performed using different Hamiltonians.

Several attempts to treat the electrons in the path integral formulation also - instead of using wave functions as in the *ab initio* path integral family - were published [245, 584, 650, 1012, 1079, 1080, 1150, 1151, 1433–1435]. These approaches are eventually exact, i.e. nonadiabaticity and full electron-phonon coupling is included at finite temperatures. However, they suffer from severe stability problems ("sign problem", "Fermion problem") [247] in the limit of degenerate electrons, i.e. at very low temperateres compared to the Fermi temperature, which is the temperature range of interest for typical problems in chemistry, materials science, and biology. Recent progress on computing electronic forces from quantum Monte Carlo simulations, possibly used in conjunction with molecular dynamics propagation, was also achieved [51, 52, 267, 560, 1667].

More traditional approaches use a wave function representation for both the electrons in the ground state and for the nuclear density matrix instead of path integrals [1426–1428, 1522]. The advantage is that real-time evolution is obtained more naturally compared to path integral simulations. Reviews of such methods with the emphasis on computing the interactions "on-the-fly" are provided in Refs. [338, 408]. An approximate wave function-based quantum dynamics method, which includes several excited states and their couplings, was also devised and used [94, 96, 914, 916, 917]. A related method, which unifies a quantum wave packet dynamics with an "on-the-fly" calculation of the electronic degrees of freedom, has been proposed more recently [687]. An alternative approach to approximate quantum dynamics consists in performing instanton or semiclassical *ab initio* dynamics [95, 730]. Also, the approximate vibrational self-consistent field approach to nuclear quantum dynamics was combined with "on-the-fly" MP2 electronic structure calculations [249].

5.5 Mixed quantum/classical hybrid molecular dynamics

5.5.1 Introduction

Quantum mechanics is computationally much more complex and thus much more demanding than classical mechanics given a certain system of interest. In a seminal paper [1597] the key idea was introduced into the field of biomolecular calculations to subdivide a complex system such that only a small, but conceptually relevant part (the "hot spot") is treated based on quantum-mechanical (QM) theories. The rest of the system (the "biomatrix") is not neglected but rather approximated by parameterized molecular mechanical (MM) force fields [283, 888, 1302], as commonly used in biomolecular molecular dynamics simulations [726, 1449, 1538, 1539]. Such embedding techniques [424, 1366, 1597], which can be viewed as atomistically resolved "system-bath" approaches [1526–1529, 1605], are now commonly denoted as mixed quantum/classical, hybrid, or quantum mechanics/molecular mechanics (QM/MM) schemes, see e.g. Refs. [43, 228, 1027, 1275, 1306, 1332] for an incomplete list of reviews.

What are the key aspects in order to couple a quantum subsystem to a classical environment? Mixed quantum/classical approaches to molecular dynamics exist, where the classical limit is rigorously achieved (in the sense of letting $\hbar \to 0$) in an atomistically defined part of a quantum system, such as those based on the hydrodynamic (Bohm) [512, 513] and path integral (Feynman) [1506] formulations of quantum mechanics. Unfortunately, they are currently not useful in order to derive practical approaches in the context of *wave function-based* electronic structure methods for *many*-electron systems, so that all working QM/MM coupling schemes feature necessarily some engineering character.

One class of QM/MM approaches is fully atomistic in the sense that both the QM system and the MM system are based on particle representations, which amounts to an embedding of the QM system in a force field environment. In "additive schemes" [1332] the full QM/MM energy and thus the Hamiltonian is obtained

$$E^{\text{QM/MM}} = E^{\text{QM}}(\{\mathbf{R}_\alpha\}) + E^{\text{MM}}(\{\mathbf{R}_I\}) + E^{\text{QM-MM}}(\{\mathbf{R}_I\}, \{\mathbf{R}_\alpha\}) \quad (5.263)$$

by *adding* to the energies E^{QM} and E^{MM} that act exclusively in the QM and MM systems, respectively, a coupling term $E^{\text{QM-MM}}$, see Eq. (5.264). Atomic positions in the MM system are denoted by $\mathbf{R}_I$, whereas $\mathbf{R}_\alpha$ are the positions of nuclei in the QM system including link atoms or pseudopotential centers, see below for more details. It is noted in passing that "subtractive schemes" [1332], such as the very successful QM-Pot technique [187, 389,

1275], fully compensate double counting of interactions, which might plague additive schemes and thus require sophisticated corrections [383, 388]. This is achieved by *subtracting* the energy of a MM calculation of the isolated QM subsystem (saturated with link atoms) from the sum of the MM energy of the entire system and the QM energy of the QM subsystem. In turn, this ansatz requires elaborate MM force fields capable of describing the interactions not only in the MM part, but also *within* the QM subsystem reasonably well during the *entire* simulation, including for instance *all* stages of chemical reactions. Clearly, this is a limitation of subtractive schemes in particular in the framework of *ab initio* molecular dynamics, which is often used in exactly those cases where reliable force fields cannot be devised.

Typical valence force field energy expressions, E^{MM}, such as the ones used for biomolecular simulation [1538, 1539], consist of short-ranged bonded interactions, representing the chemical bonds according to an *a priori* determined fixed molecular topology, and nonbonded interactions. The latter are in general split into electrostatic interactions and "steric contributions", modeling Pauli repulsion at short range in order to implement self-avoidance as well as van der Waals-type dispersion attraction. The bonded interactions consist typically of stretching (distance), bending (angle), and torsional (dihedral) terms which involve a few atoms that are connected by permanent chemical bonds. Thus, a natural but arbitrary choice to partition the hybrid total energy that couples the QM and MM systems

$$E^{\text{QM-MM}} = E_{\text{b}}^{\text{QM-MM}} + E_{\text{nb}}^{\text{QM-MM}} \tag{5.264}$$

$$E_{\text{nb}}^{\text{QM-MM}} = E_{\text{es}}^{\text{QM-MM}} + E_{\text{steric}}^{\text{QM-MM}} \tag{5.265}$$

is to use the same structure. In the following sections it will be assumed that E^{QM} is the total energy obtained from the Kohn–Sham Hamiltonian H_{e}^{KS}, Eq. (2.120), in a plane wave/pseudopotential representation, and the energy function, E^{MM}, is given by a typical valence force field such as e.g. CHARMM [888], GROMOS [1302, 1540], or AMBER [283]. The handling of the short-range nonbonded term, $E_{\text{steric}}^{\text{QM-MM}}$, which is often embodied in Lennard-Jones-type potentials, generally follows the model used in E^{MM} and may be subject to re-fitting if required.

The long-range electrostatic contribution to the nonbonded interactions, $E_{\text{es}}^{\text{QM-MM}}$, can be classified according to the sophistication of the particular coupling scheme [67, 1332]. In "mechanical embedding schemes" there is no influence of the MM charge distribution on the QM system, i.e. the QM calculation is gas-phase-like without including an additional potential due to the MM atoms. Thus, the electrostatic part of the QM-MM coupling is either neglected or for instance established by assigning fixed effective charges to the QM nuclei, such as charges according to the force field used. "Electrostatic embedding" is achieved by adding a new term to the QM Hamiltonian where the electrostatic interaction of the MM atoms with the charge density of the QM system is taken into account in terms of an external charge distribution. Finally, "polarized embedding schemes" consider in addition the polarization of the MM atoms due to the QM charge density, which can be implemented either non-self-consistently or fully self-consistently.

Particular technical but nontrivial problems are encountered in plane wave-based electronic structure calculations when evaluating the electrostatic coupling term

$$E_{\text{es}}^{\text{QM-MM}} = \sum_{I \in \text{MM}} q_I \int \frac{n(\mathbf{r})}{|\mathbf{r} - \mathbf{R}_I|} \, d\mathbf{r} \qquad (5.266)$$

since the electronic charge density, n, is defined on a FFT grid in both real and reciprocal space, see Section 3.2 for a detailed exposition. The straightforward numerical evaluation of the integral in Eq. (5.266) is prohibitive since it would involve a number of operations that scales with the number of grid points times the number of MM atoms. This amounts to an enormous effort since the real space grid, typically $\sim 100^3$, is determined by the number of plane waves used and the latter is of the order of 10 000–100 000 in typical current applications. Furthermore, in this case a notorious problem, "electron spill-out" or "charge leakage", due to the lack of orthogonality and thus Pauli repulsion for the interaction of the electrons in the QM system with the nearby MM atoms is encountered [1332]. Spill-out is always present to some extent, but it is particularly severe if the very flexible plane wave basis set is used to expand the wave function. Here, the negative electronic density can easily spread out into the "vacuum" region where the positive partial charges of the MM atoms are located, instead of being kept close to the QM nuclei as is the case if small or moderately sized Gaussian basis sets without diffuse functions are used. Various coupling schemes follow different routes to cope with these problems, as outlined in more detail below for several examples.

In addition, an elegant and rigorous dual length scale approach has been proposed [1008, 1644], implemented in the PINY package [1153, 1513], and tested [1033, 1644] in first applications. In this framework *all* charge–charge interactions, i.e. those stemming from electrons, pseudopotential cores, and MM force field sites, are treated on an equal footing within a plane wave representation. To this end, the two terms that act at long range, the Hartree and local pseudopotential energies according to Section 3.2, are split artificially into short-range and long-range contributions using the standard identity $\text{erf}(\alpha r) + \text{erfc}(\alpha r) = 1$, see e.g. Eq. (3.57). Assuming that all electrons are localized in a rather small region in a large supercell, a dense plane wave grid can be used to treat the short-range part in a small cell (subject to arbitrary boundary conditions akin to standard plane wave/pseudopotential calculations), whereas a low plane wave cutoff is sufficient to evaluate the slowly varying long-range contributions. Computational efficiency and in particular linear scaling is achieved by a Cardinal B-spline interpolation scheme between the two plane wave expansions/grids, which need not be commensurate. Refer to the original literature [1008, 1644] for details and to Refs. [818, 819] for efficient implementations of such multigrid electrostatic QM/MM coupling schemes.

Finally, bonded QM–MM interactions, $E_{\text{b}}^{\text{QM−MM}}$, must be introduced if covalent bonds connecting the QM to the MM system are cut. In "link atom-based schemes" [1332] the QM part is electronically saturated by using capping atoms, often monovalent atoms such as hydrogen [1366], which introduce additional degrees of freedom that should be constrained as far as possible or even eliminated [67, 383, 424]. Alternatively, "molecular pseudopotentials" [1417, 1573] can be constructed to saturate the valence of these QM atoms at the MM boundary.

Another and actually much older [1086] class of QM/MM methods relies on the idea of replacing the "explicit" molecular environment by an "implicit" one in terms of a smooth dielectric medium or reaction field. This reduces the molecular detail and thus the influence of the MM part essentially to a cavity shape and a dielectric constant, see e.g. Refs. [1205, 1471] for background. Such "continuum models" are particularly successful when it comes to computing solvation free energies treating the solute as the QM system, possibly including an explicit solvation layer as a buffer region. However, most implementations of this sort are tailored to optimize solute *structures* embedded in an implicit solvent environment, whereas they are not suitable to perform energy-conserving molecular dynamics simulations. One out of several schemes is presented in Section 5.5.3 which couples *ab initio* molecular dynamics to continuum solvation models [404, 1283, 1322], thereby allowing fluctuations of the solute to be sampled, including chemical reactions.

Concerning the implementation of QM/MM methods, several quite different philosophies can be followed. Ideally, a modular code is written "from scratch" which has both QM and MM methodologies available on an equal footing, thus enabling common structures, tasks, and functionality to be shared as much as possible. This is the design concept, for instance, of the PINY [1153, 1513] and CP2k [287] packages. Alternatively, a script or shell approach is used where existing "stand-alone codes" can be loosely coupled via a central control program, which is the underlying principle of ChemShell [1332] among others. Finally, the "driver concept" is based on using one existing code, either an electronic structure (molecular dynamics) or force field molecular dynamics program, which is "interfaced" with the complementary code. Data exchange between the two programs can be done either by writing and reading to/from external disks or it can be handled in the core via intimate software coupling. In the following Section 5.5.2 the CP-PAW/AMBER, CPMD/GROMOS, and EGO/CPMD interfaces will be introduced in some detail in order to discuss atomistic embedding. The former two cases are typical representatives where a Car–Parrinello code is employed as the driver, while the latter uses a force field molecular dynamics package to call an electronic structure engine. The concept of combining *ab initio* molecular dynamics with embedding in a continuum solvent environment is illustrated in Section 5.5.3 using the CP-PAW/COSMO interface. Last but not least, the important issue of dealing with electronic excitations in a QM/MM-type molecular dynamics setup is addressed in Section 5.5.4.

5.5.2 Embedding in atomistic environments

5.5.2.1 CP-PAW/AMBER interface

The pioneering implementation of Car–Parrinello molecular dynamics coupled to an external MM force field is due to Blöchl and coworkers [143, 1625–1628] and has been applied successfully in the field of homogeneous catalysis [1624, 1628]. In particular, the PAW-transformed Car–Parrinello equations [142], see Section 6.2 for background information, as implemented in the CP-PAW package [288] were coupled to the AMBER force field [283] similar in spirit to Ref. [1366]. This QM/MM approach was designed to open up the possibility of studying chemical reactions of complex systems dynamically and computing reaction free energy profiles via thermodynamic integration (see Section 3.7.3) in order to probe the influence of fluctuations and thus entropy.

The basic philosophy is to use a mechanical embedding scheme, see Section 5.5.1, where the instantaneous electronic density of the QM system is mapped onto a set of Gaussian smearing functions being tied to the sites of the moving QM nuclei. The corresponding Lagrangian is obtained by supplementing the bare Car–Parrinello Lagrangian Eq. (2.56) by a force field term

$$\mathcal{L}^{\mathrm{QM/MM}} = \mathcal{L}^{\mathrm{QM}}_{\mathrm{CP}} + \mathcal{L}^{\mathrm{MM}} + \mathcal{L}^{\mathrm{QM-MM}} \tag{5.267}$$

$$= \sum_{\alpha} \frac{1}{2} M_{\alpha} \dot{\mathbf{R}}_{\alpha}^2 + \sum_{i} \mu \left\langle \dot{\phi}_i \middle| \dot{\phi}_i \right\rangle - \left\langle \Psi_0 \middle| \tilde{H}_{\mathrm{e}}^{\mathrm{KS}}(\{\mathbf{R}_{\alpha}\}) \middle| \Psi_0 \right\rangle$$

$$+ \sum_{i,j} \Lambda_{ij} \left(\langle \phi_i | \phi_j \rangle - \delta_{ij} \right)$$

$$+ \sum_{I} \frac{1}{2} M_I \dot{\mathbf{R}}_I^2 + E^{\mathrm{MM}} \left(\{\mathbf{R}_I\} \right) + E^{\mathrm{QM-MM}} \left(\{\mathbf{R}_I\}, \{\mathbf{R}_{\alpha}\} \right) \;\;,$$

$$\tag{5.268}$$

where $\tilde{H}_{\mathrm{e}}^{\mathrm{KS}}$ is the PAW Kohn–Sham Hamiltonian for a finite, isolated QM system including the capping atoms and E^{MM} are all interactions *within* the MM system. Note that the Pulay forces, see Section 2.5 for a general discussion, must be taken into account when deriving the Euler–Lagrange equations in order to allow for an energy-conserving propagation [143]. In order to ensure efficient temperature control, separate [146, 1510] Nosé–Hoover thermostats, see Section 5.2.2, were coupled to the QM and MM systems using quite disparate inertia parameters, Q^{n}, in order to prevent strong heat flow between the thermostats themselves.

The coupling of the supercell to its periodic images is eliminated using a decoupling scheme [143] where an optimized point charge representation of the QM charge density, $n(\mathbf{r}) = \sum_i f_i |\phi_i(\mathbf{r})|^2$, is generated which is also used for the mechanical embedding procedure; see Section 3.2.3 for a general discussion of such cluster boundary conditions in the framework of plane wave basis sets. The continuous charge density $n(\mathbf{r})$ of the QM system is compressed to a linear superposition of atom-centered spherical Gaussian functions

$$n^{\mathrm{G}}(\mathbf{r}) = \sum_{\alpha} q_{\alpha} G(\mathbf{r} - \mathbf{R}_{\alpha}) \tag{5.269}$$

normalized to unity. The Gaussian width parameters are fixed and more than one Gaussian, typically three or four, are used per center; in particular for charged systems different decay parameters are useful in order to efficiently account for the compensating homogeneous background charge which creates a parabolic electrostatic potential. The parameters are determined such that the Gaussian model charge density, $n^{\mathrm{G}}(\mathbf{r})$, reproduces the multipole moments of $n(\mathbf{r})$ by reproducing $n(\mathbf{G})$ in reciprocal space near the origin, which is enforced by a suitable weighting function. Finally, in the limit of infinitesimally small widths (or equivalently large distance from the QM system) the Gaussian density can be rewritten in terms of a point-charge model

$$n^{\mathrm{pc}}(\mathbf{r}) = \sum_{\alpha} q_{\alpha} \delta(\mathbf{r} - \mathbf{R}_{\alpha}) \qquad (5.270)$$

using the *same* partial charges $\{q_{\alpha}\}$; it is found for a test case that these charges are quite close to those parameterized in the AMBER force field [283]. This set of point charges $\{q_{\alpha}\}$ at the positions of the QM nuclei $\{\mathbf{R}_{\alpha}\}$ interacts with the parameterized charges $\{q_I\}$ of the MM atoms $\{\mathbf{R}_I\}$ and constitutes the electrostatic coupling, $E_{\mathrm{es}}^{\mathrm{QM-MM}}$ in Eq. (5.265), via the MM force field [283] energy expression, $E^{\mathrm{QM-MM}}(\{\mathbf{R}_I\}, \{\mathbf{R}_{\alpha}\})$, in order to allow for an energy-conserving propagation [143].

The (short-range) steric contribution to the nonbonded interactions between the QM and MM system, $E_{\mathrm{steric}}^{\mathrm{QM-MM}}$ in Eq. (5.265), is taken into account via the parameterized van der Waals terms of the employed force field [283]. The capping in order to handle bond cuts is done with hydrogen atoms [1626–1628] within the framework of the "IMOMM approach" [963], setting the bond angles and dihedral angles equal to the respective angles used to define the capping atom.

Relevant dynamical changes in the QM and MM systems typically occur on quite disparate time scales. Large-amplitude conformational motion in the biomatrix might easily take nanoseconds or longer, whereas the accessible time scale of *ab initio* molecular dynamics is of the order of picoseconds. A multiple time step scheme [1493, 1511] has been implemented in the CP-PAW/AMBER interface [1627], which allows for energy-conserving, reversible, and symplectic integration of the QM/MM Car–Parrinello equations of motion. In particular, the computationally inexpensive MM system can be propagated much faster than the QM system, thus exploring its phase space more efficiently for a given QM configuration. In addition, "oversampling" can be achieved by artificially decreasing the masses of the MM atoms, hence speeding up their dynamics with respect to the QM nuclei. Note that this amounts to introducing, artificially, an adiabatic separation between the QM and MM systems such that the generated dynamics is fictitious. Both approaches lead to a faster sampling of the MM sector of phase space, thus offering the QM system a larger variability of MM configurations which improves the computation of averages that depend crucially on large-scale MM fluctuations such as free energy profiles obtained from thermodynamic integration.

5.5.2.2 CPMD/GROMOS *interface*

The CPMD/GROMOS interface [823] was designed to be a fully Hamiltonian QM/MM coupling scheme in the sense that a Hamiltonian is introduced which, without further approximation, yields consistent Euler–Lagrange equations for energy-conserving molecular dynamics propagation, see Ref. [228] for an early review. In particular, the Car–Parrinello approach to *ab initio* molecular dynamics, see Section 2.4, in its plane wave/pseudopotential Kohn–Sham representation, see Section 3.7.1, as implemented in CPMD [696], is used in order to propagate the QM subsystem. The MM system is described with the GROMOS simulation package and the associated force field [1302, 1540], where the electrostatic interactions are treated efficiently with a particle–particle/particle–mesh (P^{3}M) algorithm [654]. Within the class of electrostatic embedding schemes according to Section 5.5.1, this QM/MM interface [823] is tailored to study the *dynamics* of complex biomolecular systems in general, but in particular also chemical reactions. At the heart of the scheme is a hierarchical approach to deal with the electrostatic interactions in combination with an empirical modification of the short-range terms that does not require a refit of the MM force field used.

The nonbonded part Eq. (5.265) of the total energy Eq. (5.264) is written, as usual, in terms of the electrostatic and steric contributions

$$E_{\text{nb}}^{\text{QM}-\text{MM}} = E_{\text{es}}^{\text{QM}-\text{MM}} + E_{\text{steric}}^{\text{QM}-\text{MM}} \tag{5.271}$$

$$= \sum_{I \in \text{MM}} q_I \int \frac{n(\mathbf{r})}{|\mathbf{r} - \mathbf{R}_I|} \, d\mathbf{r} + \sum_{I \in \text{MM}} \sum_{\alpha \in \text{QM}} v_{\text{vdW}}(|\mathbf{R}_\alpha - \mathbf{R}_I|) \, , \tag{5.272}$$

where $n(\mathbf{r})$ denotes here the total QM charge density defined on a grid in real space, which consists of valence and (Gaussian-smeared) core contributions, n^{e} and n_I^{c}, respectively, within the plane wave/pseudopotential representation according to the scheme outlined in Section 3.2.1. It should be mentioned that the QM system must be treated as a finite cluster by decoupling it from the artificial periodic images, which is achieved by the techniques explained in Section 3.2.3, in particular the Martyna-Tuckerman decoupling [929].

The steric interactions between atoms in the QM and MM subsystems are represented by the same van der Waals-type interactions v_{vdW} as those parameterized for pure MM simulations [1302, 1540]. In the electrostatic part, care must be taken in order to prevent electron spill-out from the

QM system toward positively charged MM atoms due to the lacking Pauli repulsion, see Section 5.5.1. This is achieved by replacing in Eq. (5.272) the bare Coulomb interaction $1/r_I$ with a MM atom at position $\mathbf{R}_I$

$$E_{\text{es}}^{\text{QM–MM}} = \sum_{I \in \text{MM}} q_I \int n(\mathbf{r}) \, v_I^{\text{eff}} \left(|\mathbf{r} - \mathbf{R}_I| \right) d\mathbf{r} \tag{5.273}$$

$$v_I^{\text{eff}}(r_I) = \frac{r_{\text{c}I}^m - r_I^m}{r_{\text{c}I}^{m+1} - r_I^{m+1}} \xrightarrow{r_I \to \infty} \frac{1}{r_I} \tag{5.274}$$

in the sense of a suitable regularization as $r_I \to 0$. This ansatz is similar in spirit to the construction of empirical pseudopotentials [280, 628, 629] where the divergence at the origin (caused by the missing Pauli repulsion of the valence orbitals as a result of the neglected orthogonalization to the core electrons) has to be cut off, see also Chapter 4. Here, a set of suitable parameters m and $\{r_{\text{c}I}\}$ is required, but a clear advantage from the outset is that it is not necessary to reparameterize the partial charges $\{q_I\}$ or the MM force field as such. Test calculations [823] show that $m = 4$ and $r_{\text{c}I}$ being the covalent radius of the Ith atom is a reasonable choice. As alternatives, the expressions

$$v_I^{\text{eff}}(r_I) = \frac{r_I^m}{\left(r_{\text{c}I}^{2m+2} - r_I^{2m+2} \right)^{1/2}} \tag{5.275}$$

$$v_I^{\text{eff}}(r_I) = \frac{r_I^m}{\left(r_{\text{c}I}^{2m+2} - r_I^{2m+2} \right)^{1/2}} + v_I^{\text{rep}} \exp \left[- \left(\frac{r_I}{r_{\text{c}I}} \right)^k \right] \tag{5.276}$$

have been advocated and tested in the context of QM/MM NMR chemical shift calculations [1305], which are particularly sensitive to the electronic structure close to nuclei; here m and $k = 1$ or 2 is recommended. It is noted that the additive empirical Pauli repulsion term in Eq. (5.276), $\sim v_I^{\text{rep}}$, amounts to introducing an *additional* contribution to the short-range nonbonded van der Waals interaction, $E_{\text{steric}}^{\text{QM–MM}}$. This must be canceled or at least corrected for in order to avoid double counting of this interaction, see the discussion of the EGO/CPMD interface for a closely related discussion in the framework of the "SPLAM approach" [383].

As mentioned in Section 5.5.1, the straightforward numerical evaluation of the integral in Eq. (5.273) is not possible. On the other hand, approaches where the electrostatic potential acting on the QM fragment due to the MM atoms and the corresponding potential acting on the MM atoms as a result of the QM charge distribution are not obtained consistently violate the "*actio equals reactio*" principle and thus imply total energy, linear momentum, and angular momentum non-conservation; this can be ameliorated but not solved by applying suitable correction/rescaling schemes and/or thermostatting. This is, for instance, the case in "electrostatic embedding" schemes, see Section 5.5.1, if the exact electronic charge density of the QM system, $n(\mathbf{r})$, is fitted to an approximate model density, $n^{\text{model}}(\mathbf{r})$, in order to evaluate the electrostatic influence of the QM fragment onto the MM system, whereas the reverse coupling is done via the exact density $n(\mathbf{r})$.

Instead, the idea implemented in the CPMD/GROMOS interface [823] is to start from a Hamiltonian formulation from which the additional external electrostatic potential to be included in the Kohn–Sham Hamiltonian and the one acting onto the MM atoms as a result of the QM charge distribution can be obtained consistently. To this end, the electrostatic part is first split into short-range and long-range contributions

$$E_{\text{es}}^{\text{QM}-\text{MM}} = E_{\text{es}-\text{sr}}^{\text{QM}-\text{MM}} + E_{\text{es}-\text{lr}}^{\text{QM}-\text{MM}} \tag{5.277}$$

$$= \sum_{I \in \text{NN}} q_I \int n(\mathbf{r}) v_I^{\text{eff}} \left(|\mathbf{r} - \mathbf{R}_I| \right) \, d\mathbf{r} + E_{\text{es}-\text{lr}}^{\text{QM}-\text{MM}} \;, \tag{5.278}$$

the former stemming exclusively from "nearest-neighbor" (NN) MM atoms which are located within a certain cutoff range around the QM atoms. The short-range part is evaluated exactly by summing all grid-point/MM atom contributions in the NN class explicitly. The long-range potential is computed approximately by expanding the charge density of the QM system in terms of multipoles

$$E_{\text{es}-\text{lr}}^{\text{QM}-\text{MM}} = \sum_{I \notin \text{NN}} q_I \left\{ \frac{C}{|\mathbf{R}_I - \bar{\mathbf{R}}|} + \sum_i \frac{D_i}{|\mathbf{R}_I - \bar{\mathbf{R}}|^3} \left(R_{Ii} - \bar{R}_i \right) \right.$$

$$\left. + \frac{1}{2} \sum_{i,j} \frac{Q_{ij}}{|\mathbf{R}_I - \bar{\mathbf{R}}|^5} \left(R_{Ii} - \bar{R}_i \right) \left(R_{Ij} - \bar{R}_j \right) \right\} \tag{5.279}$$

up to quadrupolar order. The coefficients $\{C[n], D_i[n], Q_{ij}[n]\}$ are the usual multipole moments computed from the charge density $n(\mathbf{r})$ of the QM subsystem with respect to the geometric center of the QM system, $\bar{\mathbf{R}} = (\bar{R}_1, \bar{R}_2, \bar{R}_3)$, where $i = 1, 2, 3$ are the Cartesian components. A refinement can be achieved within this hierarchical approach by introducing a third region between the two just mentioned where the charge density of the QM system is represented approximately via (restrained) electrostatic potential-based (R/ESP) charges centered on the QM atoms as obtained from a dynamical fitting procedure [824] or simply taken from the MM force field used. It is noted in passing that a mechanical embedding scheme applied to the *long-range* part, i.e. the replacement of $E_{\text{es}-\text{lr}}^{\text{QM}-\text{MM}}$ according to Eq. (5.279) in Eq. (5.278) by

$$\tilde{E}_{\text{es}-\text{lr}}^{\text{QM}-\text{MM}} = \sum_{I \notin \text{NN}} q_I \left\{ \sum_{\alpha \in \text{QM}} \frac{q_\alpha}{|\mathbf{R}_\alpha - \mathbf{R}_I|} \right\} \tag{5.280}$$

where q_α are here the fixed force field charges assigned to the QM nuclei at positions $\mathbf{R}_\alpha$, leads to unsatisfactory results even if quite large NN regions are used.

The forces acting on the nuclei are obtained by taking the derivative of the QM-MM coupling Hamiltonian, Eq. (5.272), with respect to the nuclear positions. The additional potential in the Kohn–Sham Hamiltonian due to the QM-MM interaction is obtained by taking the *analytical* functional derivative of the electrostatic contribution, Eq. (5.277), with respect to the charge density $n(\mathbf{r})$; note that the multipole moments are explicitly known functionals of n. Thereby, the electron density is polarized up to any multipolar order by the atoms in the NN subset of MM atoms and up to quadrupolar order by the rest of the more distant MM atoms. Taken together, this makes possible efficient energy-conserving QM/MM Car–Parrinello molecular dynamics propagation, which allows us to properly compute time-correlation functions in the microcanonical ensemble and thus response properties such as infrared spectra [1241, 1242].

Finally, the bonded term $E_{\text{b}}^{\text{QM}-\text{MM}}$ in Eq. (5.264) must be taken into account in cases where covalent bonds between the QM nuclei and MM atoms are cut. In the framework of the CPMD/GROMOS interface an intramolecular link atom approach is implemented. As an improvement to just capping with a hydrogen atom when a carbon–carbon single bond is cut, an empirically modified monovalent link atom based on a carbon pseudopotential has been introduced [1417]. The bonded interactions across this QM–MM boundary, i.e. the stretching, bending, and torsional terms, are included as parameterized in the MM force field, considering the positions of the carbon nuclei that are replaced by pseudopotentials like MM sites. More recently, a powerful scheme to derive nonlocal pseudopotentials that reproduce *molecular* properties, called optimized effective core potentials (OECP), was devised [1572] and applied in the realm of constructing capping pseudopotentials [1573]. These pseudopotentials are of the dual-space Gaussian form [525, 608] as outlined in Section 4.4 and the parameters were determined by using the charge density $n(\mathbf{r})$ as the target function, see Ref. [1573] for details. It could be shown for a test case that such a (heptavalent) OECP could replace a methyl group close to a strongly polarizing carboxyl group better than H or F capping atoms (as represented by their standard pseudopotentials) or an empirical monovalent carbon pseudopotential. What remains to be investigated is the extent of transferability into different molecular environments.

The CPMD/GROMOS approach [228, 823, 824, 1417, 1573] to QM/MM molecular dynamics has proven to be extremely successful, as demonstrated by the wealth of applications to problems in biophysics and biochemistry exemplified in Section 9.14 and to some extent also in Section 9.13. The computational overhead due to the MM part and the QM–MM interaction in comparison to a standard Car–Parrinello simulation of the naked QM system is quite small for real-life examples, typically of the order of 10–30% of the total CPU time. However, the most expensive term to evaluate and thus a possible bottleneck in the limit of large QM systems is the *short-range* electrostatic contribution, i.e. the first term in Eq. (5.278). It requires distance evaluations between every point of the real space grid and all MM atoms in the NN class of nearest neighbors. More recently, in addition to GROMOS, the AMBER force field [283] can also be used on the MM side [1221], and extensions to include excited states, see Section 5.5.4 as well as Section 5.3.6, and to compute NMR chemical shifts [1305] were devised.

5.5.2.3 EGO/CPMD *interface*

The original stimulus of the EGO/CPMD interface [383, 387] was the quest to compute most accurately vibrational and optical spectra as well as photochemical reactions of complex biomolecular systems. Thus, a sophisticated link atom scheme, the "scaled position link atom method" (SPLAM), was devised to cut bonds between the QM and MM partitions such as to keep the resulting perturbation as small as possible. This interface is based on the EGO MM molecular dynamics program suite, originally EGO-VIII [383–388] with stochastic droplet boundary conditions and later EGO-MMII [968, 969, 1450] implementing also periodic boundary conditions, which relies on the CHARMM force field [888]. Here, the MM program is used as a driver for the QM module which is the CPMD package [696] to be run in the iterative Born–Oppenheimer molecular dynamics mode, see Section 2.3, while exchanging data via files written on hard disk.

At the heart of EGO are hierarchical "structure adapted multipole methods" (SAMM) [1053, 1054], in particular "fast multiple time step" SAMM [388] and "reaction field" SAMM [968, 969], in order to evaluate electrostatics, i.e. Eq. (5.266). Conceptually, the QM fragment is just another class in this hierarchy of charge distributions, which is ordered according to distance classes. In particular, the excess electrostatic potential due to the

MM atoms acting on the QM system

$$\Phi_{es}^{QM\leftarrow MM}(\{\mathbf{R}_I\}, \mathbf{r}) = \sum_{I\in MM} \frac{q_I}{|\mathbf{r} - \mathbf{R}_I|} \qquad (5.281)$$

$$\approx \sum_{I\in NN} \epsilon(|\mathbf{r} - \mathbf{R}_I|) \frac{q_I \, \text{erf}\left[\frac{|\mathbf{r}-\mathbf{R}_I|}{R^c}\right]}{|\mathbf{r} - \mathbf{R}_I|} + \Phi_{es}^{I\notin NN}(\{\mathbf{R}_I\}, \mathbf{r})$$

$$(5.282)$$

is obtained explicitly within the innermost ($R \le 5\text{Å}$) distance class by summing Gaussian-smeared Coulomb charges, q_I, centered at the NN atoms, see Eqs. (3.39) and (3.40); $\mathbf{r}$ denotes a point on the real space density grid, $R^c = 0.8$ Å is a uniform width parameter, and $\epsilon(R)$ is a switching function acting exclusively in link atom regions to classify bonded/nonbonded interactions. The Gaussian blurring of the MM point charges improves the convergence of the wave function optimization by removing the singularities [383, 387] and ameliorates electron spill-out from the QM electron density onto positively charged NN groups. The contributions from MM atoms further outside, $\Phi_{es}^{I\notin NN}$, are grouped into several more distance classes and are obtained from modified Coulomb sums at the next level in conjunction with sophisticated nested Taylor/multipole expansions centered at $\mathbf{R}_\alpha$. This potential is added to the external potential, Eq. (2.117), of the Kohn–Sham Hamiltonian, Eq. (2.122), yielding the electrostatic energy

$$E_{es}^{QM\leftarrow MM} = \int n(\mathbf{r}) \, \Phi_{es}^{QM\leftarrow MM}(\{\mathbf{R}_I\}, \mathbf{r}) \, d\mathbf{r} \qquad (5.283)$$

and handled within CPMD.

The electrostatic influence of the QM system on the MM system is caused by the coupling of the charge density, $n(\mathbf{r})$ as discretized on the FFT grid, to all MM point charges. Since it is not possible to evaluate this electrostatic contribution explicitly, see Eq. (5.266),

$$E_{es}^{QM\rightarrow MM} = \sum_{I\in MM} q_I \int \frac{n(\mathbf{r})}{|\mathbf{r} - \mathbf{R}_I|} \, d\mathbf{r}$$

$$\approx \sum_{I\in MM} q_I \int \frac{n^{pc}(\mathbf{r})}{|\mathbf{r} - \mathbf{R}_I|} \, d\mathbf{r} \qquad (5.284)$$

$$= \sum_{I\in MM} q_I \sum_{\alpha\in QM} \frac{q_\alpha}{|\mathbf{R}_\alpha - \mathbf{R}_I|}$$

$$= \sum_{I \in \text{MM}} q_I \, \Phi_{\text{es}}^{\text{QM} \rightarrow \text{MM}} (\mathbf{R}_I, \{\mathbf{R}_\alpha\}) \qquad (5.285)$$

the QM charge density and thus the resulting electrostatic potential is expressed approximately in EGO/CPMD in terms of effective point charges, $\{q_\alpha\}$, centered at the positions of the QM nuclei, $\{\mathbf{R}_\alpha\}$, according to Eq. (5.270). The partial charges can be obtained for a given density n by several standard procedures implemented in CPMD [696]. Thus, the full electrostatic interaction in the combined system violates slightly [383, 387] Newton's Third Law, "*actio* equals *reactio*", since the electrostatic field acting in the QM and MM subsystems is not obtained from the same electronic charge density, $n(\mathbf{r}) \neq n^{\text{pc}}(\mathbf{r})$, compare Eq. (5.283) and Eq. (5.285), such that the associated forces do not exactly cancel. As a result the total energy, linear momentum, and angular momentum of the QM/MM system are no longer conserved quantities. The heat created is removed by stochastic thermostatting, i.e. Berendsen *et al.* velocity rescaling [108], whereas elaborate infinitesimal correction schemes as detailed in Ref. [387] are devised in order to restore approximately the conservation of the total linear and angular momenta during the propagation.

The short-range nonbonded contribution between atoms in the QM and MM fragments, $E_{\text{steric}}^{\text{QM} - \text{MM}}$, is evaluated according to the CHARMM force field parameterization [888]. Crucial to the EGO/CPMD interface is the accurate removal of artifacts due to introducing link atoms, which is worked out for the case of replacing a $\cdots \text{C–C} \cdots$ single bond by a terminated $\cdots \text{C–H}$ bond [383, 387]. Following earlier work [67, 424], the C–H bond axis is oriented as the C–C bond would be oriented. In addition, the bond length of the link atom, R_{CH}, and thus its position in Cartesian space is determined by a scaling procedure

$$R_{\text{CH}} = R_{\text{CH}}^{\text{e}} + \frac{k_{\text{CC}}}{k_{\text{CH}}} \left(R_{\text{CC}} - R_{\text{CC}}^{\text{e}} \right) \qquad (5.286)$$

using harmonic approximations for the bond stiffnesses and parameterized equilibrium positions for the C–C and C–H bonds in order to approximate the stretching force of the artificial C–H bond by that of the original C–C bond. This eliminates completely the additionally introduced three degrees of freedom in SPLAM. Furthermore, corresponding harmonic energy corrections

$$\Delta E_{\text{stretch}} = k_{\text{CC}} \left(1 - \frac{k_{\text{CC}}}{k_{\text{CH}}} \right) [R_{\text{CC}} - R_{\text{CC}}^{\text{e}}]^2 \qquad (5.287)$$

$$\Delta E_{\text{bend}} = (k_{\text{HCC}} - k_{\text{HCH}}) [\theta_{\text{HCC}} - \theta_{\text{HCH}}^{\text{e}}]^2 \qquad (5.288)$$

for both stretching and bending motion of the C–H bond are introduced as well as the additional interaction terms that would be present in a force field description of a $\cdots$ C–C $\cdots$ bond, but are missing in the terminated $\cdots$ C–H bond; the usual force field parameters are used throughout.

In addition to the bonded interactions, the nonbonded interactions are also indirectly affected by the capping procedure: the $\cdots$ C–H bond is polar whereas the original $\cdots$ C–C $\cdots$ bond is essentially unpolar. This creates an artificial dipole moment, which introduces spurious terms into the electrostatic potentials, $\Phi_{es}^{QM\leftarrow MM}$ and $\Phi_{es}^{QM\rightarrow MM}$ defined in Eqs. (5.282) and (5.285), and thus into the forces acting in both the QM and MM subsystems. These interactions are removed by applying approximate dipolar corrections as explained in Refs. [383, 387]. Finally, the fictitious short-range nonbonded interaction energy of the link atom with the other QM nuclei *within* the QM fragment is approximately canceled by subtracting the force field expression of this "van der Waals" contribution from the QM total energy.

Over the years the EGO/CPMD interface has been applied very successfully, in particular to theoretical infrared spectroscopy of complex solute/solvent and biomolecular systems, see e.g. Refs. [748, 1059, 1060, 1253, 1291, 1292] as well as Sections 9.5 and 9.14.

5.5.3 Embedding in continuum environments

At variance with all previously discussed approaches, the CP-PAW/COSMO interface [1322] does not treat the environmental MM part at the atomic or molecular level. Rather, the "conductor-like screening model" (COSMO) [285, 749, 750] is used to represent the environment, which is ideally a homogeneous and inert solvent, as a dielectric continuum [1205, 1471]. The CP-PAW code [288] is used which implements the PAW-transformed Car–Parrinello equations [142] as outlined in Section 6.2. The strength of this particular [1322] and related [404, 1283] QM/MM coupling schemes without explicit solvent is clearly the expedient computation of solvation properties at finite temperatures. These "implicit solvent" approaches allow us, in practice, to screen many combinations of solute molecules in various solvents essentially at the cost of rather small molecular simulations instead of full condensed phase simulations. In addition, it becomes possible to study, with only moderate computational overhead, chemical reactions in solution for those cases where solvent molecules are just spectators and not crucially involved in the ongoing chemistry, for instance as proton or electron carriers;

the borderline case of dealing with hydrogen bonding between solute and solvent requires special care.

The essential aspect of the scheme [1322] is to treat the surface charges at the cavity boundary as fictitious dynamical variables in the spirit of an extended Lagrangian framework, see Section 2.4.8. In particular, the PAW Car–Parrinello Lagrangian for the solute, $\mathcal{L}_{\text{CP}}^{\text{QM}}$, already used in Eq. (5.268) is supplemented

$$\mathcal{L}^{\text{QM/MM}} = \mathcal{L}_{\text{CP}}^{\text{QM}} + \mathcal{L}^{\text{COSMO}} \tag{5.289}$$

$$\mathcal{L}^{\text{COSMO}} = \sum_I \frac{1}{2} M_I^{\text{Q}} \dot{Q}_I^2 - G_{\text{es}}^{\text{COSMO}}\left(\{Q_I\}, \{\mathbf{R}_\alpha\}\right)$$

$$- G_{\text{steric}}\left(\{\mathbf{R}_\alpha\}\right) - \sum_I k(1 - \Theta_I)Q_I^2 \tag{5.290}$$

with the COSMO expression [749, 750] for its electrostatic interaction with the polarized dielectric, $G_{\text{es}}^{\text{COSMO}}$, which takes care of the term $E_{\text{es}}^{\text{QM}-\text{MM}}$ in Eq. (5.265). A simple expression for all non-electrostatic contributions, G_{steric}, notably the cavity formation energy as well as Pauli repulsion and dispersion interactions, which corresponds essentially to $E_{\text{steric}}^{\text{QM}-\text{MM}}$ in Eq. (5.265), is included in terms of a simple linear approximation, $G_{\text{steric}} = \sigma_0 + \sigma_1 A$, where A is the surface area of the cavity and σ_i are empirical, fixed parameters. A bonded term between the QM and MM partitions, i.e. $E_{\text{b}}^{\text{QM}-\text{MM}}$ in Eq. (5.264), does not exist in solute/solvent setups.

The fictitious variables, Q_I, with associated inertia parameters, M_I^{Q}, are discretized, scaled surface charges. They are located at the center of surface segments $\mathbf{s}_I$ of area a_I, which span the bounding surface of the solute cavity obtained by assigning spheres centered at all QM nuclei. Since the number of segments must remain constant during the simulation in the spirit of a continuous and conservative propagation, switching functions are introduced which effectively remove the charges Q_I from those segments $\mathbf{s}_I$ that are instantaneously not part of the solute's molecular surface exposed to the solvent. Note that the last term in Eq. (5.290) is a penalty function introduced to keep the instantaneously switched-off charges, $\{(1 - \Theta_I)\}$, from blowing up during propagation since they do not contribute to the energy. This allows us to write down the COSMO free energy in terms of electrostatic solute-solvent, solvent-solvent, and self-energy contributions

$$G_{\text{es}}^{\text{COSMO}} = \sum_I Q_I \Theta_I \int_V \frac{n(\mathbf{r})}{|\mathbf{r} - \mathbf{s}_I|}\, d\mathbf{r} + \frac{1}{f} \sum_{I<J} \frac{Q_I \Theta_I Q_J \Theta_J}{|\mathbf{s}_I - \mathbf{s}_J|}$$

$$+ \frac{c}{f} \sum_I \frac{Q_I^2 \Theta_I^2}{\sqrt{a_I}} \tag{5.291}$$

$$G_{\text{steric}} = \sigma_0 + \sigma_1 \sum_I a_I \Theta_I \tag{5.292}$$

as well as the nonpolar contribution in terms of a dynamically evolving solute cavity; f is a constant screening factor needed to scale the conductor-like result to finite permittivity, c is here a geometry-dependent constant, and V is the cavity volume. In addition, the bare Coulomb interaction between the segment charges, i.e. the second term in Eq. (5.291), must be regularized in the limit $|\mathbf{s}_I - \mathbf{s}_J| \to 0$ if I and J belong to different atomic spheres, see Ref. [1322] for pertinent details. Finally, the electrostatic decoupling scheme [143] is used in order to treat the solute in a plane wave representation. Thereby, charge spill-out is prevented by using the Gaussian model density, Eq. (5.269), introduced in Section 5.5.2 for the CP-PAW/AMBER interfacein order to express the charge density $n(\mathbf{r})$ in the solute-solvent interaction, i.e. the first term in Eq. (5.291).

5.5.4 QM/MM molecular dynamics involving excited states

A natural extension of QM/MM methods is the treatment of electronically excited states within the QM fragment. This opens avenues to investigate photoinduced processes occurring in solute/solvent systems, in chromophores embedded in biomatrices, as well as photoswitchable materials. In particular, both the ROKS approach to the first excited singlet state [462, 554] introduced in Section 5.3.2, as well as linear response time-dependent density functional theory in the frequency domain [232, 233, 364, 559, 910, 912] discussed in Section 5.3.6, and the nonadiabatic QM component [153, 355, 356] from Section 5.3.3 have been combined with the QM/MM interface based on CPMD/GROMOS [823].

Within standard QM/MM applications the calculation of the MM energy contribution, E^{MM}, and of the QM–MM coupling term, $E^{\text{QM−MM}}$, relies on intramolecular and van der Waals force field parameters that were optimized to describe the interaction of ground state systems, see Eqs. (5.264)–(5.265). It is therefore first of all necessary to ensure that in the case of a description of excited states, the actual excitation region is well contained deep inside the QM part, i.e. it must be well separated from the QM–MM boundary. Secondly, the most important interaction term yet to be defined is the electrostatic interaction of the classical point charges with the electronic charge density of the quantum system in the excited state

$$E_{\text{es}}^{\text{QM}-\text{MM}} = \sum_{I \in \text{MM}} q_I \int \frac{n_{\text{QM}}(\mathbf{r})}{|\mathbf{r} - \mathbf{R}_I|} \, d\mathbf{r} \ , \tag{5.293}$$

where the sum runs over all point charges q_I at the nuclear positions $\{\mathbf{R}_I\}$ of the MM atoms and n_{QM} is the full charge density of the quantum system (see also Eq. (5.266)). For electronic structure methods which do not give direct access to the charge density in the excited state, which includes all perturbation methods and also linear response TDDFT introduced in Section 5.3.6, a consistent scheme to compute $E_{\text{es}}^{\text{QM}-\text{MM}}$ has to be found; note that n_{QM} and thus Eq. (5.293) is well defined within the ROKS and ROSS approaches to the first excited state (see Section 5.3.2). This can be achieved by making use of the fact that the charge density of the QM system can be defined as the derivative of the total energy with respect to the external potential. From this relation a working expression in the case is obtained

$$n^{\text{QM}}(\mathbf{r}) = \frac{\partial E_{\text{tot}}}{\partial V_{\text{ext}}(\mathbf{r})} = n_{\text{tot}}(\mathbf{r}) + n_{\text{P}}(\mathbf{r}) \ , \tag{5.294}$$

where $n_{\text{tot}}(\mathbf{r})$ is the ground state density (including the core charge distribution) and $n_{\text{P}}(\mathbf{r})$ is the charge distribution calculated from the sum of the two perturbation density matrices $\mathbf{P}^{(x)}$ and $\mathbf{P}^{(z)}$ from Eqs. (5.159) and (5.160). One can see that it is this relaxed density $n_{\text{tot}}(\mathbf{r}) + n_{\text{P}}(\mathbf{r})$ that acts as the charge density of the system in the excited state. This definition of $E_{\text{es}}^{\text{QM}-\text{MM}}$ ensures that the calculation of the excitation energy and nuclear forces is consistent. If this QM-MM interaction term is added to the excited state Lagrangian Eq. (5.153), the additional external potential from the classical point charges appears in the definition of $\mathcal{A}$ and also in the Kohn–Sham matrix $\mathbf{H}^{\text{KS}}$ which defines the QM(LR-TDDFT)/MM method.

The ROKS/MM approach is ideally suited to investigate the interplay of structural changes of a photoexcited solute molecule [1221] or chromophore [1222] with its environment by performing *ab initio* molecular dynamics directly in the first excited state of the QM fragment. The nonadiabatic QM/MM surface-hopping implementation [153] is particularly useful for studying photochemical transformations in condensed phase environments such as photoswitchable molecules that are embedded in polymeric or liquid crystalline materials. The QM(LR-TDDFT)/MM method [1414, 1416], on the other hand, is tailored to compute solvent shifts on optical spectra. The QM/MM approach evidently not only allows one to cut down computer time dramatically, but also makes nonpolar aprotic solvents, which interact mainly through short-range Pauli repulsion in conjunction with attractive van der Waals dispersion, more readily accessible to density functional theory-based *ab initio* molecular dynamics schemes. As an additional benefit, *artificial* charge transfer from the solute to the solvent is avoided upon treating only the solute molecule as a QM fragment, hence localizing possible electronic excitations and charge transfer, whereas all solvent molecules are modeled by the MM force field. It is known that excitations which should be localized on the solute can easily spread into solvent states [114, 115, 836]. This is particularly severe since it is known that the HOMO/LUMO gap of liquid water is severely underestimated by functionals belonging to the class of the Generalized Gradient Approximation, see Eq. (2.127), which incidentally also stabilize delocalized spin densities [1557], and that ROKS as well as ROSS underestimate the S_1 energy. Of course, the MM treatment of the solvent is not a remedy for these problems if electron transfer or coupled proton-electron transfer to (or via) the solvent is a (photo)chemically relevant process [836]. Clearly, QM/MM methods involving excited states are a yet emerging branch of the growing tree of *ab initio* molecular dynamics methods.

6

Beyond norm-conserving pseudopotentials

6.1 Introduction

For the norm-conserving pseudopotentials introduced in Section 3.1.5 and discussed in detail in Section 4.2, the all-electron wave function gets replaced inside some core radius by a soft, nodeless, pseudo wave function. The crucial restriction, however, is that the pseudo wave function must have the same norm as the all-electron wave function within the chosen core radius; note that the pseudo and all-electron wave functions are identical outside the core radius. As shown previously, good transferability requires a core radius around the outermost maximum of the all-electron wave function, because only then are the charge distribution and moments of the all-electron wave function well reproduced by the pseudo wave functions. Therefore, for elements with strongly localized orbitals (like first row, 3d, and rare earth elements), the resulting pseudopotentials require large plane wave basis sets. To avoid large basis sets, compromises are often made by increasing the core radius significantly beyond the outermost maximum of the all-electron wave function. However, this does not usually lead to a satisfactory solution because the transferability is always adversely affected when the core radius is increased, and for any new chemical environment, additional tests are required to establish the reliability of such soft norm-conserving pseudopotentials. Similarly, compromises concerning the plane wave cutoff are easily made in these cases at the expense, however, of sacrificing accuracy and reliability.

An elegant solution to this problem was proposed by Vanderbilt [1548]. In his method, the norm-conservation constraint is relaxed and to make up for the resulting charge deficit, localized atom-centered augmentation charges are introduced. These augmentation charges are defined as the charge difference between the all-electron and pseudo wave functions, but

286

for convenience they are pseudized to allow an efficient treatment of the augmentation charges on a regular grid. The core radius of the pseudopotential can now be chosen close to the nearest neighbor distance, independently of the position of the maximum of the all-electron wave function. Only for the augmentation charges must a small cutoff radius be used to restore the moments and the charge distribution of the all-electron wave function accurately. The pseudized augmentation charges are usually treated on a regular grid in real space, which is not necessarily the same as the one used for the representation of the wave functions. The relation between the ultrasoft pseudopotential (USPP) method and other plane wave-based methods is discussed by Singh [1365].

A closely related method to Vanderbilt's ultrasoft pseudopotentials was introduced by Blöchl [142] (see Ref. [145] for a review). In his projector augmented-wave method (PAW) a linear transformation is defined that connects the pseudo and all-electron wave functions. Already, Blöchl discussed the similarities of his approach to the ultrasoft pseudopotentials and other mixed basis set schemes. The PAW method is in fact a more general method, and a formal derivation of the ultrasoft pseudopotentials from the PAW equations was given later by Kresse and Joubert [790].

6.2 The PAW transformation

The following derivation of the ultrasoft pseudopotentials and the PAW method follows closely Valiev and Weare [1533]. In the first step a linear transformation [142] that connects pseudo and all-electron wave functions is introduced. The PAW method is based on a formal division of the whole space Ω into distinct regions: a collection of non-overlapping spherical regions around each atom Ω_a and the remainder, the interstitial region Ω_{I}, i.e.

$$\Omega = \Omega_{\mathrm{I}} + \bigcup_a \Omega_a \ . \tag{6.1}$$

It is clear that the plane wave basis, being the ideal choice in the interstitial region Ω_{I}, will be difficult to use in order to describe the wave functions in the atomic sphere regions. In the PAW method this problem is circumvented by introducing auxiliary wave functions which satisfy the following requirements. First, the auxiliary wave function $\tilde{\phi}_i(\mathbf{r})$ can be obtained from the all-electron wave function $\phi_i(\mathbf{r})$ via an invertible linear transformation

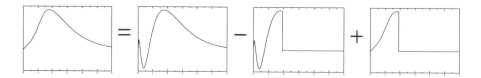

Fig. 6.1. Schematic representation of the PAW transformation. The pseudized wavefunction is constructed from the full wavefunction by subtracting the oscillatory part close to an atom and replacing it by a smooth function.

$\mathcal{T}$

$$| \tilde{\phi}_i \rangle = \mathcal{T} | \phi_i \rangle \tag{6.2}$$

$$| \phi_i \rangle = \mathcal{T}^{-1} | \tilde{\phi}_i \rangle . \tag{6.3}$$

Second, $\tilde{\phi}_i(\mathbf{r})$ is smooth, which implies that it can be represented by a plane wave basis set of a practicable size everywhere, including in particular the atomic sphere regions

$$\tilde{\phi}_i(\mathbf{r}) = \frac{1}{\sqrt{\Omega}} \sum_{\mathbf{G}} c_i(\mathbf{G}) e^{i\mathbf{G} \cdot \mathbf{r}} . \tag{6.4}$$

The first requirement ensures that the task of solving the Kohn–Sham equations can be reformulated equivalently in terms of $\tilde{\phi}_i(\mathbf{r})$, whereas the second requirement allows the entire process to be performed using the plane wave basis set.

The actual construction of $\tilde{\phi}_i(\mathbf{r})$ from a given $\phi_i(\mathbf{r})$ proceeds as follows. For each atom, one defines a finite set of local basis functions $\{\chi_\alpha^a\}$ that is expected to accurately describe the oscillating behavior of the relevant all-electron wave function $\phi_i(\mathbf{r})$ within the corresponding atomic sphere. Associated with $\{\chi_\alpha^a\}$, a set of localized projector functions $\{p_\alpha^a\}$ is introduced such that

$$\langle p_\beta^a | \chi_\alpha^a \rangle = \delta_{\alpha\beta} \tag{6.5}$$

$$p_\alpha^a(\mathbf{r}) = 0 , \quad \forall \mathbf{r} \subset \Omega_{\mathrm{I}} . \tag{6.6}$$

Using $\{\chi_\alpha^a\}$ and $\{p_\alpha^a\}$, the wave function $\phi_i(\mathbf{r})$ in the atomic sphere region can be represented as

$$\phi_i(\mathbf{r}) = \sum_\alpha c_{i,\alpha}^a \chi_\alpha^a(\mathbf{r}) + \Delta_i^a(\mathbf{r}) , \quad \forall \mathbf{r} \subset \Omega_a \tag{6.7}$$

where the coefficients $c_{i,\alpha}^a$ in this expansion are given by

$$c_{i,\alpha}^a = \langle p_\alpha^a \mid \phi_i \rangle \ . \tag{6.8}$$

The correction

$$\mid \Delta_i^a \rangle = (1 - \sum_\alpha \mid \chi_\alpha^a \rangle \langle p_\alpha^a \mid) \mid \phi_i \rangle \tag{6.9}$$

reflects the incompleteness of the set $\{\chi_\alpha^a\}$. As the size of the basis $\{\chi_\alpha^a\}$ gets larger, the local basis representation of $\phi_i(\mathbf{r})$ becomes more accurate, and $\Delta_i^a(\mathbf{r})$ goes to zero.

To define a mapping into $\{\tilde{\phi}_i(\mathbf{r})\}$, an auxiliary smooth basis set $\{\tilde{\chi}_\alpha^a\}$ is introduced subject to the following conditions. First, the basis functions $\tilde{\chi}_\alpha^a(\mathbf{r})$ are smooth, i.e. expandable in terms of the plane wave basis using a practicable plane wave cutoff everywhere and in particular in the atomic sphere regions. Second, $\tilde{\chi}_\alpha^a(\mathbf{r})$ merges differentiably into $\chi_\alpha^a(\mathbf{r})$ outside the atomic spheres, i.e

$$\tilde{\chi}_\alpha^a(\mathbf{r}) = \chi_\alpha^a(\mathbf{r}) \ , \quad \forall \mathbf{r} \subset \Omega_\mathrm{I} \ . \tag{6.10}$$

Third, both $\tilde{\chi}_\alpha^a(\mathbf{r})$ and differences $\tilde{\chi}_\alpha^a(\mathbf{r}) - \chi_\alpha^a(\mathbf{r})$ form linearly independent sets. Given these properties, a smooth wave function $\tilde{\phi}_i(\mathbf{r})$ can be obtained from $\{\tilde{\chi}_\alpha^a\}$ based on the following recipe. Inside the atomic sphere regions it is generated by replacing each occurrence of $\chi_\alpha^a(\mathbf{r})$ with $\tilde{\chi}_\alpha^a(\mathbf{r})$ in the expansion (6.7), that is

$$\tilde{\phi}_i(\mathbf{r}) = \sum_\alpha c_{i,\alpha}^a \tilde{\chi}_\alpha^a(\mathbf{r}) + \Delta_i^a(\mathbf{r}) \ , \quad \forall \mathbf{r} \subset \Omega_a \ , \tag{6.11}$$

whereas it simply coincides with $\phi_i(\mathbf{r})$

$$\tilde{\phi}_i(\mathbf{r}) = \phi_i(\mathbf{r}) \quad \forall \mathbf{r} \subset \Omega_\mathrm{I} \tag{6.12}$$

in the interstitial region. The transformation can therefore be expressed as

$$\mathcal{T} = 1 + \sum_a \sum_\alpha (\mid \tilde{\chi}_\alpha^a \rangle - \mid \chi_\alpha^a \rangle) \langle p_\alpha^a \mid \ . \tag{6.13}$$

Its inverse can be obtained as

$$\mathcal{T}^{-1} = 1 + \sum_a \sum_\alpha (\mid \chi_\alpha^a \rangle - \mid \tilde{\chi}_\alpha^a \rangle) \langle \tilde{p}_\alpha^a \mid \ , \tag{6.14}$$

where a set of smooth projector functions $\{\tilde{p}_\alpha^a\}$ is defined as

$$\langle \tilde{p}_\alpha^a \mid = \sum_\beta (p^a \mid \tilde{\chi}^a \rangle)_{\alpha\beta}^{-1} \langle p_\beta^a \mid \ . \tag{6.15}$$

It can be shown that similar to $\{p_\alpha^a\}$, the smooth projector functions $\{\tilde{p}_\alpha^a\}$ have the following properties:

$$\langle \tilde{p}_\beta^a \mid \tilde{\chi}_\alpha^a \rangle = \delta_{\alpha\beta} \tag{6.16}$$

$$\tilde{p}_\alpha^a(\mathbf{r}) = 0 , \quad \forall \mathbf{r} \subset \Omega_I . \tag{6.17}$$

Furthermore, it is straightforward to prove that

$$\langle \tilde{p}_\alpha^a \mid = \langle p_\alpha^a \mid \mathcal{T}^{-1} \tag{6.18}$$

and therefore the local basis expansion coefficients and the remainder can be represented alternatively as

$$c_{i,\alpha}^a = \langle \tilde{p}_\alpha^a \mid \tilde{\phi}_i \rangle , \tag{6.19}$$

$$\mid \Delta_i^a \rangle = (1 - \sum_\alpha \mid \tilde{\chi}_\alpha^a \rangle \langle \tilde{p}_\alpha^a \mid) \mid \tilde{\phi}_i \rangle . \tag{6.20}$$

The above two expressions show that if the basis $\{\chi_\alpha^a\}$ provides an accurate local representation for $\phi_i(\mathbf{r})$, then the smooth basis $\{\tilde{\chi}_\alpha^a\}$ provides an accurate local representation for $\tilde{\phi}_i(\mathbf{r})$ and *vice versa*. This is an important observation, since it is the objective to completely eliminate $\phi_i(\mathbf{r})$ and to seek for $\tilde{\phi}_i(\mathbf{r})$ directly.

From a practical point of view, it is the inverse transformation $\mathcal{T}^{-1}$ that plays a major role in all the applications. The expression for $\mathcal{T}^{-1}$ involves basis sets $\phi_i(\mathbf{r})$ and $\tilde{\phi}_i(\mathbf{r})$ and smooth projector functions $\{\tilde{p}_\alpha^a\}$. If desired, the projector functions $\{p_\alpha^a\}$ can be found from

$$\langle p_\alpha^a \mid = \sum_\beta (\langle \tilde{p}^a \mid \chi^a \rangle)_{\alpha\beta}^{-1} \langle \tilde{p}_\beta^a \mid . \tag{6.21}$$

In the following the approximation will be made that the local expansions are accurate and that the remainder terms can be neglected.

6.3 Expectation values

Having defined the novel PAW basis, the next step is to express matrix elements that are needed in electronic structure calculations and in molecular dynamics simulations. To this end, consider the expectation value of a general local or quasilocal operator $\mathcal{O}$ with respect to the Kohn–Sham orbitals

$$\langle O \rangle = \sum_i f_i \langle \phi_i \mid \mathcal{O} \mid \phi_i \rangle . \tag{6.22}$$

This can be rewritten in terms of the smooth wave function

$$\langle O \rangle = \sum_i f_i \langle \mathcal{T}^{-1}\tilde{\phi}_i \mid \mathcal{O} \mid \mathcal{T}^{-1}\tilde{\phi}_i \rangle$$

$$= \sum_i f_i \langle \tilde{\phi}_i \mid \mathcal{O} \mid \tilde{\phi}_i \rangle + \sum_a \sum_{\alpha\beta} \Big[\langle \tilde{\phi}_i \mid \tilde{p}_\alpha^a \rangle \left(\langle \chi_\alpha^a \mid \mathcal{O} \mid \chi_\beta^a \rangle \right.$$

$$\left. - \langle \tilde{\chi}_\alpha^a \mid \mathcal{O} \mid \tilde{\chi}_\beta^a \rangle \right) \langle \tilde{p}_\beta^a \mid \tilde{\phi}_i \rangle \Big] \ . \tag{6.23}$$

Note that the last expression cannot be derived directly using the explicit form of the transformation, but is in fact an equivalence that is based on the properties of the expansion basis. Thus, the original expression for the expectation value is seen to be split into several parts. The first part is a simple expectation value over the smooth wave function which can be calculated accurately in the plane wave basis. The second term is a one-center contribution and restricted to atomic sphere regions only. This latter term already has the same structure as the nonlocal part in a fully separable pseudopotential.

Choosing the general operator $\mathcal{O}$ to be simply the identity

$$\langle \mathbf{r} \mid \mathcal{O} \mid \mathbf{r}' \rangle = \delta(\mathbf{r} - \mathbf{r}') \tag{6.24}$$

reveals that the orthogonality properties of smooth wave functions $\{\tilde{\phi}_i\}$ are different from those of $\{\phi_i\}$. Namely, if

$$\langle \phi_i \mid \phi_j \rangle = \delta_{ij} \tag{6.25}$$

hold as usual then

$$\langle \tilde{\phi}_i \mid \mathcal{S} \mid \tilde{\phi}_j \rangle = \delta_{ij} \ , \tag{6.26}$$

which is a generalized orthogonality relation due to introducing an overlap operator which is given by

$$\mathcal{S} = 1 + \sum_a \sum_{\alpha\beta} \Big[\mid \tilde{p}_\alpha^a \rangle \left(\langle \chi_\alpha^a \mid \chi_\beta^a \rangle - \langle \tilde{\chi}_\alpha^a \mid \tilde{\chi}_\beta^a \rangle \right) \langle \tilde{p}_\beta^a \mid \Big] \ . \tag{6.27}$$

It is important to recall that this overlap depends explicitly on all nuclear positions, $\{\mathbf{R}_I\}$, by virtue of the position dependence of the local basis functions that have been introduced inside the atomic sphere regions Ω_a around each atom.

Finally, the relationship between the electron density

$$n(\mathbf{r}) = \sum_i f_i \mid \phi_i(\mathbf{r}) \mid^2 \tag{6.28}$$

and its smooth counterpart

$$\tilde{n}(\mathbf{r}) = \sum_i f_i \mid \tilde{\phi}_i(\mathbf{r}) \mid^2 \tag{6.29}$$

can be obtained by choosing

$$\mathcal{O} = \mid \mathbf{r} \rangle \langle \mathbf{r} \mid \tag{6.30}$$

for the general operator. This yields

$$n(\mathbf{r}) = \tilde{n}(\mathbf{r}) + \sum_a n^a(\mathbf{r} - \mathbf{R}_a) - \sum_a \tilde{n}^a(\mathbf{r} - \mathbf{R}_a) , \tag{6.31}$$

where one center atomic densities $n^a(\mathbf{r})$ and $\tilde{n}^a(\mathbf{r})$ are given by

$$n^a(\mathbf{r}) = \sum_i f_i \left| \sum_\alpha c_{i\alpha}^a \chi_\alpha^a(\mathbf{r}) \right|^2 \tag{6.32}$$

$$\tilde{n}^a(\mathbf{r}) = \sum_i f_i \left| \sum_\alpha c_{i\alpha}^a \tilde{\chi}_\alpha^a(\mathbf{r}) \right|^2 , \tag{6.33}$$

and the following relationships

$$n^a(\mathbf{r}) = n(\mathbf{r}) , \quad \forall \mathbf{r} \subset \Omega_a \tag{6.34}$$

$$\tilde{n}^a(\mathbf{r}) = \tilde{n}(\mathbf{r}) , \quad \forall \mathbf{r} \subset \Omega_a \tag{6.35}$$

hold for these electronic densities inside the atomic spheres Ω_a, and by construction

$$n^a(\mathbf{r}) = \tilde{n}^a(\mathbf{r}) , \quad \forall \mathbf{r} \subset \Omega_I . \tag{6.36}$$

6.4 Ultrasoft pseudopotentials

At this stage, the energy expression for Vanderbilt's ultrasoft pseudopotentials can be derived. To do so one writes the energy of a system of atoms using the results from the last section within the Kohn–Sham approach.

Thus, the electron density

$$n(\mathbf{r}) \;=\; \tilde{n}(\mathbf{r}) + \sum_a \left[n^a(\mathbf{r} - \mathbf{R}_a) - \tilde{n}^a(\mathbf{r} - \mathbf{R}_a) \right] \tag{6.37}$$

$$\tilde{n}(\mathbf{r}) \;=\; \sum_i f_i \, | \, \tilde{\phi}_i(\mathbf{r}) \, |^2 \tag{6.38}$$

$$n^a(\mathbf{r}) \;=\; \sum_i f_i \sum_{\alpha\beta} \langle \tilde{\phi}_i \, | \, \tilde{p}^a_\alpha \rangle \chi^a_\alpha(\mathbf{r}) \chi^a_\beta(\mathbf{r}) \langle \tilde{p}^a_\beta \, | \, \tilde{\phi}_i \rangle \tag{6.39}$$

$$\tilde{n}^a(\mathbf{r}) \;=\; \sum_i f_i \sum_{\alpha\beta} \tilde{\chi}^a_\alpha(\mathbf{r}) \tilde{\chi}^a_\beta(\mathbf{r}) \langle \tilde{p}^a_\beta \, | \, \tilde{\phi}_i \rangle \tag{6.40}$$

is decomposed into the various parts derived in Section 6.3. The smooth functions are orthogonal

$$\langle \tilde{\phi}_i \, | \, S \, | \, \tilde{\phi}_j \rangle = \delta_{ij} \tag{6.41}$$

with respect to the generalized overlap operator

$$S = 1 + \sum_a \sum_{\alpha\beta} \left[| \, \tilde{p}^a_\alpha \rangle \left(\langle \chi^a_\alpha \, | \, \chi^a_\beta \rangle - \langle \tilde{\chi}^a_\alpha \, | \, \tilde{\chi}^a_\beta \rangle \right) \langle \tilde{p}^a_\beta \, | \right] \, . \tag{6.42}$$

The total Kohn–Sham energy is then

$$E_{\mathrm{KS}} = \sum_i f_i \langle \tilde{\phi}_i \, | \, T + V_{\mathrm{ext}} \, | \, \tilde{\phi}_i \rangle + \sum_i f_i \sum_a \sum_{\alpha\beta} \Big[\langle \tilde{\phi}_i \, | \, \tilde{p}^a_\alpha \rangle$$

$$\times \left(\langle \chi^a_\alpha \, | \, T + V_{\mathrm{ext}} \, | \, \chi^a_\beta \rangle - \langle \tilde{\chi}^a_\alpha \, | \, T + V_{\mathrm{ext}} \, | \, \tilde{\chi}^a_\beta \rangle \right) \langle \tilde{p}^a_\beta \, | \, \tilde{\phi}_i \rangle \Big]$$

$$+ \, E_{\mathrm{Hxc}}[n(\mathbf{r})] + E_{\mathrm{ion}} \, , \tag{6.43}$$

where E_{Hxc} denotes the combined energy due to the Hartree potential and the exchange–correlation energy and E_{ion} is the electrostatic energy of the ionic cores (see Eq. (3.38) and Section 2.7.2). Next, the following new quantities:

$$Q^a_{\alpha\beta}(\mathbf{r}) \;=\; \chi^a_\alpha(\mathbf{r}) \chi^a_\beta(\mathbf{r}) - \tilde{\chi}^a_\alpha(\mathbf{r}) \tilde{\chi}^a_\beta(\mathbf{r}) \tag{6.44}$$

$$q^a_{\alpha\beta} \;=\; \int Q^a_{\alpha\beta}(\mathbf{r}) \, d\mathbf{r} \tag{6.45}$$

$$V^a_{\mathrm{ext}}(\mathbf{r}) \;=\; V^a_{\mathrm{loc}}(\mathbf{r}) + \Delta V^a_{\mathrm{ext}}(\mathbf{r}) \tag{6.46}$$

$$D^a_{\alpha\beta} \;=\; \langle \chi^a_\alpha \, | \, T + \Delta V_{\mathrm{ext}} \, | \, \chi^a_\beta \rangle - \langle \tilde{\chi}^a_\alpha \, | \, T + \Delta V_{\mathrm{ext}} \, | \, \tilde{\chi}^a_\beta \rangle \tag{6.47}$$

are introduced where $V_{\text{loc}}^a(\mathbf{r})$ is a local smooth potential and $\Delta V_{\text{ext}}^a(\mathbf{r})$ is localized within the atomic sphere regions. Using these quantities, the electron density and the overlap operator can be expressed as

$$n(\mathbf{r}) \;=\; \sum_i f_i \left[\mid \tilde{\phi}_i(\mathbf{r}) \mid^2 + \sum_a \sum_{\alpha\beta} Q_{\alpha\beta}^a(\mathbf{r}) \langle \tilde{\phi}_i \mid \tilde{p}_\alpha^a \rangle \langle \tilde{p}_\beta^a \mid \tilde{\phi}_i \rangle \right] \quad (6.48)$$

$$S \;=\; 1 + \sum_a \sum_{\alpha\beta} q_{\alpha\beta}^a \mid \tilde{p}_\alpha^a \rangle \langle \tilde{p}_\beta^a \mid \quad\quad (6.49)$$

and the Kohn–Sham energy is given by

$$E_{\text{KS}} = \sum_i f_i \langle \tilde{\phi}_i \mid -\frac{1}{2}\nabla^2 \mid \tilde{\phi}_i \rangle + \int V_{\text{loc}}(\mathbf{r}) \tilde{n}(\mathbf{r}) \, d\mathbf{r}$$

$$+ \sum_i f_i \sum_a \sum_{\alpha\beta} \langle \tilde{\phi}_i \mid \tilde{p}_\alpha^a \rangle \left(D_{\alpha\beta}^a + \int V_{\text{loc}}(\mathbf{r}) Q_{\alpha\beta}^a(\mathbf{r}) \, d\mathbf{r} \right) \langle \tilde{p}_\beta^a \mid \tilde{\phi}_i \rangle$$

$$+ E_{\text{Hxc}}[n(\mathbf{r})] + E_{\text{ion}} \; . \quad (6.50)$$

These are the working equations for the Kohn–Sham method using ultrasoft pseudopotentials.

The pseudopotentials are specified through the local pseudopotential $V_{\text{loc}}(\mathbf{r})$, the augmentation charges $Q_{\alpha\beta}^a(\mathbf{r})$ and their integrated values $q_{\alpha\beta}^a$, the nonlocal matrix elements $D_{\alpha\beta}^a$, and the projector functions $\tilde{p}_\alpha^a(\mathbf{r})$. Note that in this derivation the transition was made from a formally all-electron method to a pseudopotential treatment by assuming that the external potential is given as a norm-conserving pseudopotential. By doing so the usual approximations coming with pseudopotentials, e.g. the frozen-core approximation and the linearization of the exchange and correlation functional, were of course introduced in the ultrasoft scheme. Adding to the above equations the nonlinear core correction as introduced in Section 4.5 for norm-conserving pseudopotentials is straightforward. We refer to the literature [788, 816, 1548] for details about the various methods to determine the parameters and functions needed for ultrasoft pseudopotentials.

The only difference in the energy expression from the form with fully nonlocal pseudopotentials (see Eq. (3.71) in Section 3.4.1) is that the total charge density includes the augmentation charges $Q_{\alpha\beta}^a(\mathbf{r})$. However, for the calculation of derivatives, the special form of the metric S defined in Eq. (6.27) introduces many new terms. Starting from the extended energy functional

$$E_{\text{uspp}}(\{\phi_i\}, \Lambda, \{\mathbf{R}_I\}) = E_{\text{KS}}(\{\phi_i\}, \{\mathbf{R}_I\}) + \sum_{ij} \Lambda_{ij} \left(\langle \tilde{\phi}_i \mid \mathcal{S} \mid \tilde{\phi}_j \rangle - \delta_{ij} \right) ,$$

(6.51)

one arrives at the force expressions

$$\frac{\delta E_{\text{uspp}}}{\delta \langle \tilde{\phi}_i \mid} = f_i H_{\text{uspp}} \mid \tilde{\phi}_i \rangle + \sum_j \Lambda_{ij} \mathcal{S} \mid \tilde{\phi}_j \rangle$$

(6.52)

$$H_{\text{uspp}} = -\frac{1}{2} \nabla^2 + V_{\text{eff}}(\mathbf{r}) + \sum_a \sum_{\alpha\beta} \bar{D}_{\alpha\beta}^a \mid \tilde{p}_\alpha^a \rangle \langle \tilde{p}_\beta^a \mid$$

(6.53)

$$V_{\text{eff}}(\mathbf{r}) = \frac{\delta E_{\text{KS}}}{\delta n(\mathbf{r})} = V_{\text{loc}}(\mathbf{r}) + \int \frac{n(\mathbf{r}')}{\mid \mathbf{r} - \mathbf{r}' \mid} \, d\mathbf{r}' + V_{\text{xc}}(\mathbf{r}) ,$$

(6.54)

where $V_{\text{xc}}(\mathbf{r}) = \delta E_{\text{xc}}[n]/\delta n(\mathbf{r})$ and all the terms arising from the augmentation part of the electron density have been grouped together with the nonlocal part of the pseudopotential by defining new coefficients

$$\bar{D}_{\alpha\beta}^a = D_{\alpha\beta}^a + \int V_{\text{eff}}(\mathbf{r}) \, Q_{\alpha\beta}^a(\mathbf{r}) \, d\mathbf{r} .$$

(6.55)

The forces on the ions are calculated from the derivative of E_{uspp} with respect to the positions of the nuclei

$$\frac{\partial E_{\text{uspp}}}{\partial \mathbf{R}_I} = \frac{\partial E_{\text{KS}}}{\partial \mathbf{R}_I} + \sum_{ij} \Lambda_{ij} \left\langle \tilde{\phi}_i \left| \frac{\partial \mathcal{S}}{\partial \mathbf{R}_I} \right| \tilde{\phi}_j \right\rangle$$

(6.56)

$$\frac{\partial \mathcal{S}}{\partial \mathbf{R}_I} = \sum_{\alpha\beta} q_{\alpha\beta}^I \left\{ \left| \frac{\partial \tilde{p}_\alpha^I}{\partial \mathbf{R}_I} \right\rangle \langle \tilde{p}_\beta^I \mid + \mid \tilde{p}_\alpha^I \rangle \left\langle \frac{\partial \tilde{p}_\beta^I}{\partial \mathbf{R}_I} \right| \right\} ,$$

(6.57)

so that one obtains finally

$$\frac{\partial E_{\text{KS}}}{\partial \mathbf{R}_I} = \frac{\partial E_{\text{ion}}}{\partial \mathbf{R}_I} + \int \frac{\partial V_{\text{loc}}(\mathbf{r})}{\partial \mathbf{R}_I} \tilde{n}(\mathbf{r}) \, d\mathbf{r}$$

$$+ \int V_{\text{eff}}(\mathbf{r}) \sum_{\alpha\beta} \frac{\partial Q_{\alpha\beta}^I(\mathbf{r})}{\partial \mathbf{R}_I} \left[\sum_i f_i \langle \tilde{\phi}_i \mid \tilde{p}_\alpha^I \rangle \langle \tilde{p}_\beta^I \mid \tilde{\phi}_i \rangle \right] d\mathbf{r}$$

$$+ \sum_{\alpha\beta} \bar{D}_{\alpha\beta}^I \sum_i f_i \left[\left\langle \tilde{\phi}_i \left| \frac{\partial \tilde{p}_\alpha^I}{\partial \mathbf{R}_I} \right\rangle \langle \tilde{p}_\beta^I \mid \tilde{\phi}_i \rangle + \langle \tilde{\phi}_i \mid \tilde{p}_\alpha^I \rangle \left\langle \frac{\partial \tilde{p}_\beta^I}{\partial \mathbf{R}_I} \right| \tilde{\phi}_i \rangle \right]$$

(6.58)

for the full energy gradient in the framework of Vanderbilt's ultrasoft pseudopotentials. The additional force terms can be classified on the one hand as Pulay forces, as discussed in Section 2.5, since they can be traced back to introducing basis functions that are *localized* in the atomic sphere regions Ω_a. On the other hand, these terms can be viewed as arising from a generalized (orthonormality) constraint, as made explicit in the general formula Eq. (2.59) for the generic Car–Parrinello equations of motion (see also the discussion of Eqs. (2.111)-(2.112) in this respect).

6.5 PAW energy expression

In contrast to the ultrasoft pseudopotential derivation, the PAW approach avoids the introduction of a pseudopotential but rather works in the frozen-core approximation directly. The sum over states is therefore restricted to valence electrons, but the electronic densities always include contributions from the core electrons. In addition, the expansion functions $\chi_\alpha^a(\mathbf{r})$ have to be orthogonal to the core state on the atom (see Refs. [142, 790, 1533] for details). In the following, an all-electron treatment is assumed, although this is rarely used in practice.

The PAW method works directly with the three sets of functions $\{\chi_i^a\}$, $\{\tilde\chi_i^a\}$, and $\{\tilde p_i^a\}$. These functions, together with a local potential $V_{\text{loc}}(r)$, fully define the PAW energy expression. Similar to expectation values such as the density from Eq. (6.31), the total energy is divided into three individual terms:

$$E_{\text{KS}} = \tilde E + \sum_a E^a - \sum_a \tilde E^a \ . \tag{6.59}$$

The smooth part $\tilde E$, which is evaluated on regular grids in Fourier or real space, and the one-center contributions E^a and $\tilde E^a$, which are evaluated on radial grids in an angular momentum representation. The three contributions to E_{KS} are

$$\tilde E = \sum_i f_i \langle \tilde\phi_i \mid -\frac{1}{2}\nabla^2 \mid \tilde\phi_i \rangle$$

$$+ \int V_{\text{loc}}(\mathbf{r})\,\tilde n(\mathbf{r})\,d\mathbf{r} + E_{\text{H}}[\tilde n + \hat n] + E_{\text{xc}}[\tilde n] \tag{6.60}$$

$$E^a = \sum_i f_i \sum_{\alpha\beta} \langle \tilde\phi_i \mid \tilde p_\alpha^a \rangle \langle \chi_\alpha^a \mid -\frac{1}{2}\nabla^2 \mid \chi_\beta^a \rangle \langle \tilde p_\beta^a \mid \tilde\phi_i \rangle$$

$$+ E_{\text{H}}[n^a + n^{Za}] + E_{\text{xc}}[n^a] \tag{6.61}$$

$$\tilde{E}^a = \sum_i f_i \sum_{\alpha\beta} \langle \tilde{\phi}_i \mid \tilde{p}^a_\alpha \rangle \langle \tilde{\chi}^a_\alpha \mid -\frac{1}{2}\nabla^2 \mid \tilde{\chi}^a_\beta \rangle \langle \tilde{p}^a_\beta \mid \tilde{\phi}_i \rangle$$

$$+ \int V_{\text{loc}}(\mathbf{r})\, \tilde{n}^a(\mathbf{r})\, d\mathbf{r} + E_H[\tilde{n}^a + \hat{n}] + E_{\text{xc}}[\tilde{n}^a] \ . \qquad (6.62)$$

The potential V_{loc} is an arbitrary potential localized in the augmentation regions, i.e. inside the atomic sphere regions Ω_a. Its contribution to the total energy vanishes exactly because $\tilde{n}(\mathbf{r}) = \tilde{n}^a(\mathbf{r})$ within the atomic spheres. Since the potential contributes only if the partial wave expansion is not complete, it is used to minimize truncation errors. In addition, the point charge density n^{Za} of the nuclei and a compensation charge density $\hat{n}$ was introduced. The compensation charge density has the same multipole moments as the density $n^a + n^{Za} - \tilde{n}^a$ and is localized within the atomic regions. Whereas the division of the kinetic energy and the exchange–correlation energy is straightforward from the formulas derived for the expectation value of local operators, the derivation of the electrostatic terms is rather involved [142, 790, 1533]. In fact, another compensation charge [142] is needed in order to be able to calculate the electrostatic term in the first energy contribution solely within an energy cutoff dictated by $\tilde{\phi}_i(\mathbf{r})$.

6.6 Integrating the Car–Parrinello equations

In the following, the velocity Verlet equations will be derived for the Car–Parrinello molecular dynamics method using both ultrasoft pseudopotentials and the more general PAW transformation. The equations of motion are

$$\mu |\ddot{\phi}_i\rangle = |\varphi_i\rangle + \sum_j \Lambda_{ij}\mathcal{S}(\{\mathbf{R}_I\})|\phi_j\rangle$$

$$M_I\ddot{\mathbf{R}}_I = \mathbf{F}_I + \sum_{ij} \Lambda_{ij}\langle\phi_i|\nabla_I\mathcal{S}(\{\mathbf{R}_I\})|\phi_j\rangle \ , \qquad (6.63)$$

where the simplifying Dirac notation is used. The forces acting on the orbitals and nuclei, abbreviated $\varphi_i(\mathbf{r})$ and $\mathbf{F}_I$, respectively, are those

$$\varphi_i(\mathbf{r}) = -\frac{\delta E_{\text{KS}}}{\delta \phi^*_i(\mathbf{r})}$$

$$\mathbf{F}_I = -\frac{\partial E_{\text{KS}}}{\partial \mathbf{R}_I} \qquad (6.64)$$

derived earlier in Section 6.4, where only the tilde sign that denotes the smooth orbitals is now dropped for simplicity. The electronic force is often written in the form

$$|\varphi_i\rangle = -f_i H_{\text{uspp}}|\phi_i\rangle \ , \tag{6.65}$$

where H_{uspp} was also defined before in Eq. (6.53). In the velocity Verlet scheme, both the positions and velocities are treated explicitly. That is, one carries the information $\{\mathbf{R}_I(t), \dot{\mathbf{R}}_I(t)\}$ and $\{\phi_i(\mathbf{r}, t), \dot{\phi}_i(\mathbf{r}, t)\}$ at each time step. A prediction of the orbital positions and velocities is then made according to

$$|\dot{\bar{\phi}}_i\rangle = |\dot{\phi}_i(t)\rangle + \frac{\Delta t}{2\mu}|\varphi_i(t)\rangle$$

$$|\bar{\phi}_i\rangle = |\phi_i(t)\rangle + \Delta t|\dot{\bar{\phi}}_i\rangle \tag{6.66}$$

and similarly for the ionic positions and velocities

$$\dot{\bar{\mathbf{R}}}_I = \dot{\mathbf{R}}_I(t) + \frac{\Delta t}{2M_I}\mathbf{F}_I(t)$$

$$\bar{\mathbf{R}}_I = \mathbf{R}_I(t) + \Delta t\dot{\bar{\mathbf{R}}}_I \ . \tag{6.67}$$

The new orbital and ionic positions are then obtained from

$$|\phi_i(t + \Delta t)\rangle = |\bar{\phi}_i\rangle + \frac{\Delta t^2}{2\mu}\sum_j \Lambda_{ij}\mathcal{S}(t)|\phi_j(t)\rangle \tag{6.68}$$

and

$$\mathbf{R}_I(t + \Delta t) = \bar{\mathbf{R}}_I + \frac{\Delta t^2}{2M_I}\sum_{ij}\Lambda_{ij}\langle\phi_i(t)|\nabla_I\mathcal{S}(t)|\phi_j(t)\rangle \ , \tag{6.69}$$

respectively. The Lagrange multipliers are determined by imposing the constraint condition

$$\langle\phi_i(t + \Delta t)|\mathcal{S}(t + \Delta t)|\phi_j(t + \Delta t)\rangle = \delta_{ij} \tag{6.70}$$

so that the substitution of Eq. (6.68) into Eq. (6.70) yields the matrix equation

$$\mathbf{A} + \mathbf{XB} + \mathbf{B}^\dagger\mathbf{X}^\dagger + \mathbf{XCX}^\dagger = \mathbf{I} \ , \tag{6.71}$$

where $X_{ij} = (\Delta t^2/2\mu)\Lambda_{ij}$ and the $\mathcal{S}$-dependent matrices are given by

$$A_{ij} = \langle\bar{\phi}_i|\mathcal{S}(t + \Delta t)|\bar{\phi}_j\rangle$$

$$B_{ij} = \langle\mathcal{S}(t)\phi_i(t)|\mathcal{S}(t + \Delta t)|\bar{\phi}_j\rangle$$

$$C_{ij} = \langle\mathcal{S}(t)\phi_i(t)|\mathcal{S}(t + \Delta t)|\mathcal{S}(t)\phi_j(t)\rangle \ . \tag{6.72}$$

From the solution of this equation for the Lagrange multipliers, the new orbitals and ionic positions are obtained from Eqs. (6.68) and (6.69).

However, one can immediately see that this procedure poses a self-consistency problem, because the matrices $\mathbf{B}$ and $\mathbf{C}$ appearing in Eq. (6.71) and therefore the solution for $\mathbf{X}$ (or $\mathbf{\Lambda}$) depend on the new ionic positions $\mathbf{R}_I(t+\Delta t)$ through $\mathcal{S}(t+\Delta t)$, which in turn cannot be determined until the matrix $\mathbf{\Lambda}(t+\Delta t)$ is known. Thus, Eqs. (6.71) and (6.69) must be solved self-consistently *via* iteration. This is accomplished by first guessing a solution to Eq. (6.71) at time $t + \Delta t$ based on the solution from the two previous time steps, i.e.

$$\mathbf{X}^{(0)}(t + \Delta t) = 2\mathbf{X}(t) - \mathbf{X}(t - \Delta t) \ , \tag{6.73}$$

which is substituted into Eq. (6.69) to obtain a guess of the new ionic positions to then be used to determine the matrices in Eq. (6.71), and the procedure is iterated until a self-consistent solution is obtained. The matrix equation Eq. (6.71) must itself be solved iteratively, being quadratic in the unknown matrix $\mathbf{X}$. The details of the solution of this equation are given in Ref. [816] and will be omitted here. Despite the rapid convergence of the iterative procedure, the multiple determination of the matrix $\mathbf{X}$ in Eq. (6.71) is one of the most time-consuming parts of the calculation. Thus, each time $\mathbf{X}$ must be determined anew adds considerably to the total time per molecular dynamics step.

Once the self-consistent solution has been obtained, the new orbitals and ionic positions are calculated from Eqs. (6.68) and (6.69), and the prediction of the velocities is completed according to

$$|\mathring{\dot{\phi}}_i\rangle \ = \ |\dot{\phi}_i\rangle + \frac{\Delta t}{2\mu} \sum_j \Lambda_{ij}|\phi_j(t)\rangle + \frac{\Delta t}{2\mu}|\varphi_i(t + \Delta t)\rangle$$

$$\mathring{\dot{\mathbf{R}}}_I \ = \ \dot{\mathbf{R}}_I + \frac{\Delta t}{2M_I} \sum_{ij} \Lambda_{ij}\langle\phi_i(t)|\nabla_I \mathcal{S}(t)|\phi_j(t)\rangle$$

$$+ \frac{\Delta t}{2M_I}\mathbf{F}_I(t + \Delta t) \ . \tag{6.74}$$

The final correction step for the velocities consists of writing the new orbital and ionic velocities as

$$|\dot{\phi}_i(t + \Delta t)\rangle \ = \ |\mathring{\dot{\phi}}_i\rangle + \sum_j Y_{ij}|\phi_j(t + \Delta t)\rangle$$

$$\dot{\mathbf{R}}_I(t + \Delta t) = \dot{\mathbf{R}}_I + \frac{\mu}{M_I} \sum_{ij} Y_{ij}$$

$$\times \langle \phi_i(t + \Delta t) | \nabla_I \mathcal{S}(t + \Delta t) | \phi_j(t + \Delta t) \rangle \ . \quad (6.75)$$

The Lagrange multipliers Y_{ij} are determined by requiring that the first time derivative of the constraint condition

$$\langle \dot{\phi}_i(t) | \mathcal{S}(t) | \phi_j(t) \rangle + \langle \phi_i(t) | \mathcal{S}(t) | \dot{\phi}_j(t) \rangle$$

$$+ \sum_I \langle \phi_i(t) | \nabla_I \mathcal{S}(t) | \phi_j(t) \rangle \cdot \dot{\mathbf{R}}_I(t) = 0 \quad (6.76)$$

be satisfied at time $t + \Delta t$, Eqs. (6.75) are substituted into Eq. (6.76), and the following matrix equation for $\mathbf{Y}$ is obtained:

$$\mathbf{YP} + \mathbf{Y}^\dagger \mathbf{P}^\dagger + \mathbf{Q} + \mathbf{Q}^\dagger + \sum_I \mathbf{\Delta}^I(t) \cdot \left[\dot{\mathbf{R}}_I + \frac{\mu}{M_I} \mathrm{Tr} \left(\mathbf{Y}^\dagger \mathbf{\Delta}^I(t) \right) \right] = 0 \ , \quad (6.77)$$

where the three-component vector matrix $\mathbf{\Delta}^I$ is given by

$$\mathbf{\Delta}^I_{ij} = \langle \phi_i | \nabla_I \mathcal{S}(\{\mathbf{R}_I\}) | \phi_j \rangle \quad (6.78)$$

and the matrices $\mathbf{P}$ and $\mathbf{Q}$ are defined by

$$P_{ij} = \langle \mathcal{S}(t + \Delta t) \phi_i(t + \Delta t) | \mathcal{S}(t + \Delta t) | \phi_j(t + \Delta t) \rangle$$

$$Q_{ij} = \langle \dot{\phi}_i(t + \Delta t) | \mathcal{S}(t + \Delta t) | \phi_j(t + \Delta t) \rangle \ . \quad (6.79)$$

Note that Eq. (6.77), although linear in $\mathbf{Y}$, is most easily solved iteratively. However, because the ionic velocities in Eq. (6.75) have been used explicitly in Eq. (6.76), there is no self-consistency problem. Once Eq. (6.77) has been solved, then the new velocities can be determined via Eq. (6.75) straightforwardly. The solution for $\mathbf{Y}$ ($= \mathbf{Y}(t + \Delta t)$) can be used to obtain an initial guess for the matrix $\mathbf{X}$ in the next step *via*

$$\mathbf{X}^{(0)}(t + \Delta t) = \mathbf{X}(t) + \Delta t \mathbf{Y}(t + \Delta t) \ . \quad (6.80)$$

In the velocity Verlet algorithm as presented so far, the iteration of the position step requires several applications of the matrix $\mathbf{\Delta}^I$, which can either be stored or recalculated as needed. However, due to the size of this matrix (which is $3 \times N_{at} \times N_b^2$, where N_{at} is the number of atoms and N_b is the number of electronic states in the system), storing it for large systems is often not possible and recalculating it is expensive. To circumvent these problems, the constraint nonorthogonal orbital (CNO) method has been devised [665], which eliminates the self-consistency problem encountered when ultrasoft pseudopotentials are used. However, this method is not limited to curing the ultrasoft pseudopotential problem. The presentation in the following will be general, although the Hamiltonian will still be denoted as H_{uspp}, such that the implementation of this method for standard norm-conserving pseudopotentials will become clear as well. For notational simplicity, the special case is assumed where all occupation numbers are assumed equal ($f_i = f$, $\forall i$) but generalization is straightforward.

The basic idea of the constraint nonorthogonal orbital method is to define a new set of orbitals $\{\psi_i(\mathbf{r})\}$ which are related to the old orbitals via the transformation

$$|\phi_i\rangle = \sum_j |\psi_j\rangle T_{ji} \ . \tag{6.81}$$

Here, the matrix $\mathbf{T}$ is defined to be

$$\mathbf{T} = \mathbf{O}^{-1/2} \tag{6.82}$$

where $\mathbf{O}$ is the overlap matrix

$$O_{ij} = \langle \psi_i | \mathcal{S}(\{\mathbf{R}_I\}) | \psi_j \rangle \tag{6.83}$$

between the new orbitals with respect to the operator $\mathcal{S}$. It can be seen that because $\mathbf{O}$ is Hermitian, $\mathbf{T}$ is Hermitian as well. With the definition Eq. (6.81) for the new orbitals, it follows that the constraint condition is automatically satisfied

$$
\begin{aligned}
\langle \phi_i | \mathcal{S} | \phi_j \rangle &= \sum_{k,l} \langle \psi_k | \mathcal{S} | \psi_l \rangle T_{ki}^* T_{lj} \\
&= T_{ik} O_{kl} T_{lj} \\
&= \delta_{ij} \ .
\end{aligned}
\tag{6.84}
$$

The idea, therefore, is to use the orbitals $\{\psi_i(\mathbf{r})\}$ rather than $\{\phi_i(\mathbf{r})\}$ to formulate the Car–Parrinello molecular dynamics scheme. Note that because the generalized orthonormality condition is satisfied automatically through the definition of the orbitals $\{\psi_i\}$, it is not necessary to impose any constraint condition whatsoever on these new orbitals. In the absence of any constraint condition, the proposed molecular dynamics scheme would constitute an explicitly nonorthogonal orbital scheme. In such schemes, maintaining adiabaticity, and sometimes even stability, is not guaranteed. The use of independent Nosé–Hoover thermostats [146, 333, 651, 1063] in terms of Nosé–Hoover chains [927] can aid substantially in stabilizing the dynamics and maintaining adiabaticity. However, when the ionic motion is coupled to a thermostat, the true dynamics of the ionic degrees of freedom is no longer obtained in a strict sense.

It was therefore proposed to use an alternative method of maintaining stability and adiabaticity without sacrificing dynamical information. The above scheme is supplemented by a general set of holonomic constraint conditions

$$\sigma_\alpha[\{\psi_i\}] = 0 \qquad (6.85)$$

imposed as a means of stabilizing the dynamics. The generalized index α runs from 1 to N_c, where N_c is the total number of such constraint conditions to be specified further below. Note that because the required constraint condition is satisfied *a priori*, one has complete freedom to choose the functionals in Eq. (6.85) in any way one likes. Whatever choice one makes has no physical relevance for the dynamics. However, it is assumed that $\sigma_\alpha[\{\psi_i\}]$ is independent of atomic positions.

Based upon these considerations an *ab initio* molecular dynamics scheme is now formulated in terms of the new orbitals by proposing the following Lagrangian:

$$\mathcal{L} = \mu \sum_i \langle \dot{\psi}_i | \dot{\psi}_i \rangle + \frac{1}{2} \sum_I M_I \dot{\mathbf{R}}_I^2$$

$$- f \sum_{i,j} M_{ji} \langle \psi_i | H_{\text{uspp}} | \psi_j \rangle + \sum_\alpha \lambda_\alpha \sigma_\alpha[\{\psi_i\}] , \quad (6.86)$$

where H_{uspp} is the Kohn–Sham Hamiltonian in terms of ultrasoft pseudopotentials, Eq. (6.53), and the matrix $\mathbf{M} = \mathbf{O}^{-1}$. The potential term $f \sum_{i,j} M_{ji} \langle \psi_i | H_{\text{uspp}} | \psi_j \rangle$ is derived straightforwardly by substituting Eq. (6.81) into the original energy functional term $f \sum_{i,j} \langle \phi_i | H_{\text{uspp}} | \phi_j \rangle$. The equations of motion derived from Eq. (6.86) are

$$\mu | \ddot{\psi}_i \rangle = -f \sum_j M_{ji} H_{\text{uspp}} | \psi_j \rangle - f \sum_{j,k} \langle \psi_j | H_{\text{uspp}} | \psi_k \rangle \frac{\delta M_{kj}}{\delta \psi_i^*} + \sum_\alpha \lambda_\alpha \frac{\delta \sigma_\alpha}{\delta \psi_i^*}$$

$$M_I \ddot{\mathbf{R}}_I = -f \sum_{i,j} [M_{ji} \nabla_I \langle \psi_i | H_{\text{uspp}} | \psi_j \rangle + \nabla_I M_{ji} \langle \psi_i | H_{\text{uspp}} | \psi_j \rangle] . \quad (6.87)$$

In order to evaluate the forces appearing in Eqs. (6.87), one needs an explicit expression for the derivatives of the elements of the matrix $\mathbf{M} = \mathbf{O}^{-1}$ with respect to the parameters shown. In general, given a matrix $\mathbf{A}$ which depends on some parameter λ, the derivative of $\mathbf{A}^{-1}$ with respect to λ is given by

$$\frac{d\mathbf{A}^{-1}}{d\lambda} = -\mathbf{A}^{-1}\frac{d\mathbf{A}}{d\lambda}\mathbf{A}^{-1} \;, \tag{6.88}$$

which can be applied straightforwardly when the matrix $\mathbf{A}$ and its dependence on λ are known explicitly. Using Eq. (6.88) and Eq. (6.83) for the matrix $\mathbf{O} = \mathbf{M}^{-1}$, one can show that the derivatives appearing in Eqs. (6.87) are given explicitly by

$$\frac{\delta M_{kj}}{\delta \psi_i^*} = -\sum_l M_{ki} M_{lj} \mathcal{S}|\psi_l\rangle$$

$$\nabla_I M_{kj} = -\sum_{l,m} M_{kl} M_{mj} \langle \psi_l | \nabla_I \mathcal{S} | \psi_m \rangle \;. \tag{6.89}$$

The substitution of Eqs. (6.89) into Eqs. (6.87) yields force expressions that could be used in actual computations. However, it is found that it is computationally more convenient to re-express the forces on the orbitals $\{\psi_i(\mathbf{R})\}$ in terms of the orthonormal orbitals $\{\phi_i(\mathbf{r})\}$, which is accomplished with the help of Eq. (6.81). For example, the electronic force terms are re-expressed as follows:

$$\sum_j M_{ji} H_{\mathrm{uspp}}|\psi_j\rangle = \sum_{j,k} T_{jk} T_{ki} H_{\mathrm{uspp}}|\psi_j\rangle$$

$$= \sum_k H_{\mathrm{uspp}}|\phi_k\rangle T_{ki} \tag{6.90}$$

and

$$\sum_{j,k} \frac{\delta M_{kj}}{\delta \psi_i^*} \langle \psi_j | H_{\mathrm{uspp}} | \psi_k \rangle = -\sum_{j,k,l} M_{ki} M_{lj} \mathcal{S}|\psi_l\rangle \langle \psi_j | H_{\mathrm{uspp}} | \psi_k \rangle$$

$$= -\sum_{j,k,l,m,n} T_{km} T_{mi} T_{ln} T_{nj} \mathcal{S}|\psi_l\rangle \langle \psi_j | H_{\mathrm{uspp}} | \psi_k \rangle$$

$$= -\sum_{m,n} \mathcal{S}|\phi_n\rangle \langle \phi_n | H_{\mathrm{uspp}} | \phi_m \rangle T_{mi} \tag{6.91}$$

so that the electronic force may be expressed as

$$|\bar{\varphi}_i\rangle = -f \sum_j \left[H_{\mathrm{uspp}}|\phi_j\rangle - \sum_k \mathcal{S}|\phi_k\rangle \langle \phi_k | H_{\mathrm{uspp}} | \phi_j \rangle \right] T_{ji} \tag{6.92}$$

where the notation $|\bar{\varphi}_i\rangle$ is used here to distinguish this force from the electronic force $|\varphi_i\rangle$ appearing in Eqs. (6.63). The ionic force may be worked out in the same way with the result

$$M_I \ddot{\mathbf{R}}_I = -f \sum_i \langle \phi_i | \nabla_I H_{\text{uspp}} | \phi_i \rangle - f \sum_{ij} \langle \phi_j | \nabla_I S | \phi_j \rangle \langle \phi_j | H_{\text{uspp}} | \phi_i \rangle \ , \quad (6.93)$$

where the first term is just the usual ionic force that was denoted $\mathbf{F}_I$ in Eq. (6.63).

The advantage of expressing the forces in terms of the orthonormal orbitals is immediately apparent from Eqs. (6.92) and (6.93). The forces appearing in these equations are of the same form as those in the original formulation given in Eqs. (6.63), except that the matrix Λ_{ij} has been replaced by $\langle \phi_i | H_{\text{uspp}} | \phi_j \rangle$. The additional difference is the rotation of the electronic forces in Eq. (6.92) by the matrix $\{T_{ji}\}$. Therefore, the conversion from an orthonormal to the constraint nonorthogonal orbital scheme is straightforward, requiring practically no modification of the force calculation. The equations of motion of this method can be written compactly in the form

$$\mu|\ddot{\psi}_i\rangle = |\bar{\varphi}_i\rangle + \sum_\alpha \lambda_\alpha \frac{\delta\sigma_\alpha}{\delta\psi_i^*}$$

$$M_I \ddot{\mathbf{R}}_I = \mathbf{F}_I + f \text{Tr}\left(\mathbf{\Delta}^I \mathbf{H}_{\text{uspp}}\right) \ , \quad (6.94)$$

where $\mathbf{H}_{\text{uspp}}$ denotes the matrix with elements $\{\langle \phi_i | H_{\text{uspp}} | \phi_j \rangle\}$. These are the basic Car–Parrinello equations of motion of the constraint nonorthogonal orbital method. The important point to stress is that the constraint conditions no longer depend on the ionic positions and additionally that the ionic forces no longer depend on the Lagrange multipliers. The self-consistency problem has been eliminated, so that the need for an iterative solution to the constraint problem is no longer required. Also, the matrix $\mathbf{\Delta}^I$ is used only once to calculate the force and then can be discarded. Therefore, there is no longer a need to store this matrix, a fact which reduces the memory required to simulate large systems.

In choosing an explicit form for the constraint functional introduced in Eq. (6.85), the interest is specifically in functionals for which the invariance of the Lagrangian under unitary transformations of the original orbitals $\{|\psi_i\rangle\}$ in the space of occupied states

$$|\psi_i'\rangle = \sum_j U_{ij} |\psi_j\rangle \quad (6.95)$$

is preserved. The first choice along these lines is an orthonormality constraint of the form

$$\langle \psi_i | \psi_j \rangle - \delta_{ij} = 0 \ , \tag{6.96}$$

which for a system of N_b electronic states yields the usual $N_c = N_b^2$ constraint conditions and will require N_b^2 Lagrange multipliers. If there are M basis functions, then the calculation of these constraint functions scales as $N_b^2 M$. Substituting Eq. (6.96) into the equations of motion Eq. (6.94) yields

$$\mu |\ddot{\psi}_i\rangle = |\bar{\varphi}_i\rangle + \sum_{i,j} \Lambda_{ij} |\psi_j\rangle$$

$$M_I \ddot{\mathbf{R}}_I = \mathbf{F}_I + f \mathrm{Tr}\left(\mathbf{\Delta}^I \mathbf{H}_{\mathrm{uspp}}\right) \tag{6.97}$$

for this case. The constrained equations of motion Eqs. (6.97) take the form of the standard Car–Parrinello equations, but maintain in addition the imposed unitary symmetry property. The integration of these equations by the velocity Verlet algorithms follows the standard procedure as outlined previously. With the choice Eq. (6.96), the number of overlap matrices which must be computed is reduced compared to the orthonormal orbital method described before, which makes the constraint nonorthogonal orbital method more efficient than the standard orthonormal orbital approach.

The second example of Eq. (6.85) to be considered is the length or norm constraint condition expressed as

$$\langle \psi_i | \psi_i \rangle - 1 = 0 \ , \tag{6.98}$$

which constitutes only N_b constraint conditions and therefore has significantly better scaling than condition Eq. (6.96). When substituted into Eq. (6.94), the resulting equations take the form

$$\mu |\ddot{\psi}_i\rangle = |\bar{\varphi}_i\rangle + \lambda_i |\psi_i\rangle$$

$$M_I \ddot{\mathbf{R}}_I = \mathbf{F}_I + f \mathrm{Tr}\left(\mathbf{\Delta}^I \mathbf{H}_{\mathrm{uspp}}\right) \tag{6.99}$$

such that this scheme also preserves the unitary symmetry property. The numerical integration of this set of equations, together with the constraint condition Eq. (6.98), will be discussed in the context of the velocity Verlet integrator in which the constraint condition

$$\langle \psi_i | \dot{\psi}_i \rangle = 0 \tag{6.100}$$

is also enforced. The orbitals at time $t + \Delta t$ are determined from those at t from

$$|\psi_i(t + \Delta t)\rangle = |\check{\psi}_i\rangle + x_i |\psi_i(t)\rangle \ , \tag{6.101}$$

where $x_i = (\Delta t^2 / 2\mu)\lambda_i$ and

$$|\breve{\psi}_i\rangle = |\psi_i(t)\rangle + \Delta t |\dot{\psi}_i(t)\rangle + \frac{\Delta t^2}{2\mu} |\bar{\varphi}_i(t)\rangle \ . \tag{6.102}$$

The ionic positions at $t + \Delta t$ are given by

$$\mathbf{R}_I(t + \Delta t) = \mathbf{R}_I(t) + \Delta t \dot{\mathbf{R}}_I(t) + \frac{\Delta t^2}{2M_I} \left[\mathbf{F}_I(t) + f \text{Tr} \left(\mathbf{\Delta}^I(t) \mathbf{H}_{\text{uspp}}(t) \right) \right] \tag{6.103}$$

and the Lagrange multipliers in Eq. (6.101) are determined by requiring that the constraint condition Eq. (6.98) be satisfied at $t + \Delta t$. When this condition is imposed, the result is a set of N_{b} independent quadratic equations for the x_i's, which take the form

$$x_i^2 + 2x_i + a_i = 0 \tag{6.104}$$

where

$$a_i = \langle \breve{\psi}_i | \breve{\psi}_i \rangle - 1 \ . \tag{6.105}$$

In deriving Eq. (6.104), the fact has been exploited that the constraint conditions Eqs. (6.98) and (6.100) are satisfied at t and the fact that $|\psi_i(t)\rangle$ is also orthogonal to the force $|\bar{\varphi}_i(t)\rangle$ which can be seen directly from Eq. (6.92). Of the two solutions to Eq. (6.104), we choose the one for which $x_i \to 0$ as $a_i \to 0$, which corresponds to

$$x_i = -1 + \sqrt{1 - a_i} \ . \tag{6.106}$$

The orbital velocities at $t + \Delta t$ are given by

$$|\dot{\psi}_i(t + \Delta t)\rangle = |\breve{\dot{\psi}}_i\rangle + y_i |\psi_i(t + \Delta t)\rangle \ , \tag{6.107}$$

where

$$|\breve{\dot{\psi}}_i\rangle = |\dot{\psi}_i(t)\rangle + \frac{\Delta t}{2\mu} \left[|\bar{\varphi}_i(t)\rangle + |\bar{\varphi}_i(t + \Delta t)\rangle \right] + \frac{x_i}{\Delta t} |\psi_i(t)\rangle \tag{6.108}$$

and similarly, the ionic velocities are given by

$$\dot{\mathbf{R}}_I(t + \Delta t) = \dot{\mathbf{R}}_I(t) + \frac{\Delta t}{2M_I} \left[\mathbf{F}_I(t) + \mathbf{F}_I(t + \Delta t) \right.$$

$$\left. + f \text{Tr} \left(\mathbf{\Delta}^I(t) \mathbf{H}_{\text{uspp}}(t) + \mathbf{\Delta}^I(t + \Delta t) \mathbf{H}_{\text{uspp}}(t + \Delta t) \right) \right] \ . \tag{6.109}$$

To determine y_i in Eq. (6.107), the constraint condition Eq. (6.100) is enforced at $t + \Delta t$, which yields

$$y_i = -\langle \psi_i(t + \Delta t) | \breve{\dot{\psi}}_i \rangle \ . \tag{6.110}$$

The simplicity of this second method to define the constraints makes it appealing. However, the constraint conditions Eqs. (6.98) and (6.100) alone are not enough to ensure stability in that they do not prevent the vectors $\{|\psi_i\rangle\}$ from becoming linearly dependent. As the vectors begin to overlap, the off-diagonal elements of the matrix $\langle\psi_i|\psi_j\rangle$ begin to grow. This problem can be circumvented by exploiting the unitary symmetry property. Let the unitary transformation diagonalize the matrix $\langle\psi_i|\psi_j\rangle$. The rotation in Eq. (6.95) can be performed without altering the dynamics. When this rotation is carried out, the matrix $\langle\psi_i|\psi_j\rangle$ will be diagonal, and a subsequent enforcement of the norm constraint condition Eq. (6.98) in the *next* time step will bring this matrix very close to the unit matrix. This operation can be performed when the largest off-diagonal element of the matrix $\langle\psi_i|\psi_j\rangle$ exceeds a specified tolerance. Generally, this tolerance can be chosen large ($\sim 10^{-3}$) such that the diagonalization and rotation, both $N_b^2 M$ operations, need only be carried out infrequently. Implementing this procedure requires some care since after a rotation is performed, the orbitals $\{\psi_i(t)\}$ with which one begins the next time step do not satisfy Eq. (6.98). Therefore, in calculating the Lagrange multiplier for the orbitals $\{\psi_i(t+\Delta t)\}$ for the step immediately following the rotation, Eq. (6.104) will need to be replaced by

$$c_i x_i^2 + 2b_i x_i + a_i = 0 \qquad (6.111)$$

where

$$
\begin{aligned}
c_i &= \langle\psi_i(t)|\psi_i(t)\rangle \\
b_i &= \langle\breve{\psi}_i|\psi_i(t)\rangle
\end{aligned}
\qquad (6.112)
$$

so that the solution for x_i is

$$x_i = \frac{1}{c_i}\left[-b_i + \sqrt{b_i^2 - a_i c_i}\right] \qquad (6.113)$$

which vanishes when $a_i = 0$. Eq. (6.111) need only be used after a rotation step, otherwise x_i is calculated from Eq. (6.104).

Finally, it is noted again that the constraint nonorthogonal orbital method is not limited to the case of ultrasoft pseudopotentials or the PAW method. It can just as easily be applied to the standard norm-conserving pseudopotentials as well. For this case, one does not use the overlap matrix as given in Eq. (6.83), but rather substitutes the unit operator for the operator $\mathcal{S}$ yielding

$$O_{ij} = \langle\psi_i|\psi_j\rangle \ . \qquad (6.114)$$

The constraint nonorthogonal orbital method with the orthonormality constraint condition Eq. (6.96) applied reduces to the usual Car–Parrinello scheme in the special case when norm-conserving pseudopotentials are used.

7

Computing properties

7.1 Adiabatic density-functional perturbation theory: Hessian, polarizability, NMR

7.1.1 Introduction

The total energy and charge density are basic quantities of density functional theory and give access to a wide range of experimental observables. Most of these quantities can be obtained as derivatives of the energy or density with respect to external perturbations, as collected in Table 7.1 for a selection of relevant response properties. As a simple example, the force exerted on a given nucleus is given by the negative derivative of the total energy of the system with respect to the position of this nucleus. The calculation of such energy derivatives can be done by finite-difference methods. The total energy is computed at slightly different values of the external perturbation and the derivative of the total energy curve with respect to the small disturbance is calculated numerically. Concerning the force, this just amounts to displacing the nucleus of interest by small amounts along the three Cartesian coordinates with respect to its equilibrium position as determined earlier from a geometry optimization. Although this is a very convenient and most straightforward method, recent practice has shown that perturbative techniques within the framework of density functional theory are much more powerful compared to numerical methods. Such techniques are very similar to the treatment of perturbations within Hartree–Fock theory, i.e. the coupled perturbed Hartree–Fock formalism [837]. These techniques were discovered and rediscovered within density functional theory many times. They are based either on the Sternheimer equation, Green's functions, sum-over-states techniques, or on the Hylleras variational technique. Generalizations to arbitrary order of perturbations, based on the "$2n + 1$ theorem" of perturbation theory and on generalized Sternheimer equations, were also

309

Table 7.1. *List of some properties related to derivatives of the total energy. Here, ϵ_α is an external electrical field, x_i a Cartesian displacement of a nuclei, B_α a magnetic field, and m_i the nuclear magnetic moment.*

Derivative	Observable
$\frac{d^2 E}{d\epsilon_\alpha d\epsilon_\beta}$	polarizability
$\frac{d^2 E}{dx_i dx_j}$	harmonic force constants
$\frac{d^2 E}{dx_i d\epsilon_\alpha}$	dipole derivatives: infrared intensities
$\frac{d^3 E}{dx_i d\epsilon_\alpha d\epsilon_\beta}$	polarizability derivative: Raman intensities
$\frac{d^2 E}{dB_\alpha dB_\beta}$	magnetizability
$\frac{d^2 E}{dm_i dB_\beta}$	nuclear magnetic shielding tensor

proposed. The complete literature is not given here but can be found in the publications of the Baroni and Gonze groups [79, 535–537].

7.1.2 Coupled perturbed Kohn–Sham equations

In this section, the basic working equations underlying density functional perturbation theory (DFPT) are derived. Consider the extended Kohn–Sham energy functional

$$\mathcal{E}^{\text{KS}}[\{\phi_i\}] = E^{\text{KS}}[\{\phi_i\}] + \lambda E^{\text{pert}}[\{\phi_i\}] + \sum_{ij} \Lambda_{ij} \left(\langle \phi_i \mid \phi_j \rangle - \delta_{ij} \right) , \quad (7.1)$$

which includes an energy functional E^{pert} due to an external perturbation of strength λ. According to the usual perturbation approach, all quantities are expanded in powers of λ in the form

$$X(\lambda) = X^{(0)} + \lambda X^{(1)} + \lambda^2 X^{(2)} + \cdots , \quad (7.2)$$

where X can be the Kohn–Sham energy E^{KS}, the Kohn–Sham orbitals $\phi_i(\mathbf{r})$, the electron density $n(\mathbf{r})$, the Lagrange multipliers $\mathbf{\Lambda}$, or the Kohn–Sham

Hamiltonian H_{KS}. The upper index in parentheses indicates the order of the perturbation and (0) stands for the unperturbed system defined by $\lambda = 0$.

Because the Kohn–Sham energy satisfies a variational principle under constraints, it is possible to derive a constraint variational principle for the $2n$th-order derivative of the energy with respect to the nth-order derivative of the orbitals $\phi_i(\mathbf{r})$ [536, 1363]. If the expansion of the wave function up to an order of $n - 1$ is known, then the variational principle for the $2n$th-order derivative of the energy is given by

$$E^{(2n)} = \min_{\phi_i^{(n)}} \left(\mathcal{E}^{\mathrm{KS}} \left[\sum_k \lambda^k \phi_i^{(k)} \right] \right)^{(2n)} \tag{7.3}$$

under the constraint, in the parallel transport gauge

$$\sum_{k=0}^{n} \langle \phi_i^{(n-k)} \mid \phi_j^{(k)} \rangle = 0 \tag{7.4}$$

for all occupied states i and j. The requirement that the constraint condition, i.e. orthogonality in the present case, is fulfilled at every order in the perturbation was used to derive the above equations. It turns out that there is a certain degree of arbitrariness in the definition of the Lagrange multipliers. By adding additional conditions that leave the energy invariant, a special gauge for the Lagrange multipliers is chosen. The best known gauge is the canonical gauge for the zeroth-order wave function in Hartree–Fock and Kohn–Sham theory. In this gauge the matrix of Lagrange multipliers is diagonal. For the perturbation treatment at hand, the parallel transport gauge defined above is most convenient. The explicit expressions for $E^{(2n)}$ can be derived by introducing Eq. (7.1) into Eq. (7.3). For zeroth order one gets back the Kohn–Sham equations

$$H_{\mathrm{KS}}^{(0)} \mid \phi_i^{(0)} \rangle = \sum_j \Lambda_{ij}^{(0)} \mid \phi_j^{(0)} \rangle \; , \tag{7.5}$$

and for second order the following expression is obtained:

$$E^{(2)} \left[\{\phi_i^{(0)}\}, \{\phi_i^{(1)}\} \right] = \sum_{ij} \langle \phi_i^{(1)} \mid H_{\mathrm{KS}}^{(0)} \delta_{ij} + \Lambda_{ij}^{(0)} \mid \phi_j^{(1)} \rangle$$

$$+ \sum_i \left\{ \langle \phi_i^{(1)} \mid \frac{\delta E^{\mathrm{pert}}}{\delta \langle \phi_i \mid} + \frac{\delta E^{\mathrm{pert}}}{\delta \mid \phi_i \rangle} \mid \phi_i^{(1)} \rangle \right\}$$

$$+ \frac{1}{2} \int \int \mathcal{K}(\mathbf{r}, \mathbf{r}') \, n^{(1)}(\mathbf{r}) \, n^{(1)}(\mathbf{r}') \, d\mathbf{r} \, d\mathbf{r}'$$

$$+ \int \frac{d}{d\lambda} \frac{\delta E_{\text{Hxc}}}{\delta n(\mathbf{r})} \bigg|_{n^{(0)}} n^{(1)}(\mathbf{r}) \, d\mathbf{r} + \frac{1}{2} \frac{d^2 E_{\text{Hxc}}}{d\lambda^2} \bigg|_{n^{(0)}} , \quad (7.6)$$

where the first-order wave functions are varied under the constraints

$$\langle \phi_i^{(0)} \mid \phi_j^{(1)} \rangle = 0 , \quad (7.7)$$

for all occupied states i and j. The zeroth-order density and the first-order perturbed density are given by

$$n^{(0)}(\mathbf{r}) = \sum_i f_i \mid \phi_i^{(0)}(\mathbf{r}) \mid^2 \quad (7.8)$$

$$n^{(1)}(\mathbf{r}) = \sum_i f_i [\phi_i^{(0)\star}(\mathbf{r})\phi_i^{(1)}(\mathbf{r}) + \phi_i^{(1)\star}(\mathbf{r})\phi_i^{(0)}(\mathbf{r})] , \quad (7.9)$$

respectively. Further, the zeroth-order Kohn–Sham Hamiltonian

$$H_{\text{KS}}^{(0)}[\{\phi_i^{(0)}\}] = \frac{1}{2}\nabla^2 + V_{\text{ext}}^{(0)}(\mathbf{r}) + \int \frac{n^{(0)}(\mathbf{r}')}{\mid \mathbf{r} - \mathbf{r}' \mid} d\mathbf{r}' + \frac{\delta E_{\text{xc}}}{\delta n(\mathbf{r})} \bigg|_{n^{(0)}} , \quad (7.10)$$

the combined Hartree and exchange–correlation energy

$$E_{\text{Hxc}}[n] = \frac{1}{2} \int \int \frac{n^{(0)}(\mathbf{r})n^{(0)}(\mathbf{r}')}{\mid \mathbf{r} - \mathbf{r}' \mid} d\mathbf{r}' d\mathbf{r} + E_{\text{xc}}[n] , \quad (7.11)$$

and the second-order energy kernel

$$\mathcal{K}(\mathbf{r}, \mathbf{r}') = \frac{\delta^2 E_{\text{Hxc}}[n]}{\delta n(\mathbf{r})\delta n(\mathbf{r}')} \bigg|_{n^{(0)}} . \quad (7.12)$$

have been introduced above.

Since the second-order energy $E^{(2)}$ is variational with respect to $\phi_i^{(1)}$ the Euler–Lagrange, or in this case self-consistent Sternheimer equations [537, 1399] can be deduced

$$\mathcal{Q}\sum_j \left(H_{\text{KS}}^{(0)}\delta_{ij} + \Lambda_{ij}^{(0)} \right) \mathcal{Q}\mathcal{P}_c \mid \phi_j^{(1)} \rangle = -\mathcal{Q}H_{\text{KS}}^{(1)} \mid \phi_i^{(0)} \rangle , \quad (7.13)$$

where $\mathcal{Q}$ is the projector upon the unoccupied states

$$\mathcal{Q} = 1 - \sum_i \mid \phi_i^{(0)} \rangle \langle \phi_i^{(0)} \mid \quad (7.14)$$

and the first-order Hamiltonian $H_{\text{KS}}^{(1)}$ is given by

$$H_{\text{KS}}^{(1)} = \frac{\delta E^{\text{pert}}}{\delta \langle \phi_i \mid} + \int \mathcal{K}(\mathbf{r}, \mathbf{r}') \, n^{(1)}(\mathbf{r}') \, d\mathbf{r}' + \frac{d}{d\lambda} \frac{\delta E_{\text{Hxc}}}{\delta n(\mathbf{r})} \bigg|_{n^{(0)}} . \quad (7.15)$$

Equation (7.13) can either be considered as a set of equations for $\{\phi_i^{(1)}\}$ that have to be solved self-consistently, or as one linear system of size $N_b \times M$, where N_b is the number of occupied orbitals and M the number of basis functions.

The evaluation of the kernel function $\mathcal{K}(\mathbf{r}, \mathbf{r}')$ requires the second functional derivative of the exchange and correlation functionals. In the case of LDA functionals this is easily done, however, its evaluation is more complex for gradient-corrected functionals [381]. The integral to be calculated is

$$\int \frac{\delta^2 E_{\mathrm{xc}}[n, |\nabla n|]}{\delta n(\mathbf{r})\delta n(\mathbf{r}')}\bigg|_{n^{(0)}(\mathbf{r})} n^{(1)}(\mathbf{r}')d\mathbf{r}' =$$

$$\int \frac{\delta V_{\mathrm{xc}}[n, |\nabla n|](\mathbf{r})}{\delta n(\mathbf{r}')}\bigg|_{n^{(0)}(\mathbf{r})} n^{(1)}(\mathbf{r}')d\mathbf{r}' \ , \quad (7.16)$$

where $E_{\mathrm{xc}}[n, |\nabla n|]$ is a (semi-)local functional that depends on the density n and its gradient $|\nabla n|$ at position $\mathbf{r}$. For this calculation the exchange–correlation potential has to be considered to depend only parametrically on $\mathbf{r}$. With such an exchange–correlation functional of the general form

$$E_{\mathrm{xc}}[n] = \int F_{\mathrm{xc}}(n(\mathbf{r}), |\nabla n|(\mathbf{r})) \ d\mathbf{r} \ , \quad (7.17)$$

where F_{xc} is just a function of $n(\mathbf{r})$ and $|\nabla n|(\mathbf{r})$, the associated potential becomes

$$V_{\mathrm{xc}}[n, |\nabla n|](\mathbf{r}) = \frac{\partial F_{\mathrm{xc}}(\mathbf{r})}{\partial n(\mathbf{r})} - \nabla \frac{\partial F_{\mathrm{xc}}(\mathbf{r})}{\partial |\nabla n(\mathbf{r})|} \frac{\nabla n(\mathbf{r})}{|\nabla n(\mathbf{r})|} \quad (7.18)$$

at point $\mathbf{r}$. The functional derivative $\delta F/\delta f$ of a functional $F[f]$ is defined as follows:

$$\int \frac{\delta F}{\delta f(\mathbf{r})} u(\mathbf{r}) \ d\mathbf{r} = \frac{d}{d\varepsilon} F[f(\mathbf{r}) + \varepsilon u(\mathbf{r})]\bigg|_{\varepsilon=0} \ , \quad (7.19)$$

with ε being an infinitesimal dimensionless parameter, which leads to

$$\int \frac{\delta^2 E_{\mathrm{xc}}[n, |\nabla n|]}{\delta n(\mathbf{r})\delta n(\mathbf{r}')}\bigg|_{n^{(0)}(\mathbf{r})} n^{(1)}(\mathbf{r}')d\mathbf{r}' =$$

$$\frac{d}{d\varepsilon} V_{\mathrm{xc}}[n + \varepsilon n^{(1)}, |\nabla n + \varepsilon \nabla n^{(1)}|]\bigg|_{\varepsilon=0, n=n^{(0)}} =$$

$$\frac{\partial}{\partial n} V_{\mathrm{xc}}[n, |\nabla n|]n^{(1)} + \frac{\partial}{\partial \nabla n} V_{\mathrm{xc}}[n, |\nabla n|]\nabla n^{(1)} \quad (7.20)$$

when applied to this exchange–correlation potential. Using furthermore the identity $\partial/\partial\nabla n = \partial/\partial|\nabla n| \cdot (\nabla n/|\nabla n|)$, Eq. (7.20) results in

$$\int \frac{\delta^2 E_{\mathrm{xc}}[n, |\nabla n|]}{\delta n(\mathbf{r})\delta n(\mathbf{r}')}\bigg|_{n^{(1)}(\mathbf{r})} n^{(1)}(\mathbf{r}')d\mathbf{r}' =$$

$$\left\{ \frac{\partial^2 F_{\mathrm{xc}}}{\partial n^2}n^{(1)} + \frac{\partial^2 F_{\mathrm{xc}}}{\partial n\partial|\nabla n|}\frac{\nabla n}{|\nabla n|}\nabla n^{(1)} \right.$$

$$- \nabla\left[\frac{\partial^2 F_{\mathrm{xc}}}{\partial n\partial|\nabla n|}\frac{\nabla n}{|\nabla n|}n^{(1)} \frac{\partial^2 F_{\mathrm{xc}}}{\partial|\nabla n|^2}\frac{\nabla n}{|\nabla n|}\frac{\nabla n}{|\nabla n|}\nabla n^{(1)} \right.$$

$$\left. \left. + \frac{\partial F_{\mathrm{xc}}}{\partial|\nabla n|}\left(\frac{\nabla n^{(1)}}{|\nabla n|} - \frac{\nabla n}{|\nabla n|^3}\nabla n \cdot \nabla n^{(1)} \right) \right] \right\}\bigg|_{n=n^{(0)}} \quad, \quad (7.21)$$

which is the final analytic expression. The evaluation of these terms is rather cumbersome and alternatively, a finite difference approach can be used

$$\frac{1}{2}\int\int n^{(1)}(\mathbf{r}')\frac{\delta^2 E_{\mathrm{xc}}}{\delta n(\mathbf{r}')\delta n(\mathbf{r})}n^{(1)}(\mathbf{r}) \; d\mathbf{r}d\mathbf{r}'$$

$$= \frac{1}{2}\lim_{\varepsilon\to 0}\frac{1}{\varepsilon^2}\left[E_{\mathrm{xc}}\left[n^{(0)}(\mathbf{r}) + \varepsilon n^{(1)}(\mathbf{r})\right] \right.$$

$$\left. + E_{\mathrm{xc}}\left[n^{(0)}(\mathbf{r}) - \varepsilon n^{(1)}(\mathbf{r})\right] - 2E_{\mathrm{xc}}\left[n^{(0)}(\mathbf{r})\right] \right] \quad (7.22)$$

and therefore

$$V_{\mathrm{xc}}^{(1)} = \lim_{\varepsilon\to 0}\frac{V_{\mathrm{xc}}\left(n^{(0)} + \varepsilon n^{(1)}\right) - V_{\mathrm{xc}}\left(n^{(0)} - \varepsilon n^{(1)}\right)}{2\varepsilon} \quad . \quad (7.23)$$

In Eq. (7.23), accuracy requires the use of a small ε, but on the other hand, numerical stability favors a large ε. In practice, a reasonable compromise can be achieved with a value of $\varepsilon \approx 10^{-3}$.

A comparison of an analytic implementation and the numerical differentiation has been done by Egli and Billeter [381]. At variance with former findings [284], it is concluded in the more recent work that the implementation of the analytical formulas is numerically more stable without loss of performance.

7.1.3 Nuclear Hessian

Harmonic vibrational frequencies of a system can be calculated as the eigenvalues of the dynamical matrix

$$D_{I\alpha,J\beta} = \frac{1}{\sqrt{M_I M_J}} \frac{\partial^2 E_{\mathrm{KS}}}{\partial R_{I\alpha} \partial R_{J\beta}} \ , \tag{7.24}$$

where the different ions are labeled $I, J = 1, \ldots, N_{\mathrm{at}}$ and the Cartesian coordinates with index $\alpha, \beta = 1, 2, 3$. This can be calculated using density functional perturbation theory within the general framework outlined in the previous section. To this end, one has to consider $3N_{\mathrm{at}}$ perturbations, one for each small displacement of the atom I from its equilibrium position $\mathbf{R}_I$ in the direction α. Thus, the perturbative functional in Eq. (7.1) is given by

$$E^{\mathrm{pert}}[\{\phi_i\}] = \frac{\partial E_{\mathrm{ion}}}{\partial R_{I\alpha}} + \sum_k f_k \int \phi_k^*(\mathbf{r}) \frac{\partial V_{\mathrm{ext}}(\mathbf{r})}{\partial R_{I\alpha}} \phi_k(\mathbf{r}) \, d\mathbf{r} \ , \tag{7.25}$$

where V_{ext} describes the ionic Coulomb potential assuming norm-conserving pseudopotentials of the sort

$$V_{\mathrm{ext}}(\mathbf{r}) = V_{\mathrm{loc}}(\mathbf{r}) + \sum_L | p_L \rangle \omega_L \langle p_L | \ , \tag{7.26}$$

with only one projector per angular momentum channel. This form of separable nonlocal pseudopotential was introduced in Chapter 3, see Eq. (3.31) and is discussed in detail in Chapter 4, see Eq. (4.49). The first term in Eq. (7.25) is due to the ionic interaction and can be calculated easily from the expressions for E_{ion}, see Section 3.2. For the simple case of a single projector the general perturbation functional is

$$E^{\mathrm{pert}}[n] = \frac{\partial E_{\mathrm{ion}}}{\partial R_{I\alpha}} + \int \frac{\partial V_{\mathrm{loc}}(\mathbf{r})}{\partial R_{I\alpha}} n(\mathbf{r}) \, d\mathbf{r}$$

$$+ \sum_L \sum_k f_k \left(\left\langle \phi_k \left| \frac{\partial p_L}{\partial R_{I\alpha}} \right\rangle \omega_L \langle p_L | \phi_k \rangle + \langle \phi_k | p_L \rangle \omega_L \left\langle \frac{\partial p_L}{\partial R_{I\alpha}} \middle| \phi_k \right\rangle \right) \ .$$

The electronic structure term of the dynamic matrix can be calculated using the Hellmann–Feynman theorem generalized to nonlocal potentials

$$\frac{\partial E_{\mathrm{KS}}}{\partial R_{I\alpha}} = \frac{\partial E_{\mathrm{ion}}}{\partial R_{I\alpha}} + \sum_k f_k \left\langle \phi_k^{(0)} \middle| \frac{\partial V_{\mathrm{ext}}}{\partial R_{I\alpha}} \middle| \phi_k^{(0)} \right\rangle \tag{7.27}$$

and taking the second derivative

$$\frac{\partial^2 E^{KS}}{\partial R_{I\alpha} \partial R_{J\beta}} = \frac{\partial^2 E_{ion}}{\partial R_{I\alpha} \partial R_{J\beta}} + \sum_k \left(\left\langle \phi_k^{(0)} \left| \frac{\partial^2 V_{ext}}{\partial R_{I\alpha} \partial R_{J\beta}} \right| \phi_k^{(0)} \right\rangle \right.$$

$$\left. + \left\langle \phi_{k,I\alpha}^{(1)} \left| \frac{\partial V_{ext}}{\partial R_{J\beta}} \right| \phi_k^{(0)} \right\rangle + \left\langle \phi_k^{(0)} \left| \frac{\partial V_{ext}}{\partial R_{J\beta}} \right| \phi_{k,I\alpha}^{(1)} \right\rangle \right) . \quad (7.28)$$

This formalism allows us to compute the dynamical matrix analytically within the framework of density functional perturbation theory.

7.1.3.1 Selected eigenmodes of the Hessian

Harmonic vibrational frequencies and normal modes are calculated from the dynamical matrix. If only part of the eigenvalues are needed [319, 435, 436, 1207], an iterative diagonalization scheme, e.g. Lanczos [827] or Davidson [310] diagonalization could be used. This might be useful in geometry optimization or in the location of saddle points as well as when only certain types of vibrations are of interest, for instance surface modes or hydrogen stretching modes. Iterative diagonalization schemes are based on matrix vector multiplication

$$\mathbf{a} = \mathbf{H}\mathbf{q} \quad (7.29)$$

in quite general terms. This is of special interest when the calculation of the matrix $\mathbf{H}$ can be avoided and only an algorithm for the application of a vector to the matrix has to be available.

Consider a collective displacement [435]

$$\mathbf{q} = \sum_{I,\alpha} q_{I\alpha} \mathbf{e}_{I\alpha} , \quad (7.30)$$

where the indices I and α run over all atoms and coordinates x, y, z respectively, and $\mathbf{e}_{I\alpha}$ is a $3N_{at}$ Cartesian unit vector. The matrix vector multiplication then becomes

$$\mathbf{a}_{J\beta} = \sum_{I,\alpha} \frac{\partial^2 E_{KS}}{\partial \mathbf{R}_{J\beta} \partial \mathbf{R}_{I\alpha}} q_{I\alpha} = \frac{\partial^2 E_{KS}}{\partial \mathbf{R}_{J\beta} \partial q} . \quad (7.31)$$

This can be calculated with density functional perturbation theory, which can be used to calculate the response of the Kohn–Sham orbitals to the collective displacement $q_{I,\alpha}$

$$\mathbf{a}_{J\beta} = \sum_{I,\alpha} q_{I\alpha} \frac{\partial^2 E_{\mathrm{ion}}}{\partial R_{I\alpha} \partial R_{J\beta}} + \sum_k f_k \sum_{I\alpha} q_{I\alpha} \left\langle \phi_k^{(0)} \right| \frac{\partial^2 V_{\mathrm{ext}}}{\partial R_{I\alpha} \partial R_{J\beta}} \left| \phi_k^{(0)} \right\rangle$$

$$+ \sum_k \left(\left\langle \phi_{k,q}^{(1)} \right| \frac{\partial V_{\mathrm{ext}}}{\partial R_{J\beta}} \left| \phi_k^{(0)} \right\rangle + \left\langle \phi_k^{(0)} \right| \frac{\partial V_{\mathrm{ext}}}{\partial R_{J\beta}} \left| \phi_{k,q}^{(1)} \right\rangle \right) . \quad (7.32)$$

7.1.4 Polarizability

An extension of density functional perturbation theory [1171] also allows us to handle cases where the perturbation cannot be expressed in a Hamiltonian form. One such case [1170] is an external electric field which couples with the electric polarization $\mathbf{P}^{\mathrm{el}}$ in a periodic system. Using the modern theory of polarization [740, 1210–1212, 1375] in the Γ-point-only sampling of the Brillouin zone (see Section 5.3.6), this perturbation can be written [1329] in terms of the Berry phase

$$\varphi_\mu = \mathrm{Im} \log \det \mathbf{Q}^{(\mu)} , \quad (7.33)$$

using the Kohn–Sham orbitals, where the matrix $\mathbf{Q}^{(\mu)}$ is defined as

$$\mathbf{Q}_{ij}^{(\mu)} = \langle \phi_i | \exp\left[i\mathbf{G}_\mu \cdot \mathbf{r}\right] | \phi_j \rangle \quad (7.34)$$

and $\mathbf{G}_\mu$ is the smallest vector in a periodically repeated cubic cell in the direction μ, see also Section 7.2.2. This formula is in principle valid in the limit of an infinitely large cell, but in a non-conducting material this is a good approximation even with relatively small supercells and Eq. (7.33) can also be generalized to cells of arbitrary shape [109]. Using the Berry phase the μ-component of the polarization is given by (cell volume Ω)

$$P_\mu^{\mathrm{el}} = \frac{2\,|e|}{|G_\mu|\,\Omega} \varphi_\mu , \quad (7.35)$$

which induces a perturbation in the Kohn–Sham functional of the type

$$\lambda E^{\mathrm{pert}}\left[\{|\phi_i\rangle\}\right] = -\sum_\nu E_\nu^{\mathrm{ext}} \frac{2\,|e|}{|G_\nu|\,\Omega} \mathrm{Im} \log \det \mathbf{Q}^{(\nu)} \quad (7.36)$$

via the coupling of the polarization to the external electric field $\mathbf{E}^{\mathrm{ext}}$ in the dipole approximation, see Section 5.3.6. In this case the perturbative parameter is the field component E_ν^{ext}, and thus

$$|\phi_i\rangle \simeq |\phi_i^{(0)}\rangle - E_\nu^{\mathrm{ext}} |\phi_i^{\nu\,(1)}\rangle . \quad (7.37)$$

The derivative $\delta E^{\mathrm{pert}} / \delta \langle \phi_i^{(0)} |$ and its conjugate ket can be evaluated using the formula for the derivative of a matrix $\mathbf{A}$ with respect to a generic variable x

$$\frac{d}{dx} (\ln \det \mathbf{A}) = \sum_{ij} \frac{dA_{ij}}{dx} A_{ji}^{-1}. \qquad (7.38)$$

The perturbative term in the case of polarization is therefore

$$\frac{2|e|}{|\mathbf{G}_\nu|} \mathrm{Im} \left[\sum_{ij} \left(\left\langle \phi_i^{\nu\,(1)} \left| \exp\left[i\mathbf{G}_\nu \cdot \mathbf{r}\right] \right| \phi_j^{(0)} \right\rangle \right. \right.$$

$$\left. \left. + \left\langle \phi_i^{(0)} \left| \exp\left[i\mathbf{G}_\nu \cdot \mathbf{r}\right] \right| \phi_j^{\nu\,(1)} \right\rangle \right) \left(\mathbf{Q}_{ji}^{(\nu)} \right)^{-1} \right]. \qquad (7.39)$$

Using this term one can calculate the first-order correction to the set of Kohn–Sham orbitals $\{\phi_i^{(1)}\}$. This allows us to evaluate the induced polarization and thus the dipole moment in the μ-direction

$$\delta P_\mu^{\mathrm{el}} \;=\; \frac{2|e|}{|\mathbf{G}_\mu|\Omega} \, \delta\phi_\mu$$

$$= \; -\sum_\nu \frac{2|e|}{|\mathbf{G}_\nu|\Omega} \mathrm{Im} \left[\sum_{ij} \left(\left\langle \phi_i^{\nu\,(1)} \left| \exp\left[i\mathbf{G}_\mu \cdot \mathbf{r}\right] \right| \phi_j^{(0)} \right\rangle \right. \right.$$

$$\left. \left. + \left\langle \phi_i^{(0)} \left| \exp\left[i\mathbf{G}_\mu \cdot \mathbf{r}\right] \right| \phi_j^{\nu\,(1)} \right\rangle \right) \left(\mathbf{Q}_{ji}^{(\nu)} \right)^{-1} \right] E_\nu^{\mathrm{ext}} \qquad (7.40)$$

as well as the polarizability tensor $\boldsymbol{\alpha}$

$$\alpha_{\mu\nu} = \frac{2|e|}{|\mathbf{G}_\nu|\Omega} \mathrm{Im} \left[\sum_{ij} \left(\left\langle \phi_i^{\nu\,(1)} \left| \exp\left[i\mathbf{G}_\mu \cdot \mathbf{r}\right] \right| \phi_j^{(0)} \right\rangle \right. \right.$$

$$\left. \left. + \left\langle \phi_i^{(0)} \left| \exp\left[i\mathbf{G}_\mu \cdot \mathbf{r}\right] \right| \phi_j^{\nu\,(1)} \right\rangle \right) \left(\mathbf{Q}_{ji}^{(\nu)} \right)^{-1} \right] \qquad (7.41)$$

according to its definition $\alpha_{\mu\nu} = -\partial P_\mu^{\mathrm{el}}/\partial E_\nu^{\mathrm{ext}}$ (see also Section 5.3.6).

7.1.5 NMR chemical shifts

The *ab initio* calculation of chemical shifts has become more and more popular, and over the years many methods have been developed in the quantum chemistry community to perform such computations. Good reviews [623, 731] of the various approaches and recent developments in this field are available. One major problem that appears in these calculations is the choice of the gauge. While being in principle a cyclic variable, the choice of gauge can significantly affect the results in an actual calculation. To minimize this effect, several solutions have been proposed. In the GIAO method (Gauge Including Atomic Orbitals [347]), one transforms the gauge of the basis set functions to the position of their nuclei, whereas in the IGLO method (Individual Gauges for Localized Orbitals [808]), the gauges of the final wave functions are transformed to their centers of charge. The CSGT method (Continuous Set of Gauge Transformations [733]) finally defines a gauge which depends on the position where the induced current is to be calculated. However, there is another issue that restricts the applicability of the existing implementations of these methods to isolated systems. The Hamiltonian which represents the magnetic field contains the position operator. In an extended system, which would typically be treated under periodic boundary conditions, this operator is ill-defined as discussed in general terms in Section 7.2.2. In particular, the position operator and therefore the perturbation Hamiltonian operator do not have any periodicity as would be required for periodic boundary conditions. Practical solutions to both the gauge and position operator problems will be presented in the following, after having defined the chemical shift and magnetic susceptibility tensors.

Mauri *et al.* [975, 976] have presented a formalism which allows the calculation of chemical shifts and other magnetic properties in extended systems using periodic boundary conditions. This formulation is based on a magnetic field which is modulated in space. To return to the experimental situation of a homogeneous field, the limit of infinite modulation wavelength is evaluated numerically, which is achieved by using a small but finite wave vector. An alternative method for extended systems in periodic boundary conditions proposed by Sebastiani and Parrinello [1304] takes advantage of the exponential decay properties of localized Wannier orbitals [764, 1594] (see Section 7.2) and treats these localized orbitals as virtually independent. For the gauge problem, a particular variant of the CSGT method [733] mentioned above is adapted to these localized orbitals.

7.1.5.1 Chemical shifts and susceptibilities

When a magnetic field is applied to a medium, it induces a current due to the modification of the electronic ground state. This electronic current distribution induces an additional inhomogeneous magnetic field. The chemical shift tensor is defined as the proportionality factor between the induced and the externally applied magnetic field at the positions of the nuclei

$$\sigma(\mathbf{R}) = \frac{\partial \mathbf{B}^{\mathrm{ind}}(\mathbf{R})}{\partial \mathbf{B}^{\mathrm{ext}}} \quad . \tag{7.42}$$

The induced field is determined by the total electronic current $\mathbf{j}(\mathbf{r})$ through

$$\mathbf{B}^{\mathrm{ind}}(\mathbf{r}) = \frac{\mu_0}{4\pi} \int \frac{\mathbf{r}' - \mathbf{r}}{|\mathbf{r}' - \mathbf{r}|^3} \times \mathbf{j}(\mathbf{r}') \, d\mathbf{r}' \quad , \tag{7.43}$$

where μ_0 is the vacuum permeability. In periodic systems also the current density will be periodic and one can evaluate Eq. (7.43) in reciprocal space from the Fourier transform of the current

$$\mathbf{B}^{\mathrm{ind}}(\mathbf{G} \neq 0) = -\mu_0 \, i \, \frac{\mathbf{G}}{|\mathbf{G}|^2} \times \mathbf{j}(\mathbf{G}) \quad . \tag{7.44}$$

The $\mathbf{G} = \mathbf{0}$ component of the field depends on the bulk magnetic susceptibility tensor, $\boldsymbol{\chi}$, and on the shape of the sample

$$\mathbf{B}^{\mathrm{ind}}(\mathbf{G} = 0) = \kappa \, \boldsymbol{\chi} \, \mathbf{B}^{\mathrm{ext}} \quad , \tag{7.45}$$

where the numerical prefactor is given by $\kappa = 2/3$ for the special case of a spherical system. The bulk susceptibility $\boldsymbol{\chi}$ can also be expressed as a function of the orbital electronic current as

$$\boldsymbol{\chi} = \frac{\mu_0}{2\Omega} \frac{\partial}{\partial \mathbf{B}^{\mathrm{ext}}} \int_{\Omega} \mathbf{r} \times \mathbf{j}(\mathbf{r}) \, d\mathbf{r} \quad , \tag{7.46}$$

where the integral is over one unit cell of volume Ω. A single cell is sufficient since the integral is invariant under translations of any lattice vector $\mathbf{L}$ because of

$$\mathbf{j}(\mathbf{r} + \mathbf{L}) = \mathbf{j}(\mathbf{r}) \qquad \text{and} \qquad \int_{\Omega} \mathbf{j}(\mathbf{r}) \, d\mathbf{r} = 0 \quad . \tag{7.47}$$

Therefore, the integral over the complete sample can be written as the sum of integrals over unit cells, and all these integrals are equal. The molar susceptibility is related to $\boldsymbol{\chi}$ through $\chi^{\mathrm{m}} = \Omega N_{\mathrm{A}} \chi$, with Avogadro's number N_{A}.

The standard procedure to obtain the orbital electronic current density $\mathbf{j}$ is perturbation theory. The field $\mathbf{B}^{\mathrm{ext}}$ is represented by a vector potential $\mathbf{A}$ satisfying $\mathbf{B}^{\mathrm{ext}} = \nabla \times \mathbf{A}(\mathbf{r})$. A typical choice for $\mathbf{A}$ in the case of a homogeneous magnetic field is

$$\mathbf{A}(\mathbf{r}) = -\frac{1}{2} \, (\mathbf{r} - \mathbf{R}^{\mathrm{go}}) \times \mathbf{B}^{\mathrm{ext}} \tag{7.48}$$

with a cyclic variable $\mathbf{R}^{go}$, the "gauge origin". The perturbation Hamiltonians at first and second order in the field strength are given by

$$H_{KS}^{(1)} = \frac{-e}{m_e} \mathbf{p} \cdot \mathbf{A}(\mathbf{r}) \tag{7.49}$$

$$H_{KS}^{(2)} = \frac{e^2}{2m_e} \mathbf{A}(\mathbf{r}) \cdot \mathbf{A}(\mathbf{r}), \tag{7.50}$$

with the momentum operator $\mathbf{p}$ and the charge e and mass m_e of the electron, see also Section 5.3.6. The first-order perturbation gives rise to a correction in the electronic ground state

$$\phi = \phi^{(0)} + B_\mu^{ext} \phi^{(1)} \tag{7.51}$$

with respect to the unperturbed system. This correction $\phi^{(1)}$ is responsible for the induced current, which can be obtained as

$$\mathbf{j}(\mathbf{r}') = \frac{e^2}{m_e} \mathbf{A}(\mathbf{r}') |\phi^{(0)}(\mathbf{r}')|^2 + \frac{-e}{m_e} \langle \phi^{(0)} | \left[\mathbf{p} | \mathbf{r}' \rangle \langle \mathbf{r}' | + | \mathbf{r}' \rangle \langle \mathbf{r}' | \mathbf{p} \right] | \phi^{(1)} \rangle . \tag{7.52}$$

7.1.5.2 The gauge origin problem

The current density, Eq. (7.52), written in terms of the single orbital contributions $\mathbf{j}_k$, can be separated into dia- and paramagnetic terms

$$\mathbf{j}(\mathbf{r}') = \sum_k \mathbf{j}_k(\mathbf{r}') = \sum_k \mathbf{j}_k^{dia}(\mathbf{r}') + \mathbf{j}_k^{para}(\mathbf{r}')$$

$$\mathbf{j}_k^{dia}(\mathbf{r}') = \frac{e^2}{m_e} \mathbf{A}(\mathbf{r}') |\phi_k^{(0)}(\mathbf{r}')|^2$$

$$\mathbf{j}_k^{para}(\mathbf{r}') = \frac{-e}{m_e} \langle \phi_k^{(0)} | \left[\mathbf{p} | \mathbf{r}' \rangle \langle \mathbf{r}' | + | \mathbf{r}' \rangle \langle \mathbf{r}' | \mathbf{p} \right] | \phi_k^{(1)} \rangle . \tag{7.53}$$

Both contributions individually depend on the gauge, whereas the total current $\mathbf{j}$ is gauge-independent. However, the two contributions are large numbers and have opposite signs. For the present choice of the vector potential, Eq. (7.48), $\mathbf{A}(\mathbf{r})$ is linear in the gauge origin $\mathbf{R}^{go}$. Therefore, the diamagnetic current $\mathbf{j}_k^{dia}$ grows linearly in $\mathbf{R}^{go}$ so that $\mathbf{j}_k^{para}$ must compensate for this in order to fulfill the invariance of the total current. Thus, for large distances $|\mathbf{r} - \mathbf{R}^{go}|$, the total current density $\mathbf{j}$ results from the cancellation of two large terms.

Many techniques have been developed to minimize this problem for isolated molecules [347, 733, 808]. For periodic systems, probably the most natural approach is the so-called "$\mathbf{R}^{go} = \mathbf{r}$"-variant of the CSGT method [733]. For each point $\mathbf{r}'$ in space, the current density is calculated with the gauge origin $\mathbf{R}^{\text{go}}$ being set equal to $\mathbf{r}'$. This method makes the diamagnetic part vanish analytically

$$\mathbf{j}_k^{\text{dia}}(\mathbf{r}') = \mathbf{0} \; , \tag{7.54}$$

such that cancellations of large numbers no longer occur. In practice, the current is computed as

$$\mathbf{j}_k(\mathbf{r}') = \frac{-e}{m_{\text{e}}} \langle \phi_k^{(0)} | \left(\mathbf{p}|\mathbf{r}'\rangle\langle\mathbf{r}'| + |\mathbf{r}'\rangle\langle\mathbf{r}'|\mathbf{p} \right) \left[|\phi_k^{\mathbf{r}\times\mathbf{p}}\rangle - \mathbf{r}' \times |\phi_k^{\mathbf{p}}\rangle \right] \cdot \mathbf{B}^{\text{ext}} \; . \tag{7.55}$$

Here, $|\phi_k^{\mathbf{r}\times\mathbf{p}}\rangle$ and $|\phi_k^{\mathbf{p}}\rangle$ are the first-order perturbation orbitals for the special perturbation Hamiltonians

$$|\phi_k^{\mathbf{r}\times\mathbf{p}}\rangle \;\;\mapsto\;\; H_{\text{KS}}^{(1)} = \mathbf{r} \times \mathbf{p} \tag{7.56}$$

$$|\phi_k^{\mathbf{p}}\rangle \;\;\mapsto\;\; H_{\text{KS}}^{(1)} = \mathbf{p} \; . \tag{7.57}$$

This formulation avoids actually calculating distinct orbitals $\phi^{(1)}$ for each point $\mathbf{r}'$ in space. Denoting the perturbation theory Green's function by

$$\mathcal{G}_{lk} = - \left(H_{\text{KS}}^{(0)} \delta_{kl} - \langle \phi_k^{(0)} | H_{\text{KS}}^{(0)} | \phi_l^{(0)} \rangle \right)^{-1} \; , \tag{7.58}$$

one can formally express the first-order perturbation orbitals for an arbitrary perturbation operator $\mathcal{O}$ as

$$|\phi_k^{\mathcal{O}}\rangle = \sum_l \mathcal{G}_{kl} \, \mathcal{O} \, |\phi_l^{(0)}\rangle \tag{7.59}$$

where $\mathcal{O}$ is either $\mathbf{p}$ or $\mathbf{r} \times \mathbf{p}$. By expanding Eq. (7.59) in the basis of the unperturbed unoccupied orbitals, one would obtain the well-known sum-over-states expression for the first-order perturbed wave function.

This Green's function formulation is not used in the actual calculation, but rather one performs a variational energy minimization, see Section 7.1.2. Equation (7.59) serves only as a compact notation to obtain a closed expression for the current density

$$\mathbf{j}_k(\mathbf{r}') = \frac{-e}{m_{\text{e}}} \sum_l \langle \phi_k^{(0)} | \left(\mathbf{p}|\mathbf{r}'\rangle\langle\mathbf{r}'| + |\mathbf{r}'\rangle\langle\mathbf{r}'|\mathbf{p} \right)$$

$$\left[\mathcal{G}_{kl} \, (\mathbf{r} \times \mathbf{p}) \, |\phi_l^{(0)}\rangle - \mathcal{G}_{kl} \, (\mathbf{r}' \times \mathbf{p}) \, |\phi_l^{(0)}\rangle \right] \cdot \mathbf{B}^{\text{ext}} \; . \tag{7.60}$$

In this formulation, it becomes apparent that any simultaneous translations of the relative origin for the operator $\mathbf{r}$ and the gauge $\mathbf{R}^{\text{go}} = \mathbf{r}'$ automatically cancel each other out. In particular, the current is invariant under arbitrary orbital-specific translations $\mathbf{d}_l$

$$\mathbf{j}_k(\mathbf{r}') = \frac{-e}{m_e} \sum_l \langle \phi_k^{(0)} | \left(\mathbf{p} | \mathbf{r}' \rangle \langle \mathbf{r}' | + | \mathbf{r}' \rangle \langle \mathbf{r}' | \mathbf{p} \right)$$

$$\left[\mathcal{G}_{kl} \left((\mathbf{r} - \mathbf{d}_l) \times \mathbf{p} \right) | \phi_l^{(0)} \rangle - \mathcal{G}_{kl} \left((\mathbf{r}' - \mathbf{d}_l) \times \mathbf{p} \right) | \phi_l^{(0)} \rangle \right] \cdot \mathbf{B}^{\text{ext}} \; . \quad (7.61)$$

The CSGT gauge still leaves the freedom to translate the coordinate system individually for each orbital according to Eq. (7.61).

A straightforward application of Eq. (7.61) would be too expensive computationally. In fact, it would require one inversion of the Hamiltonian for each real space mesh point $\mathbf{r}'$. However, the second term of $\mathbf{j}_k$ can be rewritten as

$$-\frac{-e}{m_e} \sum_l \langle \phi_k^{(0)} | \left(\mathbf{p} | \mathbf{r}' \rangle \langle \mathbf{r}' | + | \mathbf{r}' \rangle \langle \mathbf{r}' | \mathbf{p} \right) \mathcal{G}_{kl} \, (\mathbf{r}' - \mathbf{d}_l) \times \mathbf{p} \, | \phi_l^{(0)} \rangle \cdot \mathbf{B}^{\text{ext}}$$

$$= -\frac{-e}{m_e} \sum_l \langle \phi_k^{(0)} | \left(\mathbf{p} | \mathbf{r}' \rangle \langle \mathbf{r}' | + | \mathbf{r}' \rangle \langle \mathbf{r}' | \mathbf{p} \right) (\mathbf{r}' - \mathbf{d}_k) \times \mathcal{G}_{kl} \, \mathbf{p} \, | \phi_l^{(0)} \rangle \cdot \mathbf{B}^{\text{ext}}$$

$$+ \Delta \mathbf{j}_k(\mathbf{r}') \; , \quad (7.62)$$

where

$$\Delta \mathbf{j}_k(\mathbf{r}') = -\frac{-e}{m_e} \sum_l \langle \phi_k^{(0)} | \left(\mathbf{p} | \mathbf{r}' \rangle \langle \mathbf{r}' | + | \mathbf{r}' \rangle \langle \mathbf{r}' | \mathbf{p} \right)$$

$$\mathcal{G}_{kl} \, (\mathbf{d}_k - \mathbf{d}_l) \times \mathbf{p} \, | \phi_l^{(0)} \rangle \cdot \mathbf{B}^{\text{ext}} \; . \quad (7.63)$$

The evaluation of the first term of Eq. (7.62) can be done at the computational cost of one total energy calculation, while $\Delta \mathbf{j}_k$ requires one such calculation per electronic state k. At first sight, the sum $\Delta \mathbf{j} = \sum_k \Delta \mathbf{j}_k$ seems to be equal to zero, since the inner operator is antisymmetric in the k, l indices. However, since the momentum operators in Eq. (7.63) do not commute with the Green's function, $\Delta \mathbf{j}$ does not vanish unless all $\mathbf{d}_l$ are equal.

7.1.5.3 The position operator problem

In order to compute NMR chemical shifts with the procedure outlined above, the orbitals are first localized by means of a unitary rotation in the occupied subspace. The rotation is chosen such that the spatial extension of the wave

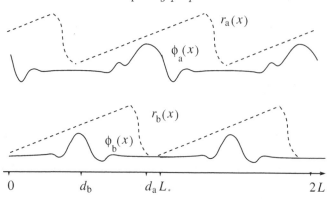

Fig. 7.1. Two localized orbitals with their corresponding position operators. Figure adapted from Ref. [1304]

functions is minimal, yielding maximally localized Wannier functions, see Section 7.2. It can be shown that in an insulator the resulting localized wave functions decay exponentially [764]. If the unit cell is chosen such that the lattice parameter is larger than the decay length, the orbital is significantly different from zero only within a limited region of the cell, and it practically vanishes everywhere else. The next step is to assign individual virtual cells to these Wannier orbitals. The virtual cells are chosen such that for the corresponding wave function, the cell borders are located in regions of space where the orbital density is close to zero. Then, the position operator is defined normally running from $-L/2$ to $+L/2$ inside the virtual cell. At the borders, it makes a smooth transition back from $+L/2$ to $-L/2$, yielding a sawtooth shape as depicted in Fig. 7.1. This jump is not sharp, in order to avoid components of very high frequency in the operator. As a consequence of this definition, the position operator now matches the periodic boundary conditions, since it is identical in every virtual cell and all its replicas.

This localization transformation does not affect the chosen orbital gauge. If it did, additional terms in the perturbation Hamiltonian and the wave function orthonormality relations, as in the IGLO method, would appear. The crucial difference is that no individual orbital gauge origins are used; the gauge is always "$\mathbf{R}^{go} = \mathbf{r}'$". Instead, an individual reference system is defined for both $\mathbf{r}$ and $\mathbf{R}^{go}$ simultaneously, as described by the relative origins $\mathbf{d}_k$ in Eq. (7.61).

The problem that arises for this construction is that the new operator has a completely unphysical shape around the borders of the virtual cell. By choosing the virtual cells as described above, the unphysical transitions lie in those regions of space where the wave function vanishes. As a consequence, the problematic part of the operator is only applied where it has no effect. Hence, the sawtooth shape of the position operator as indicated by Fig. 7.1 is a reasonable approximation as long as the orbitals are sufficiently localized. This represents a certain restriction for this method, namely, it is required that the decay length is significantly smaller than the lattice constant of the simulation box. Only in such a case can the virtual cell be chosen with its borders in a region of vanishing density. It follows that for a system with truly delocalized orbitals, like a metal, this approach is not applicable. In such systems, the decay of the Wannier orbitals is only algebraic and the necessary cell size would by far exceed the computationally tractable cell volume.

7.1.5.4 Density functional perturbation theory

At this stage, density functional perturbation theory as introduced in Section 7.1 can be used in order to calculate the response orbitals due to the external magnetic perturbation $\mathbf{B}^{\mathrm{ext}}$. The derivation has strong analogies with other variational schemes used in quantum chemistry. In particular, it is similar to the stationary perturbation theory by Kutzelnigg [810], which is used in the IGLO implementation. The starting point of the derivation is the functional for the second-order energy of the system, Eq. (7.6), which is variational in the first-order perturbation wave functions $\phi^{(1)}$

$$E^{(2)} = \sum_{kl} \langle \phi_k^{(1)} | H_{\mathrm{KS}}^{(0)} \delta_{kl} - \lambda_{kl} | \phi_l^{(1)} \rangle$$

$$+ \sum_k \left[\langle \phi_k^{(1)} | H_{\mathrm{KS}}^{(1)} | \phi_k^{(0)} \rangle + \langle \phi_k^{(0)} | H_{\mathrm{KS}}^{(1)} | \phi_k^{(1)} \rangle \right]$$

$$+ \frac{1}{2} \int \frac{\delta^2 E_{\mathrm{Hxc}}[n^{(0)}]}{\delta n(\mathbf{r}) \delta n(\mathbf{r}')} n^{(1)}(\mathbf{r}) n^{(1)}(\mathbf{r}') \, d\mathbf{r}' \, d\mathbf{r} \quad . \quad (7.64)$$

In the present case where the perturbation is a magnetic field, the energy functional simplifies considerably because the first-order density vanishes analytically everywhere. The reason is that the perturbation Hamiltonian and the first-order wave functions are purely imaginary, and thus, the two terms in the response density cancel. The matrix element of the magnetic perturbation Hamiltonian, Eq. (7.49), is given in the position representation by

$$\langle \mathbf{r} | H_{\mathrm{KS}}^{(1)} | \mathbf{r}' \rangle = i \, \frac{-e\hbar}{2m_{\mathrm{e}}} \, \delta(\mathbf{r} - \mathbf{r}') \, (\mathbf{r} - \mathbf{R}^{\mathrm{go}}) \times \mathbf{B}^{\mathrm{ext}} \cdot \nabla \quad . \quad (7.65)$$

It is purely imaginary so that with real wave functions and a necessarily real energy, the first-order orbitals $\phi_k^{(1)}$ must be purely imaginary, too. Hence, the first-order density vanishes analytically for magnetic perturbations and the energy functional, Eq. (7.64), simplifies to

$$E^{(2)} = \sum_{kl} \langle \phi_k^{(1)} | H_{KS}^{(0)} \delta_{kl} - \lambda_{kl} | \phi_l^{(1)} \rangle$$

$$+ \sum_k \left[\langle \phi_k^{(1)} | H_{KS}^{(1)} | \phi_k^{(0)} \rangle + \langle \phi_k^{(0)} | H_{KS}^{(1)} | \phi_k^{(1)} \rangle \right] . \quad (7.66)$$

The stationarity condition on the energy can be written as an inhomogeneous system of coupled equations

$$\sum_l \left(H_{KS}^{(0)} \delta_{kl} - \lambda_{kl} \right) | \phi_l^{(1)} \rangle = -H_{KS}^{(1)} | \phi_k^{(0)} \rangle \quad (7.67)$$

for the $\phi_k^{(1)}$.

7.1.5.5 Pseudopotential correction

In pseudopotential/plane wave calculations no core orbitals are taken into account and the valence wave functions have an incorrect shape in the core region (see Chapter 4). The chemical shift, on the other hand, is extremely sensitive to precisely that region of space because the interaction between nuclear spin and current is proportional to $1/r^2$. Thus, it is not clear *a priori* whether a pseudopotential implementation can give meaningful results at all. However, the contribution of the core orbitals to the chemical shift is often almost constant for hydrogen and first row elements with respect to the chemical environment of the atom. In a recent investigation of Gregor *et al.* [552], it has been shown that this property can be exploited to correct for the frozen-core approximation. It turns out that a simple additive constant is sufficient to reproduce the all-electron shieldings satisfactorily in many cases.

Finally, the gauge including projector augmented wave (GIPAW) method by Pickard and Mauri [1148], which is an extension of Blöchl's PAW method (see Chapter 6), solves the pseudopotential problem elegantly within the plane wave framework. As the PAW approach is not invariant with respect to gauge origin transformation in the presence of magnetic fields, it was necessary to introduce magnetic field-dependent phase factors. These phase factors are chosen in the spirit of the gauge including atomic orbital (GIAO) method.

7.2 Wannier functions: dipole moments, IR spectra, atomic charges

7.2.1 Introduction

In the independent-particle approximation the electronic ground state of a periodic system can be labeled naturally according to Bloch's theorem: single-particle wave functions (orbitals) are assigned a quantum number $\mathbf{k}$ for the crystal momentum together with a band index n as introduced in Section 3.1.2. However, alternative representations are available to this widely used choice. In the Wannier representation [1594], a real-space picture of localized orbitals assigns as quantum numbers the lattice vector $\mathbf{R}$ of the cell where the orbital is localized, together with an orbital index n.

Wannier functions are a powerful tool [959] in the study of the electronic and dielectric properties of materials. They are the condensed matter equivalent of localized molecular orbitals [184] and as such provide an insightful picture of the nature of chemical bonding, otherwise missing from the Bloch picture of extended orbitals; the latter translate into the set of canonical molecular orbitals for finite systems. By transforming the occupied electronic manyfold into a set of maximally localized Wannier functions (MLWFs), it becomes possible to obtain an enhanced understanding of chemical coordination and bonding properties. This is achieved *via* an analysis of factors such as changes in shape or symmetry of the maximally localized Wannier functions or changes in the locations of their centers of charge. The charge center of a maximally localized Wannier function provides a kind of classical correspondence for the location of an electron (or electron pair) in a quantum-mechanical insulator. This analogy is extended further by the modern theory of bulk polarization developed by Vanderbilt, Resta, and others [740, 1210–1212, 1375] (see also Section 5.3.6), which relates the vector sum of the centers of the Wannier functions to the macroscopic polarization of a crystalline insulator.

7.2.2 Position operator in periodic systems

The position operator in periodic systems has been investigated in detail by Resta [1211]. The position operator within the Schrödinger representation acts by multiplying the wave function by the space coordinate. This applies only to the bound eigenstates of a finite system which belong to the class of square-integrable wave functions. However, in condensed matter theory one usually considers a large system within periodic boundary conditions, and the position operator in its common form becomes meaningless. For the sake of simplicity, only the one-dimensional case will be dealt with in this presentation. The Hilbert space of the single-particle wave functions is defined by the condition $\phi(x+L) = \phi(x)$, where L is the imposed periodicity, chosen to be large with respect to atomic dimensions. An operator maps any vector of the given space into another vector belonging to the same space. The multiplicative position operator x is not a legitimate operator when periodic boundary conditions are adopted for the state vectors, since $x\phi(x)$ is not a periodic function whenever $\phi(x)$ is periodic. On the other hand, any periodic function of x is a legitimate multiplicative operator. This is the case, e.g., for the nuclear potential acting on the electrons.

Since the position operator is ill-defined, so is its expectation value, whose observable effects in condensed matter are related to macroscopic polarization (i.e. dipole moment density). For the crystalline case, the problem of dielectric polarization has been solved [740, 1210] using a Berry phase [1329]. It is an observable which cannot be cast as the expectation value of any operator, being instead a gauge-invariant phase of the wave function. Through this theory one arrives at defining the expectation value of the position in an extended quantum system within periodic boundary conditions. Among the most relevant features, the expectation value is defined modulo L, and the operator is no longer a single-particle operator: it acts as a genuine many-body operator on the periodic wave function of N electrons.

As a starting point serves the simpler case where periodic boundary conditions are not chosen and the N-particle ground state wave function, Ψ_0, goes to zero exponentially outside a bounded region in space. One may safely use the operator $\hat{X} = \sum_{i=1}^{N} x_i$, and define the position expectation value as

$$\langle X \rangle = \langle \Psi_0 \mid \hat{X} \mid \Psi_0 \rangle = \int x\, n(x)\, dx \ , \qquad (7.68)$$

where $n(x)$ is the one-particle density and x_i the position (operator) of the ith particle in one dimension. The value $\langle X \rangle$ scales with the system size and the quantity of interest is the dipole moment per unit length, which coincides with macroscopic polarization. The expectation value of this same operator cannot be evaluated if the wave function obeys periodic boundary conditions.

However, when periodic boundary conditions are adopted the position expectation value cannot be defined according to the straightforward generalization of Eq. (7.68) but through

$$\langle X \rangle = \frac{L}{2\pi} \text{Im} \ln \langle \Psi_0 \mid \exp\left[i\frac{2\pi}{L}\hat{X}\right] \mid \Psi_0 \rangle, \qquad (7.69)$$

which implies that the expectation value $\langle X \rangle$ is thus defined only modulo L. The right-hand side of Eq. (7.69) is not simply the expectation value of an operator, the form as the imaginary part of a logarithm is essential. Furthermore, its main ingredient is the expectation value of a multiplicative operator, $\exp\left[i\,2\pi \hat{X}/L\right]$. It is important to realize that this is a genuine many-body operator. An operator is defined to be one-body whenever it is the sum of N identical operators, acting on each electronic coordinate separately. In order to express the expectation value of such an operator the full many-body wave function is not needed. Instead, knowledge of the reduced density matrix $\rho(x, x')$ is sufficient.

The electronic polarization corresponding to this definition of the position expectation value is

$$P^{\text{el}} = \lim_{L \to \infty} \frac{-e}{2\pi} \text{Im} \ln \langle \Psi_0 \mid \exp\left[i\frac{2\pi}{L}\hat{X}\right] \mid \Psi_0 \rangle \ , \tag{7.70}$$

where $-e$ is the electron charge. For the special case of a system of non-interacting particles one can furthermore simplify the formulas. Suppose one deals with a crystalline system of lattice constant a where periodic boundary conditions over M linear cells are imposed. There are M equally spaced Bloch vectors in the reciprocal cell $[0, 2\pi/a)$,

$$q_s = \frac{2\pi}{Ma}s \ , \qquad\qquad s = 0, 1, \ldots, M - 1 \tag{7.71}$$

and the size of the periodically repeated system is $L = Ma$. The one-body orbitals can be chosen to have the Bloch form

$$\phi_{q_s,m}(x + \tau) = \exp\left[iq_s\tau\right]\phi_{q_s,m}(x) \ , \tag{7.72}$$

where $\tau = la$ is a lattice translation, and m is a band index. There are N/M occupied bands in a Slater determinant wave function, which is written as

$$|\Psi_0\rangle = \mathcal{A} \prod_{m=1}^{N/M} \prod_{s=0}^{M-1} \phi_{q_s,m}(x) \ , \tag{7.73}$$

where $\mathcal{A}$ is the antisymmetrizer. Next, a new set of Bloch orbitals is defined by

$$\tilde{\phi}_{q_s,m}(x) = \exp\left[-i\frac{2\pi}{L}x\right]\phi_{q_s,m}(x) \tag{7.74}$$

and the expectation value of Eq. (7.69) is recast as

$$\langle X \rangle = -\frac{L}{2\pi}\text{Im} \ln \langle \Psi_0 \mid \tilde{\Psi}_0 \rangle \ , \tag{7.75}$$

where $| \tilde{\Psi}_0 \rangle$ is the Slater determinant of the $\tilde{\phi}$'s. The overlap among two determinants is equal to the determinant of the overlap matrix among the orbitals

$$\langle X \rangle = -\frac{L}{2\pi} \text{Im ln det } \mathbf{S} \ , \tag{7.76}$$

where

$$S_{sm,s'm'} = \int_0^L \phi^*_{q_s,m}(x) \exp\left[-i\frac{2\pi}{L}x\right] \phi_{q_{s'},m'}(x) \, dx \ . \tag{7.77}$$

Owing to the orthogonality properties of the Bloch functions, the overlap matrix elements vanish except when $q_{s'} = q_s + 2\pi/L$, that is for $s' = s + 1$. The determinant of the $N \times N$ matrix can be factorized into M small determinants

$$\text{det } \mathbf{S} = \prod_{s=0}^{M-1} \text{det } \mathbf{S}(q_s, q_{s+1}) \ , \tag{7.78}$$

where for the elements of the small overlap matrix the notation

$$S_{m,m'}(q_s, q_{s+1}) = \int_0^L \phi^*_{q_s,m}(x) \exp\left[-i\frac{2\pi}{L}x\right] \phi_{q_{s'},m'}(x) \, dx \tag{7.79}$$

is used and $\phi^*_{q_M,m}(x) = \phi^*_{q_0,m}(x)$, the so-called periodic gauge, is implicitly understood. Using Eq. (7.70) one gets

$$P^{\text{el}} = -\frac{e}{2\pi} \lim_{L\to\infty} \text{Im ln} \prod_{s=0}^{M-1} \text{det } \mathbf{S}(q_s, q_{s+1}) \tag{7.80}$$

as the final result for the polarization.

In *ab initio* molecular dynamics simulation the periodicity of the wave functions and the potential are taken to be the same. This further approximation, i.e. using the Γ-point only, amounts to the case $M = 1$ and is justified by the rather large unit cells needed anyway for molecular dynamics. In this approximation the formula for the cell dipole per unit length becomes

$$P^{\text{el}} = \frac{e}{2\pi} \text{Im ln det } \mathbf{S} \ , \tag{7.81}$$

where the matrix $\mathbf{S}$ is now defined using the Kohn–Sham orbitals ϕ_n

$$S_{n,m} = \int_0^L \phi^*_n(x) \exp\left[-i\frac{2\pi}{L}x\right] \phi_m(x) \, dx \ . \tag{7.82}$$

The polarization formula for the system of non-interacting electrons has therefore been used as an approximation for the expectation value in the interacting system.

7.2.3 Localization functionals

Wannier functions are defined in terms of a unitary transformation performed on occupied Bloch orbitals [1594]. One major problem in a practical calculation is their non-uniqueness. This is a result of the indeterminacy of the Bloch orbitals, which are, in the case of a single band, only determined up to a phase factor and, in the multi-band case, up to an arbitrary unitary transformation among all occupied orbitals at every point in the Brillouin zone. As proposed by Marzari and Vanderbilt [960], one can resolve this non-uniqueness by requiring that the total spread of the localized functions should be minimal. This criterion is in close analogy with the Boys-Foster method [184, 456] for finite systems, where one uses the spread defined through the conventional position operator. The periodic technique has been applied successfully to crystal systems and to small molecules within a general **k**-point scheme [960, 1373]. An extension to disordered systems within the Γ-point approximation was also worked out [1355]. In the following, the focus will be the Γ-point approximation only, which is of particular interest when one would like a localized orbital picture within the framework of *ab initio* molecular dynamics. Upon minimization of the spread functional the appropriate unitary transformation to the localized orbitals has to be calculated.

In Resta's treatment [1211], the fundamental object for studying localization of an electronic state within Born-von Kármán boundary conditions is the dimensionless complex number

$$z = \int_0^L \exp\left[i2\pi x/L\right] |\phi(x)|^2 \, dx \; , \tag{7.83}$$

which is a special case of the quantity defined earlier in Eq. (7.82). Here, L is the linear dimension and $\phi(x)$ denotes the wave function. Using this number and by considering the definition of the spread of the wave function to be $\Omega = \langle x^2 \rangle - \langle x \rangle^2$, where $\langle \cdots \rangle$ denotes an expectation value, Resta has shown that to order $O(1/L^2)$ the functional for the spread in one dimension is

$$\Omega = \frac{1}{(2\pi)^2} \ln |z|^2 \; . \tag{7.84}$$

In three dimensions the following dimensionless complex number [109] is studied within Born-von Kármán boundary conditions:

$$z_I = \int_V \exp\left[i\mathbf{G}_I \cdot \mathbf{r}\right] |\phi(\mathbf{r})|^2 d\mathbf{r} \; . \tag{7.85}$$

Here, I labels a general reciprocal lattice vector, $\mathbf{G}_I = l_I \mathbf{b}_1 + m_I \mathbf{b}_2 + n_I \mathbf{b}_3$, where $\mathbf{b}_\alpha$ are the primitive reciprocal lattice vectors, the integers l, m, and n are the Miller indices, $\phi(\mathbf{r})$ denotes the wave function, and V is the volume.

One must now define an appropriate function of the z_I's that gives the three-dimensional spread in the case of an arbitrary simulation cell. In a molecular dynamics simulation the cell parameters (primitive lattice vectors) to describe systems of general symmetry are given by $\mathbf{a}_1$, $\mathbf{a}_2$, and $\mathbf{a}_3$. As already used at several places, it is convenient to form a matrix of these cell parameters, $\mathbf{h} = [\mathbf{a}_1, \mathbf{a}_2, \mathbf{a}_3]$, where the volume of the simulation cell is given by the determinant of $\mathbf{h}$ (see Section 3.1.1). It is also very useful to define scaled coordinates, $\mathbf{s} = \mathbf{h}^{-1} \cdot \mathbf{r}$, that are located within the unit cube (see Eq. (3.2)). For systems of general symmetry one can compute the reciprocal space vectors with the knowledge of the matrix of cell parameters. Thus, the Ith reciprocal lattice vector is

$$\mathbf{G}_I = 2\pi \left(\mathbf{h}^{-1}\right)^{\mathrm{T}} \cdot \mathbf{g}_I \ , \tag{7.86}$$

where the superscript T denotes transposition and $\mathbf{g}_I = (l_I, m_I, n_I)$ is the Ith Miller index (see Section 3.1.1). This expression is substituted into Eq. (7.85) and the definition of $\mathbf{r}$ is used to obtain

$$z_I = \det\mathbf{h} \int_0^1 \exp\left[i2\pi\mathbf{g}_I^{\mathrm{T}} \cdot \mathbf{s}\right] \ |\phi(\mathbf{h} \cdot \mathbf{s})|^2 \ . \tag{7.87}$$

Note that the exponential is independent of any coordinate system. Following Resta [1211] one can write the electron density in terms of a superposition of localized density and its periodic images

$$|\phi(\mathbf{h} \cdot \mathbf{s})|^2 = \sum_{\mathbf{m}=-\infty}^{\infty} n_{\mathrm{loc}}(\mathbf{h} \cdot \mathbf{s} - \mathbf{h} \cdot \mathbf{s}_0 - \mathbf{h} \cdot \mathbf{m}) \ , \tag{7.88}$$

where $\mathbf{m}$ is a vector of integers and $\mathbf{h} \cdot \mathbf{s}_0$ is the center of the distribution such that

$$\int_{-\infty}^{\infty} \mathbf{h} \cdot \mathbf{s}\, n_{\mathrm{loc}}(\mathbf{h} \cdot \mathbf{s})ds = 0 \ . \tag{7.89}$$

Using the Poisson summation formula [865], Eq. (7.87) is rewritten as

$$z_I = \exp\left[i2\pi\mathbf{g}_I^{\mathrm{T}} \cdot \mathbf{s}_0\right] \hat{n}_{\mathrm{loc}}(-2\pi\mathbf{g}_I^{\mathrm{T}} \cdot \mathbf{h}^{-1}) \ , \tag{7.90}$$

where $\hat{n}_{\mathrm{loc}}$ denotes the Fourier transform of n_{loc}. Furthermore, since n_{loc} is considered to be localized, its Fourier transform is smooth in reciprocal space and thus one can be assured that it is well represented about $g_I = 0$.

Expanding $\hat{n}_{\text{loc}}(-2\pi \mathbf{g}_I^{\text{T}} \cdot \mathbf{h}^{-1})$ to second order one obtains

$$\hat{n}_{\text{loc}}(-2\pi \mathbf{g}_I^{\text{T}} \cdot \mathbf{h}^{-1}) = 1 + \sum_{\alpha} g_{\alpha,I} \frac{\partial \hat{n}_{\text{loc}}}{\partial g_{\alpha,I}} \Big|_{g_I = 0}$$

$$+ \frac{1}{2} \sum_{\alpha,\beta} g_{\alpha,I} g_{\beta,I} \frac{\partial^2 \hat{n}_{\text{loc}}}{\partial g_{\alpha,I} \partial g_{\beta,I}} \Big|_{g_I = 0} + \cdots \quad , \quad (7.91)$$

where the second term is zero given the imposed condition $\langle \mathbf{h} \cdot \mathbf{s} \rangle = 0$. Thus, one is left with

$$\hat{n}_{\text{loc}}(-2\pi \mathbf{g}_I^{\text{T}} \cdot \mathbf{h}^{-1}) = 1 - \frac{(2\pi)^2}{2} V \sum_{\alpha,\beta} g_{\alpha,I} g_{\beta,I}$$

$$\times \int_{-\infty}^{\infty} s_\alpha s_\beta \, n_{\text{loc}}(\mathbf{h} \cdot \mathbf{s}) \, ds \quad (7.92)$$

and combining Eq. (7.92) with Eq. (7.90) one gets

$$1 - |z_I| = V \frac{(2\pi)^2}{2} \sum_{\alpha,\beta} g_{\alpha,I} g_{\beta,I} \int_{-\infty}^{\infty} s_\alpha s_\beta \, n_{\text{loc}}(\mathbf{h} \cdot \mathbf{s}) \, ds \quad . \quad (7.93)$$

Keeping in mind that $\int_{-\infty}^{\infty} \mathbf{h} \cdot \mathbf{s} \, n_{\text{loc}}(\mathbf{h} \cdot \mathbf{s}) ds = 0$, one can define the spread of the electronic distribution for the case of a general box through

$$\langle \mathbf{r}^2 \rangle - \langle \mathbf{r} \rangle^2 = \left\langle (\mathbf{h} \cdot \mathbf{s})^2 \right\rangle = \sum_{\alpha,\beta} \mathcal{G}_{\alpha\beta} V \int_{-\infty}^{\infty} s_\alpha s_\beta \, n_{\text{loc}}(\mathbf{h} \cdot \mathbf{s}) \, ds \quad . \quad (7.94)$$

Here, $\mathcal{G}_{\alpha\beta} = \sum_{\mu} h_{\alpha\mu}^{\text{T}} h_{\mu\beta}$ can be thought of as a metric tensor to describe the corresponding distances in the unit cube according to Eq. (3.3).

Equation (7.94) shows exactly how the length scales are built into the spread through the metric tensor. From direct comparison of Eq. (7.93) and Eq. (7.94) one can see that for supercells of general symmetry linear combinations of $g_{\alpha,I} g_{\beta,I}$ have to be chosen that reproduce the metric tensor, $\mathcal{G}_{\alpha\beta}$. However, as stated earlier, $g_{\alpha,I}$ are dimensionless numbers. Thus, an appropriate generalization

$$\mathcal{G}_{\alpha\beta} = \sum_{I} \omega_I g_{\alpha,I} g_{\beta,I} \quad (7.95)$$

takes the form of a sum rule; ω_I are the "weights" with the appropriate dimensions to be determined later. Thus, it should also be clear that $\mathcal{G}_{\alpha\beta}$ will have at most six independent entries (for triclinic symmetry) and thus a maximum of six weights are needed. Finally, this expression is generalized to more than one state, $|\phi\rangle \to |\phi_n\rangle$ and the desired expression for the spread Ω in a supercell of general symmetry is

$$\Omega = \frac{2}{(2\pi)^2} \sum_n \sum_I \omega_I (1 - |z_{I,n}|) + O(2\pi \mathbf{g}_I^T \cdot \mathbf{h}^{-1})^2$$

$$z_{I,n} = \int_V \exp\left[i\mathbf{G}_I \cdot \mathbf{r}\right] |\phi_n(\mathbf{r})|^2 \, d\mathbf{r} \ , \tag{7.96}$$

where Eq. (7.95) determines the set $\{\mathbf{G}_I\}$.

At this point it is useful to make contact with other spread formulas that are present in the current literature. Following Resta's derivation one finds the formula [1211]

$$\Omega = -\frac{1}{(2\pi)^2} \sum_n \sum_I \omega_I \log |z_{I,n}|^2 \ , \tag{7.97}$$

according to the present notion with $z_{I,n}$ defined as above. This expression is obtained by inserting Eq. (7.92) into Eq. (7.90), taking the log of the absolute value, and expanding to consistent order. Silvestrelli [1347], on the other hand, uses (again, in this notation)

$$\Omega = \frac{1}{(2\pi)^2} \sum_n \sum_I \omega_I (1 - |z_{I,n}|^2) \ , \tag{7.98}$$

with a similar definition for $z_{I,n}$. Obviously, Eq. (7.98) is obtained from Eq. (7.97) by an expansion of the logarithm. At first glance, it seems confusing that there are different definitions for the spread. Admittedly, one has to keep in mind that all formulas are only valid up to the order given in Eq. (7.96). Thus, although different, they are consistent and there is no fundamental reason to choose one definition of the spread over another.

One can also derive a general expression for the expectation value of the periodic position operator for computing the center of the localized function. It is recalled that for a cubic simulation supercell the expectation value of the position operator is given as

$$r_{\alpha,n} = -\frac{L}{2\pi} \operatorname{Im} \log z_{\alpha,n} \tag{7.99}$$

$$z_{\alpha,n} = \int_V \exp\left[i\mathbf{g}_\alpha \cdot \mathbf{r}\right] |\phi_n(\mathbf{r})|^2 \, d\mathbf{r} \ , \tag{7.100}$$

where $\mathbf{g}_1 = (1,0,0)$, $\mathbf{g}_2 = (0,1,0)$, and $\mathbf{g}_3 = (0,0,1)$. Again, the salient feature of Eq. (7.100) is that the expectation value of the exponential is invariant with respect to the choice of cell. Thus, a general equation for the expectation value of the position operator in supercells of arbitrary symmetry is

$$r_{\alpha,n} = -\sum_\beta \frac{h_{\alpha\beta}}{2\pi} \text{Im} \, \log z_{\alpha,n} \, . \tag{7.101}$$

Finally, the weights ω_I have to be determined as defined in the sum rule Eq. (7.95) for supercells of general symmetry. As mentioned, the metric $\mathcal{G}$ will contain at most six independent entries as defined by the case of the lowest symmetry, i.e. triclinic. Thus, Eq. (7.95) is a linear set of six equations with six unknowns and one has the freedom to choose the six Miller indices, $\mathbf{g}_I$ for the linear combination. For computational convenience of computing z_i one chooses $\mathbf{g}_1 = (1,0,0)$, $\mathbf{g}_2 = (0,1,0)$, $\mathbf{g}_3 = (0,0,1)$, $\mathbf{g}_4 = (1,1,0)$, $\mathbf{g}_5 = (1,0,1)$, $\mathbf{g}_6 = (0,1,1)$. With this choice of $\mathbf{g}_i$ the explicit system of equations based on Eq. (7.95) takes a simple form and yields the following set of general weights:

$$\omega_1 = g_{11} - g_{12} - g_{13}$$

$$\omega_2 = g_{22} - g_{12} - g_{23}$$

$$\omega_3 = g_{33} - g_{13} - g_{23}$$

$$\omega_4 = g_{12}$$

$$\omega_5 = g_{13}$$

$$\omega_6 = g_{23} \, , \tag{7.102}$$

which is the last ingredient needed to compute the expectation value of the position operator in a periodic system of general symmetry.

7.2.4 Localization methods

7.2.4.1 Generalized localization procedure

The mathematical problem which defines the localization procedure is to find the unitary transformation $\mathbf{U}$ of the orbitals

$$|\tilde{\phi}_n\rangle = \sum_i U_{in} |\phi_i\rangle \tag{7.103}$$

that simultaneously minimizes the spread functional, Ω. In order to arrive at a general expression it is convenient to introduce a generalized form

$$\Omega = \sum_n \sum_I f(|z_{I,n}|^2)$$

$$z_{I,n} = \langle \phi_n | O^I | \phi_n \rangle , \tag{7.104}$$

where f and O^I denote an appropriate function and operator, respectively. If one neglects the weights and constants in favor of simplicity, one can obtain the different spread functionals of the last section, defined through Eqs. (7.96), (7.97), and (7.98) by setting

$$O^I = \exp[i\mathbf{G}_I \cdot \mathbf{r}]$$

$$f_1(|z_{I,n}|^2) = \sqrt{|z_{I,n}|^2} = |z_{I,n}|$$

$$f_2(|z_{I,n}|^2) = \log(|z_{I,n}|^2)$$

$$f_3(|z_{I,n}|^2) = |z_{I,n}|^2 , \tag{7.105}$$

where the values of the index I range at most from one to six. It is important to notice that maximizing Eq. (7.104) is equivalent to minimizing one of the spread functionals Eqs. (7.96), (7.97), or (7.98).

The actual calculation of maximally localized WFs or maximally localized MOs within the presented localization procedure is relatively simple. First, one takes the output of a conventional electronic structure calculation (i.e. Bloch orbitals and canonical molecular orbitals in the periodic and finite cases, respectively), chooses a spread functional, and solves for the unitary transformation producing the orbitals that maximize Eq. (7.104).

One must now focus on the computation of the localization transformation $\mathbf{U}$ that rotates the orbitals suitably. To ensure a maximally localized function an efficient solution of

$$\frac{\partial \Omega}{\partial U_{ij}} = 0 \tag{7.106}$$

has to be devised where $\mathbf{U}$ is considered to be real since the Γ-point approximation is invoked. There are two principal alternatives for parameterizing this unitary transformation. First, as a direct product of elementary plane rotations and second, as the exponential of an antisymmetric matrix. The first parameterization scheme, discussed in the next subsection, amounts to the well-known Jacobi optimization procedure for finding eigenvalues of general matrices. The second parameterization choice of $\mathbf{U}$ is based on the exponential alternative that will be investigated in Section 7.2.4.3.

7.2.4.2 Orbital rotations

The traditional method in quantum chemistry for computing localized molecular orbitals is the method of two-by-two orbital rotations first introduced by Edmiston and Ruedenberg [380]. The basic idea of the method is to tackle the problem of finding $\mathbf{U}$ by performing a sequence of consecutive two-by-two rotations among all pairs of orbitals. The elementary step consists of a plane rotation where two orbitals i and j are rotated through an angle γ. To proceed, an optimal angle is selected to ensure that the spread functional, as defined in Eq. (7.104), is iteratively maximized. The transformed expectation values are denoted $\tilde{z}_{I,i/j}$ and are obtained as

$$\tilde{z}_{I,i} = \cos(\gamma)z_{I,i} + \sin(\gamma)z_{I,j} , \quad \tilde{z}_{I,j} = -\sin(\gamma)z_{I,i} + \cos(\gamma)z_{I,j} . \quad (7.107)$$

Thus, by combining Eq. (7.107) with Eq. (7.104) it is straightforward to calculate the change in the functional value, $\Delta\Omega$ as a function of γ. The most natural way to obtain the optimal angle which maximizes the change in the functional value is to compute the derivative of $\Delta\Omega$ with respect to γ, set it to zero, and solve for γ. This is precisely the way the method of orbital rotations is implemented, and an explicit calculation yields

$$\tan(4\gamma) = -\frac{a}{b} , \quad (7.108)$$

$$a = \text{Re}[Y_{ij}(\bar{Y}_{ii} - \bar{Y}_{jj})] , \quad b = |Y_{ij}|^2 - \frac{1}{4}|Y_{ii} - Y_{jj}|^2 ,$$

where $Y_{ij} = \sum_I z_{I,ij}$, $z_{I,ij} = \langle \phi_i | O^I | \phi_j \rangle$ and Re denotes the real part and $\gamma + n\pi/4$ are the solutions of Eq. (7.108) corresponding to maxima and minima. For a maximum the condition

$$\frac{\partial^2\Omega}{\partial\gamma^2} = 16\left(b\cos(4\gamma) - a\sin(4\gamma)\right) < 0 \quad (7.109)$$

has to be fulfilled.

There is one restriction to this scheme: Eq. (7.108) is only valid in the case $f_3(x) = x$ as given in Eq. (7.105). In the other cases, i.e. $f_1(x) = \sqrt{x}$ and $f_2(x) = \log(x)$, no analogous formula is derivable. The reason is that the explicit solution of $\partial\Delta\Omega/\partial\gamma = 0$ with respect to γ seems not analytically tractable. Nevertheless, one can still implement the method of orbital rotations in the above cases by a numerical maximization of $\Delta\Omega$ as a function of γ using derivative information.

7.2.4.3 Exponential representation

The ansatz $|\tilde{\phi}_n\rangle = \sum_i U_{in}|\phi_i\rangle$, where $\mathbf{U}$ is a unitary matrix, leads to the transformed expectation value

$$\tilde{z}_{I,n} = \sum_{ij} U_{in}^\dagger U_{jn} z_{I,ij} \ . \tag{7.110}$$

As discussed above, one parameterizes $\mathbf{U} = \exp[\mathbf{A}]$ as the exponential of an antisymmetric matrix and calculates the gradient with respect to $\mathbf{A}$. Using the chain rule the gradient splits into two contributions:

$$\frac{\partial \Omega}{\partial A_{ij}} = \sum_{st} \frac{\partial \Omega}{\partial U_{st}} \frac{\partial U_{st}}{\partial A_{ij}} = \sum_{I,n} \sum_{st} \frac{\partial f(|z_{I,n}|^2)}{\partial U_{st}} \frac{\partial U_{st}}{\partial A_{ij}} \ . \tag{7.111}$$

It is worth noting that only the first contribution depends on the type of spread functional and its evaluation is straightforward:

$$\frac{\partial f(|z_{I,n}|^2)}{\partial U_{st}} = f'(|z_{I,n}|^2) \frac{\partial |z_{I,n}|^2}{\partial U_{st}} \tag{7.112}$$

$$= 2 \, f'(|z_{I,n}|^2) \left(\sum_i U_{in} \delta_{tn} z_{I,is} \right)$$

$$\times \left(\sum_{kl} U_{kn} U_{ln} \bar{z}_{I,kl} \right) + \text{c.c.} \ , \tag{7.113}$$

where $\bar{z}$ denotes the complex conjugate (**c.c.**) of z and f' is the derivative of f. Combining Eq. (7.111) and Eq. (7.112), a general form for the gradient of the spread functional is obtained as

$$\frac{\partial \Omega}{\partial A_{ij}} = \sum_{st} M_{st} \frac{\partial U_{st}}{\partial A_{ij}} \ , \tag{7.114}$$

where $\mathbf{M} = \partial \Omega / \partial U_{st} = \partial f(|z_{I,n}|^2)/\partial U_{st}$ is defined via Eq. (7.112).

The calculation of $\partial U_{st}/\partial A_{ij}$ is more subtle. One has to calculate the derivative of a matrix function, here the exponential function, $\mathbf{U} = \exp[\mathbf{A}]$ with respect to $\mathbf{A}$. This can be done by writing the matrix function in an alternative way using a complex contour integral [489]

$$\frac{\partial \mathbf{U}}{\partial A_{ij}} = \frac{\partial \exp[\mathbf{A}]}{\partial A_{ij}} \tag{7.115}$$

$$= \frac{1}{2\pi i} \frac{\partial}{\partial A_{ij}} \oint \exp[z](z\mathbf{1} - \mathbf{A})^{-1} dz$$

$$= \frac{1}{2\pi i} \oint \exp[z](z\mathbf{1} - \mathbf{A})^{-1} (\mathbf{1}^{ij} - \mathbf{1}^{ji}) (z\mathbf{1} - \mathbf{A})^{-1} dz$$

$$= \mathbf{R}^{\dagger} \frac{1}{2\pi i} \oint \exp[z](z\mathbf{1} - \mathbf{\Lambda})^{-1}\mathbf{R}(\mathbf{1}^{ij} - \mathbf{1}^{ji})\mathbf{R}^{\dagger}(z\mathbf{1} - \mathbf{\Lambda})^{-1} dz\ \mathbf{R}\ ;$$

here $\mathbf{1}$ denotes the identity matrix, $(\mathbf{1}^{ij})_{kl} = \delta_{ki}\delta_{lj}$, $\mathbf{R}$ is the eigenvector matrix of $\mathbf{A}$ with eigenvalues λ_k, and $\Lambda_{kl} = \lambda_k \delta_{kl}$. Carrying out the integration over z one obtains

$$\frac{1}{2\pi i} \oint \frac{\exp[z]}{(z - \lambda_k)(z - \lambda_l)} dz = \begin{cases} \exp[\lambda_k], & \lambda_k = \lambda_l, \\ \frac{\exp[\lambda_k] - \exp[\lambda_l]}{\lambda_k - \lambda_l}, & \lambda_k \neq \lambda_l. \end{cases} \tag{7.116}$$

Performing some simple algebraic transformations, Eq. (7.114) becomes

$$\frac{\partial \Omega}{\partial A_{ij}} = \mathrm{Tr}\left[\mathbf{M}^{\mathrm{T}}\mathbf{R}^{\dagger}\{\mathbf{C}^{ij}, \mathbf{B}\}\mathbf{R}\right]$$

$$= (\mathbf{R}^{\dagger}\{\mathbf{R}\mathbf{M}^{\mathrm{T}}\mathbf{R}^{\dagger}, \mathbf{B}\}\mathbf{R})_{ji} - (\mathbf{R}^{\dagger}\{\mathbf{R}\mathbf{M}^{\mathrm{T}}\mathbf{R}^{\dagger}, \mathbf{B}\}\mathbf{R})_{ij}\ , \tag{7.117}$$

where the matrix B_{ij} is defined through Eq. (7.116), $\{\mathbf{C}^{ij}, \mathbf{B}\}$ denotes a component-by-component matrix multiplication, and $\mathbf{C}^{ij} = \mathbf{R}(\mathbf{1}^{ij} - \mathbf{1}^{ji})\mathbf{R}^{\dagger}$. The final transformation in Eq. (7.117) is verified by inserting the explicit definition of the matrix $\mathbf{1}^{ij}$. With the above scheme the desired gradient is obtained analytically and in addition one is able to combine the iterative localization procedure with gradient methods developed to accelerate convergence.

7.2.5 Wannier functions in Car–Parrinello simulations

In a Car–Parrinello molecular dynamics simulation the Kohn–Sham orbitals are propagated together with the nuclei using coupled equations of motion derived from a Lagrangian function. When a unitary transformation is applied to the orbitals and their velocities, the values of the Lagrangian and the forces on the nuclei are not changed. Therefore, the same nuclear trajectory is generated. This property was used before in Section 6.6 to develop algorithms based on nonorthogonal orbitals.

Techniques from gauge-field theory are employed in an alternative formulation by Tuckerman *et al.* [670, 1464] that relies on a powerful covariant generalization of the Car–Parrinello Lagrangian. Using the Dirac gauge-fixing method, a family of Car–Parrinello molecular dynamics schemes can be derived. In the "Wannier gauge" a method can be derived that allows maximally localized Wannier orbitals to be generated dynamically as the calculation proceeds, i.e. "on-the-fly" instead of localizing the propagated Bloch orbitals *a posteriori*. An approximate algorithm for integrating the equations of motion that is stable and maintains orbital locality was developed based on the exact equations. The resulting algorithm has similar characteristics for performance and accuracy as the one described in the following section.

An efficient implementation [743] of the calculation of maximally localized Wannier functions is needed in order to make their calculation along a molecular dynamics trajectory feasible. The serial Jacobi algorithm outlined in this section can readily be ported to parallel computer architectures. A Jacobi plane rotation only involves two columns and, therefore, there are disjoint operations which can be executed in parallel. By using special orderings, the number of independent operations can be kept maximal at each step of the algorithm. Many parallel orderings have been proposed in the literature (see Ref. [1661] and references therein). The ring Jacobi ordering with alternating forward and backward steps [379, 1661] was chosen in the implementation in the CPMD code. In order to minimize the number of communication steps, the matrices are distributed on individual processors in block form. A single Jacobi rotation in the parallel algorithm is then generalized to plane rotations between blocks [529]. Two other methods [573, 670] that allow for an efficient calculation of Wannier centers in parallel have been reported in the literature.

In order to illustrate the performance of this algorithm the Wannier function centers (WFCs) and their spread, i.e. the first and second moment of the Wannier orbitals, respectively, have been calculated at an arbitrary step within a Car–Parrinello molecular dynamics simulation of 32 water molecules at ambient conditions. The gradient of the spread functional Ω with respect to an infinitesimal rotation can be calculated as

$$G_{ij} = \left.\frac{\partial \Omega}{\partial U_{ij}}\right|_{\mathbf{U}=0} = 4\sum_{\alpha}[z_{\alpha,ij}(z_{\alpha,ii} - z_{\alpha,jj})] \qquad (7.118)$$

and the absolute value of the maximum element of matrix $\mathbf{G}$ is used to

Table 7.2. *Effect of convergence criteria in the calculation of maximally localized Wannier functions on the accuracy of their first (centers) and second (spread) moments for a system of 32 water molecules. Errors are calculated with respect to the results obtained for a maximum gradient of 10^{-8}. The first line (BO) refers to a calculation where orbitals were quenched to the Born–Oppenheimer surface first (again using a maximum gradient of 10^{-8}), whereas the other lines and the reference calculation used orbitals from a Car–Parrinello molecular dynamics simulation; see text for discussion.*

| Accuracy | Center | | Spread | |
	Mean error	Max error	Mean error	Max error
10^{-8} BO	$2.1 \cdot 10^{-4}$	$2.1 \cdot 10^{-3}$	$6.6 \cdot 10^{-4}$	$2.0 \cdot 10^{-3}$
10^{-2}	$3.7 \cdot 10^{-5}$	$5.7 \cdot 10^{-4}$	$3.3 \cdot 10^{-5}$	$1.2 \cdot 10^{-4}$
10^{-3}	$1.2 \cdot 10^{-5}$	$2.0 \cdot 10^{-4}$	$1.1 \cdot 10^{-5}$	$5.0 \cdot 10^{-5}$
10^{-4}	$2.0 \cdot 10^{-6}$	$5.5 \cdot 10^{-5}$	$1.6 \cdot 10^{-6}$	$1.1 \cdot 10^{-5}$
10^{-6}	$1.3 \cdot 10^{-8}$	$8.4 \cdot 10^{-7}$	$7.8 \cdot 10^{-9}$	$1.0 \cdot 10^{-6}$

monitor the progress of optimization; the Jacobi method was used throughout the calculations. The accuracies of the Wannier function center and spread values are compiled in Table 7.2. The first row lists the accuracy of the calculated values from orbitals taken directly from the Car–Parrinello simulation with respect to fully optimized Kohn–Sham orbitals at the same geometry. The former values will depend on the parameters chosen for the Car–Parrinello simulation (time step, electron mass) and are related directly to the fictitious kinetic energy. The values given in Table 7.2 can be considered typical for a Car–Parrinello simulation. The maximum error in the position of the Wannier function center is a direct measure for the uncertainty in the value for molecular dipole moments. In the other rows errors are listed for less stringent convergence criteria. As can be seen, even for a rather loose convergence criteria of 10^{-2}, errors are smaller than the uncertainty introduced by the fact that the system is not exactly on the Born–Oppenheimer surface.

In Fig. 7.2 the development of the maximum value of the gradient G_{ij} is shown as a function of Car–Parrinello molecular dynamics steps since the last computation of maximally localized Wannier functions. The gradient increases linearly with a rather small slope and even after ten steps it is still below 0.02. From the data reported in Table 7.2 one can estimate that the

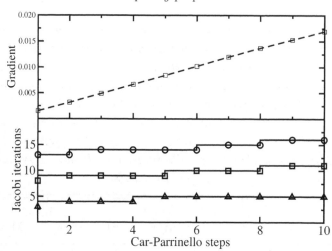

Fig. 7.2. Upper panel: Development of the gradient for orbitals within a typical Car–Parrinello molecular dynamics simulation of 32 water molecules starting from a set of maximally localized Wannier functions. Lower panel: Number of Jacobi sweeps needed to re-optimize maximally localized Wannier functions to an accuracy of 10^{-4} (triangles), 10^{-6} (squares), 10^{-8} (circles); see text for discussion.

Wannier function centers are at this point still accurate to about 10^{-3} Bohr. In the lower panel of Fig. 7.2 the number of Jacobi sweeps to reach a certain accuracy is shown. As the Jacobi method converges exponentially and the gradient only increases linearly, the number of sweeps is almost constant.

One can make use of these findings in setting up an efficient protocol for Car–Parrinello molecular dynamics with maximally localized Wannier functions. Starting with a set of maximally localized Wannier functions, a number of Car–Parrinello molecular dynamics steps are performed. At a certain sampling rate, typically five to ten Car–Parrinello steps, the dipole moment is calculated from the Berry phase and a single Jacobi sweep is carried out and the computed dipole moment and Wannier function centers are stored for later analysis. Using this protocol the gradient of the maximally localized Wannier functions has been monitored over a Car–Parrinello molecular dynamics simulation with the CPMD package [696] of 1 ps using 32 water molecules. As can be seen in Fig. 7.3, the gradient of the maximally localized Wannier functions is rather stable and stays below a value of 0.006, which is sufficient for an accurate calculation of molecular dipole moments.

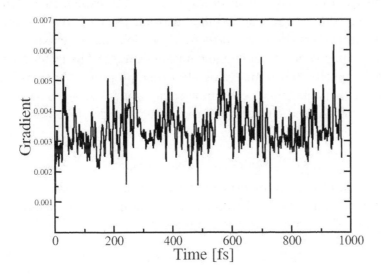

Fig. 7.3. Gradient of the orbitals in a typical Car–Parrinello molecular dynamics simulation of 32 water molecules. Every fifth MD step (i.e. about every 0.5 fs) a single Jacobi sweep was applied; see text for discussion.

7.2.6 Applications: dipole moments, infrared spectra, and atomic charges

Many papers have appeared since the mid to end 1990s in the literature that derive molecular properties from Wannier functions. In the area of chemistry, maximally localized Wannier functions are used as a tool to understand the nature of chemical bonding whereas in physics, they are used, for instance, as compact descriptors of both local and global dielectric properties or to analyze and compute electric currents.

At this stage, a general note of caution is in order. In most cases such Wannier function analyses are carried out in the pseudopotential/plane wave framework of Chapter 3, where only part of the electrons is considered explicitly. This is handled straightforwardly when norm-conserving pseudopotentials as introduced in Chapter 4 are used since the norm-conserving pseudo orbitals carry the full valence charge when integrated over full space. Thus, the total electronic charge is obtained by adding the charge density of the pseudo orbitals to the charge of the core electrons that is hidden in the pseudopotential. As a result, analyzing electronic properties in terms of norm-conserving pseudo orbitals, for instance using the transformation to Wannier orbitals, is meaningful as long as the underlying pseudopotential approximation does not break down as such. As explained in Chapter 6, this is very different in Vanderbilt's ultrasoft pseudopotential scheme where the norm-conservation property from Section 4.2 is relaxed at the expense of introducing additional augmentation charges. The latter have to be considered explicitly when analyzing electronic properties, such as Wannier orbitals. This turns out to be cumbersome in many cases, such that a simple remedy often used consists in carrying out single-point analyses of representative configurations by using norm-conserving pseudopotentials.

7.2.6.1 *Molecular dipole moments*

In the framework of density functional theory within the Kohn–Sham approach (assuming spin-restricted Kohn–Sham orbitals $\{\phi_k\}$ for simplicity), the polarization of the system is approximated by

$$M_\alpha^{\text{el}} = \frac{2e}{|\mathbf{G}_\alpha|}\text{Im} \ln \det \mathbf{Z}_\alpha \ , \tag{7.119}$$

as a generalization of the one-dimensional case Eq. (7.81) with α denoting the Cartesian components of the electronic contribution $\mathbf{M}^{\text{el}}$ to the dipole moment vector $\mathbf{M}$ of the total system. Here, the matrices $\mathbf{Z}_\alpha$ are defined according to Eq. (7.82) in terms of matrix elements of the Kohn–Sham orbitals

$$(\mathbf{Z}_\alpha)_{kl} = \langle \phi_k \mid \exp\left[-i\mathbf{G}_\alpha \cdot \mathbf{r}\right] \mid \phi_l \rangle \ , \tag{7.120}$$

where the indices k and l run over all occupied orbitals. The nuclear contribution must be added to get the total dipole moment of the system, which is obtained trivially from the positions and charges of the nuclei taking the effective core charges into account according to the specific pseudopotentials used.

For the analysis of a molecular liquid or solid it is most convenient if the electronic contribution $\mathbf{M}^{\text{el}}$ to the total dipole moment of the system can be written, at least to a good approximation, as a sum of molecular dipole moments $\{\boldsymbol{\mu}_I^{\text{el}}\}$

$$\mathbf{M}^{\text{el}} \approx \sum_I \boldsymbol{\mu}_I^{\text{el}} \tag{7.121}$$

that are ascribed to individual molecules I (or even to molecular fragments of larger molecules or supramolecular aggregates in more extreme cases). An optimal result would be achieved if the three $\mathbf{Z}_\alpha$ matrices could be diagonalized using the same transformation. In this case the determinant in Eq. (7.119) would reduce to a product and the separation would be exact.

However, the operators $\exp[-i\mathbf{G}_\alpha \cdot \mathbf{r}]$ do not commute and a simultaneous diagonalization is therefore not possible. Still, it is possible to proceed along the lines of these ideas, which is outlined in the following.

The set of maximally localized Wannier functions $\{w_k(\mathbf{r})\}$ that can be obtained according to Eq. (7.103) from the set of occupied Bloch Kohn–Sham orbitals $\{\phi_l(\mathbf{r})\}$ using the localization functionals and methods from Sections 7.2.3 and 7.2.4, respectively, by carrying out a suitable unitary transformation $\mathbf{U}$

$$w_k(\mathbf{r}) = \sum_l U_{kl}\phi_l(\mathbf{r}) \tag{7.122}$$

can be reinterpreted as a unique basis to produce "maximally diagonal" sets of matrices. On the other hand, the expectation value of the position operator according to Eq. (7.100) is given by

$$\mathbf{r}_k = -\frac{L}{2\pi} \operatorname{Im} \ln \mathbf{z}_k \ , \tag{7.123}$$

which is just the center of a given maximally localized Wannier function where k labels the different Wannier orbitals and thus electrons. With this definition the electronic part of the total dipole moment of the supercell can be approximated by

$$\mathbf{M}^{\mathrm{el}} \approx -2e \sum_k \mathbf{r}_k \ , \tag{7.124}$$

where the sum includes all occupied states, and molecular dipole moments can be defined as

$$\boldsymbol{\mu}_I^{\mathrm{el}} = -2e \sum_{k \in I} \mathbf{r}_k \ , \tag{7.125}$$

where the sum runs over all Wannier function centers that can be associated with molecule I. This is possible as Wannier function centers for closed shell molecules tend to be located close to atoms or bond centers and can therefore be attributed uniquely to a molecule; this includes also lone pair electrons. It is worth pointing out that on an operational level, Eq. (7.124) implies that the electronic dipole moment vector (and according to Eq. (7.125) also its decomposition into individual molecular contributions) can be obtained, in the Wannier representation, as a simple sum of position vectors $\mathbf{r}_k$ multiplied by the appropriate charges (i.e. $-1e$ and $-2e$ in spin-polarized and spin-restricted calculations, respectively), like for classical point particles!

A more general definition of a molecular dipole moment within a condensed system would only require that maximally localized Wannier functions on different molecules do not overlap. This can be achieved by using the definition

$$\mu_I^{\text{el}} = \sum_\alpha \frac{2e}{|\mathbf{G}_\alpha|} \text{Im} \ln \det \mathbf{Z}_\alpha^I \ , \tag{7.126}$$

where the matrices $\mathbf{Z}^I$ are molecular submatrices for each molecule I. Tests have shown that molecular dipole moments calculated using the definitions in Eqs. (7.125) and (7.126) are almost identical for molecules without delocalized electrons.

7.2.6.2 Solute infrared absorption spectra

The very same decomposition of the electronic charge into molecular contributions as worked out in Section 7.2.6.1 can also be used to derive a method that allows the total infrared absorption spectrum of a molecular condensed phase system to be decomposed into molecular components. As a special case, this allows us to compute on firm ground the infrared absorbance of a solute molecule, such as a chromophore in solution, which allows us to investigate solvent shifts.

Denoting the molecular dipole moment of the solute, calculated using Wannier function centers, by $\boldsymbol{\mu}_0$ and the molecular dipole moments of the solvent molecules by $\boldsymbol{\mu}_I$, the dipole time autocorrelation function needed for the calculation of the infrared spectra (see Section 5.4.5) can be decomposed into three parts [476]:

$$C(t) = C_{\text{SS}}(t) + C_{\text{MM}}(t) + C_{\text{SM}}(t) \ , \tag{7.127}$$

where the full system time-correlation function

$$C(t) = \langle \mathbf{M}(t) \cdot \mathbf{M}(0) \rangle \tag{7.128}$$

based on the total dipole moment $\mathbf{M}$, which includes both the electronic and nuclear contributions, Eqs. (5.244)–(5.246), and the partial time-correlation functions

$$C_{\text{SS}}(t) = \sum_{IJ}^{\text{solvent}} \langle \boldsymbol{\mu}_I(t) \cdot \boldsymbol{\mu}_J(0) \rangle \tag{7.129}$$

$$C_{\text{MM}}(t) = \langle \boldsymbol{\mu}_0(t) \cdot \boldsymbol{\mu}_0(0) \rangle \tag{7.130}$$

$$C_{\text{SM}}(t) = \sum_I^{\text{solvent}} \left(\langle \boldsymbol{\mu}_I(t) \cdot \boldsymbol{\mu}_0(0) \rangle + \langle \boldsymbol{\mu}_0(t) \cdot \boldsymbol{\mu}_I(0) \rangle \right) \tag{7.131}$$

have been introduced. The term $C_{\text{SS}}(t)$ should yield the infrared spectra of the solvent after Fourier transformation, whereas the terms $C_{\text{MM}}(t)$ and

$C_{SM}(t)$ contribute to the spectrum of the solute molecule. Whether or not the cross-correlation term $C_{SM}(t)$ should be included in the analysis of the solvent spectrum has been answered differently by different authors [476, 478, 671]. Finally, care has been taken to correct approximately for neglected quantum effects on the motion of the nuclei according to the background provided in Section 5.4.5.

This approach clearly relies on the separation of the charge density into a sum of molecular contributions. Although intuitively appealing, all the problems that are potentially related to such an additive approach are carried over to the spectra. However, the alternative of a direct determination of the spectra by subtracting a normalized spectrum from a pure solvent simulation from a spectrum determined with the full system time-correlation function is not feasible. Although this would be consistent with the way most experimental spectra are obtained, it would confront us with an enormous signal-to-noise ratio. The problem is especially serious because of the small system sizes and short simulation times that are presently accessible in *ab initio* molecular dynamics simulations. However, efficient parallelization techniques such as those introduced in Chapter 8, in conjunction with powerful massively parallel computers, will certainly improve the situation in the future.

7.2.6.3 Atomic charges

The calculation of atomic charges from Wannier function centers is based on a series of approximations. It is assumed that the total charge density can be divided into molecular charge densities using Wannier functions as explained earlier. Furthermore, the electrostatic potential is reliably represented outside the molecule by unit charges at the Wannier function centers. This approximation and extensions using higher multipoles have been investigated [1262] in great detail. Finally, the reduction of the total charge distribution to a set of partial atomic charges can only be achieved in a non-unique fashion. The method presented [743] is closely related to the D-RESP procedure [824].

Consider a molecule of N_{at} atoms with charges Z_A and atomic positions $\mathbf{R}_A$. The electronic distribution of the molecule is described by M Wannier function centers with charges $-q_w$ at positions $\mathbf{r}_a$. The charge q_w is fixed and has a value of one for the spin-polarized case and a value of two in spin-restricted calculations. It is assumed that the molecules are neutral

$$\sum_{A}^{N_{at}} Z_A - M q_w = 0 \; , \tag{7.132}$$

but this restriction can easily be lifted. One is now looking for a set of charges q_A that reproduce the electrostatic potential of the molecule as closely as possible. The electrostatic potential of the molecule derived from the Wannier function centers is defined as

$$V(\mathbf{r}) = \sum_A \frac{Z_A}{|\mathbf{R}_A - \mathbf{r}|} - \sum_a \frac{q_w}{|\mathbf{r}_a - \mathbf{r}|} \ . \tag{7.133}$$

In order to determine the parameters $\{q_A\}$, this potential is sampled at many positions $\mathbf{r}_i$ outside the molecule and optimized with respect to the charges $\{q_A\}$ within a least-squares fit. In addition, reference charges $\{q_A^0\}$ are added in order to stabilize the optimization. This is achieved by a quadratic term in $q_A - q_A^0$ with a given weight w. The zeroth and first moments of the resulting charge distribution, i.e. the total charge and the total dipole moments, will be enforced exactly.

Imposing the constraints on the moments by four Lagrange multipliers λ, ϵ_x, ϵ_y, and ϵ_z, the function to minimize is given by

$$\Omega(\{q_A\}, \lambda, \{\epsilon_\alpha\}) = \sum_i \left(\sum_A \frac{Z_A}{|\mathbf{R}_A - \mathbf{r}_i|} - \sum_a \frac{q_w}{|\mathbf{r}_a - \mathbf{r}_i|} - \sum_A \frac{q_A}{|\mathbf{R}_A - \mathbf{r}_i|} \right)^2$$

$$+ w \sum_A (q_A - q_A^0)^2 - \lambda \sum_A q_A - \sum_{\alpha=x,y,z} \epsilon_\alpha \left(\sum_A q_A r_A^\alpha - \mu_\alpha \right) , \tag{7.134}$$

where the first sum includes all sampling points $\{\mathbf{r}_i\}$ chosen. The variation of Ω with respect to the unknowns

$$\frac{\partial \Omega}{\partial q_A} = 0 \ , \qquad \frac{\partial \Omega}{\partial \lambda} = 0 \ , \qquad \frac{\partial \Omega}{\partial \epsilon_\alpha} = 0 \tag{7.135}$$

results in the equations

$$\sum_B \sum_i \frac{q_B}{|\mathbf{R}_B - \mathbf{r}_i|} \frac{2}{|\mathbf{R}_A - \mathbf{r}_i|} + 2wq_A - \lambda - \sum_{\alpha=x,y,z} \epsilon_\alpha r_A^\alpha =$$

$$2wq_A^0 - \sum_i \left(\sum_a \frac{q_w}{|\mathbf{r}_a - \mathbf{r}_i|} - \sum_C \frac{Z_C}{|\mathbf{R}_C - \mathbf{r}_i|} \right) \frac{2}{|\mathbf{R}_A - \mathbf{r}_i|} \tag{7.136}$$

$$\sum_B q_B = 0 \tag{7.137}$$

$$\sum_B q_B r_B^\alpha = \mu_\alpha \qquad \text{for } \alpha = x, y, z \tag{7.138}$$

being satisfied for all $A = 1, \ldots, N_{at}$. This system of linear equations with dimension $N_{at} + 4$ can easily be solved. Provided a suitable choice for the parameter w and the reference charges has been made, results will only be slightly dependent on the sampling points $\mathbf{r}_i$ [824].

8

Parallel computing

8.1 Introduction

Ab initio molecular dynamics calculations call for substantial computer resources. Memory and CPU time requirements make it necessary to run projects on the largest computers available. Today and in the foreseeable future, these high-end resources are provided exclusively by parallel computers. There are many different types of parallel architectures available, differing in their memory access system and their communication system. In addition, widely different performances are seen for the amount of data that can be sent from one processor to another per unit time (bandwidth) and the minimal time needed to send a small message (latency). Furthermore, different parallel programming paradigms are supported. In order to have a portable code that can be used on most of the current computer architectures, CPMD was programmed using standard communication libraries and making no assumption about the topology of the processor network and memory access system. Since this approach is also the basis of other general-purpose *ab initio* molecular dynamics codes, it will serve in this book as the specific case in order to explain in general how to design and code up a highly efficient parallel program.

Minimizing the communication was the major goal in the implementation of the parallel plane wave code in CPMD. Therefore, the algorithms had to be adapted to the distributed data model chosen. The most important decisions concern the data distribution of the largest arrays in the calculation, which are the ones holding information on the wave function, such as e.g. orbital positions and velocities. Several distribution strategies can be envisaged and were used before [189, 190, 240, 274, 1618, 1619].

First, the data are distributed over the bands [1618]. Each processor holds all expansion coefficients of an electronic band locally. Although rather straightforward to implement, this choice has severe drawbacks. The number of bands is usually of the same magnitude as the number of processors. This leads to a severe load-balancing problem that can only be avoided for certain magic numbers, namely if the number of bands needed is a multiple of the number of CPUs available. In other words, changing the number of available CPUs in the course of a study, or changing the system while keeping the number of CPUs fixed, can only be done at the expense of an enormous waste of resources. Furthermore, this approach requires performing three-dimensional Fourier transforms locally. The memory requirements for the Fourier transform only increase linearly with system size, but their prefactor is very big and a distribution of these arrays is desirable. In addition, all parts of the program that do not contain loops over the number of bands have to be parallelized using another scheme, leading to additional communication and synchronization overheads.

Second, the data are distributed over the Fourier space components and the real space grid is also distributed [189, 190, 240, 274]. This scheme allows for a straightforward parallelization of all parts of the program that involve loops over the Fourier components or the real space grid. Only a few routines are not covered by this scheme. The disadvantage is that all three-dimensional Fourier transforms require communication.

Third, it is possible to use a combination of the above two schemes [1619]. This leads to the most complicated approach, as only a careful arrangement of algorithms avoids the disadvantages of the other schemes while still keeping their advantages.

Fourth, it is possible to distribute the loop over **k**-points. As most calculations only use a limited number of **k**-points and many even sample exclusively the Γ-point, this method is of limited use only. However, combining the distribution of the **k**-points with one of the other methods mentioned above might result in a very efficient approach.

The CPMD program is parallelized using the distribution in Fourier and real space. The data distribution is held fixed during a calculation, i.e. static load-balancing is used. In all parts of the program where the distribution of the plane waves does not apply, an additional parallelization over the number of atoms or bands is used. However, the data structures involved are replicated on all processors. This is the basic parallelization scheme that has been supplemented with more advanced techniques for massively parallel computers. Once the basic methods are discussed in the following sections, the advanced techniques can be explained in Section 8.4.

A special situation exists for the case of path integral calculations (see Section 5.4), where an inherent parallelization over the Trotter slices is present.

The problem is "embarrassingly parallel" in this variable and perfect parallelism can be observed on all types of computers, even on workstation farms (for instance connected by GigaBit only) or clustered supercomputers (called "meta-computing"). In practice, the parallelization over the Trotter slices will be combined with one of the schemes mentioned above, allowing for good results even on massively parallel machines with several thousand processors. Similar cases are encountered for other types of calculations, like linear response calculations with several independent perturbations and for optimization methods based on the nudged elastic band method [631, 632] where loosely coupled independent calculations are required.

8.2 Data structures

In addition to the variables used in the serial version as introduced in Section 3.8, local copies have to be defined which will be indexed by a superscript indicating the processor number. The total number of processors is P and each processor p has a certain number of plane waves, atoms, electronic bands, and real space grid points assigned:

N_{at}^p	number of atoms on processor p
N_p^p	number of projectors on processor p
N_b^p	number of electronic bands or states on processor p
N_{PW}^p	number of plane waves on processor p
N_D^p	number of plane waves for density/potential on processor p
N_x^p, N_y, N_z	number of grid points in x, y, and z direction on processor p
$N^p = N_x^p N_y N_z$	total number of grid points on processor p.

Note in particular that the real space grid is only distributed over the x coordinates. This decision is related to the performance of the Fourier transform, which will be discussed in more detail in the following sections. The distribution algorithm for atoms, projectors, and bands just divides the total number of these quantities into equal chunks based on their arbitrary numbering. The algorithms that use these parallelization schemes do not play a major role in the overall performance of the program (at least for systems accessible with the computers available today and in the near future), and small imperfections in load-balancing can be ignored.

Data structures that are replicated on all processors:

$r(3, N_{at})$ nuclear positions

$v(3, N_{at})$ nuclear velocities

$f(3, N_{at})$ nuclear forces

$fnl(N_p, N_b)$ overlap of projectors and bands

$smat(N_b, N_b)$ overlap matrices between bands.

Data structures that are distributed over all processors:

$g(3, N_{PW}^p)$ plane wave indices

$ipg(3, N_{PW}^p)$ mapping of **G**-vectors (positive part)

$img(3, N_{PW}^p)$ mapping of **G**-vectors (negative part)

$rhog(N_{PW}^p)$ densities (n, n_c, n_{tot}) in Fourier space

$vpot(N_{PW}^p)$ potentials (V_{loc}, V_{xc}, V_H) in Fourier space

$n(N_x^p, N_y, N_z)$ densities (n, n_c, n_{tot}) in real space

$v(N_x^p, N_y, N_z)$ potentials (V_{loc}, V_{xc}, V_H) in real space

$vps(N_D^p)$ local pseudopotential

$rpc(N_D^p)$ core charges

$pro(N_{PW}^p)$ projectors of nonlocal pseudopotential

$eigr(N_D^p, N_{at})$ structure factors

$dfnl(N_p, N_b^p, 3)$ derivative of fnl

$cr(N_{PW}^p, N_b)$ bands in Fourier space

$cv(N_{PW}^p, N_b)$ velocity of bands in Fourier space

$cf(N_{PW}^p, N_b)$ forces of bands in Fourier space.

Several different goals should be achieved in the distribution of the plane waves over processors. All processors should hold approximately the same number of plane waves. If a plane wave for the wave function cutoff is on a certain processor, the same plane wave should be on the same processor for the density cutoff. The distribution of the plane waves should be such that at the beginning or end of a three-dimensional Fourier transform no additional communication is needed.

To achieve all of these goals, the following heuristic algorithm [274] is used. The plane waves are ordered into so-called "pencils" and each pencil holds all plane waves with the same g_y and g_z components. The number of pencils on a processor is proportional to the work for the first step in the three-dimensional Fourier transform. The pencils are numbered according to the total number of plane waves that are part of it. Pencils are distributed over processors in a "round robin" fashion, switching directions after each round. This is done first for the wave function cutoff. For the density cutoff the distribution is carried over, and all new pencils are distributed according

Table 8.1. *Distribution of plane waves and "pencils" in parallel runs using different numbers P of processors. The example is for 32 water molecules at a density of 1 g/cm³ in a cubic box with a volume of 6479.0979 Bohr³ and a 70 Rydberg cutoff for the wave functions.*

P	Wave function cutoff				Density cutoff			
	Plane waves		Pencils		Plane waves		Pencils	
	max	min	max	min	max	min	max	min
1	32 043	32 043	1 933	1 933	256 034	256 034	7 721	7 721
2	16030	16013	967	966	128 043	127 991	3 859	3 862
4	8 016	8 006	484	482	64 022	63 972	1 932	1 929
8	4011	4000	242	240	32 013	31 976	966	964
16	2 013	1 996	122	119	16 011	15 971	484	482
32	1 009	994	62	59	8 011	7 966	242	240
64	507	495	32	29	4 011	3 992	122	119
128	256	245	16	14	2 006	1 996	62	59

to the same algorithm. Experience shows that this algorithm leads to good results for the load-balancing on both levels, the total number of plane waves and the total number of pencils. Table 8.1 shows the different distributions for an example of 32 water molecules at 70 Rydberg cutoff. Even for 128 processors the algorithm is capable to find a distribution with an excellent load-balancing. Special care has to be taken for the processor that holds the $\mathbf{G} = \mathbf{0}$ component. This component has to be treated separately from the nonzero components in the calculation of the overlaps and the particular processor that holds this component is called $p0$.

8.3 Computational kernels

There are three communication routines mostly used in the parallelization of the CPMD code. All of them are collective communication routines, meaning that all processors are involved. This also implies that synchronization steps are performed during the execution of these routines. Occasionally, other communication routines have to be used (e.g. in the output routines for the collection of data), but they do not appear in the basic computational kernels. The three essential routines are the `Broadcast`, `GlobalSum`, and `MatrixTranspose`. In the `Broadcast` routine, data are sent from one processor (px) to all other processors

$$x^p \leftarrow x^{\mathrm{px}} \ . \tag{8.1}$$

In the `GlobalSum` routine a data item is replaced on each processor by the sum over this quantity on all processors

$$x^p \leftarrow \sum_p x^p \ . \tag{8.2}$$

The `MatrixTranspose` changes the distribution pattern of a matrix, e.g. from row distribution to column distribution

$$x(p,:) \leftarrow x(:,p) \ . \tag{8.3}$$

On a parallel computer with P processors, a latency time of t_{L} (the time needed for the, first data to arrive) and a bandwidth of B (data transmitted per unit time), the time spent in the basic routines is

`Broadcast`	$\log_2[P]\{t_{\mathrm{L}} + N/B\}$
`GlobalSum`	$\log_2[P]\{t_{\mathrm{L}} + N/B\}$
`MatrixTranspose`	$Pt_{\mathrm{L}} + N/(PB),$

where it is assumed that the amount of communicated data N is constant. The time needed in `Broadcast` and `GlobalSum` will increase with the logarithm of the number of processors involved. The time for the matrix transposition scales for one part linearly with the number of processors. Once this part is small, then the latency part will be dominant and increase linearly. Besides load-balancing problems, the communication routines will limit the maximum speedup that can be achieved on a parallel computer for a given problem size.

Once the distribution of the data structures is done, the parallelization of the computational kernels is in most cases easy. In the `StructureFactor` and `Rotation` routines, the loop over the plane waves N_{D} has to be replaced by N_{D}^p. The routines performing inner products have to be adapted for the $\mathbf{G} = \mathbf{0}$ term and the global summation of the final result:

```
MODULE DotProduct
IF (p == p0) THEN
    ab = A(1) * B(1) + 2 * sdot(2 * (N_D^p - 1),A(2),1,B(2),1)
ELSE
    ab = 2 * sdot(2 * N_D^p,A(1),1,B(1),1)
END IF
```

```
CALL GlobalSum[ab]


MODULE Overlap
CALL SGEMM('T','N',Nb,Nb,2*NpPW,2,&
          & ca(1,1),2*NpPW,cb(1,1),2*NpPW,0,smat,Nb)
IF (p == p0) CALL SDER(Nb,Nb,-1,ca(1,1),2*NpPW,&
          & cb(1,1),2*NpPW,smat,Nb)
CALL GlobalSum[smat]
```

Similarly, the overlap part of the FNL routine has to be changed and the loops restricted to the local number of plane waves:

```
MODULE FNL
FOR i=1:Np,M
   IF (MOD(lp(i),2) == 0) THEN
      FOR j=0:M-1
         pf = -1**(lp(i+j)/2)
         FOR k=1:NpPW
            t = pro(k) * pf
            er = REAL[eigr(k,iat(i+j))]
            ei = IMAG[eigr(k,iat(i+j))]
            scr(k,j) = CMPLX[t * er,t * ei]
         END
      END
   ELSE
      FOR j=0:M-1
         pf = -1**(lp(i+j)/2+1)
         FOR k=1:NpPW
            t = pro(k) * pf
            er = REAL[eigr(k,iat(i+j))]
            ei = IMAG[eigr(k,iat(i+j))]
            scr(k,j) = CMPLX[-t * ei,t * er]
         END
      END
   END IF
   IF (p == p0) scr(1,0:M-1) = scr(1,0:M-1)/2
   CALL SGEMM('T','N',M,Nb,2*NpPW,2,&
             & scr(1,0),2*NpPW,cr(1,1),2*NpPW,0,fnl(i,1),Np)
END
```

```
CALL GlobalSum[fnl]
```

The routines that need the most changes are the ones that include Fourier transforms. Due to the complicated breakup of the plane waves, a new mapping has to be introduced. The map `mapxy` ensures that all pencils occupy contiguous memory locations on each processor:

```
MODULE INVFFT
scr1(1:N_x,1:N^D_pencil) = 0
FOR i=1:N^p_D
   scr1(ipg(1,i),mapxy(ipg(2,i),ipg(3,i))) = rhog(i)
   scr1(img(1,i),mapxy(img(2,i),img(3,i))) = CONJG[rhog(i)]
END
CALL ParallelFFT3D("INV",scr1,scr2)
n(1:N^p_x,1:N_y,1:N_z) = REAL[scr2(1:N^p_x,1:N_y,1:N_z)]


MODULE FWFFT
scr2(1:N^p_x,1:N_y,1:N_z) = n(1:N^p_x,1:N_y,1:N_z)
CALL ParallelFFT3D("FW",scr1,scr2)
FOR i=1:N^p_D
   rhog(i) = scr1(ipg(1,i),mapxy(ipg(2,i),ipg(3,i)))
END
```

Due to the mapping of the y and z directions in Fourier space onto a single dimension, input and output arrays of the parallel Fourier transform do have different shapes:

```
MODULE Density
rho(1:N^p_x,1:N_y,1:N_z) = 0
FOR i=1:N_b,2
   scr1(1:N_x,1:N^PW_pencil) = 0
   FOR j=1:N^p_PW
      scr1(ipg(1,i),mapxy(ipg(2,i),ipg(3,i))) = &
            & c(j,i) + I * c(j,i+1)
      scr1(img(1,i),mapxy(img(2,i),img(3,i))) = &
            & CONJG[c(j,i) + I * c(j,i+1)]
   END
   CALL ParallelFFT3D("INV",scr1,scr2)
   rho(1:N^p_x,1:N_y,1:N_z) = rho(1:N^p_x,1:N_y,1:N_z) + &
```

```
        & REAL[scr2(1:N_x^p,1:N_y,1:N_z)]**2 + &
        & IMAG[scr2(1:N_x^p,1:N_y,1:N_z)]**2
END


MODULE VPSI
FOR i=1:N_b,2
   scr1(1:N_x,1:N_pencil^PW) = 0
   FOR j=1:N_PW^p
      scr1(ipg(1,i),mapxy(ipg(2,i),ipg(3,i))) = &
           & c(j,i) + I * c(j,i+1)
      scr1(img(1,i),mapxy(img(2,i),img(3,i))) = &
           & CONJG[c(j,i) + I * c(j,i+1)]
   END
   CALL ParallelFFT3D("INV",scr1,scr2)
   scr2(1:N_x^p,1:N_y,1:N_z) = scr2(1:N_x^p,1:N_y,1:N_z) * &
      & vpot(1:N_x^p,1:N_y,1:N_z)
   CALL ParallelFFT3D("FW",scr1,scr2)
   FOR j=1:N_PW^p
      FP = scr1(ipg(1,i),mapxy(ipg(2,i),ipg(3,i))) &
           & + scr1(img(1,i),mapxy(img(2,i),img(3,i)))
      FM = scr1(ipg(1,i),mapxy(ipg(2,i),ipg(3,i))) &
           & - scr1(img(1,i),mapxy(img(2,i),img(3,i)))
      fc(j,i) = f(i) * CMPLX[REAL[FP],IMAG[FM]]
      fc(j,i+1) = f(i+1) * CMPLX[IMAG[FP],-REAL[FM]]
   END
END
```

The parallel Fourier transform routine can be built from a multiple one-dimensional Fourier transform and a parallel matrix transpose. As mentioned above, only one dimension of the real space grid is distributed in the CPMD code. This allows us to combine the transforms in the y and z direction to a series of two-dimensional transforms. The handling of the plane waves in Fourier space breaks the symmetry and two different transpose routines are needed, depending on the direction. All the communication is done in the routine `ParallelTranspose`. This routine consists of a part where the coefficients are gathered into matrix form, the parallel matrix transpose, and a final part where the coefficients are put back according to the mapping used:

```
MODULE ParallelFFT3D(tag,a,b)
IF (tag == "INV") THEN
   CALL MLTFFT1D(a)
   CALL ParallelTranspose("INV",b,a)
   CALL MLTFFT2D(b)
ELSE
   CALL MLTFFT2D(b)
   CALL ParallelTranspose("FW",b,a)
   CALL MLTFFT1D(a)
END IF
```

All other parts of the program use the same patterns for the parallelization as the ones shown in this section.

8.4 Massively parallel processing

Two types of limitations can be encountered when trying to run a parallel code on a computer: increasing the number of processors working on a problem will no longer lead to a faster calculation, or the memory available is not sufficient to perform a calculation, independently of the number of processors available. The first type of limitation is related to bad load-balancing or the computation becomes dominated by the non-scaling part of the communication routines. Load balancing problems in the CPMD code are almost exclusively due to the distribution of the real space arrays. Note that only the x coordinate is distributed and that there are typically of the order of 100 grid points in each direction. Figure 8.1 shows the maximal theoretical speedup for a calculation with a real space grid of dimension 100. The steps are due to the load-balancing problems initiated by the granularity of the problem (the dimension is an integer value). Note that no further speedup can be achieved once 100 processors are reached. The dotted curve in Fig. 8.1 shows measured speedups obtained from actual calculations of a realistic system using the CPMD code including all contributions of full pseudopotential/plane wave density functional calculations. This comparison clearly demonstrates that the load-balancing problem in the Fourier transforms affects the performance of this special example. Where these steps appear and how severe the performance losses are depends, of course, on the system under consideration.

In order to overcome this limitation a method based on processor groups has been implemented into the code. For the two most important routines where the real space grid load-balancing problem appears, the calculation

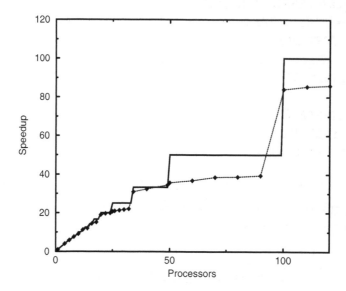

Fig. 8.1. Maximal theoretical speedup for a calculation with a real space grid of dimension 100 (solid line). Effective speedup obtained with CPMD for a 32 water molecule system with an energy cutoff of 70 Rydberg and a real space grid of dimension 100 (dotted line with diamonds)

of the charge density and the application of the local potential, a second level of parallelism is introduced. The processors are arranged into a two-dimensional grid and groups are built according to the row and column indices. Each processor is a member of its column group (`colgrp`) and its row group (`rowgrp`). In a first step a data exchange in the column group ensures that all the data needed to perform Fourier transforms within the row groups are available. Then each row group performs the Fourier transforms independently and in the end another data exchange in the column groups rebuilds the original data distribution. This scheme (shown in the pseudo code below for the density calculation) needs roughly double the amount of communication. Advantages are the improved load-balancing for the Fourier transforms and the bigger data packages in the matrix transposes. The number of plane waves in the row groups (N_{PW}^{pr}) is calculated as the sum over all local plane waves in the corresponding column groups.

```
MODULE Density
rho(1:Nˣᵖʳ,1:Nᵧ,1:N_z) = 0
```

```
FOR i=1:N_b,2*Pc
  CALL ParallelTranspose(c(:,i),colgrp)
  scr1(1:N_x,1:N_{pencil,r}^{PW}) = 0
  FOR j=1:N_{PW}^{pr}
    scr1(ipg(1,i),mapxy(ipg(2,i),ipg(3,i))) = &
        & c(j,i) + I * c(j,i+1)
    scr1(img(1,i),mapxy(img(2,i),img(3,i))) = &
        & CONJG[c(j,i) + I * c(j,i+1)]
  END
  CALL ParallelFFT3D("INV",scr1,scr2,rowgrp)
  rho(1:N_x^{pr},1:N_y,1:N_z) = rho(1:N_x^{pr},1:N_y,1:N_z) + &
    & REAL[scr2(1:N_x^p,1:N_y,1:N_z)]**2 + &
    & IMAG[scr2(1:N_x^p,1:N_y,1:N_z)]**2
END
CALL GlobalSum(rho,colgrp)
```

The use of processor groups (task groups) as shown in Table 8.2 using an IBM Blue Gene/L computer allows one to access efficiently massively parallel computers with more than $P = 1000$ processors even for rather small systems [661]. Even though the parallel efficiency decreases, there are still sizeable speedups possible which brings down the time per *ab initio* molecular dynamics step to something like 1 to 10 seconds for problems that belong currently to the small- and medium-size categories, respectively. This, in turn, allows us to generate throughput of the order of 100 and 10 picoseconds of *ab initio* trajectory per week, respectively, when using such massively parallel platforms. These realistic examples imply furthermore that nanosecond time scales will be accessible in the near future also on the level of *ab initio* simulations.

The effect of the non-scalability of the global communication used in CPMD is analyzed in Fig. 8.2 using a Cray T3E/600 (300MHz Dec Alpha EV56 processors, torus network) computer. This example shows the percentage of time used in the global communication routines, GlobalSum and Broadcast, and the time spent in the parallel Fourier transforms for a system of 64 silicon atoms with an energy cutoff of 12 Rydberg. It can clearly be seen that the global sums and broadcasts do not scale and therefore become more important the more processors are used. The Fourier transforms, on the other hand, scale nicely for this range of processors. The point where communication becomes dominant depends both on the size of the system and on the performance ratio of communication versus CPU.

Finally, the memory available on each processor may become a bottleneck

Table 8.2. *Two realistic examples on the performance of taskgroups for massively parallel processing using an IBM Blue Gene/L computer [661]; note that one node comprises two processors. Bulk system of 32 water molecules at 1 g/cm³ density and a liquid/vapor interface of 120 methanol molecules with a large vacuum region. Note that the results for the water system were obtained with an improved version of the code compared to Ref. [661]*

Number of nodes	Number of task groups	Time/step in seconds	Parallel efficiency
Water			
8	1	5.74	100%
16	1	2.88	100%
32	4	1.65	87%
64	4	0.87	82%
128	8	0.46	78%
256	8	0.33	54%
512	8	0.23	39%
Methanol			
128	1	75.2	100%
256	1	36.0	104%
512	1	29.5	64%
512	2	18.0	105%
1024	1	14.9	63%
1024	2	14.1	67%
1024	4	10.6	89%

for large computations. The replicated data approach for some arrays adapted in the implementation of the code poses limits on the system size that can be processed on a given type of computer. In the outline given in this chapter, there are two types of arrays that scale quadratically in system size that are replicated: the overlap matrix of the projectors with the wave functions (fnl) and the overlap matrices of the wave functions themselves (smat). The fnl matrix is involved in two types of calculations, where the parallel loop goes either over the bands or the projectors. In order to avoid communication, two copies of the array are kept on each processor. Each

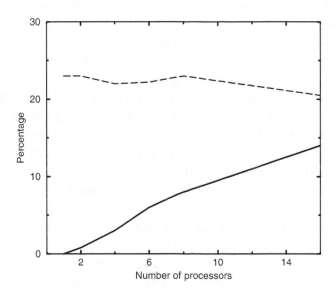

Fig. 8.2. Percentage of total CPU time spent in global communication routines (solid line) and in Fourier transform routines (dashed line) for a system of 64 silicon atoms on a Cray T3E/600 computer.

copy holds the data needed in one of the distribution patterns. This scheme needs only a small adaptation of the code described above. The distribution of the overlap matrices (`smat`) causes some more problems. In addition to the adaptation of the overlap routine, also the matrix multiply routines needed for the orthogonalization step have to be done in parallel. Although there are libraries available for these tasks, the complexity of the code is increased considerably.

There is an attractive alternative to the complicated schemes outlined above for computers built up from shared-memory multi-processor nodes [661, 662]. Using loop-level parallelism in a shared-memory environment allows for an increase in performance with only minor additional memory requirements. Shared-memory parallelization on the loop level is achieved by using OpenMP compiler directives and multi-threaded libraries (BLAS and FFT) if available. Compiler directives have been used to ensure parallelization of all longer loops (those that depend on the number of plane waves or the number of grid points in real space), and to avoid parallelization on the shorter ones. This type of parallelization is independent from the MPI parallelization and can be used alone or combined with the distributed-memory approach. Tests on various shared-memory computers have shown that an efficient parallelization up to 16 processors can be achieved. It does not come as a surprise that loop-level parallelism is very effective in CPMD. The code has a vectorization degree of above 99% and reaches routinely more than 75% efficiency on vector processors.

The combined approach is especially interesting for the following reasons. The shared-memory parallelization is also effective on the serial parts of the distributed-memory scheme, e.g. determination of rotation matrices in the constraint algorithm. For a given total number of processors P, the number of tasks involved in the distributed-memory parallelization can easily be decreased by one order of magnitude. This reduces drastically the impact of the latency in the all-to-all communications and allows us to obtain good scaling behavior for up to thousands of processors and enhances, in addition, the performance on loosely coupled clusters.

In order to demonstrate the performance of CPMD for such calculations, a set of supercell crystalline systems derived from the silicon carbide (SiC) unit cell and containing an increasing number of atoms ranging from 216 to 1000 have been used as characterized in Table 8.3. The systems are named *In* (where $n = 3$, 4, and 5) to indicate the rank n of the supercell considered; *I3*, for example, means a system consisting of a $3 \times 3 \times 3$ supercell of the original simple cubic cell of SiC (which contains 4 carbon atoms and 4 silicon atoms). Capability calculations and the demonstration of TeraFlop/s performance with CPMD have been carried out on the HPCx system in Daresbury (UK). This IBM p690 series computer consisted of 40 switched Regatta frames, each of them logically partitioned into four parts to give a cluster of 160 SMP (symmetric multiprocessing) partitions (8 Power4 1.3 GHz processors per partition) for a total of 1280 processors. These partitions are interconnected with a double-plane Colony switch having a latency of 21 μs and a bidirectional bandwidth of 720 MB/s. This supercomputer had a peak performance of ≈ 6.6 TeraFlop/s and a demonstrated LINPACK performance of ≈ 3.2 TeraFlop/s. When assessing the performance of the CPMD code one has to consider that a large effort has been made to minimize the operation count. Therefore any paper-and-pencil estimate of the number of operations during a single *ab initio* molecular dynamics step will overestimate the number of operations. With the help of a hardware performance monitor, the actual number of floating point operations could be determined directly. In the following, the number of operations used to calculate the performance was that derived from the runs on a single processor, whenever possible for memory reasons, otherwise the actual operation count was used. The maximum overhead introduced by the parallelization in the operation

Table 8.3. *Size of the three systems derived from cubic silicon carbide used for the tests [662].*

System	Number of atoms	Number of plane waves	FFT grid
I3	216	477 534	$128 \times 128 \times 128$
I4	512	1 131 630	$168 \times 168 \times 168$
I5	1 000	2 209 586	$256 \times 256 \times 256$

Table 8.4. *Systems I3, I4, and I5 from Table 8.3 on switched Regatta frames [662].*

System	Number of processors	MPI tasks	SMP threads	Time/step in seconds	Performance in GigaFlop/s
I3	1024	128	8	4.3	160
I4	672	64	8	20.1	380
	1280	160	8	10.6	703
I5	512	64	8	99.5	563
	1024	256	4	71.9	780
	1024	128	8	56.3	1017
	1232	154	8	52.1	1087

count, i.e. the number of operations with 1024 processors minus the number of operations with one processor, is less than 4%.

The results of these benchmark calculations of the SiC test system from Table 8.3 on an IBM p690 series computer are shown in Table 8.4 and in Fig. 8.3. The number of MPI tasks (and therefore the number of nodes) for the runs on the switched Regatta frames have been chosen in order to match as closely as possible the number of planes in the FFT grid. In order to test the limit of the parallelism in the CPMD code, one run has been performed on 32 switched frames. A rather low parallel efficiency ($\approx 15\%$) is measured for going from one frame to 32 frames. However, this can really be considered a limiting case. From the analysis of the time spent in the different parts of the code, one can see that $\approx 50\%$ of the total time is spent in the all-to-all communication of the FFT and that of this time, $\approx 90\%$ is latency bound.

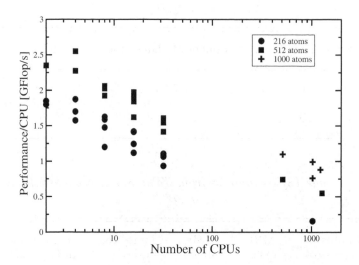

Fig. 8.3. Performance in GigaFlop/s per CPU for the three different SiC systems [662] from Table 8.3 on an IBM p690 series computer with up to 1232 CPUs.

For system $I4$ one finds a parallel efficiency of $\approx 36\%$ going from one frame to 40 frames and an almost perfect scaling from 21 to 40. The overall parallel efficiency (i.e. from one processor to 1280 processors) was 20%. The same observations as for system $I3$ can be made. More than 50% of the time is spent in the all-to-all communication related to the FFT, and this time is bound by latency. For system $I5$ there is no reference to determine the global parallel efficiency, but one obtains $\sim 20\%$ of the peak performance and because the single-processor calculations for the other systems give a maximum efficiency of $\sim 40\%$, one can reasonably expect a parallel efficiency of $\sim 45\%$. Moreover, going from 512 to 1024 processors, one recovers a parallel efficiency close to 90%. A very important point to highlight is the usefulness of the mixed approach. It helps in reducing the number of MPI tasks in all-to-all communications, partly hiding the latency of non-optimal switches. This is evident when comparing the results of the two runs with 1024 processors. The one with 256 MPI tasks and 4 SMP threads per task is 30% slower than the one with 128 MPI tasks and 8 SMP threads per task. A run with 1024 MPI tasks would yield almost no performance increase over a 256 processor run, owing to load-balancing and communication problems.

The same set of calculations is summarized in Fig. 8.3. The performance in GFlop/s per CPU is shown as a function of the total number of CPUs used in the calculation. Due to the parallel overhead, a decrease in performance for each system when going to more CPUs is seen. For a given number of CPUs, the performance increases when going to larger simulation systems. There are two different effects that cause this behavior. For larger systems the parallel overhead becomes important only for larger numbers of CPUs and the larger data structures on a single CPU allow for a more efficient use of the local resources.

Parallel computers have been instrumental in the success of *ab initio* molecular dynamics. Only with the computer power available through parallel computers are the impressive range of applications possible. On the other hand, the *ab initio* molecular dynamics codes with their excellent parallelization have provided landmark calculations that boosted the adaptation of parallel computers in science. The parallelization techniques outlined in this chapter will allow for an efficient usage of the current and at least next generation of parallel computers. This allows the *ab initio* molecular dynamics community to stay at the forefront of high-performance computing, and leaves enough time to develop new strategies for future generations of computer architectures.

Part III

Applications

9

From materials to biomolecules

9.1 Introduction

Ab initio molecular dynamics has been called a "virtual matter laboratory" [510] and even a "virtual laboratory" [934]. This notion is justified in view of the obvious parallels to experiments, being performed typically in the real laboratory. In the virtual lab, ideally, a system is prepared in some initial state and then evolves according to the basic laws of (quantum-statistical) physics – without the need for experimental input. The trajectory thus generated by *ab initio* molecular dynamics can be analyzed in arbitrary detail, including the dynamics of the electronic structure, which hence offers deep insights into the occuring processes. As a third step, the initial state and/or the external conditions such as for instance the temperature or composition of the system can be varied, but the system can also be exposed to light, electrical current, hydrostatic pressure, or uniaxial mechanical forces. Thus, not only the investigated matter as such is virtual, but also the apparatus used to manipulate matter, for instance a laser beam or an atomic force microscope, is fully represented "in silico". It is clear to every practitioner that this viewpoint is highly idealistic for more than one reason, but still this philosophy allows one to compute observables with predictive power and is at the same time the reason behind the broad application range and versatility of *ab initio* simulations. Furthermore, progress in the general availability of powerful computer hardware makes it easier as time goes on to follow the "virtual lab avenue" to achieve scientific progress.

It is also evident from the exploding number of papers dealing with *ab initio* molecular dynamics, see for instance Fig. 1.2, that a truly comprehensive and exhaustive survey of applications cannot be given in a meaningful way. Instead, the strategy chosen here is to provide the reader with an extensive, albeit by far incomplete collection of references that try to cover as

371

much of the broad scope of this approach as possible - instead of discussing in depth the physics or chemistry of only a few specific applications. These references are supposed to serve as pointers for the practitioner or newcomer to locate specific applications. In addition, they can be used to find more recent literature via citation databases.

In order to help the reader, the selected articles are organized in terms of various subfields. Furthermore, a separate section is included that is devoted to papers where novel analysis methods to compute properties have been introduced, used for the first time, or reviewed/explained in detail. The underlying selection of articles is based on a general literature search in order to suppress personal preferences as much as possible. A rather good "snapshot" of the field's state-of-the-art in 2005 can be gleaned from the Special Issue "Parrinello Festschrift: From Physics via Chemistry to Biology" [38]. Finally, a continuously updated list of articles, where specifically the CPMD *ab initio* molecular dynamics program package [696] has been used, can be found at its website http://www.cpmd.org/.

9.2 Solids, minerals, materials, and polymers

The very first application of Car–Parrinello molecular dynamics [222] dealt with bulk silicon, one of the basic materials in the semiconductor industry. Classic solid-state applications of this technique focus on the properties of crystals, such as those of CuCl where anharmonicity and off-center displacements of the Cu along the (111) directions were found to be important to describe the crystal structure as a function of temperature and pressure [127]. The A7 to simple cubic transformation in As was investigated using *ab initio* molecular dynamics at constant pressure [302]. Chemical bonding of elemental lithium has been studied using orbital and ELF analysis from the molecular regime to solids and surfaces [1245]. By applying external pressure, the hydrogen sublattice was found to undergo amorphization in $Mg(OH)_2$ and $Ca(OH)_2$ – a phenomenon that was interpreted in terms of frustration [1201].

Various properties of molecular solids, such as solid nitromethane [1503], crystalline nitric acid trihydrate [1413], solid benzene [988], stage-1 alkali-graphite intercalation compounds [609, 610], or the one-dimensional intercalation compound $2HgS \cdot SnBr_2$ [1254], were determined based on first principles. The molecular solid HBr undergoes various phase transitions upon compression. The dynamical behavior of one of these phases, disordered HBr-I, could be clarified using *ab initio* molecular dynamics [676]. Using *ab initio* constant-pressure simulations, phase transitions in solid H_2S have been studied including the calculation of theoretical IR spectra and analysis of the chemical bonding [1239]. Several HCl hydrate crystals have been studied in terms of free energy profiles for proton transfer and infrared spectra and compared to high-concentration liquid acid simulations [1346]. The structure as well as the associated infrared and inelastic neutron scattering (INS) spectra of the Zundel cation, $H_5O_2^+$, in crystalline $H_5O_2^+ClO_4$ environment were simulated using Car–Parrinello molecular dynamics which led to an assignment of the complex vibrations [1565]. Properties of solid cubane C_8H_8 were obtained in constant-pressure simulations and compared to experiment [1213].

Structure, phase transitions, and short-time dynamics of magnesium silicate perovskites were analyzed in terms of *ab initio* trajectories [1595]. The mechanism for the pressure-induced transformation of cristobalite to stishovite and post-stishovite phases has been obtained from constant-pressure *ab initio* molecular dynamics simulations [757]. The displacement of an oxygen atom in pure alpha quartz was studied via first principles molecular dynamics [157]. These simulations show that when an O atom in a Si-O-Si bridge is moved away from its original equilibrium position, a new stable energy minimum can be reached. Depending on the spin state and charge of the system, this minimum can give rise to either a threefold oxygen or to an unsaturated Si atom carrying a dangling bond. Quartz solubility and speciation of silica in aqueous fluid up to 1300°C and 20 kbar has been modeled based on the chain reaction formalism [501]. Metadynamics is a method to accelerate the exploration of free energy surfaces of complex systems, see Section 5.2.4, which is well suited to induce phase transitions [921] in solids and minerals as reviewed in Ref. [920]. Using its *ab initio* version, the high-pressure phase transition in the deuterated lithium hydroxide crystalline state has been studied in the isobaric/isothermal ensemble [1094]. Polytypic stacking-fault structures intermediate between the perovskite and post-perovskite phases have been investigated with *ab inito* metadynamics [1078].

In the quest to explore alternative gate dielectrics, first principles simulations within the local density approximation were used to investigate the electronic properties of the ZrO_2/Si and $ZrSiO_4$/Si interfaces [1168]. How the inclusion of nitrogen in a silica matrix changes its dielectric constant and the underlying mechanisms was elucidated [438]. The properties of fully relaxed vacancies in GaAs were investigated in Ref. [815]. Sliding of grain boundaries in aluminum as a typical ductile metallic material was generated and analyzed in terms of atomistic rearrangements [1023]. Microfracture in a sample of amorphous silicon carbide was induced by uniaxial strain and found to induce Si segregation at the surface [481]. A first principles derived approach is developed to study finite-temperature properties of $Pb(Zr_{1-x}Ti_x)O_3$ solid solutions near the morphotropic phase boundary, including their piezoelectric properties [89]. Using the *ab initio* path integral technique [931, 946, 948, 1507], see Section 5.4, the preferred sites of hydrogen and muonium impurities in crystalline silicon [1014, 1015], or the proton positions in $HCl \cdot nH_2O$ crystalline hydrates [1575] could be located. The proton diffusion process in imidazole-based molecular crystals, which are new candidate materials for fuel cell membranes, was simulated using dynamical first principles techniques [668, 1042].

Classical proton diffusion in crystalline silicon at high temperatures was an early application to the dynamics of atoms in solids [196] and continues to be of interest, for instance in amorphous silicon [1410]. The classical diffusion of hydrogen in crystalline GaAs was followed in terms of diffusion paths [1589] and photoassisted reactivation of H-passivated Si donors in GaAs was simulated based on first principles [1016]. Oxygen diffusion in p-doped silicon can be enhanced by adding hydrogen to the material, an effect that could be rationalized by simulations [220]. *Ab initio* dynamics also helped to quantify the barrier for the diffusion of atomic oxygen in a model silica host [592]. Diffusion of the self-interstitial in silicon has been studied at several temperatures in the range from 700 to 1200°C, thus determining the most frequent migration mechanisms and computing the migration energy and the Arrhenius prefactor [1264]. The microscopic mechanism of the proton diffusion in protonic conductors, in particular Sc-doped $SrTiO_3$ and Y-doped $SrCeO_3$, is studied via *ab initio* molecular dynamics, where it is found that covalent OH bonds are formed during the process [1339]. Ionic diffusion in a ternary superionic conductor was obtained by *ab initio* dynamics [1607]. The dynamic properties of mobile Cu ions in the superionic conductor CuI are studied by performing a population analysis as a function of time, which shows that the time evolution of local bonding plays an important role in high ionic conductivity [1337]. Using similar methods, the migration of Ag in the superionic conductor Ag_2Se was studied [1338]. Defect-induced self-diffusion in sodium is analyzed in terms of various defect mechanisms obtained via direct simulation, which indicate that the interstitial is a significant contributor for self-diffusion at temperatures close to melting, thus providing a possible explanation for the bending of the Arrhenius plot at high temperatures [1368]. Ferro/paraelectricity and isotope effects in hydrogen-bonded potassium dihydrogen phosphate, KH_2PO_4, crystals (KDP) have been studied [777, 778].

Even more complex dynamical processes occurring in condensed matter can be studied by *ab initio* dynamical methods. The early stages of nitride growth on cubic silicon carbide including wetting were modeled by depositing nitrogen atoms on the Si-terminated SiC(001) surface [480]. The radiation-induced formation of H_2^* defects in c-Si via vacancies and self-interstitials was simulated by *ab initio* molecular dynamics [399]. Proton motion and isomerization pathways of a complex photochromic molecular crystal composed of 2-(2,4-dinitrobenzyl)pyridine dyes was generated by *ab initio* methods [466]. Two distinct mechanisms of ammonium ion rotation in crystalline NH_4F have been characterized using *ab initio* calculations: in one all the cations rotate in phase, while in the other the rotating ions are isolated from each other by non-rotated ions [16].

The properties of polymers and macromolecular materials have also been investigated with *ab initio* simulations. Early applications of semiempirical ZDO molecular dynamics [1587] were devoted to defects in conducting polymers, in particular to solitons, polarons, and alkali doping in polyacetylene [1587, 1588] as well as to muonium implanted in *trans* and *cis* polyacetylene [439]. More recent are calculations of Young's modulus for crystalline polyethylene [579], soliton dynamics in positively charged polyacetylene chains [252], the spontaneous appearance of trans-gauche defects close to the melting temperature of polyethylene [1326], charge localization in doped polypyrrole [281], chain rupture of polyethylene chains under tensile load [1265], the influence of a knot on the strength of a polymer strand [1267], cis-polyacetylene under external stress [1220], or ion diffusion in polyethylene oxide [1098]. *Ab initio* molecular dynamics simulations were performed to study the motion of single metal atoms and atom clusters of Cu and Ta in polymers to gain an insight into their diffusion mechanisms and characteristics, showing that jumps between cavities inside the polymer are crucial and that cross-linking within the polymers does not significantly affect the scenario [304]. A multiscale modeling approach is used [326] to study the properties of a polycarbonate melt near a nickel surface, as a model system for the interaction of polymers with metal surfaces, showing strong chemisorption of chain ends, resulting in significant modifications of the melt composition when compared to an inert wall. Using similar methods, benzene adsorption onto Ni(111) and Au(111) has been simulated [1293]. Electronic, optical, and structural properties of the three base forms of polyaniline polymers have been investigated using the Car–Parrinello method with periodic isolated infinite chains [242] and a crystalline environment [243, 244]. By monitoring the time evolution of the geometry of the pernigraniline base polymer, evidence for oscillating quinoid-phenylene character of each chain ring is given [243, 244].

9.3 Surfaces, interfaces, and heterogeneous catalysis

A large number of dynamical studies dealing with surfaces, growth processes, and with adsorbed atoms and molecules possibly involving surface chemical reactions have appeared over the years. Specific to *ab initio* molecular dynamics is its capability to describe not only weak physisorption but also strong interactions leading to chemical bonds between the adsorbate and the surface (chemisorption), as well as dynamical processes on (and of) surfaces including reactions [1175]. Such dynamical first principles approaches to surface chemical reactions, and in particular to computational heterogeneous catalysis, expanded greatly in the 1990s and have established a mature subfield [511, 555, 697]. The interaction of hydrogen with clean and adsorbate-covered metal and semiconductor surfaces has been reviewed in Ref. [556]. *Ab initio* calculations of surface vibrational motion (phonons) can be obtained from frozen-phonon, linear-response, and molecular dynamics approaches as discussed and compared in Ref. [469] in the realm of semiconductor surfaces. Most such *ab initio* dynamical simulations are performed in the framework of slab boundary conditions introduced in Section 3.2.3. The convergence properties of the "quasi-periodic slab approach" to chemisorption studies have been compared to the "finite cluster approach" using CO on Cu(100) as a test system [1454] and dissociation of water on the defective (001) surface of PbS (galena) [1634].

The surface of liquid silicon has been studied, yielding a pronounced layering similar to that observed in low-temperature liquid metals like Ga and Hg, which is traced back to directional bonding of Si atoms at the surface that propagates into the bulk [401]. The surface dynamics of the (7×7) reconstructed Si(111) surface has been studied at finite temperatures [1577]. Various properties of the clean (0001) α-quartz surface have been determined [1215]. Investigations of dynamical and thermodynamic properties of sp-bonded metals, in particular the Al(110) surface close to its premelting point and the free surface of liquid Na, carried out in the framework of the Born–Oppenheimer molecular dynamics approach in the ensemble density functional formulation are reviewed in Ref. [1586].

Typical examples of first principles molecule/surface studies treat for instance C_2H_2, C_2H_4, and trimethylgallium adsorbates on the GaAs(001)-(2×4) surface [543], thiophene on the catalytically active $MoS_2(010)$ [1202] or RuS_2 [1369] surfaces, small molecules on a nitric acid monohydrate crystal surface [1475], CO on Si(001) [678], various molecules on TiO_2 [81, 1317], the structure of chemisorbed acetylene on the Si(001)-(2×1) surface [991], chemisorption of quinizarin on α-Al_2O_3 [464, 465], sulfur on

Si(100) at various coverages [1660], or sulfuric acid adsorbed on ZrO_2(101) and ZrO_2(001) [575]. The behavior of ammonia adsorbed on MgO(100) was investigated in Ref. [828]. The adsorption of amino acid species such as glycine, methionine, serine, and cysteine on partially hydroxylated rutile (100) and (110) wet surfaces including well-known defects was simulated by Car–Parrinello molecular dynamics [830]. Adsorption sites on the octahedral and tetrahedral surfaces of the kaolinite group of clay minerals have been studied using water and acetic acid molecules [1519].

Applications in the traditional surface science field include investigations of the diffusion of Si adatoms on a double-layer stepped Si(001) surface [739], the diffusion of a single Ga adatom on the GaAs(100)-c(4×4) surface [857], molecular motion of NH_3 on MgO [831], the dynamics of the (0001) face of hexagonal ice [901], the transition from surface vibrations to liquid-like diffusional dynamics of the Ge(111) surface [1437], molecular vs. dissociative adsorption of water layers on MgO(100) as a function of coverage [1072], the adsorption of HCl on the (0001) face of hexagonal ice [902], or desorption processes of D_2 from Si(100) [557]. The adsorption and reaction of water on a free amorphous silica surface has been studied with Car–Parrinello molecular dynamics leading to an exothermic formation of two silanol groups at a two-membered ring on this SiO_2 surface [1010]. Thermal contraction, the formation of adatom-vacancy pairs, and finally premelting was observed in *ab initio* simulations of the Al(110) surface at temperatures up to 900 K [962]. Hydroxylation and dehydroxylation reactions at the surfaces of amorphous silica and brucite, $Mg(OH)_2$, have been studied by Car–Parrinello molecular dynamics simulations. A particular topological defect on the amorphous silica surface, the two-membered (2M) silicon ring, is shown to react fast with water thus confirming the experimental assignment of the most reactive surface sites [964].

Early studies of chemical reactions on surfaces include those of the dissociation of an H_2O molecule on MgO [831, 832], dissociation of Cl_2 on GaAs(110) [903], or chlorine adsorption and reactions on Si(100) [315]. The initial stages of the oxidation reaction of a Mg(0001) surface by direct attack of molecular O_2 was simulated dynamically [203], including the penetration of the oxidation layer into the bulk. Similarly, the growth of an oxide layer was generated on a Si(100) surface [1521]. The oxidation of CO on Pt(111) [14, 1656] and the reaction $HCl + HOCl \rightarrow H_2O + Cl_2$ as it occurs on an ice surface [874] have been studied. The adsorption and initial dehydrogenation of methanol on Pt(111) was examined using *ab initio* molecular dynamics, revealing that in solution the methanol molecule orients with the -CH_3 group toward the surface, in contrast to the vacuum adsorption structure [973]. Chemisorption and physical adsorption of H_2O at the (100) and (110) surfaces of the rutile modification of TiO_2 have been studied using the Car–Parrinello method [829]. Spontaneous dissociation was observed in the simulation when a water molecule was inserted into a vacancy on the partially reduced (100) surface, whereas on the partially reduced (110) surface no spontaneous dissociation took place but hydroxylated configurations were stable. The energy dependence of ion-surface collision reactions leading to film growth has been investigated through *ab initio* molecular dynamics for collisions between Al^+ and a gibbsite surface in terms of reaction scenarios [1228] and microscopic insights into the local structural disorder upon impact [1229]. Collision reactions between cyano radical (CN) and dimethylacetylene (C_4H_6) were investigated using *ab initio* molecular dynamics showing nonreactive collision, incorporation, and substitution reaction channels [1438].

CO oxidation on Ru(0001) at low coverages has been investigated, confirming that Ru is very inactive for CO oxidation under UHV conditions, the activation of the chemisorbed O atom from the initial hcp hollow site (the most stable site) to the bridge site being the crucial step for the reaction [1655]. Heterogenous catalysts are often poisoned, which was for instance studied in the case of hydrogen dissociation on the Pd(100) surface in the presence of adsorbed sulfur layers [558]. The dynamics of NO_x species adsorbed on BaO(100) has been investigated [191] with *ab initio* molecular dynamics simulations at a temperature of 300°C. For both nitrites and nitrates, diffusion events between anion sites are observed, including a large number of possible adsorption configurations. These findings support the use of spillover mechanisms often postulated in mechanistic models of catalysts based on the NO_x storage and reduction concept. The diffusion pathways of O_2 and vacancies on a defective rutile TiO_2(110) surface have been studied [1467] in terms of molecular and dissociated adsorption using *ab initio* string molecular dynamics [725]. CO oxidation on a TiO_2(110) supported gold film has been investigated [876]. Structural and charging effects on bonding and catalyzed oxidation of CO on gold nanoclusters pinned by F-centers on the MgO(001) surface has been studied [581, 1649]. Coverage of the catalytically used ZnO surface with water leads to complex superstructures and interfacial dynamics [368, 998, 999]. Through first principles molecular dynamics the low-temperature oxidation of the Si(001) surface from the initial adsorption of an O_2 molecule to the formation of a native oxide layer is studied [271], where boron and phosphorus impurities are found to catalyze oxidation.

Atomic processes leading to growth have been studied using *ab initio* simulations: homoepitaxial crystal growth on Si(001), the low-temperature dynamics of Si(111)-(7×7) [1456], and homoepitaxial SiC growth [1214]. Similarly, it has been uncovered that the transition temperature for graphitization of flat and stepped diamond (111) surfaces depends sensitively on the type of surface [736]. An atomistic model for the nucleation of aluminum oxide on the Al(111) surface is derived from first principles molecular dynamics simulations: the process begins with the dissociative adsorption of O_2 molecules on the metal surface before the O atoms are spontaneously incorporated underneath the topmost Al surface layer, initiating the nucleation of the oxide far below the saturation coverage of one (1×1) O adlayer [270]. Qualitatively different growth mechanisms of the Cu/α-$Al_2O_3(0001)$ interface have been found by first principles molecular dynamics as a function of the transition metal coverage and temperature of the anhydrous vs. partially hydroxylated oxide surface [1273]. Si-doped (110) GaAs cross-sectional surfaces are investigated using first principles calculations with the aim of identifying simple defect configurations [365].

The Pd(100)/water, Pd(100)/O/water, and Si(111)/water interfaces were simulated based on *ab initio* molecular dynamics [755, 1531]. Water covering the surface of a microscopic model of muscovite mica is found to form a two-dimensional network of hydrogen bonds, called two-dimensional ice, on that surface [1073]. The adsorption of water on the $TiO_2(110)$ surface has been studied with *ab initio* molecular dynamics up to the bilayer regime of a fully hydrated surface, providing evidence for efficient proton transfer pathways along chain-like wire structures [1653] which seem to become less favored upon growth of further layers [1654]. The interaction of water and H_2S with the pyrite (100) surface was investigated [1401, 1402], as well as the adsorption properties of glycine at the water/pyrite interface at extreme thermodynamic conditions [155, 1047, 1156]. The adsorption of water on Pt(111) has been studied using *ab initio* molecular dynamics, yielding a well-ordered molecular bilayer on the surface [992]. The H-terminated InP(100)/water interface [492], Cu(110)/water and Ag(111)/water interfaces [691, 692], and dynamics and reactions of hydrated α-alumina surfaces [611] have been analyzed. Coadsorbed hydroxyl and water species on a Rh(111) surface have been studied by dynamical *ab initio* methods, and the possibility of a structural diffusion of the OH^- defect on the surface through a Grotthuss-like mechanism is proposed [1561]. The ice-to-water phase transition was observed in an *ab initio* molecular dynamics study of pressure-induced melting of a thin ice film confined between two parallel metal surfaces [1331]. The interface between the [001] face of crystalline aluminum and the coexisting liquid has been studied in an *ab initio* molecular dynamics simulation using an orbital-free density functional description of the electronic structure [711]. The liquid–vapor interface of pure water has been studied by large-scale Car–Parrinello molecular dynamics simulations [802, 935], where it could be shown that a certain system size is necessary in order to produce an intrinsically stable interface [1560].

The metal–organic interface of Pd–porphyrin and perylene monolayers on Au(111) was analyzed using an *ab initio* approach [825]. Self-assembled monolayers of thiolates on Cu(111) feature a metal-to-molecule charge redistribution and are found to be very mobile at room temperature, which generates a significant nonuniform diffuse background to Bragg scattering [773]. An interesting possibility is to compute the tip-surface interactions in atomic force microscopy as, for example, done for a neutral silicon tip interacting with an InP(110) surface [1469] or Si(111) [1140, 1142]. The relaxed atomic structure of a model ceramic/metal interface, {222}MgO/Cu was simulated, including lattice constant mismatch, using first principles local-density functional theory plane wave/pseudopotential methods with about 400 heavy atoms [99]. The liquid–vapor interfaces of neat lithium and sodium near their respective triple points [533], as well as those of two binary liquid alloys [534], $Na_{0.3}K_{0.7}$ and $Li_{0.4}Na_{0.6}$, have been studied with *ab initio* simulations using samples of 2000-3000 particles in a slab geometry with periodic boundary conditions. In the latter two cases, the total ionic density distributions along the normal to the interface display some layering with a virtually pure monolayer of the lower surface tension component located outermost at the interface [534].

9.4 Mechanochemistry and molecular electronics

Mechanically induced chemistry, in short "mechanochemistry" [793, 934], is a novel field where in general the interplay of chemistry and mechanics, and in particular mechanical activation of covalent bonds by externally applied forces, is studied [126, 461]. Constrained *ab initio* molecular dynamics simulations, see Section 3.7.3, are an ideal tool in order to study these effects. The influence of tensile stress on the properties of polymers, including mechanically induced chain ruptures and its mechanisms, have been studied [10, 1220, 1265–1267] using *ab initio* molecular dynamics methodology. The properties of siloxane elastomers under tensile stress have been simulated, in comparison to single-molecule AFM experiments [463]. Crucial here is the stress-induced chemical bond breaking occurring in the high-force regime, where a description of the molecular electronic structure is essential in order to determine the rupture mechanism.

Car–Parrinello molecular dynamics simulations, mimicking quasistatically AFM experiments, have demonstrated [792] that pulling a single thiolate molecule anchored on a stepped gold surface at room temperature does not preferentially break the sulfur–gold chemical bond, but leads to the formation of a monoatomic gold nanowire followed by breaking a gold–gold bond with a rupture force in the range of about 1–2 nN. This observation can be used to mechanochemically [793] modify clusters and surfaces by using such a "molecular nanohook". The behavior of copper is significantly different from gold in that there is a propensity to break the anchored molecule apart by cleaving the carbon–sulfur bond thus leaving chemisorbed S atoms on the surface [772].

Junctions consisting of organic molecules that are clamped between gold nanocontacts [1562] have been studied. The mechanical properties of suspended gold nanowires as such, pure gold chains, and those including non-metal impurity atoms have been simulated using *ab initio* molecular dynamics [642, 855, 856, 1067, 1068]. The molecular mechanism for the functionality of lubricant additives, pressure-induced changes of cross-linkers in chemical networks, has been studied on the nanoscale using *ab initio* molecular dynamics [1037]. The chemomechanical response of triphosphates and zinc phosphates to changes in pressure and temperature has been simulated up to maximum values of about 20 GPa and 1000 K in order to mimic roughly the extreme conditions to which phosphates may be exposed during their role as engine antiwear films [1038]. In these systems, atoms undergo pressure-induced changes in coordination number, which is partially reversible upon decompression, and feature proton transfer reactions.

Ab initio techniques were also used to investigate the role of tip–surface interactions in noncontact atomic force microscopy (AFM) [1139, 1141], atomic-scale image formation in noncontact AFM on metallic surfaces [342], and image formation in AFM of a GaAs(110) surface in terms of tip morphology/image contrast [732]. In the latter case, the effect of tip morphology on the image contrast was investigated by considering three different tip apexes on a Si tip: the Si apex with a half-filled dangling bond, the Ga apex with an empty dangling bond, and the As apex with a fully-filled dangling bond. It is shown that the dangling-bond state of the tip apex has a significant effect on the image contrast. AFM operating in the contact mode [704] has been studied, as well as the nanomanipulation capability of dynamic surface force microscopy [343]. Time-resolved dynamical STM imaging was performed for a wet ZnO surface where water molecules are found to autodissociate, thereby hydroxylating the surface and leading to an intrinsic averaging of the STM images which needs to be taken into account when interpreting measured images [368]. Point defects on ZnO in thermody-namic equilibrium with a gas phase such as F-centers and dimer vacancies were characterized by STM imaging as well as by scanning tunneling spectroscopy (STS) [779].

Theoretical and in particular computational molecular electronics [306, 1058] obtained a tremendous boost by various atomistic implementations of conductance calculations. For instance, the evolution of the structure and conductance of an Al nanowire subject to a tensile stress has been studied [706]. The calculations show the correlation between discontinuous changes in the force (associated with changes in the bonding structure of the nanowire) and abrupt modifications of the conductance as the nanowire develops a thinner neck [706]. The issue of forces that are induced by electric currents through molecules has been addressed as well [1566]. Light-induced conduction switching of photochromic molecules such as dithienylethene and azobenzene, suspended between leads, has been studied using molecular dynamics methods [861, 1652]. Maximally localized Wannier functions (see Section 7.2) have been used in order to compute [212] ballistic quantum conductance in metallic single-walled nanotubes [852] within the framework of Landauer's theory of equilibrium currents [306]. The additional contribution of inelastic collisions leads to a quantum dissipative current that has been worked out [205, 206, 493] and applied to an *ab initio* treatment of a monolayer of benzene dithiolate molecules between gold electrodes [494] relevant to molecular electronics devices.

9.5 Water and aqueous solutions

Molecular liquids and in particular associated ones certainly belong to the traditional realm of molecular dynamics simulations. The accurate description of water has been and continues to be a challenge [658, 1371] for both experiment and simulations due to the directional nature and the weakness of the hydrogen bonds in conjunction with its autodissociation properties. This leads, in particular in its fluid phases, to delicate association phenomena that are energetically in between (strong) covalent bonding and (weak) van der Waals interactions, the spontaneous formation of charge defects, and thus proton transfer [939]. Initiated by the pioneering *ab initio* simulation of water [817], this associated liquid turned out to play a crucial role in assessing both the power and limitations of *ab initio* molecular dynamics for simulating liquids. Thus, a rapidly growing number of investigations on water using similar methods has been emerging ever since, see for instance Refs. [26, 53, 165, 327, 411–413, 561, 618, 658, 693, 780, 804, 899, 900, 981,

1090, 1160, 1298, 1301, 1330, 1354, 1357, 1358, 1367, 1387, 1470, 1553] for a nevertheless severely incomplete list. Most of these investigations differ in minor technical aspects from the very first simulation [817], but importantly they can rely on a massive improvement of more data quality due to the gratifying development of computational power. In addition, electronic properties of water (and ice) have been investigated with great success, such as band structure and electronic density of states [656, 1160], various X-ray absorption spectra [238, 638, 1074], and optical absorption spectra [580]. More recently, the development of *ab initio* Monte Carlo simulations [804, 981–983], not only in the canonical and isobaric-isothermal but also in the Gibbs ensemble [468], paved the way to a direct investigation of the vapor-liquid coexistence of water [982, 983] and other liquids [782].

More intricate properties of pure liquid water, such as the autodissociation mechanism $H_2O(aq) \rightarrow H^+(aq) + OH^-(aq)$, its pH and p$K$ value have also been investigated [497, 1385, 1482, 1483]. Closely related is the study of the dissociation products themselves, i.e. the $H^+(aq)$ and $OH^-(aq)$ solvation complexes in liquid water, which impacts on such issues as the structural (Grotthuss) diffusion mechanism and the Eigen and Zundel solvation motifs [54, 55, 262, 694, 695, 955, 1494–1496, 1501, 1509, 1662]; see Refs. [936, 939] for reviews. The *ab initio* path integral technique, see Section 5.4, has been applied to study quantum effects on the structure of aqueous solutions [955, 957, 1509] and later of neat liquid water itself [261]. Proton transfer along short chains of water molecules ("water wires") in vacuum [987, 1260] and water alignment as well as proton transport through water-filled carbon nanotubes [323–325, 898] have been studied using *ab initio* molecular dynamics approaches. Such protonated water clusters can also be treated in more complex environments such as the membrane protein bacteriorhodopsin [613, 970, 1241, 1242]. The density and temperature dependence of proton diffusion in water have been studied in order to provide a key to understanding the acceleration of synthetic organic reactions in supercritical water and, more generally, in the framework of green chemistry [156]. Related is the issue of the dissociation of protic acids such as HCl in water [813, 814] and in highly concentrated HCl solutions [1345] or concentrated aqueous HF [1344]. Highly concentrated basic solutions of KOH have been analyzed in terms of structure, dynamics, and infrared spectra [1662].

Supercritical water [162, 163, 452] and water under extreme pressure and temperature conditions [241, 322, 527, 531, 1298–1300] is accessible using essentially the same methodology as for studying water at ambient conditions. The hydrated electron in water, $e^-(aq)$, at ordinary and supercritical conditions shows that in the former case, the electron cleaves an ellipsoidal cavity in the hydrogen bond network in which six H_2O molecules form the solvation shell whereas at supercritical conditions, instead, the electrons localize in pre-existing cavities in the discontinuous network [160].

Since chemical reactions often occur in aqueous media, the solvation properties of water are of utmost importance so that after the first study [954] of this kind the hydration properties of various ions [27, 28, 64, 113, 200, 655, 674, 675, 866, 867, 887, 1049, 1076, 1112, 1178, 1195, 1196, 1376, 1421, 1583, 1616] and solvated molecules [49, 50, 476–478, 813, 814, 858, 873, 1025, 1406, 1407, 1643] were successfully investigated by applying *ab initio* molecular dynamics to aqueous liquids [751]. The influence of the self-interaction error inherent to many functionals used to simulate liquids has been investigated for the hydroxyl radical in water, $OH^{\bullet}(aq)$, by applying semi-empirical self-interaction correction schemes [1557]. The important issue of the relative location of the HOMO in aqueous solution was addressed for several cations and anions from the microsolvation limit to the bulk extrapolation [655]. The various pK_a values of aqueous pentaoxyphosphoranes, $P(OH)_5(aq)$, at ambient conditions have been studied using coordination-constrained *ab initio* simulations [311, 358]. The redox properties of aqueous systems have been studied in great detail using dedicated *ab initio* molecular dynamics techniques [148–152, 657, 1448, 1453]. Using *ab initio* metadynamics, see Section 5.2.4, proton transfer of aqueous formic acid was shown to be water-mediated as deprotonation and reprotonation by means of a proton wire was observed [850]. *Ab initio* simulations of aqueous solvation of ethanol and ethylene reveal that the dipole moment of ethanol increases from 1.8 to 3.1 Debye upon solvation, while the apolar ethylene molecule attains an average dipole moment of 0.5 Debye in addition to featuring π-hydrogen-bonded solvation shell water molecules [1537]. Water-acetonitrile mixtures have been investigated focusing on the properties of hydrogen-bonded aggregates revealing dipole-bound dimer-like configurations [66]. The acidic properties of an equimolar HF/water mixture have been studied [1362] in terms of solvation complexes such as H_3O^+, $H_5O_2^+$, HF_2^-, and FH^-OH_2.

Hydrophobic hydration and anomalously fast diffusion of a H atom in water at ambient conditions have been studied using Car–Parrinello molecular dynamics and ELF analysis (see Section 10.2) of the electronic structure [745]. The observed anomalously fast diffusion of H atoms (and other small hydrophobic species such as H_2 or He) is explained in terms of a topological diffusion mechanism driven by shape fluctuations of the hydrogen-bonded cavity that hosts the solute [745]. Hydrophilic and hydrophobic interactions of wet hydroxyl- and methyl-terminated alkanethiol monolayers have been studied [1353]. The structure of the "Al$_{13}$ polymer", thought to be a major component of the aluminum chlorohydrate polymer system, has been studied both in the gas phase and in aqueous solution using Car–Parrinello molecular dynamics [1158, 1159], which shows that this polymer is stable on the picosecond time scale in an aqueous environment. A dimethyl sulfoxide/water mixture was studied with Car–Parrinello simulation techniques: a threefold coordination at the DMSO oxygen, methyl group hydrogen-oxygen atom contacts, Russel structures, and DMSO-$(H_2O)_2$ clusters are observed [742]. Using advanced techniques based on the analysis of Wannier functions, see Sections 7.2.6 and 10.2, solvent effects on electronic properties have been extracted in such a DSMO/water mixture [743]. Electronic properties of aqueous solutions, such as the UV absorption properties of Cu^+ and Ag^+ ions in water [112], have been studied using time-dependent density functional theory response techniques introduced in Section 5.3.6. Spectral properties of complex solutes in water can be treated efficiently via QM/MM approaches [1221], see Sections 5.5 and 5.3.6, as exemplified by recent studies of acetone [1221, 1416], aminocoumarins [1414, 1416], p-benzoquinone [1060], phosphate ions [748], or formaldehyde [1291, 1292] in ambient water, whereas all-QM approaches [112, 114, 115, 836] allow us to study nontrivial electronic interactions between solute and solvent.

9.6 Non-aqueous liquids and solutions

Many other molecular liquids exist that share the phenomenon of association with water being the best studied prototype system. Partial reviews on the subject of *ab initio* simulations as applied to hydrogen-bonded liquids can be found in the literature [541, 932, 936, 939, 1381, 1384]. Liquid HF, for instance, is a strongly associated liquid which features short-lived hydrogen-bonded zig-zag chains [1236]. Its microscopic structure as well as its static and dynamic properties have been simulated at several thermodynamic states [781], including the gas phase at liquid–vapor coexistence conditions [782]. Akin to water, charge defects such as excess protons in liquid HF introduce interesting dynamical phenomena which have been studied [1199, 1200] using the *ab initio* path integral technique outlined in Section 5.4.2 in order to include quantum effects. The thermodynamic and structural properties of liquid HF under pressure for different temperatures ranging between 500 and 3000 K have also been studied, leading to different mechanisms of proton transfer [893]. Liquid HCl at 310 K has been analyzed in detail concerning the hydrogen bonding and aggregate structuring of the liquid [366], but more importantly its dielectric susceptibility has been computed as a function of frequency [367]. The dissociation and recombination of hydrochloric acid in bulk liquid glycerol under ambient conditions was studied with *ab initio* simulations [1663].

The structure of liquid ammonia at 273 K was investigated with a variety of techniques so that the limitations of using classical nuclei, simple point charge models, small systems, and various density functionals could be assessed [346]. The *ab initio* simulated solvation behavior of "unbound electrons" in liquid ammonia at 260 K was found to be consistent with the physical picture extracted from experiment [331–334]. The strongly metallic Li/NH$_3$ liquid at saturation concentration has been analyzed in comparison to stretched pure NH$_3$ using the free energy functional and Wannier orbital analysis to decouple the unbound electrons [255]. *Ab initio* Monte Carlo simulations based on Hartree–Fock theory have been performed [600] for liquid ammonia at 277 K and 1 atm. A clear influence of three-body contributions has been detected at short distances upon comparison to simulations based on the pair potential approximation derived from the same electronic structure approach [600]. However, the hydrogen bond in the isolated ammonia dimer, (NH$_3$)$_2$, is known to be strongly bent (by about 20°), whereas all standard functionals including GGAs, hybrid functionals, meta-GGAs, as well as available force field models yield a (close to) linear hydrogen bond arrangement. Thus, a special parameterization of an HCTH-type functional, HCTH/407+, has been devised in order to cure that problem [164]. By applying this tailored functional to liquid ammonia, it could be demonstrated explicitly that the solvation shell of the ammonia molecule in the liquid phase is dominated by steric packing effects and not so much by directional hydrogen bonding interactions. In addition, the propensity of ammonia molecules to form bifurcated and multifurcated hydrogen bonds in the liquid phase is found to be negligibly small [164].

Another associated liquid, methanol, was simulated at 300 K using an adaptive finite-element method [1491], in conjunction with Born–Oppenheimer molecular dynamics [1488] as well as with the standard Car–Parrinello approach [598]. In agreement with experimental evidence, the majority of the molecules are found to be engaged in short linear hydrogen-bonded chains with some branching points [1488]. The microscopic mechanism of the proton transport process in liquid methanol has been simulated at 300 K where it is found that the defect structure associated with an excess proton is a hydrogen-bonded cationic chain whose length generally exceeds the average chain length in pure liquid methanol [1032]. Various properties of liquid formic acid have been determined from Car–Parrinello molecular dynamics [65]. Properties of neat liquid formamide (HCONH$_2$)

have been studied by Born–Oppenheimer *ab initio* molecular dynamics using an adaptive finite-element method [1487]. The molecular ionic liquid 1,3-dimethylimidazolium chloride has been investigated using Car–Parrinello molecular dynamics [199]. Car–Parrinello molecular dynamics simulations of supercritical carbon dioxide (sc-CO_2) have been performed at temperatures of about 320 K and density of ≈ 0.7 g/cc in order to understand its microscopic structure and dynamics [1263]. Blue-shifted intermolecular hydrogen bonds have been studied using Car–Parrinello simulations at 100 K using fluoroform dissolved in liquid carbon monoxide [1219]. Molecules in complex but chemically inert molecular solvents, i.e. those that interact mainly via electrostatic and van der Waals interactions, can be simulated efficiently by QM/MM approaches, see Section 5.5, using a force field MM model for the solvent as applied in the case of acetone and aminocoumarins in acetonitrile [1414, 1416].

Ab initio molecular dynamics is also an ideal tool to study other complex fluids, in particular those with partial covalency, metallic fluids, molten salts, as well as their transformations as a function of temperature, pressure, or concentration. *Ab initio* molecular dynamics of dilute [451, 1316] and concentrated [1350] molten $K_x \cdot (KCl)_{1-x}$ mixtures were performed at 1300 K entering the metallic regime by crossing the insulator–metal transition. The properties of water-free KF $\cdot$ nHF melts depend crucially on polyfluoride anions $H_m F_{m+1}^-$ and solvated K^+ cations. *Ab initio* simulations allow for a direct comparison of these complexes in the liquid, gaseous, and crystalline phase [1574]. The changes of the measured structure factor of liquid sulfur as a function of temperature can be rationalized on the atomistic level by various chain and ring structures that can be analyzed statistically in *ab initio* molecular dynamics simulations [1485]. Liquid $GeSe_2$ is characterized by strong chemical bonds that impose a structure beyond the usual very short distances due to network formation [966, 967]. The viscosity of liquid CdTe has been computed from *ab initio* molecular dynamics using the Stokes–Einstein relation [759]. Zintl alloys such as liquid NaSn [1314] or KPb [1319] have very interesting bonding properties that manifest themselves in strong temperature and concentration dependences of their structure factors (including the appearance of the so-called first sharp diffraction peak [1320]) or electric conductivities. Atomic structures of liquid alloys $Al_{80}Mn_{20}$ and $Al_{80}Ni_{20}$ have been calculated by first principles molecular dynamics simulations [698]. Pronounced short-range ordering is found in both systems, but the atomic magnetic moments play a key role in determining the short-range arrangement in case of the Mn system [698]. Constant pressure first principles molecular dynamics simulations were carried out to study structural and electronic properties of the polymeric phase of liquid phosphorus at high pressures: around 1 GPa atoms are connected by p-type covalent bonds and feature a Peierls distortion as in liquid arsenic [1028, 1029]. Later, it has been argued that the local density approximation used is not adequate and that chains of linked and opened up ("butterfly") P_4 molecules may serve as a seed triggering the transition from the molecular to the network phase [502]. The liquid–liquid transition observed in phosphorus at high temperature and high pressure is shown to be a transition from a nonmetallic molecular liquid composed of stable tetrahedral P_4 molecules to metallic polymeric liquid [1321]. Under ambient conditions, pure antimony pentafluoride (SbF5) is a strongly associated liquid which has been investigated by Car–Parrinello molecular dynamics [1198]. A very strong ionic character is found for the Sb–F bond and a pronounced tendency for molecular oligomerization via a barrierless, diffusion-limited process.

Metals are ideal systems to investigate the metal–insulator transition upon expansion of the liquid [128, 789] or upon melting [313]; see Section 9.8 for applications in the realm of extreme thermodynamic conditions. Liquid copper was simulated at 1500 K where structural and dynamical data were found to be in excellent agreement with experimental [1111]. The direct coexistence of solid and liquid aluminum has been simulated [21] at zero pressure using systems containing up to 1000 atoms for 15 ps. The local structure of liquid tellurium at about 1100 K and at continuously varied densities has been investigated [1393]. *Ab initio* Monte Carlo simulations have been performed to determine the equilibrium properties of liquid lithium (and lithium clusters) at different temperatures [1592]. In particular, density functional methods were employed to calculate the potential energy change for each proposed change of configuration, which was then accepted or rejected according to the Metropolis Monte Carlo scheme. Transport coefficients of liquid metals can also be obtained from first principles molecular dynamics using the Green–Kubo formalism [1348, 1392]. The microscopic mechanism of the semiconductor-metal transition in liquid As_2Se_3 could be rationalized in terms of a structural change as found in *ab initio* simulations performed as a function of temperature and pressure [1341]. The temperature dependence of various properties of the liquid $In_{20}Sn_{80}$ alloy has been studied from about 800 to 1200 K with a focus on coordination [1658]. The III–V semiconductors, such as GaAs, assume metallic behavior when melted, whereas the II–VI semiconductor CdTe does not. The different conductivities could be traced back to pronounced structural dissimilarities of the two systems in the melt [515].

9.7 Glasses and amorphous systems

Related to the simulation of dynamically disordered fluid systems are investigations of amorphous or glassy materials which are subject to static disorder in conjunction with extremely slow dynamics. In view of the severe limitations on system size and time scale, and thus on the accessible correlation lengths and times, *ab initio* molecular dynamics can only provide fairly local information in this sense. Thus, it is often used together with force field molecular dynamics or Monte Carlo simulations employed to prepare suitable initial configurations. Within these inherent constraints the microscopic structure of amorphous selenium [646], the structure of phosphorus–selenium glasses [1323], glassy As_2Se_3 [860], tetrahedral amorphous carbon [487, 908, 980], hydrogenated amorphous carbon [134], hydrogenated silicon with a five-atom void [374], the amorphization of silica [1610], vitreous silica [101], boron doping in amorphous Si:H [405], as well as the Raman spectrum [1110] and dynamic structure factor [1113] of quartz glass and their relation to short-range order could be studied. The structural properties of glassy and liquid sodium tetrasilicate, $Na_2Si_4O_9$, obtained from Car–Parrinello molecular dynamics were compared to those obtained for the same samples by classical molecular dynamics simulations using an empirical interaction potential [681, 682]. Similarly, for an amorphous silica surface [1011] it has been found that although the model potential used was able to correctly reproduce the surface on the length scale beyond approximately 5Å, it is necessary to use an *ab initio* simulation method to reliably predict the structure at small length scales. It could be shown that specific differences are related to the modifications of the atomic charges due to the introduction of Na atoms, which are not taken into account in the classical simulations. The calcium aluminosilicate [CaO–Al_2O_3–SiO_2] melt is compared to a silica melt in terms of the local structures [102]. Femtosecond laser pulse-induced structural changes in vitreous silica and their role in changing the refractive index of the glass have been investigated using *ab initio* molecular dynamics simulation methods based on finite-temperature density functional theory [1318]. The microstructure of supercooled nickel has been analyzed in terms of orientational order and three-dimensional pair analysis techniques showing the importance of fivefold symmetry local structures [700]. In the case of undercooled zirconium liquids, the local order is more complex than the icosahedral one as traditionally suggested [699].

The properties of supercooled CdTe were compared to the behavior in the liquid state in terms of its local structure [516]. Defects in amorphous $Si_{1-x}Ge_x$ alloys generated by *ab initio* annealing were found to explain ESR spectra of this system [738]. The infrared spectrum of a sample of amorphous silicon was obtained and found to be in quantitative agreement with experimental data [316]. Pressure-induced structural transformations in hydrogenated silicon clusters [922, 923, 1021, 1022] have been investigated using the constant-pressure technique for finite, nonperiodic systems as outlined in Section 5.2.3. The CO_2 insertion into a model of argon-bombarded porous SiO_2 was studied [1188]. In particular, the electronic properties of amorphous GaN were investigated using *ab initio* methods [1409]. In a search for improved gate dielectric materials, realistic models of amorphous ZrO_2 were generated in a "melt-and-quench" fashion using *ab initio* MD and subsequently analyzed in terms of coordination statistics, vibrational, and dielectric properties [1659].

Larger systems and longer annealing times are accessible after introducing additional approximations into the first principle treatment of the electronic structure that underlies *ab initio* molecular dynamics. Using such density functional tight-binding (DFTB) methods [1315], a host of different amorphous carbon nitride samples with various stoichiometries and densities could be generated and characterized in terms of trends [1603]. Similarly, the pressure-induced glass-to-crystal transition in condensed sodium was investigated [41] and two structural models of amorphous GaN obtained at different densities were examined in terms of their electronic structure [1409].

9.8 Matter at extreme conditions

A strong advantage of *ab initio* simulations is their predictive power also at extreme thermodynamic conditions, an area where molecular dynamics relying on fitted potential models optimized for ambient conditions might encounter severe difficulties. At the same time, extreme conditions are a challenge to experiment, so that *ab initio* molecular dynamics offers a unique avenue to investigate and predict the properties beyond ambient conditions. Thus, high pressures and/or high temperatures such as those found in the Earth's core, on other planets, or on stars can easily be achieved in the "virtual laboratory" [510, 934]. This opens up the possibility of studying not only properties but also phase transformations and chemical reactions at these conditions [116]. Furthermore, conditions of geophysical and astrophysical interest can be produced in the real laboratory, using techniques based on diamond anvil cells, shock waves, or lasers for instance. The limitations of these experimental approaches are, however, not so much related to generating the extreme conditions as one might expect, but rather to actually measuring meaningful observables.

In the virtual laboratory this type of information is accessible and the melting of diamond at high pressure [484], the phase transformation from the antiferromagnetic insulating δ-O_2 phase to a nonmagnetic metallic molecular ζ-O_2 phase [1325], dense fluid oxygen [1005], the phase diagram of carbon at high pressures and temperatures [567, 1593], as well as transformations of molecular systems such as methane [30], carbon monoxide [110], nitrogen [971], CO_2 [571, 1324, 1446], H_2O ice [100, 103–107, 119, 591, 843–846, 1170, 1364, 1455, 1484], solid hydrogen [129, 130, 649, 747, 763, 1492], the phase diagram of water and ammonia up to 7000 K and 300 GPa [241, 527, 531], and solid $Ar(H_2)_2$ molecular solids [111] under pressure could be probed. The pioneering *ab initio* molecular dynamics studies of hot fluid hydrogen [15], including computations of the sound velocity of dense hydrogen at conditions on Jupiter [17], stimulated further investigations into that subject. Liquid deuterium between four- and sixfold compression and temperatures of 5000 and 10 000 K has been studied [482]. In this regime the liquid goes continuously from a dissociation/recombination regime, where a substantial proportion of atoms form D_2 complexes, to a scattering regime, where mostly atoms are present [482]. The equation of state of dense deuterium is computed in the density range of $0.67 \leq \rho \leq 1.60$ g/cm^3, where excellent agreement between path integral Monte Carlo and Car–Parrinello molecular dynamics is found [1004]. The equation of state of liquid deuterium up to eightfold compression and temperatures between 2000 and 20 000 K has been computed [177] with excellent agreement with gas gun shock wave measurements.

Along similar lines, properties of a liquid Fe–S mixture under Earth's core conditions [23], the viscosity of liquid iron [314, 1392], the melting of hcp iron at high pressure based on the full potential linear muffin tin orbital method [92], etc. were investigated at extreme state points. Constant pressure *ab initio* molecular dynamics has been used to investigate the mechanisms of compression of liquid SiO_2, which becomes denser than quartz at a pressure of about 6 GPa. This high compressibility is traced to medium-range changes in the topology of the atomic network due to the creation of topological defects [1479]. The optical properties of shock-compressed silica up to a pressure of 1200 GPa have been determined using *ab initio* molecular dynamics simulations [839]. A significant rise in conductivity and reflectivity as both the pressure and temperature increased is found, which is traced back to the dissociation of molecular systems in the SiO_2 fluid [839]. The melting curve of MgO in direct solid/liquid coexistence has been simulated up to 135 GPa, which is the relevant pressure at the core mantle boundary [22]. A review on *ab initio* simulations relevant to minerals at conditions found in the Earth's mantle is provided in Ref. [1614], whereas *ab initio* studies relevant to Earth's inner and outer core, i.e. iron alloys at pressure of ca. 300 GPa and temperatures of about 5000–6000 K are reviewed in Ref. [24].

Ab initio molecular dynamics of laser-induced non-thermal heating of silicon [1349, 1352] and graphite [1356] gave fascinating insights into the dynamics of electronically strongly excited matter. The structure and the electrical conductivity of warm dense gold are computed during the first picoseconds after a short-pulse laser illumination showing that the ions remain in their initial fcc structure for several picoseconds, despite electron temperatures ranging from a few to several electronvolts after excitation [979]. *Ab initio* molecular dynamics calculations were performed for the equation of state of aluminum, spanning condensed matter and dense plasma regimes [1420]. Electronic exchange and correlation are included with either a zero- or finite-temperature local density approximation potential, and above the Fermi temperature a final state pseudopotential has been used to describe thermally excited ion cores.

Zinc complexation has been simulated in hot hydrothermal chloride brines [601], which are highly concentrated aqueous $ZnCl_2$ solutions with temperatures of up to 300°C [601]. Adsorption, desorption, and chemical reactivity properties of glycine at the surface of pyrite in contact with hot and pressurized water, mimicking the conditions in deep sea near hydrothermal vents, were studied with Car–Parrinello methods [155, 1156], including defects and free energy surfaces for desorption [1047].

9.9 Clusters, fullerenes, and nanotubes

Investigations of clusters made of silicon and simple metal atoms were among the first applications of *ab initio* molecular dynamics. Here, the feasibility to conduct finite-temperature simulations and in particular the possibility to search globally for minima turned out to be instrumental for success [8, 71, 647, 648, 1177, 1231, 1233, 1313], see e.g. Refs. [34, 70, 715] for reviews. Such investigations focus more and more on more complex clusters such as those with varying chemical composition [336, 337, 341, 437, 620, 797, 798, 1232]. Cluster melting is also accessible on an *ab initio* footing [176, 791, 1243, 1244, 1256] and molecular clusters, complexes, or cluster aggregates are also being actively investigated [137, 263, 265, 1083, 1240, 1246, 1458, 1459, 1508, 1581, 1582, 1647]. Quantum effects on the structural and electronic properties of metal nanoclusters made from light nuclei, such as Li_n clusters, have been investigated [320, 1243, 1244] using *ab initio* path integrals, see Section 5.4. Photoelectron spectroscopy of aluminum cluster anions in the range from ten to a hundred atoms has been investigated using Born–Oppenheimer molecular dynamics [9]. Pressure-induced structural transformations in hydrogenated silicon clusters [922, 923, 1021, 1022] have been investigated using the constant-pressure technique for finite, nonperiodic systems as outlined in Section 5.2.3. Molecular microsolvation clusters have been investigated in great detail with particular emphasis on aqueous solvation of ions, complexes, and small molecules [496, 655, 722, 805, 886, 942, 944, 993, 1044, 1468].

Type III–V semiconductor clusters embedded in sodalite minerals show quantum confinement and size effects that can be rationalized by *ab initio* simulations [198, 1478]. Supported clusters such as Cu_n on an $MgO(100)$ surface are found to diffuse by "rolling" and "twisting" motions with very small barriers [1045]. The diffusion of protonated helium clusters in various sodalite cages was generated using *ab initio* dynamics [434]. Photo-induced structural changes in Se chains and rings were generated by a vertical HOMO → LUMO excitation and monitored by *ab initio* dynamics [653].

With the discovery and production of finite carbon assemblies, *ab initio* investigations of the properties of fullerenes [35, 37, 1084], the growth process of nanotubes [121, 140, 256], or the electrical conductivity of nanowires [76, 582] became of great interest. The early stages of single-walled carbon nanotube growth on top of metal nanoparticles have been simulated dynamically showing that an sp^2-bonded cap is formed on an iron catalyst, following the diffusion of carbon atoms from hydrocarbon precursors on the nanoparticle surface [1192]. Spontaneous formation of a GaP fullerene cage has been found in *ab initio* molecular dynamics simulations starting from a bulk fragment [1476]. Modeling irradiation with ultrashort, very intense laser pulses, nonthermal melting of fullerite has been found in terms of C_{60} cage opening, leading to coalescence and formation of a fluid-like phase of small chains of twofold coordinated carbon [488]. *Ab initio* molecular dynamics simulations have been used to study the subpicosecond chemistry and dynamics of hyperthermal $O(^3P)$ collisions with single-walled carbon nanotubes, with a particular focus on insertion and substitutional doping [897]. Tube curvature effects are found to result in an increase of the epoxide binding energy with a decrease in tube diameter, whereas no noticeable effects of tube diameter on insertion were found.

9.10 Complex and fluxional molecules

In most cases the structure of molecules, clusters, or molecular aggregates can be described faithfully in terms of the concept of an "equilibrium structure", which is the global minimum of the relevant potential energy surface, and isomer structures being local minima. However, there are many examples where this traditional, static view is misleading in the sense that fluctuations, i.e. thermal fluctuations, zero-point vibrations or quantum tunneling, lead to qualitative changes that cannot be taken into account by *a posteriori* corrections. Thus, other well-defined concepts like the "most probable structure" or the "average structure", taking into account the influence of fluctuation effects, need to be considered [758] and in some extreme cases only a dynamical perspective can provide satisfactory insights [941]. Molecules where these structures deviate significantly from the equilibrium structure are often called fluxional or floppy molecules. *Ab initio* simulations are an ideal tool in order to include these effects *a priori* in the theoretical description and to investigate their impact [758].

The first examples where taking fluxionality into account has been demonstrated to be vital in order to understand the systems are the ground state of protonated methane (CH_5^+) [947, 952] including some isotopologues [951] and of protonated acetylene ($C_2H_3^+$) [758, 949] which were shown to be subject to constant dynamical rearrangements due to unavoidable quantum-mechanical fluctuation effects. The related dynamical exchange of atoms in these molecules can also be excited by thermal fluctuations [56, 178, 758, 796, 950, 1486]; a dynamical animation of such a scrambling process is available in Ref. [941]. In addition, it was shown that CH_5^+ is three-center two-electron bonded and that this bonding topology does not change qualitatively in the presence of strong quantum motion [953]. The fluxional behavior of the protonated ethane molecular ion $C_2H_7^+$ was investigated by *ab initio* molecular dynamics as well [377]. Another example of this class of molecules is the organometallic compound $C_2H_2Li_2$, which has an unexpected ground state structure that was found only after careful *ab initio* simulated annealing [1235]. In addition, this complex shows at high temperatures intramolecular hydrogen migration that is mediated via a lithium hydride subunit [1235]. The neutral and ionized SiH_5 and Si_2H_3 species display a rich dynamical behavior which was seen during *ab initio* molecular dynamics simulations [532]. The lithium pentamer Li_5 was found to perform pseudorotational motion on a time scale of picoseconds or faster at temperatures as low as 77 K [505]. The pseudorotation mechanism in SF_4 was investigated using first principles molecular dynamics where the chemical shieldings of the fluorine atoms were computed along the trajectory [1120].

Quantum effects in complex molecular systems can be considered within the framework of *ab initio* path integrals as introduced in Section 5.4. The very first such study focused on H_5^+ [946] followed by *ab initio* quantum simulations of CH_5^+ [947, 951–953], $C_2H_3^+$ [949], $H_5O_2^-$ and $H_3O_2^-$ [264, 1508], $H^+ \cdot (H_2O)_n$ [946, 1581, 1582], and Li_8 [320, 1243, 1244, 1599]. More recently, isotope effects on the hydrogen bond in $NH_4^+ \cdots CBeH_2$ and on small molecules such as ammonia, H_2O, H_3O^+, $(H_2O)_2^-$ in the gas phase have been studied as well using *ab initio* path integral technology [612, 1333, 1336, 1429–1432]. Similarly, unsaturated hydrocarbons such as benzene or ethylene have been investigated with respect to nuclear quantum effects [170–172, 877, 1179, 1185–1187, 1294, 1295] using more approximate, non-self-consistent methods. Implanted muons in organic molecules (benzene, 3-quinolyl nitronyl nitroxide, para-pyridyl nitronyl nitroxide, phenyl nitronyl nitroxide, and para-nitrophenyl nitronyl nitroxide) were investigated using approximate *ab initio* path integral simulations that include the strong quantum broadening of the muonium [1534, 1535]. Using *ab initio* instanton dynamics, the inversion splitting of the NH_3, ND_3, and PH_3 molecules due to the umbrella mode was estimated [730]. Similarly, a semiclassical *ab initio* dynamics approach was used to compute the tunneling rate for intramolecular proton transfer in malon aldehyde [95]. Even the influence of heavy-atom quantum effects on intramolecular proton transfer in malon aldehyde at room temperature can be quantified using constrained *ab initio* path integrals [1506].

Protonated water clusters are a challenge to both simulation and experiment. *Ab initio* simulated annealing can be used to explore the potential energy landscape and to locate various minima, such as for instance done for protonated water clusters [1601]. The gas phase IR spectrum of the $O \cdots H \cdots O$ fragment of $H_5O_2^+$ and its perdeuterated analogue were calculated using *ab initio* classical molecular dynamics based on MP2 forces [1563, 1564], including also the spectroscopically relevant adduct with Ar atoms [1274] and quantum effects have been assessed as well [1508]. One-dimensional protonated water clusters form chains or "water wires" which have been studied by *ab initio* simulation techniques in vacuum [987, 1260] and in confining structures such as nanotubes [323–325, 898] and bacteriorhodopsin [613, 970, 1241, 1242].

Ab initio molecular dynamics can also be used in order to optimize large and complex molecular systems. The determination of the structure of a RNA duplex including its hydration water [660], investigations of geometry and electronic structure of porphyrins and porphyrazines [826], and the simulation of a bacteriochlorophyll model crystal [904] are some applications to large molecules. Similarly, the "carboplatin" complex [1474] – a drug with large ligands – as well as the organometallic complex Alq3 [296] – an electroluminescent material used in organic light-emitting diodes – were investigated with respect to structural, dynamical, and electronic properties. Molecular dynamics simulations of trimethylaluminum $Al(CH_3)_3$ have been carried out in order to investigate the properties of the gas-phase dimer [68]. Photoisomerization intermediates generated by vertical excitation of the silver trimer anion complex have been simulated using CASSCF molecular dynamics [703]. The structures and vibrational frequencies of tetrathiafulvalene in different oxidation states were probed by *ab initio* molecular dynamics [729].

9.11 Chemical reactions and transformations

Early applications of *ab initio* molecular dynamics were devoted to reactive scattering of small molecules in the gas phase such as $CH_2 + H_2 \rightarrow CH_4$ [1590] or $H^- + CH_4 \rightarrow CH_4 + H^-$ [853]. The "on-the-fly" approach can be compared to classic trajectory calculations on very accurate global potential energy surfaces. This was, for instance, done for the well-studied exothermic exchange reaction $F + H_2 \rightarrow HF + H$ in Ref. [299]. Other gas phase reactions studied were $Li(2p) + H_2 \rightarrow LiH(^1\Sigma) + H(^1S)$ in Ref. [914], $F + C_2H_4 \rightarrow C_2H_3F + H$ in Ref. [175], $2O_3 \rightarrow 3O_2$ in Ref. [371], $F^- + CH_3Cl \rightarrow CH_3F + Cl^-$ in Ref. [1425], hydroxyl radical with nitrogen dioxide radical [348], formaldehyde radical anion with CH_3Cl in Ref. [1640], the reduction of $OH^\bullet$ with 3-hexanone [467] and other radical reactions in the atmosphere [458], or the hydrolysis (or solvolysis, S_N2 nucleophilic substitution) of methyl chloride with water [5, 6]. *Ab initio* molecular dynamics suggests that the thermal decomposition of 5-nitro-1-hydrogen-tetrazole might be the main contribution to the N_2-releasing process of 5-nitro-tetrazole's decomposition by uncovering three distinct reaction channels [1591]. Unimolecular thermal dissociation of acetic acid has been studied, and three dissociation channels were identified, in particular internal rotations were found to facilitate H migration among functional groups [875]. An *ab initio* molecular dynamics study of the S_N2 reaction $Cl^- + CH_3Br \rightarrow CH_3Cl + Br^-$ has been performed using Blue Moon sampling [1193]. Additional impact studies show that, depending on impact velocity, recrossing of the barrier can occur. Thermal effects on the $ClCH_2CN + Cl$ S_N2-reaction at 300 K have been studied by *ab initio* molecular dynamics where the mechanistic role of the cyano-substituent is explained by the formation of a hydrogen bond in a pre-reactive complex [1095]. *Ab initio* simulations [1641] reveal distinctive features of three types of mechanisms passing through the S_N2-like transition state for the reaction of $CH_2O^{\cdot-}$ with CH_3Cl. The four possible mechanisms of ring closure in dioxin formation from chlorophenols are studied using *ab initio* molecular dynamics in terms of free energy barriers [403]. Based on 1200 trajectories, the two photodissociation channels of formic acid, $HCOOH \rightarrow H_2O + CO$ and $HCOOH \rightarrow CO_2 + H_2$, have been studied [806]. In the heterolysis rearrangement of protonated pinacolyl alcohol, $(CH_3)_3C\text{-}CH(CH_3)\text{-}OH_2^+$, the stepwise pathway revealed by dynamical *ab initio* simulations at finite temperatures is found to be qualitatively different from the concerted mechanism predicted by the intrinsic reaction coordinate (IRC) approach [29]. These results imply that transition states and IRCs may have only limited importance with respect to the actual mechanism of chemical reactions, which offers new interpretations of organic reactivity where dynamical effects can play pivotal roles [29].

In addition to allowing us to study complex gas-phase chemistry, *ab initio* molecular dynamics also opened up the avenue to simulate condensed-phase chemistry at finite temperatures [39]. The crucial aspect hereby is the possibility of including the condensed phase environment at the explicit atomistic level in the simulation of chemical transformations. In particular, this allowed us for the first time to access "wet chemistry" directly in the "virtual lab" beyond using continuum solvation models. Some applications out of this emerging field are the cationic polymerization of 1,2,5-trioxane [295, 297], the initial steps of the dissociation of HCl in water [813, 814], the formation of sulfuric acid by letting SO_3 react in liquid water [990], or the acid-catalyzed addition of water to formaldehyde [989]. Reactive solvation of the ethyl radical cation has been studied in ambient water [1018]. Hydrocarbon reactivity has been investigated in the superacid SbF_5/HF, with emphasis on protonation of simple alkanes [1197]. Furthermore, gas phase and condensed phase reaction mechanisms can be compared most directly to probe the influence of the solvent. An important class of problems are systems where the solvent is not an inert spectator but plays an active role instead, as often enountered in solvent-mediated reactions. A convincing case study among many are Car–Parrinello simulations of the reaction of boron trichloride, BCl_3, with ammonia which have nicely shown that the mechanism in liquid ammonia differs markedly from the two-step gas-phase mechanism [1208].

The *ab initio* Car–Parrinello-type variant of the metadynamics technique [667, 821] described in Section 5.2.4 is an efficient novel tool in order to study complex chemical reactions in a few-dimensional subspace of generalized collective variables that spans the reaction coordinate as reviewed in Ref. [393]. The metadynamics method has already been applied to various chemical reactions and transformations in chemistry and chemical physics, see for instance Refs. [269, 392, 499, 668, 675, 720, 721, 850, 1094, 1403, 1404]. Using this technique, the Lewis-acid catalyzed hydrosilylation of phenylacetylene has been studied in the gas phase with a focus on the detailed reaction mechanism including the role of the catalyst [1666]. The free energy landscape of the E2 and S_N2 reaction channels of the well-known $CH_3CH_2F + F^-$ reaction have been explored [391], and the mechanism of the formation of β-lactone from epoxide and CO catalyzed by Co carbonyl has been simulated [1404]. Methods for accelerating chemical reactions based on bias potentials that depend on the system's HOMO-LUMO gap are constructed and applied to complex chemical reactions such as ethylene isomerization, $2 + 2$ cycloaddition of ethylene, and disrotatory cyclization of butadiene [1584], see also Ref. [1036].

Proton transfer is an elementary process which is at the heart of many phenomena of broad interest and thus has implications in many fields. Intramolecular proton transfer was studied in malon aldehyde [95, 1506, 1623], a Mannich base [406], and formic acid dimers [1013, 1532]. Targeted *ab initio* molecular dynamics has been used to analyze the reaction mechanism of double proton transfer in the formic acid dimer at room temperature [909]. The pioneering *ab initio* molecular dynamics simulations of proton and hydroxyl diffusion in liquid water were reported in the mid-1990s [1494–1496]. Intimately related to this problem is the autodissociation mechanism of pure water at ambient conditions [497, 1385, 1482, 1483]. Recently it became possible to study proton motion including nuclear quantum effects [666, 955, 957, 1199, 1200, 1508, 1509] by using the *ab initio* path integral technique [931, 946, 948, 1507], see Section 5.4.2. Protonation of nitric acid has been studied [679] in terms of intermediates and transition state structures to find that the dominant channel is $HNO_3 + H^+ \rightarrow NO_2^+ + H_2O$. *Ab initio* molecular dynamics simulations have been used to study the proton transfer reaction that converts neutral into zwitterionic glycine in aqueous solution in terms of potential of mean force associated with the direct intramolecular proton transfer event in glycine, thus giving insights into the free energy difference between the two species and the associated free energy barrier of this process [858].

Ab initio molecular dynamics also allows chemical reactions to take place in solid phases, in particular if a constant-pressure methodology is used [116], see Section 5.2. For instance, solid-state reactions such as pressure-induced transformations of methane [30], propene [1039], and carbon monoxide [110] or the polymerization [117] and amorphization [116] of acetylene were investigated. Solid-state reactions that are most relevant to extreme conditions are collected in Section 9.8.

More recently, photoinduced reactions of molecules also became accessible within the framework of *ab initio* dynamics, such as for instance the *cis–trans* photoisomerization in ethylene [93], excited state dynamics in conjugated polymers [138], bond breaking in the S_8 ring [1340], or transformations of diradicales [426, 427]. Further studies, in particular those relying on the ROKS and linear response time-dependent density functional theory approaches, are presented in Section 9.13.

9.12 Homogeneous catalysis and zeolites

The polymerization of olefines is an important class of chemical reactions on the industrial scale. In the light of such applications the detailed understanding of these reactions might lead to the design of novel catalysts. Driven by such stimulations, several catalysts were investigated in detail such as metal alkyles [1445], platinum–phospine complexes [286], or Grubbs' ruthenium–phosphine complexes [1], metallocenes [1630]. In addition, elementary steps of various catalytic processes were the focus of *ab initio* molecular dynamics simulations. Among those are chain branching and termination steps in polymerizations [1630], ethylene metathesis [1], "living polymerization" of isoprene with ethyl lithium [1237], Ziegler–Natta heterogenous polymerization of ethylene [158, 159, 161], Reppe carbonylation of Ni–CH=CH$_2$ using Cl(CO)$_2$ [40], or Sakakura–Tanaka functionalization [906]. Similar to the situation in a real laboratory, side reactions can occur also in the virtual laboratory, such as e.g. the β-hydrogen elimination as an unpredicted reaction path [907]. A digression on using finite-temperature *ab initio* dynamics in homogeneous catalysis research can be found in Ref. [1629]. Pioneering dynamical QM/MM Car–Parrinello studies [143, 1627, 1628], using the PAW transformation [142, 145] outlined in Section 6.2, focused on homogeneous catalysis [1624, 1628]. Valuable insights into the mechanism and stereocontrol of propene polymerization with sterorigid metallocences have been obtained based on Car–Parrinello simulations [710]. The enantioselective palladium-catalyzed hydrosilylation of styrene has been simulated using hybrid QM/MM Car–Parrinello molecular dynamics [892]. Rhodium-catalyzed hydroformylation of ethene has been simulated by probing the activity of several catalyst variants [514]. Using the powerful metadynamics technique from Section 5.2.4, the role of the catalyst in the Lewis-acid catalyzed hydrosilylation of phenylacetylene has been studied [1666]. Similarly, the mechanism of the $Co(CO)_4^-$ catalyzed β-lactone synthesis from epoxide and CO has been simulated [1404].

Zeolites often serve as catalysts as well, and are at the same time ideal candidates for finite-temperature *ab initio* simulations in view of their chemical complexity [479]. A host of different studies [213, 442, 443, 445–447, 576, 705, 1296, 1297, 1327, 1328, 1457, 1580] contributed greatly to the understanding of these materials and the processes occurring therein, such as the initial stages of the methanol to gasoline conversion [1580]. The oxidation of NO_2^- to NO^{3-} by O_2 is studed in a sodalite cage by using the combined Blue Moon ensemble and Car–Parrinello dynamics approaches [448]. Intermolecular electronic excitation transfer, EET, of chlorine host molecules inside the noncrossing channels of the zeolite bikitaite has been studied [444] using the ROKS excited state methodolgy from Section 5.3.2.

9.13 Photophysics and photochemistry

The field of first principles molecular dynamics involving excited states [96, 350, 355, 915] became effectively accessible in the mid to late 1990s and the "on-the-fly" treatment of nonadiabatic transitions in terms of both the forces and nonadiabatic couplings has been included only recently in 2002 for the first time [356]. Some examples are investigations of the *cis-trans* photoisomerization in ethylene [93], excited state dynamics in conjugated polymers [138], bond breaking in the S_8 ring [1340], transformations of diradicals [426, 427], excited state intramolecular proton transfer in the $S_2(\pi\pi^\star)$ state in malon aldehyde [278], or the photodynamics of azobenzene in its first excited state [1472]; see Ref. [578] for an assessment of the "multiple spawning method" [96, 915] in comparison to Tully's fewest switches surface hopping [1515–1517] and exact quantum dynamics in terms of model systems. The equation-of-motion coupled cluster (EOM-CCSD) method has been used "on-the-fly" in order to include explicitly electron correlation in excited state dynamics [61]. The processes induced by photoexcitation of $Cl^-(H_2O)_3$ and $I^-(H_2O)_3$ microsolvation clusters have been studied as well using *ab initio* molecular dynamics [770].

Access to excited state dynamics of truly complex molecular systems including liquid-state environments and QM/MM coupling schemes has again been achieved in the framework of density functional theory, in particular relying on the ROKS approach [462] and variants thereof, see Section 5.3.2. The first test application was the well-known photoinduced isomerization of an isolated formaldimine molecule (CH_2NH) in its first excited state, S_1, without including any nonadiabatic effects [462]. Early applications of ROKS were devoted to the *cis–trans* photoisomerization of retinal [1019, 1020], as reviewed in Ref. [1020]. The excited state structures and excitation energies from the ROKS and ROSS approaches, see Section 5.3.2, have been systematically compared to those from linear response TDDFT for a set of medium-sized molecules with different characteristic lowest excitations [1077]. Similarly, the generalized ROKS scheme has been applied to study and assess systematically the $\pi - \pi^\star$ transition in polyenes, cyanines, and protonated imines [554]. This approach has been applied to the photoinduced *cis–trans* isomerization of butadiene and to the conrotatory ring opening of cyclohexadiene [1061]. The performance of ROKS for a photoinduced *cis–trans* isomerization in the first excited state of formaldimine has been compared to multi-reference post-Hartree–Fock and TDDFT calculations along both the minimum energy path and realistic finite-temperature trajectories [353]. Using the ROKS methodology the photodissociation mechanism of diiodomethane (CH_2I_2) in acetonitrile solution has been studied with *ab initio* molecular dynamics simulations showing how the iso-diiodomethane photoproduct can be formed [1075]. Various nucleobases have been studied that way in the gas and liquid phases with a focus on their photostabilities [833, 834]. Intermolecular EET has been studied in a wire of chlorine molecules hosted in a zeolite channel using ROKS [444]. ROKS excited state simulations were used to explain the unidirectional nature of a complex light-driven intramolecular rotation process in a chiral molecule [553]. In the realms of materials science, photoinduced transformations of small rings in amorphous silica have been investigated by Car–Parrinello molecular dynamics [361] and by the first principles metadynamics approach introduced in Section 5.2.4 in terms of reaction mechanisms [360].

Simulations of photoinduced processes in complex systems including nonadiabatic transitions and radiationless decay via conical intersections, became accessible due to advances in *ab initio* molecular dynamics, see Sections 5.3.3 and 5.5.4. Excited state proton transfer (ESPT) and internal conversion in o-hydroxybenzaldehyde have been studied [352] using *ab initio* molecular dynamics beyond the Born–Oppenheimer approximation [356] based on a Kohn–Sham description of the electronic ground state coupled via Tully's surface-hopping to a ROKS treatment of the S_1 state. Using the same nonadiabatic simulation technique, photophysical and photochemical properties of nucleobases, in particular excited state proton transfer and radiationless decay, has been studied in the gas phase [351, 835] as well as in a microsolvation environment and in bulk aqueous solution [836]. More recently, the *ab initio* surface-hopping idea has been combined with explicit time-dependent Kohn–Sham theory to yield an Ehrenfest molecular dynamics scheme, see Sections 2.2 and 5.3.5. Such techniques have been applied to study the photoinduced electron injection from a molecular chromophore into a TiO_2 surface and the excited state relaxation of the green fluorescent protein (GFP) chromophore [289, 369, 370]. Conical intersection seams of molecular systems are complex low-dimensional geometric objects in high-dimensional space, but methods have been devised and applied in order to map them using constrained molecular dynamics [820] and to characterize them by computing nonadiabatic coupling vectors [135, 353].

Time-dependent density functional response theory has been implemented in a plane wave/pseudopotential *ab initio* molecular dynamics framework [354, 357, 659] as introduced in Section 5.3.6. These technical advances allowed us to compute (vertical and adiabatic) electronic excitation spectra, gradients in particular electronically excited states, and nonadiabatic coupling vectors between electronic states of complex molecules not only in isolation but also in solution. Issues such as the thermal broadening of the photoabsorption spectrum of formamide at room temperature [357] and finite temperature effects on the optical absorption properties of the $Ru^{2+}(aq)$ complex [113] can be addressed. A method to include in addition quantum-mechanical fluctuation effects, such as zero-point motion and tunneling of light nuclei, has been worked out by combining *ab initio* path integrals and the linear response approach to time-dependent density functional theory [320], which yields at the same time unambiguously the formal classical limit. The solvent shift and intensity enhancement of the $^1A_2$ $n \rightarrow \pi^\star$ electronic transition in acetone has been studied using TDDFT theory in the linear response framework [114]. The influence of Hartree–Fock exchange in linear response TDDFT has been assessed for charge transfer excitations in solvated molecular systems [115], in particular the $n \rightarrow \pi^\star$ transition in acetone using the B3LYP and PBE0 hybrid functionals in comparison to BLYP. Using dynamical TDDFT methods the photophysical properties of the s-tetrazine molecule have been investigated in the gas phase, in microsolvation clusters, and in the bulk solution to show that the solvent shifts can be explained by the polarization of the Kohn–Sham orbitals of the solute [1076].

The combination of excited state methods, such as ROKS or linear-response TDDFT, with a QM/MM coupling allows us to simulate directly the dynamics of photoprocesses occurring in quite complex environments. Prime examples are chromophores in biomatrices or in molecular solvents, which are both treated at the force field (MM) level only. Such combined techniques were used to study optical properties of acetone [1221, 1416] and aminocoumarins [1414, 1416], not only in water but also in more complicated solvents such as acetonitrile. This approach is not only computationally economical, as the simulation in solution is only marginally more demanding than the associated gas phase simulation, but also avoids problems of artificial mixing of solute and solvent orbitals and thus spurious excitations involving charge transfer due to the deficiencies of the underlying (semi-local) density functionals [114, 115]. Protein effects in relation to the rhodopsin chromophore, retinal, have also been studied within this mixed technique [1223]. The optical response of the green fluorescent protein (GFP) has been studied within a combined QM/MM and time-dependent density functional theory approach [911]. More biologically relevant examples are presented in Section 9.14.

9.14 Biophysics and biochemistry

Applications of *ab initio* molecular dynamics to molecules and processes of interest in life sciences [36, 226, 228, 1306] began to emerge in the mid to late 1990s with a significant boost due to the development of QM/MM approaches that are suitable to conduct stable, energy-conserving molecular dynamics simulations [143, 383, 388, 823, 1628], see Section 5.5. Early applications in biomolecular science were investigations of the crystal structure of a fully hydrated RNA duplex (sodium guanylyl-3'-5'-cytidine nonahydrate) [660], the simulation of a bacteriochlorophyll model crystal [904], structure models for the cytochrom P450 enzyme family [1308–1310], nanotubular polypeptides [227], a synthetic biomimetic model of galactose oxidase [1234], interconversion pathways of the protonated β-ionone Schiff base [1460], or of the binding properties of small molecules of physiological relevance such as O_2, CO or NO to iron-porphyrines and its complexes [1248–1252]. Cyclic AMP response element binding protein (CREB) is involved in the activation of transcriptional DNA machinery by binding to the coactivator CREB-binding protein (CBP). *Ab initio* simulations methods were used to investigate the dynamics and energetics of a relatively large, fully hydrated model complex representing pSer133 and its counterparts of the CBP domain [1126]. The reduction reactions for nitrogen fixation at Sellmann-type dinuclear model complexes have been studied with the help of Car–Parrinello simulation techniques [744], including photochemical activation processes [1206]. The workings of nitrogenase itself have been studied allowing for non-collinear magnetic structures of all iron centers [728]. The protonation state of the compound II intermediate (Cpd II) of the catalase reaction cycle has been studied using Car–Parrinello molecular dynamics [1247].

Ab initio molecular dynamics was used to study the gas-phase conformational dynamics of an alanine dipeptide analogue [1600]. It is found that conformational transformation between C5 and C7eq occurs on the picosecond time scale, whereas classical molecular dynamics using popular force fields does not yield a transition even after nanoseconds. The formation of a glycosidic bond in liquid water has been simulated, using constraint techniques, in order to extract a reaction scheme based on a detailed analysis of the electronic structure using Wannier functions and ELF analysis [1406, 1407]. Solute-solvent charge transfer in a peptide solvated in liquid water has been studied [1128] by Car–Parrinello simulations using Wannier function charge decomposition, as outlined in Section 7.2.6. Amino acids are important ingredients as they are the building blocks of polypeptides, which in turn form channels and pores for ion exchange. Motivated by their ubiquity, glycine and alanine as well as some of their oligopeptides and helical (periodic) polypeptides were studied in great detail [727]. Deprotonation of a histidine residue has been studied in aqueous solution using constrained Car–Parrinello molecular dynamics [683]. *Ab initio* molecular dynamics simulations have been used to study the proton transfer reaction that converts neutral into zwitterionic glycine in aqueous solution in terms of potential of mean force associated with the direct intramolecular proton transfer event in glycine, thus giving insights into the free energy difference between the two species and the associated free energy barrier of this process [858]. The adsorption properties of glycine at the water/pyrite interface, which is relevant for protein synthesis in the framework of the "iron-sulfur world" prebiotic scenario, were studied with Car–Parrinello methods [155, 1156] including metadynamics [1047] to explore desorption free energy surfaces.

Hybrid Car–Parrinello QM/MM calculations [228], see Section 5.5, were used to investigate the reaction mechanism of hydrolysis of a common beta-lactam substrate, cefotaxime, by a monozinc beta-lactamase, which suggests a fundamental role for an active site water in the catalytic mechanism binding the zinc ion in the first step of the reaction [1127]. Using similar techniques, the peptide hydrolysis reaction catalyzed by HIV-1 protease has been simulated [1147]. Key steps of the enzymatic reaction of caspase-3 were also studied. The hydrolysis of the acyl-enzyme complex is described with the BLYP functional, whereas the protein and the solvent are treated using the GROMOS96 force field [1415]. The structure and binding of cisplatin to DNA in aqueous solution were investigated where the platinated moiety is treated at the density functional level and the biomolecular frame with the AMBER force field [1378]. Using QM/MM Car–Parrinello simulations the hydrogen bonding pattern at the binding site of the complexes of HIV virus type-1 AP and the eukaryotic endothiapepsin and penicillopepsin has been investigated [1568]. Decarboxylation of orotidine $5'$-monophosphate (Omp) to uridine $5'$-monophosphate by orotidine $5'$-monophosphate decarboxylase (ODCase) is clarified based on molecular dynamics simulations [1194]. Also, the optical spectral properties of biomolecular complexes can be treated efficiently via QM/MM approaches [1060, 1414, 1416]. Based on suitable QM/MM schemes, various aspects of bacteriorhodopsin have been studied [613–615, 851]. In particular, theoretical infrared spectroscopy of biomolecular systems became possible in the framework of dynamical QM/MM simulations such as investigations of the ubiquinones Q(A) and Q(B) in the bacterial photosynthetic reaction center of *Rhodobacter sphaeroides* [1059] and the very anharmonic IR response of protonated water networks in bacteriorhodopsin [970, 1241, 1242]. The structural and dynamical properties of the four α-helix bundle Due Ferri 1 (DF1), which is a generic mimic of diiron proteins, have been explored using QM/MM Car–Parrinello simulations [891, 1100].

Proton transport through water wires is an important biophysical process in the chemiosmotic theory for biochemical ATP production. A pioneering Car–Parrinello study [1261] focused on proton transfer along a hydrogen-bonded chain of water molecules in a polyglycine analogue of the helical ion channel Gramicidin A (gA). Using the *ab initio* path integral technique [931, 946, 948, 1507], see Section 5.4, the properties of linear water wires with an excess proton were studied at room temperature [987]. Using classical nuclei instead, proton transfer along such wires in vacuum [1260] and in carbon nanotubes [323–325, 898] has been studied as well. Using dynamical QM/MM methods, excess protons in hydrogen-bonded water local-area networks ("WLANs") in bacteriorhodopsin have been investigated [970, 1241, 1242]. Hypothetical hydroxide and proton migration along the linear water chain in aquaporin GlpF from *Escherichia coli* were studied by QM/MM Car–Parrinello molecular dynamics simulations [709]. The dynamical flexibility and proton transfer in the bridged binuclear structural motif in the active site of arginase has been probed by *ab initio* molecular dynamics [684, 685].

Understanding the basic chemistry of the interaction of radicals with nucleobases is imperative when trying to elucidate the potential effects of radiation on DNA. Possible mechanisms for the gas-phase damage of guanine in the presence of an OH radical indicated that dehydrogenation of guanine (to yield a guanine radical and water) is spontaneous and most favored at all hydrogen sites except at C8; spontaneous hydroxylation at C8 and C4 is also found in accordance with experimental findings [1043]. The interaction of DNA bases with an OH radical in liquid water is investigated using Car–Parrinello molecular dynamics [1635], which indicates that the specific mechanisms of the initial phase of DNA damage are different in thymine and guanine. Using the "Earth Simulator" the mechanism of electron hole (positive charge) localization in a radical cation Z-DNA crystal has been simulated to show that at room temperature structural deformation does not provide an efficient localization mechanism; instead, evidence is given for the importance of changes in the protonation state for stabilizing the radical defect [500], see also Ref. [498]. Photophysics and photochemistry of several nucleobases have been studied [833, 834], also including explicitly the possibility of nonradiative decay from the first excited state to the ground state [351, 835]. Explicit excitation into the first electronically excited state and the subsequent nonadiabatic $S_1 \leftrightarrow S_0$ dynamics of guanine in liquid water have been studied [836] beyond the Born–Oppenheimer approximation by using the nonadiabatic Car–Parrinello scheme presented in Section 5.3.3.

Pioneering studies of aspects of the first step in the vision process, the 11-*cis* to all-*trans* photoisomerization of models of the rhodopsin chromophore, used standard Car–Parrinello molecular dynamics methodology [131–133, 197, 1125]. Later, an explicit treatment of the first excited state, S_1, within the ROKS scheme, see Section 5.3.2, has been used [1019, 1020] including also QM/MM S_1 studies [1222], see Section 5.5.4, An overview of early *ab initio* molecular dynamics approaches to photobiochemistry can be found in Ref. [1020]. Using QM/MM Car–Parrinello simulation, see Section 5.5, solvent and protein effects on the structure and dynamics of the rhodopsinchromophore, retinal, have been studied [1223]. *Ab initio* molecular dynamics using restricted Hartree–Fock theory have been performed to study interconversion pathways (bond rotation and ring inversion) of the protonated beta-ionone Schiff base [1460]. Raman and infrared frequencies as well as the corresponding intensities have been computed [1477] for DsRed model chromophores including *ab initio* molecular dynamics in the S_1 excited state using the linear response TDDFT gradients introduced in Section 5.3.6. Optical properties of GFP have been studied within a combined QM/MM and time-dependent density functional theory approach [911]. Also, the optical spectral properties of biomolecular complexes in solution can be treated efficiently via QM/MM approaches such as aminocoumarins [1414, 1416] and p-benzoquinone [1060].

10

Properties from *ab initio* simulations

10.1 Introduction

After having established *ab initio* simulation methods as such in the sense of being able to generate molecular dynamics trajectories based on first principles forces, enormous progress has been achieved, starting in the mid-1990s, in order to analyze them. These approaches must, of course, transcend the usual techniques used in classical simulations since the analysis can and should take advantage of the availability of the electronic structure information that is generated along the trajectories in addition to just knowing the positions and momenta of the nuclei as in force field-based simulations. In the following sections, a selection of publications is presented where such specific techniques were applied for the first time, assessed, reviewed, or compared to other approaches.

10.2 Boys–Wannier, population, ELF, and Fukui electronic structure analyses

It is well known from quantum chemistry that chemical bonding [807, 809, 812] is difficult to discuss in terms of the canonical orbitals only. Many routes have been followed in order to understand bonding in specific systems beyond looking at the molecular electron density difference with reference to non-interacting atoms [602]. An old idea (going back to Hund, 1931, according to Kutzelnigg [807, 812]) is that of using a suitable unitary transformation of the set of canonical orbitals to localized orbitals [380]. Several localization proceedures have been introduced for molecules, for instance the criteria proposed by Boys [184], Foster–Boys [456] or Pipek–Mezey [1154]. In simple cases the localized molecular orbitals (LMOs) produced can be related directly to Lewis-type structures in terms of localized bonds and lone pairs. The generalization of the localization idea

407

from finite to periodic systems is not straightforward at all, but the problem has been solved in terms of the maximally localized Wannier functions [960] as explained in Section 7.2; generalizations to metallic-like systems with entangled band and fractionally occupied bands are also available [1374, 1466]. This implies that the well-known analysis of molecules or clusters in terms of Boys LMOs can be carried over to the electronic structure and bonding in condensed matter systems where this is called Wannier analysis [960, 1347, 1355, 1357, 1664, 1669]; see Ref. [959] for an authorative review.

Going beyond the static analysis of selected configurations, this technique allows one to follow how the electrons "flow" during chemical reactions, thus making possible the extraction of typical reaction mechanism schemes [1406, 1407] on which organic textbooks rely heavily. An animation of the behavior of the dynamical three-center two-electron bond in the fluxional molecule CH_5^+ in terms of its LMOs can be found in Ref. [941]. In addition, electronic currents within molecules [11] as a result of chemical reactions can be extracted that put the concepts of concerted vs. nonconcerted reactions, as amply discussed in the context of cycloadditions, onto a firm theoretical basis. A step beyond the original scheme is the introduction of effective molecular orbitals [656, 1583] that are localized on individual molecules, for instance in a molecular liquid. This opens up possibilities to decompose bands of disordered condensed matter into thermal fluctuations in the orbital energies and electronic broadening due to intermolecular coupling [656].

Of course, standard population analyses using orbital based charge partitioning schemes such as Mulliken [1041], Löwdin [881], or Davidson [309, 622, 1218] charges as well as Mayer bond orders [978] can be computed during a dynamical *ab initio* simulation, see for instance Refs. [1337, 1338]. When using plane waves or other originless basis sets, see Section 2.8, this approach requires first a projection onto a localized auxiliary basis set [1270, 1271, 1311, 1312]. This approach has been implemented from early on in the CPMD package [696]. In particular, a (minimal) basis of Slater-type atomic orbitals or the pseudo atomic wave function of the atomic reference state used in the generation of the pseudopotential can be employed, see Section 3.6.1, followed by the population analyses using this optimized auxiliary basis.

Alternative electronic structure analysis methods are based on electron density as such rather than on the orbitals used to represent the charge density [59]. This leads to decomposition schemes that work in real space rather than in Hilbert space. Bader's "atoms in molecules" analysis [59] has also been investigated in the context of pseudopotential charge densities [630] and applied, for instance, to the controversial case of chemical bonding in a silicon analogue of alkynes [1152]. Another useful tool is the electron localization function (ELF) [87, 1276, 1277], which turned out to be a convenient approach to classify the electronic structure using topological concepts [60, 204, 1360]; see Ref. [1278] for a review and http://www.cpfs.mpg.de/ELF/ for a very useful online resource on the concepts, basics, and usage of ELF. Most importantly, systems that cannot be easily characterized in terms of LMOs, and thus not easily understood using localized bonds, are amenable to ELF analysis [769, 1243, 1245, 1278, 1360, 1509]. An extreme example is the fluxional behavior of CH_5^+, where quantum-mechanical fluctuations induce delocalization phenomena and thus fluctuating bonds [953] (not to be confused with resonance structures known from conjugated π-systems). Furthermore, the ELF approach is useful to study the evolution of chemical bonding from small molecules, clusters to condensed matter within a unifying framework [1245], or to investigate molecule/metal junctions [1240, 1246]. A generalization to open-shell and spin-polarized systems is available [768] and has been used to investigate a radical impurity in solution [745]. The extension to time-dependent processes including excited states has been worked out more recently in the framework of time-dependent density functional theory [207, 394].

Chemical reactivity indices are a generalization of the frontier orbital (HOMO/LUMO) concept which is particularly lucid in the framework of density functional theory [266, 495, 1102]. For instance, the concept of hardness/softness can be formulated in terms of (electronic) Fukui functions as site-specific reactivity indices and thus local measures of electrophilic and nucleophilic attack. They probe locally the electronic response of a system to small changes in the total number of electrons within a fixed nuclear configuration, and indeed reduce to the LUMO and HOMO densities in lowest order, i.e. when the density and thus the Kohn–Sham orbitals are not relaxed upon this perturbation. These functions can be conveniently evaluated numerically within the framework of density-functional perturbation theory, see Section 7.1. Implementing this approach in conjunction with *ab initio* molecular dynamics the electronic structure of aqueous solutions of Na^+ and Ag^+ was investigated [1583]. In particular, the nuclear Fukui functions were found to be very sensitive to the chemical nature of the component species, giving for Ag^+ a susceptibility 3.5 times the value for a H_2O molecule while the result for Na^+ is more than a factor of four smaller compared to a solvent molecule. The electronic structure of the solution is further characterized by construction of effective molecular orbitals and

energies. This analysis reveals that the effective highest occupied molecular orbital (HOMO) of the hard cation, Na^+, remains buried in the valence
bands of the solvent, whereas the HOMO of Ag^+ is found to mix with the
lone pair electrons of its four ligand H_2O molecules to form the (global)
HOMO of the solution. Electronic Fukui functions have been used in order
to compare the site-specific reactivity of an amino acid anchored in different
adsorption modes at a water/mineral interface to that of the surface atoms
themselves [1156].

Finally, as a note of caution, the reader is reminded that all these analyses
refer only to those electrons that are explictly included in the simulations,
i.e. the valence orbitals in the case of pseudopotential approaches discussed
in Chapter 4. Furthermore, particular attention has to be payed when using
non-norm-conserving schemes such as Vanderbilt's ultrasoft pseudopotentials by including that part of the "valence" charge density in the analysis
which is represented by the augmentation charges introduced in order to re-
lax the norm-conservation requirement, see Chapter 6 for details. Note that
schemes exist in order to re-introduce approximately in the atomic core regions the full electronic structure in terms of all-electron orbitals within the
frozen-core approach [490]. This is required in calculations of X-ray absorption and diffraction, NMR chemical shifts or hyperfine splittings among
others [997, 1148, 1536], which depend sensitively on the electronic structure at the nuclear cores. An elegant and general approach [637] to this
reconstruction can be achieved using the atomic orbitals which are obtained
within the PAW transformation due to Blöchl, see Section 6.2.

10.3 Dipole moments, infrared and Raman spectroscopy

Within the framework of force field molecular dynamics it is only possible
to compute vibrational power spectra or vibrational density of states. In
ab initio molecular dynamics, however, it is possible to compute both the
nuclear and electronic contributions to the total dipole moment (or bulk
polarization) of a system as outlined in Section 7.2. This, in turn, allows
us to compute the corresponding time autocorrelation function from which
the finite-temperature infrared spectrum, including properly intensities and
anharmonicities as well as mode-coupling and broadening effects, can be
obtained according to Section 5.4.5. The pioneering computation of the
infrared spectrum of liquid water [1354] and solid ice under pressure [119]
was possible thanks to the evaluation of the electronic bulk polarization
by means of the Berry-phase formulation (see Section 7.2). Furthermore,
the total polarization can be decomposed into the sum of contributions

stemming from the electronic charges that are assigned to the centroids of the Wannier localized orbitals, see Section 7.2.6. This observation allows us to compute in an approximate but well-defined way the contribution of individual molecular entities or subgroupings thereof to the total dipole moment (or bulk polarization). Using this idea led to the computation of the "molecular dipole moment" of individual H_2O molecules in liquid water, including the corresponding probability distribution function at room temperature [1357–1359]. Along the same lines it is possible to decompose, approximately, infrared spectra in terms of individual contributions. A decomposition scheme of complex spectra into linear contributions was applied to understand the origin of the broad infrared absorption of aqueous acids ("Zundel continua") [669, 671, 672]. The accumulated progress in the development of *ab initio* molecular dynamics methods for the computation of infrared absorption spectra in condensed phase molecular systems is reviewed and illustrated by a detailed account of an application [476] to aqueous uracil [477]. Methods to extract (quasi-) harmonic frequencies and normal modes from finite-temperature simulations have also been developed, including decompositions in terms of irreducible representations or local modes [761, 905, 942, 944, 1291, 1292].

Even within *ab initio* molecular dynamics the infrared spectra are computed in the framework of a classical–mechanical time evoluation of the nuclei. A step beyond the classical approximation is to use *ab initio* centroid molecular dynamics [956, 1119], which generates a quasi-classical time evolution of the nuclear dynamics as explained in Section 5.4.3. It has been demonstrated [1183] that this approach is clearly superior to any quantum correction factor scheme applied to classical lineshape functions, see Section 5.4.5. After a pioneering feasibility study of *ab initio* centroids [956], a detailed investigation of the vibrational dynamics of diatomic molecules, H_2 and HF, is presented in Ref. [1082]. Finite-temperature Raman spectra have been calculated from *ab initio* molecular dynamics of highly compressed ice as a function of hydrostatic pressure [1170].

10.4 Magnetism, NMR and EPR spectroscopy

Techniques have been developed and applied in order to compute magnetic susceptibilites and NMR chemical shifts of condensed matter systems using periodic boundary conditions [975–977, 1650]. The calculation of NMR chemical shifts and magnetic susceptibilities in systems under periodic boundary conditions based on simulted trajectories can assign the chemical shifts to individual atoms in experimental spectra [1303]. This approach has been generalized in the context of QM/MM NMR chemical shift calculations [1305]. First principles molecular dynamics of liquid oxygen was performed where the magnetic structure evolves according to a generalized density functional scheme allowing for noncollinear spin configurations and thus noncollinear magnetism [1070, 1071]. The calculated magnetic structure factor shows antiferromagnetic correlations between molecules in the first shell, transient configurations in which the molecular moments are aligned in an antiferromagnetic fashion are found as well as the occurrence of long-lived O_4 molecular units [1070, 1071]. Chemical shifts in solution were computed for various metal solvation complexes in water [201, 202], such as ^{51}V and ^{59}Co.

The influence of hydrogen-bond fluctuations on the g-tensor of benzo-semiquinone in aqueous solutation has been evaluated using Car–Parrinello molecular dynamics [49, 50]. First principles calculations of the EPR g-tensor in extended periodic systems are discussed in Ref. [317]. The temperature dependence of hyperfine coupling constants of the D_3O and H_3O radicals in the gas phase has been investigated with dynamical *ab initio* methods [1424]. Hyperfine interactions in aqueous solution of Cr^{3+} were studied in the framework of *ab initio* molecular dynamics [1645], taking into account core spin-polarization effects in the framework of pseudopotential calculations [1646].

10.5 Electronic spectroscopy and redox properties

Various versions of the time-dependent density functional linear response method for the computation of electronic excitation spectra, gradients in electronically excited states, and nonadiabatic coupling vectors have been implemented in a plane wave/pseudopotential *ab initio* molecular dynamics formalism [354, 357, 659]. The pioneering implementation has been applied to the thermally broadened photoabsorption spectrum of formamide at room temperature [357]. Such methods have been used to simulate the finite temperature absorption band of the Ru^{2+} hexahydrate coordination complex in aqueous solution in comparison to the corresponding $Ru^{2+}(H_2O)_6$ complex in vacuo, revealing that bulk solvation has a negligible effect on the position and shape of the absorption profile [113]. A method for including both quantum and thermal fluctuation effects on photoabsorption spectra of finite systems has been worked out by combining the *ab initio* path integral treatment of the nuclei with the linear response approach to time-dependent density functional theory (see Section 5.4.4) and applied to the

Li_8 cluster [320]. Using the classical approximation, line broadening effects have been studied in charged metal clusters [1035].

The photoelectron spectra of the Li_4F_4 cluster have been obtained at 200 and 500 K by evaluating the transition dipole autocorrelation function [620]. X-ray absorption spectra (XAS, NEXAFS) of water have been computed based on *ab initio* molecular dynamics simulations [238, 638, 1074] with particular focus on the full and half core hole (FCH, HCH) approaches [239] to computing inner-shell (core level) excitations within the (plane wave/pseudopotential) GGA approach. Going beyond the GGA approach is achieved by using the GW approximation on top of the GGA calculation and finally solving the Bethe–Salpeter equation. The resulting GGA + GWA + BSE combination allowed one to compute the optical absorption spectrum [580] (i.e. the imaginary part of the dielectric function) of hexagonal ice in convincing agreement with experiment. A procedure for calculating Auger decay transition rates, including effects of core–hole excited state dynamics [1436], has been implemented and applied to the normal and first resonant Auger processes of gas phase water ($O(1s) \rightarrow 4a(1)$ process) in comparison to high-resolution experiments.

The electrochemical properties of aqueous redox reaction systems such as $Ag^{2+} + Cu^+ \rightarrow Ag^+ + Cu^{2+}$, Ru^{2+}/Ru^{3+}, $RuO_4^{2-} + MnO_4^- \rightarrow RuO_4^- + MnO_4^{2-}$, or Ag^+/Ag^{2+} have been studied in great detail. This has been possible using specifically tailored *ab initio* molecular dynamics techniques [148–152, 657, 1448, 1453] including a dedicated grand canonical ensemble approach based on an explicit description of two redox states [1453].

10.6 X-ray diffraction and Compton scattering

A direct calculation of the coherent X-ray scattering spectrum [780] of liquid water under ambient conditions has been carried out using an all-electron implementation of the Gaussian augmented plane wave (GAPW) method [287, 1552]. Molecular dynamics can be used to sample fluctuations and thus generate broadening of X-ray diffraction data beyond the Debye-Waller approximation based on static structure models. The influence of finite-temperature dynamical disorder effects on X-ray diffraction has been demonstrated to be crucial for ice under extreme compression [107] when compared to experiment. Thiolate SAMs on copper surfaces have been shown to be mobile at room temperature, thus leading to a broad but nonuniform background to Bragg scattering due to a significant diffuse contribution to the coherent structure factor [773].

X-ray absorption spectra of water have been computed based on *ab initio* molecular dynamics simulations [238, 239, 638, 1074]. Compton scattering profiles were computed using Wannier functions and applied to analyze hydrogen bonds in various environments in terms of covalency [1225, 1226].

10.7 External electric fields, scanning probe imaging, conductivity, and currents

A static homogeneous electric field can be applied in density functional calculations using periodic boundary conditions and a nonlocal energy functional depending on the applied field within an *ab initio* molecular dynamics scheme [1372, 1530]. The reliability of the method is demonstrated in the case of bulk MgO for the Born effective charges, and the high- and low-frequency dielectric constants, which is evaluated by performing damped molecular dynamics in an electric field, thus avoiding the calculation of the dynamical matrix. Application of this method to vitreous silica shows good agreement with experiment and illustrates its potential for systems of large size. Using such methods the real and imaginary parts of the dielectric susceptibility of an associated liquid, hydrogen chloride at 313 K, have been studied as a function of frequency by evolving the system directly in a homogeneous electric field [367]. A scheme for performing *ab initio* supercell calculations of charged slabs at constant electron chemical potential rather than at constant number of electrons has been introduced for applications in electrochemistry [885].

Computational approaches to scanning probe microscopies at the atomic scale have been reviewed thoroughly [644]. The operation of the STM on metallic surfaces from the tunneling to the contact regime has been explored with a combination of first-principles total energy methods and a calculation of the electronic currents based on nonequilibrium Keldish–Green's function techniques, including forces and atomic relaxations [139]. The time evolution of STM images computed along an *ab initio* molecular dynamics trajectory has been evaluated and shown to lead to nontrivial averaging effects [368]. Atomic force microscopy (AFM) has been studied in both the noncontact mode [342, 732, 1139, 1141] used for imaging and in contact mode [704], where covalent tip-surface interactions play a role and can be exploited for nanoscale manipulation [343].

The electronic contribution to the conductivity is accessible in dynamical simulations (within the linear response approximation or first-order perturbation theory) via the Green–Kubo relations that rely on the current-current autocorrelation function [375] or via the Kubo–Greenwood approach which is a sum-over-states approach using the momentum operator matrix elements of the ground state [449, 483] or including excited states [1348, 1350, 1351]. Computing currents and thus electron transport in molecular wire junctions and nanostructures is a challenge to first principles approaches [1058]. Maximally localized Wannier functions (see Section 7.2) have been used in order to compute [212] ballistic quantum conductance in metallic single-walled nanotubes [852] within the framework of Landauer's theory of equilibrium currents [306]. The nonballistic contribution of inelastic scattering via collissons leads to a quantum dissipative current that has been worked out [205, 206, 493] and applied to an *ab initio* treatment of a monolayer of benzene dithiolate molecules between gold electrodes [494] relevant to molecular electronics devices. The issue of additional forces and torques that are induced by the flow of electric currents through molecules has been addressed [1566]. Light-induced changes of the electron transport properties of photochromic molecules with two isomeric state conformations have been studied using molecular dynamics methods [861, 1652].

11

Outlook

Standard *ab initio* molecular dynamics, which is based on ground state adiabatic dynamics using Kohn–Sham density functional theory in the microcanonical ensemble with classical nuclei, is by now, more than 20 years after its emergence [222], a well-established and mature simulation technology. There are many powerful and "easy-to-use" computer packages available that are able to perform *ab initio* molecular dynamics simulations (for instance ABINIT [2], CASTEP [234], CONQUEST [282], CPMD [696], CP2k [287], CP-PAW [288], DACAPO [303], FHI98md [421], NWChem [1069], ONETEP [1085], PINY [1153], PWscf [1172], SIESTA [1343], S/PHI/nX [1377], or VASP [1559] to name but a few), which are opening up new avenues for a broad range of applications. In academic research these codes are not only used by specialist groups as in the early to mid-1990s, but more and more in experimental groups in order to complement ongoing experiments - or even to allow for any meaningful physical interpretation of the measured data in the first place. In addition, *ab initio* simulations are becoming increasingly popular in industrial R&D as well. Analysis of databases such as those underlying Fig. 1.2 clearly shows that an increasing number of companies seem to be interested in this methodology. Researchers affiliated to Bayer, Corning, DSM, Dupont, Exxon, Ford, Hitachi, Hoechst, Kodak, NEC, Philips, Pirelli, Shell, Toyota, Xerox and others refer in their research publications to the Car–Parrinello paper [222] or use *ab initio* molecular dynamics in their work. All of this certainly contributed to the fact, which can be quantified by citation analyses such as the one shown in Fig. 1.2, that this utmost efficient *and* elegant Car–Parrinello approach to *ab initio* computer simulation [222] became a success story [1093, 1216].

Furthermore, the mode of most applications has shifted principally in the new millennium away from feasibility and proof-of-principle studies at the cutting edge toward systematic and large-scale investigations. This means that properties and processes can now be studied comprehensively by varying several control parameters such as temperature, pressure, chemical composition, etc. simultaneously, as known for a long time from traditional computer simulations. At the same time these new possibilities allow one to check more carefully than ever before for statistical as well as systematic errors in the results, for instance by studying the same system using more than one electronic structure method by employing, for example, different density functionals. This dramatic progress also implies that the cutting-edge examples given in Chapter 9 will already become standard applications within a couple of years and much of the sophisticated analysis techniques from Chapter 10 will be used routinely by then as well!

Bibliography

[1] O. M. Aagaard, R. J. Meier, and F. Buda. Ruthenium-catalyzed olefin metathesis: A quantum molecular dynamics study. *Journal of the American Chemical Society*, 120:7174–7182, 1998.

[2] ABINIT. Ref. [539]; distributed under the terms of the GNU General Public Licence; See http://www.abinit.org/.

[3] P. S. Addison. *The Illustrated Wavelet Transform Handbook*. Institute of Physics Publishing, Bristol, 2002.

[4] P. Ahlrichs, R. Everaers, and B. Dünweg. Screening of hydrodynamic interactions in semidilute polymer solutions: A computer simulation study. *Physical Review E*, 64:040501(R), 2001.

[5] M. Aida, H. Yamataka, and M. Dupuis. *Ab initio* molecular dynamics simulations on the hydrolysis of methyl chloride with explicit consideration of three water molecules. *Chemical Physics Letters*, 292:474–480, 1998.

[6] M. Aida, H. Yamataka, and M. Dupuis. *Ab initio* MD simulations of a prototype of methyl chloride hydrolysis with explicit consideration of three water molecules: a comparison of MD trajectories with the IRC path. *Theoretical Chemistry Accounts*, 102:262–271, 1999.

[7] H. Akai and P. H. Dederichs. A simple improved iteration scheme for electronic-structure calculations. *Journal of Physics C*, 18:2455–2466, 1985.

[8] J. Akola and M. Manninen. Aluminum–lithium clusters: First-principles simulation of geometries and electronic properties. *Physical Review B*, 65:245424, 2002.

[9] J. Akola, M. Manninen, H. Häkkinen, U. Landman, X. Li, and L.-S. Wang. Aluminum cluster anions: Photoelectron spectroscopy and *ab initio* simulations. *Physical Review B*, 62:13216–13228, 2000.

[10] D. Aktah and I. Frank. Breaking bonds by mechanical stress: When

do electrons decide for the other side? *Journal of the American Chemical Society*, 124:3402–3406, 2002.

[11] D. Aktah, D. Passerone, and M. Parrinello. Insights into the electronic dynamics in chemical reactions. *Journal of Physical Chemistry A*, 108:848–854, 2004.

[12] A. Alavi. Private communication.

[13] A. Alavi. Path integrals and *ab initio* molecular dynamics. In K. Binder and G. Ciccotti, eds., *Monte Carlo and Molecular Dynamics of Condensed Matter Systems*, chapter 25, pages 648–666. Italian Physical Society SIF, Bologna, 1996.

[14] A. Alavi, P. Hu, T. Deutsch, P. L. Silvestrelli, and J. Hutter. CO oxidation on Pt(111): An ab initio density functional theory study. *Physical Review Letters*, 80:3650–3653, 1998.

[15] A. Alavi, J. Kohanoff, M. Parrinello, and D. Frenkel. *Ab Initio* molecular dynamics with excited electrons. *Physical Review Letters*, 73:2599–2602, 1994.

[16] A. Alavi, R. M. Lynden-Bell, and R. J. C. Brown. The pathway to reorientation in ammonium fluoride. *Chemical Physics Letters*, 320:487–491, 2000.

[17] A. Alavi, M. Parrinello, and D. Frenkel. *Ab-initio* calculation of the sound-velocity of dense hydrogen – implications for models of Jupiter. *Science*, 269:1252–1254, 1995.

[18] M. Albrecht, A. Shukla, M. Dolg, P. Fulde, and H. Stoll. A Hartree–Fock *ab initio* band-structure calculation employing Wannier-type orbitals. *Chemical Physics Letters*, 285:174–179, 1998.

[19] B. J. Alder and T. E. Wainwright. Phase transition for a hard sphere system. *Journal of Chemical Physics (Letters to the Editor)*, 27:1208–1209, 1957.

[20] M. M. G. Alemany, M. Jain, L. Kronik, and J. R. Chelikowsky. Real-space pseudopotential method for computing the electronic properties of periodic systems. *Physical Review B*, 69:075101, 2004.

[21] D. Alfe. First-principles simulations of direct coexistence of solid and liquid aluminum. *Physical Review B*, 68:064423, 2003.

[22] D. Alfe. Melting curve of MgO from first-principles simulations. *Physical Review Letters*, 94:235701, 2005.

[23] D. Alfe and M. J. Gillan. First-principles simulations of liquid Fe-S under Earth's core conditions. *Physical Review B*, 58:8248–8256, 1998.

[24] D. Alfe, M. J. Gillan, L. Vocadlo, J. Brodholt, and G. D. Price. The

ab initio simulation of the Earth's core. *Philosophical Transactions of the Royal Society London, Series A*, 360:1227–1244, 2002.

[25] M. P. Allen and D. J. Tildesley. *Computer Simulation of Liquids*. Clarendon Press, Oxford, 1987 and 1990.

[26] M. Allesch, E. Schwegler, F. Gygi, and G. Galli. A first principles simulation of rigid water. *Journal of Chemical Physics*, 120:5192–5198, 2004.

[27] S. Amira, D. Spangberg, and K. Hermansson. Distorted five-fold coordination of Cu^{2+}(aq) from a Car–Parrinello molecular dynamics simulation. *Physical Chemistry Chemical Physics*, 7:2874–2880, 2005.

[28] S. Amira, D. Spangberg, V. Zelin, M. Probst, and K. Hermansson. Car–Parrinello molecular dynamics simulation of Fe^{3+}(aq). *Journal of Physical Chemistry B*, 109:14235–14242, 2005.

[29] S. C. Ammal, H. Yamataka, M. Aida, and M. Dupuis. Dynamics-driven reaction pathway in an intramolecular rearrangement. *Science*, 299:1555–1557, 2003.

[30] F. Ancilotto, G. L. Chiarotti, S. Scandolo, and E. Tosatti. Dissociation of methane into hydrocarbons at extreme (planetary) pressure and temperature. *Science*, 275:1288–1290, 1997.

[31] H. C. Andersen. Molecular-dynamics simulations at constant pressure and/or temperature. *Journal of Chemical Physics*, 72:2384–2393, 1980.

[32] H. C. Andersen. RATTLE - A velocity version of the SHAKE algorithm for molecular-dynamics calculations. *Journal of Computational Physics*, 52:24–34, 1983.

[33] H. L. Anderson. Scientific uses of the MANIAC. *Journal of Statistical Physics*, 43:731–748, 1986.

[34] W. Andreoni. Computer simulations of small semiconductor and metal clusters. *Zeitschrift für Physik D*, 19:31–36, 1991.

[35] W. Andreoni. Computational approach to the physical chemistry of fullerenes and their derivatives. *Annual Review of Physical Chemistry*, 49:405, 1998.

[36] W. Andreoni. Density-functional theory and molecular dynamics: A new perspective for simulations of biological systems. *Perspectives in Drug Discovery and Design*, 9-11:161–167, 1998.

[37] W. Andreoni and A. Curioni. Ab initio approach to the structure and dynamics of metallofullerenes. *Applied Physics A*, 66:299–306, 1998.

[38] W. Andreoni, D. Marx, and M. Sprik. Parrinello Festschrift: From

physics via chemistry to biology (Special Issue). *ChemPhysChem*, 6(9):1671-1947, September 2005. DOI: 10.1002/cphc.200590020.

[39] W. Andreoni, D. Marx, and M. Sprik. A tribute to michele parrinello: From physics via chemistry to biology. *ChemPhysChem*, 6:1671–1676, 2005. Editorial of the "Parrinello Festschrift" [38].

[40] F. De Angelis, N. Re, A. Sgamellotti, A. Selloni, J. Weber, and C. Floriani. A dynamical density functional study of CO migration in the Reppe carbonylation. *Chemical Physics Letters*, 291:57–63, 1998.

[41] M. I. Aoki and K. Tsumuraya. *Ab initio* molecular-dynamics study of pressure-induced glass-to-crystal transitions in the sodium system. *Physical Review B*, 56:2962–2968, 1997.

[42] H. Appel and E. K. U. Gross. Static and time-dependent many-body effects via density-functional theory. In J. Grotendorst, D. Marx, and A. Muramatsu, eds., *Quantum Simulations of Complex Many-Body Systems: From Theory to Algorithms*, pages 255–268. John von Neumann Institute for Computing, Forschungszentrum Jülich, 2002. See http://www.theochem.rub.de/go/cprev.html.

[43] J. Aqvist and A. Warshel. Simulation of enzyme-reactions using valence-bond force-fields and other hybrid quantum–classical approaches. *Chemical Reviews*, 93:2523–2544, 1993.

[44] T. A. Arias. Multiresolution analysis of electronic structure: Semi-cardinal and wavelet bases. *Reviews of Modern Physics*, 71:267–311, 1999.

[45] T. A. Arias, M. C. Payne, and J. D. Joannopoulos. *Ab initio* molecular dynamics: Analytically continued energy functionals and insights into iterative solutions. *Physical Review Letters*, 69:1077–1080, 1992.

[46] T. A. Arias, M. C. Payne, and J. D. Joannopoulos. *Ab initio* molecular-dynamics techniques extended to large-length-scale systems. *Physical Review B*, 45:1538–1549, 1992.

[47] E. Artacho, D. Sánchez-Portal, P. Ordejón, A. Garcia, and J. M. Soler. Linear-scaling ab-initio calculations for large and complex systems. *physica status solidi (b)*, 215:809–817, 1999.

[48] N. W. Ashcroft and N. D. Mermin. *Solid State Physics*. Saunders College Publishing, Philadelphia, 1976.

[49] J. R. Asher, N. L. Doltsinis, and M. Kaupp. *Ab initio* molecular dynamics simulations and g-tensor calculations of aqueous benzosemiquinone radical anion: Effects of regular and "T-stacked" hydrogen bonds. *Journal of the American Chemical Society*, 126:9854–9861, 2004.

[50] J. R. Asher, N. L. Doltsinis, and M. Kaupp. Extended Car–Parrinello molecular dynamics and electronic g-tensors study of benzosemiquinone radical anion. *Magnetic Resonance in Chemistry*, 43:S237–S247, 2005.

[51] R. Assaraf and M. Caffarel. Computing forces with quantum Monte Carlo. *Journal of Chemical Physics*, 113:4028–4034, 2000.

[52] R. Assaraf and M. Caffarel. Zero-variance zero-bias principle for observables in quantum Monte Carlo: Application to forces. *Journal of Chemical Physics*, 119:10536–10552, 2003.

[53] D. Asthagiri, L. R. Pratt, and J. D. Kress. Free energy of liquid water on the basis of quasichemical theory and *ab initio* molecular dynamics. *Physical Review E*, 68:041505, 2003.

[54] D. Asthagiri, L. R. Pratt, and J. D. Kress. *Ab initio* molecular dynamics and quasichemical study of H^+(aq). *Proceedings of the National Academy of Sciences of the United States of America*, 102:6704–6708, 2005.

[55] D. Asthagiri, L. R. Pratt, J. D. Kress, and M. A. Gomez. Hydration and mobility of HO^-(aq). *Proceedings of the National Academy of Sciences of the United States of America*, 101:7229–7233, 2004. See also Los Alamos Natl. Lab. Tech. Rep. LA-UR-02-7006 (2002), available from http://www.arxiv.org/abs/physics/0211057.

[56] O. Asvany, P. Kumar, B. Redlich, I. Hegemann, S. Schlemmer, and D. Marx. Understanding the infrared spectrum of bare CH_5^+. *Science*, 309:1219–1222, 2005. DOI:10.1126/science.1113729; S. Borman. Chemistry Highlights 2005. *Chemistry and Engineering News*, 83(51):15-20, December 19 (2005).

[57] G. B. Bachelet, D. R. Hamann, and M. Schlüter. Pseudopotentials that work - From H to Pu. *Physical Review B*, 26:4199–4228, 1982. Erratum: [58].

[58] G. B. Bachelet, D. R. Hamann, and M. Schlüter. Erratum: Pseudopotentials that work - From H to Pu. *Physical Review B*, 29:2309–2309, 1984. Original article: [57].

[59] R. F. W. Bader. *Atoms in Molecules - A Quantum Theory*. Clarendon Press, Oxford, 2003.

[60] R. F. W. Bader, S. Johnson, T.-H. Tang, and P. L. A. Popelier. The electron pair. *Journal of Physical Chemistry*, 100:15398–15415, 1996.

[61] K. K. Baeck and T. J. Martinez. *Ab initio* molecular dynamics with equation-of-motion coupled-cluster theory: Electronic absorption spectrum of ethylene. *Chemical Physics Letters*, 375:299–308, 2003.

[62] R. Baer. Non-adiabatic couplings by time-dependent density functional theory. *Chemical Physics Letters*, 364:75–79, 2002.

[63] R. Baer and R. Gould. A method for *ab initio* nonlinear electron-density evolution. *Journal of Chemical Physics*, 114:3385–3392, 2001.

[64] I. Bakó, J. Hutter, and G. Pálinkás. Car–Parrinello molecular dynamics simulation of the hydrated calcium ion. *Journal of Chemical Physics*, 117:9838–9843, 2002.

[65] I. Bakó, J. Hutter, and G. Pálinkás. Car–Parrinello molecular dynamics simulation of liquid formic acid. *Journal of Physical Chemistry A*, 110:2188–2194, 2006.

[66] I. Bakó, T. Megyes, and G. Pálinkás. Structural investigation of water–acetonitrile mixtures: An *ab initio*, molecular dynamics and X-ray diffraction study. *Chemical Physics*, 316:235–244, 2005.

[67] D. Bakowies and W. Thiel. Hybrid models for combined quantum mechanical and molecular mechanical approaches. *Journal of Physical Chemistry*, 100:10580–10594, 1996.

[68] S. Balasubramanian, C. J. Mundy, and M. L. Klein. Trimethylaluminum: A computer study of the condensed phases and the gas dimer. *Journal of Physical Chemistry B*, 102:10136–10141, 1998.

[69] A. Baldereschi. Mean-value point in Brillouin zone. *Physical Review B*, 7:5212–5215, 1973.

[70] P. Ballone and W. Andreoni. Density functional theory and Car–Parrinello molecular dynamics for metal clusters. In W. Ekardt, ed., *Metal Clusters*. Wiley, New York, 1999.

[71] P. Ballone, W. Andreoni, R. Car, and M. Parrinello. Equilibrium structures and finite temperature properties of silicon microclusters from *ab initio* molecular-dynamics calculations. *Physical Review Letters*, 60:271–274, 1988.

[72] P. Bandyopadhyay, S. Ten-no, and S. Iwata. Structures and photoelectron spectroscopies of $Si_2C_2^-$ studied with *ab initio* multicanonical Monte Carlo simulation. *Journal of Physical Chemistry A*, 103:6442–6447, 1999.

[73] J. A. Barker. A quantum-statistical Monte Carlo method; path integrals with boundary conditions. *Journal of Chemical Physics*, 70:2914–2918, 1979.

[74] R. N. Barnett, H.-P. Cheng, H. Häkkinen, and U. Landman. Studies of excess electrons in sodium–chloride clusters and of excess protons in water clusters. *Journal of Physical Chemistry*, 99:7731–7753, 1995.

[75] R. N. Barnett and U. Landman. Born–Oppenheimer molecular-dynamics simulations of finite systems – Structure and dynamics of $(H_2O)_2$. *Physical Review B*, 48:2081–2097, 1993.

[76] R. N. Barnett and U. Landman. Cluster-derived structures and conductance fluctuations in nanowires. *Nature*, 387:788–791, 1997.

[77] R. N. Barnett, U. Landman, and A. Nitzan. Dynamics and spectra of a solvated electron in water clusters. *Journal of Chemical Physics*, 89:2242–2256, 1988.

[78] R. N. Barnett, U. Landman, A. Nitzan, and G. Rajagopal. Born–Oppenheimer dynamics using density-functional theory - Equilibrium and fragmentation of small sodium clusters. *Journal of Chemical Physics*, 94:608–616, 1991.

[79] S. Baroni, S. de Gironcoli, A. Dal Corso, and P. Giannozzi. Phonons and related crystal properties from density-functional perturbation theory. *Reviews of Modern Physics*, 73:515–562, 2001.

[80] S. Baroni and P. Giannozzi. Towards very large-scale electronic-structure calculations. *Europhysics Letters*, 17:547–552, 1992.

[81] S. P. Bates, M. J. Gillan, and G. Kresse. Adsorption of methanol on $TiO_2(110)$: A first-principles investigation. *Journal of Physical Chemistry B*, 102:2017–2026, 1998.

[82] M. H. Beck, A. Jäckle, G. A. Worth, and H.-D. Meyer. The multi-configuration time-dependent Hartree (MCTDH) method: a highly efficient algorithm for propagating wavepackets. *Physics Reports*, 324:1–105, 2000.

[83] T. L. Beck. Real-space mesh techniques in density-functional theory. *Reviews of Modern Physics*, 72:1041–1080, 2000.

[84] A. D. Becke. Density-functional exchange-energy approximation with correct asymptotic-behavior. *Physical Review A*, 38:3098–3100, 1988.

[85] A. D. Becke. Density-functional thermochemistry. 3. The role of exact exchange. *Journal of Chemical Physics*, 98:5648–5652, 1993.

[86] A. D. Becke. A new mixing of Hartree–Fock and local density-functional theories. *Journal of Chemical Physics*, 98:1372–1377, 1993.

[87] A. D. Becke and K. E. Edgecombe. A simple measure of electron localization in atomic and molecular systems. *Journal of Chemical Physics*, 92:5397–5403, 1990.

[88] A. D. Becke and M. R. Roussel. Exchange holes in inhomogeneous systems - A coordinate-space model. *Physical Review A*, 39:3761–3767, 1989.

[89] L. Bellaiche, A. Garcia, and D. Vanderbilt. Finite-temperature properties of $Pb(Zr_{1-x}Ti_x)O_3$ alloys from first principles. *Physical Review Letters*, 84:5427–5430, 2000.

[90] L. Bellaiche and K. Kunc. All-electron calculations with plane waves in solid lithium hydride. *International Journal of Quantum Chemistry*, 61:647–656, 1997.

[91] L. Bellaiche and K. Kunc. Core effects in lithium hydride. *Physical Review B*, 55(8):5006–5014, Feb 1997.

[92] A. B. Belonoshko, R. Ahuja, and B. Johansson. Quasi-*ab initio* molecular dynamic study of Fe melting. *Physical Review Letters*, 84:3638–3641, 2000.

[93] M. Ben-Nun and T. J. Martinez. *Ab initio* molecular dynamics study of *cis–trans* photoisomerization in ethylene. *Chemical Physics Letters*, 298:57–65, 1998.

[94] M. Ben-Nun and T. J. Martinez. Nonadiabatic molecular dynamics: Validation of the multiple spawning method for a multidimensional problem. *Journal of Chemical Physics*, 108:7244–7257, 1998.

[95] M. Ben-Nun and T. J. Martinez. Semiclassical tunneling rates from *ab initio* molecular dynamics. *Journal of Physical Chemistry A*, 103:6055–6059, 1999.

[96] M. Ben-Nun and T. J. Martinez. *Ab initio* quantum molecular dynamics. *Advances in Chemical Physics*, 121:439–512, 2002.

[97] P. Bendt and A. Zunger. New approach for solving the density-functional self-consistent-field problem. *Physical Review B*, 26:3114–3137, 1982.

[98] P. Bendt and A. Zunger. Simultaneous relaxation of nuclear geometries and electric charge densities in electronic structure theories. *Physical Review Letters*, 50:1684–1688, 1983.

[99] R. Benedek, A. Alavi, D. N. Seidman, L. H. Yang, D. A. Muller, and C. Woodward. First principles simulation of a ceramic/metal interface with misfit. *Physical Review Letters*, 84:3362–3365, 2000.

[100] M. Benoit, M. Bernasconi, P. Focher, and M. Parrinello. New high-pressure phase of ice. *Physical Review Letters*, 76:2934–2936, 1996.

[101] M. Benoit, S. Ispas, P. Jund, and R. Jullien. Model of silica glass from combined classical and *ab initio* molecular-dynamics simulations. *European Physical Journal B*, 13:631–636, 2000.

[102] M. Benoit, S. Ispas, and M. E. Tuckerman. Structural properties of molten silicates from *ab initio* molecular-dynamics simulations: Comparison between $CaO–Al_2O_3–SiO_2$ and SiO_2. *Physical Review B*, 64:224205, 2001.

[103] M. Benoit and D. Marx. The shapes of protons in hydrogen bonds depend on the bond length. *ChemPhysChem*, 6:1738–1741, 2005.

[104] M. Benoit, D. Marx, and M. Parrinello. Quantum effects on phase transitions in high-pressure ice. *Computational Materials Science*, 10:88–93, 1998.

[105] M. Benoit, D. Marx, and M. Parrinello. Tunnelling and zero-point motion in high-pressure ice. *Nature*, 392:258–261, 1998. See also [1455].

[106] M. Benoit, D. Marx, and M. Parrinello. The role of quantum effects and ionic defects in high-density ice. *Solid State Ionics*, 125:23–29, 1999.

[107] M. Benoit, A. H. Romero, and D. Marx. Reassigning hydrogen-bond centering in dense ice. *Physical Review Letters*, 89:145501, 2002.

[108] H. J. C. Berendsen, J. P. M. Postma, W. F. van Gunsteren, A. DiNola, and J. R. Haak. Molecular-dynamics with coupling to an external bath. *Journal of Chemical Physics*, 81:3684–3690, 1984.

[109] G. Berghold, C. J. Mundy, A. H. Romero, J. Hutter, and M. Parrinello. General and efficient algorithms for obtaining maximally localized Wannier functions. *Physical Review B*, 61:10040–10048, 2000.

[110] S. Bernard, G. L. Chiarotti, S. Scandolo, and E. Tosatti. Decomposition and polymerization of solid carbon monoxide under pressure. *Physical Review Letters*, 81:2092–2095, 1998.

[111] S. Bernard, P. Loubeyre, and G. Zerah. Phase transition in $Ar(H_2)_2$: A prediction of metallic hydrogen organized in lamellar structures. *Europhysics Letters*, 37:477–482, 1997.

[112] L. Bernasconi, J. Blumberger, M. Sprik, and R. Vuilleumier. Density functional calculation of the electronic absorption spectrum of Cu^+ and Ag^+ aqua ions. *Journal of Chemical Physics*, 121:11885–11899, 2004.

[113] L. Bernasconi and M. Sprik. Time-dependent density functional theory description of on-site electron repulsion and ligand field effects in the optical spectrum of hexaaquoruthenium(II) in solution. *Journal of Physical Chemistry B*, 109:12222–12226, 2005.

[114] L. Bernasconi, M. Sprik, and J. Hutter. Time dependent density functional theory study of charge-transfer and intra-molecular electronic excitations in acetone–water systems. *Journal of Chemical Physics*, 119:12417–12431, 2003.

[115] L. Bernasconi, M. Sprik, and J. Hutter. Hartree–Fock exchange in time dependent density functional theory: Application to charge

transfer excitations in solvated molecular systems. *Chemical Physics Letters*, 394:141–146, 2004.

[116] M. Bernasconi, M. Benoit, M. Parrinello, G. L. Chiarotti, P. Focher, and E. Tosatti. *Ab initio* simulation of phase transformations under pressure. *Physica Scripta*, T66:98–101, 1996.

[117] M. Bernasconi, G. L. Chiarotti, P. Focher, M. Parrinello, and E. Tosatti. Solid-state polymerization of acetylene under pressure: *Ab initio* simulation. *Physical Review Letters*, 78:2008–2011, 1997.

[118] M. Bernasconi, G. L. Chiarotti, P. Focher, S. Scandolo, E. Tosatti, and M. Parrinello. First-principle constant-pressure molecular-dynamics. *Journal of Physics and Chemistry of Solids*, 56:501–505, 1995.

[119] M. Bernasconi, P. L. Silvestrelli, and M. Parrinello. *Ab initio* infrared absorption study of the hydrogen-bond symmetrization in ice. *Physical Review Letters*, 81:1235–1238, 1998.

[120] B. J. Berne, G. Ciccotti, and D. F. Coker, eds. *Classical and Quantum Dynamics in Condensed Phase Simulations*. World Scientific, Singapore, 1998.

[121] J. Bernholc, C. Brabec, M. B. Nardelli, A. Maiti, C. Roland, and B. I. Yakobson. Theory of growth and mechanical properties of nanotubes. *Applied Physics A*, 67:39–46, 1998.

[122] J. Bernholc, E. L. Briggs, C. Bungaro, M. B. Nardelli, J. L. Fattebert, K. Rapcewicz, C. Roland, W. G. Schmidt, and Q. Zhao. Large-scale applications of real-space multigrid methods to surfaces, nanotubes, and quantum transport. *physica status solidi (b)*, 217:685–701, 2000.

[123] J. Bernholc, N. O. Lipari, and S. T. Pantelides. Scattering-theoretic method for defects in semiconductors. 2. Self-consistent formulation and application to the vacancy in silicon. *Physical Review B*, 21:3545–3562, 1980.

[124] J. Bernholc, J.-Y. Yi, and D. J. Sullivan. Structural transitions in metal-clusters. *Faraday Discussions*, 92:217–228, 1991.

[125] B. Bernu and D. M. Ceperley. Path integral Monte Carlo. In J. Grotendorst, D. Marx, and A. Muramatsu, eds., *Quantum Simulations of Complex Many-Body Systems: From Theory to Algorithms*, pages 51–61. John von Neumann Institute for Computing, Forschungszentrum Jülich, Jülich, 2002. See http://www.theochem.rub.de/go/cprev.html.

[126] M. K. Beyer and H. Clausen-Schaumann. Mechanochemistry: The mechanical activation of covalent bonds. *Chemical Reviews*,

105:2921–2948, 2005. See Note on p. 2922: "From a different direction, the term mechanochemistry has recently been introduced in quantum molecular dynamics simulations of the pulling of gold nanowires in atomic force microscopy (AFM).[11]" which is Ref. [793] in this bibliography.

[127] S. R. Bickham, J. D. Kress, L. A. Collins, and R. Stumpf. Ab initio molecular dynamics studies of off-center displacements in CuCl. *Physical Review Letters*, 83:568–571, 1999.

[128] S. R. Bickham, O. Pfaffenzeller, L. A. Collins, J. D. Kress, and D. Hohl. *Ab initio* molecular dynamics of expanded liquid sodium. *Physical Review B*, 58:R11813–R11816, 1998.

[129] S. Biermann, D. Hohl, and D. Marx. Proton quantum effects in high pressure hydrogen. *Journal of Low Temperature Physics*, 110:97–102, 1998.

[130] S. Biermann, D. Hohl, and D. Marx. Quantum effects in solid hydrogen at ultra-high pressure. *Solid State Communications*, 108:337–341, 1998.

[131] A. Bifone, H. J. M. de Groot, and F. Buda. Ab initio molecular dynamics of retinals. *Chemical Physics Letters*, 248:165–172, 1996.

[132] A. Bifone, H. J. M. de Groot, and F. Buda. *Ab initio* molecular dynamics of rhodopsin. *Pure and Applied Chemistry*, 69:2105–2110, 1997.

[133] A. Bifone, H. J. M. de Groot, and F. Buda. Energy storage in the primary photoproduct of vision. *Journal of Physical Chemistry B*, 101:2954–2958, 1997.

[134] M. M. M. Bilek, D. R. McKenzie, D. G. McCulloch, and C. M. Goringe. *Ab initio* simulation of structure in amorphous hydrogenated carbon. *Physical Review B*, 62:3071–3077, 2000.

[135] S. R. Billeter and A. Curioni. Calculation of nonadiabatic couplings in density-functional theory. *Journal of Chemical Physics*, 122:034105, 2005.

[136] K. Binder and G. Ciccotti, eds. *Monte Carlo and Molecular Dynamics of Condensed Matter Systems*. Italian Physical Society SIF, Bologna, 1996.

[137] G. Bischof, A. Silbernagl, K. Hermansson, and M. Probst. Quantum chemical study of the molecular dynamics of hydrated Li^+ and Be^{2+} cations. *International Journal of Quantum Chemistry*, 65:803–816, 1997.

[138] E. R. Bittner and D. S. Kosov. Car–Parrinello molecular dynamics on

excited state surfaces. *Journal of Chemical Physics*, 110:6645–6656, 1999.

[139] J. M. Blanco, C. Gonzalez, P. Jelinek, J. Ortega, F. Flores, and R. Perez. First-principles simulations of STM images: From tunneling to the contact regime. *Physical Review B*, 70:085405, 2004.

[140] X. Blase, A. De Vita, J.-C. Charlier, and R. Car. Frustration effects and microscopic growth mechanisms for BN nanotubes. *Physical Review Letters*, 80:1666–1669, 1998.

[141] P. E. Blöchl. Generalized separable potentials for electronic-structure calculations. *Physical Review B*, 41:5414–5416, 1990.

[142] P. E. Blöchl. Projector augmented-wave method. *Physical Review B*, 50:17953–17979, 1994.

[143] P. E. Blöchl. Electrostatic decoupling of periodic images of plane-wave-expanded densities and derived atomic point charges. *Journal of Chemical Physics*, 103:7422–7428, 1995.

[144] P. E. Blöchl. Second-generation wave-function thermostat for *ab initio* molecular dynamics. *Physical Review B*, 65:104303, 2002.

[145] P. E. Blöchl, C. J. Först, and J. Schimpl. Projector augmented wave method: *ab initio* molecular dynamics with full wave functions. *Bulletin of Materials Science*, 26:33–41, 2003.

[146] P. E. Blöchl and M. Parrinello. Adiabaticity in first-principles molecular dynamics. *Physical Review B*, 45:9413–9416, 1992.

[147] S. Blügel. *First principles calculations of the electronic structure of magnetic overlayers on transition metal surfaces.* PhD thesis, Rheinisch-Westfälische Technische Hochschule (RWTH), Aachen, 1988.

[148] J. Blumberger, L. Bernasconi, I. Tavernelli, R. Vuilleumier, and M. Sprik. Electronic structure and solvation of copper and silver ions: A theoretical picture of a model aqueous redox reaction. *Journal of the American Chemical Society*, 126:3928–3938, 2004.

[149] J. Blumberger and M. Sprik. *Ab initio* molecular dynamics simulation of the aqueous Ru^{2+}/Ru^{3+} redox reaction: The Marcus perspective. *Journal of Physical Chemistry B*, 109:6793–6804, 2005.

[150] J. Blumberger and M. Sprik. Quantum versus classical electron transfer energy as reaction coordinate for the aqueous Ru^{2+}/Ru^{3+} redox reaction. *Theoretical Chemistry Accounts*, 115:113–126, 2006.

[151] J. Blumberger, Y. Tateyama, and M. Sprik. *Ab initio* molecular dynamics simulation of redox reactions in solution. *Computer Physics Communications*, 169:256–261, 2005.

[152] J. Blumberger, I. Tavernelli, M. L. Klein, and M. Sprik. Diabatic free energy curves and coordination fluctuations for the aqueous Ag^+/Ag^{2+} redox couple: A biased Born–Oppenheimer molecular dynamics investigation. *Journal of Chemical Physics*, 124:064507, 2006.

[153] M. Böckmann, N. L. Doltsinis, and D. Marx. Nonadiabatic QM/MM *ab initio* molecular dynamics for condensed phase applications. Unpublished, 2005.

[154] M. Bockstedte, A. Kley, J. Neugebauer, and M. Scheffler. Density-functional theory calculations for poly-atomic systems: Electronic structure, static and elastic properties and *ab initio* molecular dynamics. *Computer Physics Communications*, 107:187–222, 1997.

[155] C. Boehme and D. Marx. Glycine on a wet pyrite surface at extreme conditions. *Journal of the American Chemical Society*, 125:13362–13363, 2003.

[156] M. Boero, T. Ikeshoji, and K. Terakura. Density and temperature dependence of proton diffusion in water: A first-principles molecular dynamics study. *ChemPhysChem*, 6:1775–1779, 2005.

[157] M. Boero, A. Oshiyama, and P. L. Silvestrelli. E′ centers in alpha quartz in the absence of oxygen vacancies: A first-principles molecular-dynamics study. *Physical Review Letters*, 91:206401, 2003.

[158] M. Boero, M. Parrinello, and K. Terakura. First principles molecular dynamics study of Ziegler–Natta heterogeneous catalysis. *Journal of the American Chemical Society*, 120:2746–2752, 1998.

[159] M. Boero, M. Parrinello, and K. Terakura. Ziegler–Natta heterogeneous catalysis by first principles computer experiments. *Surface Science*, 438:1–8, 1999.

[160] M. Boero, M. Parrinello, K. Terakura, T. Ikeshoji, and C. C. Liew. First-principles molecular-dynamics simulations of a hydrated electron in normal and supercritical water. *Physical Review Letters*, 90:226403, 2003.

[161] M. Boero, M. Parrinello, K. Terakura, and H. Weiss. Car–Parrinello study of Ziegler–Natta heterogeneous catalysis: Stability and destabilization problems of the active site models. *Molecular Physics*, 100:2935–2940, 2002.

[162] M. Boero, K. Terakura, T. Ikeshoji, C. C. Liew, and M. Parrinello. Hydrogen bonding and dipole moment of water at supercritical conditions: A first-principles molecular dynamics study. *Physical Review Letters*, 85:3245–3248, 2000.

[163] M. Boero, K. Terakura, T. Ikeshoji, C. C. Liew, and M. Parrinello.

Water at supercritical conditions: A first principles study. *Journal of Chemical Physics*, 115:2219–2227, 2001.

[164] A. D. Boese, A. Chandra, J. M. L. Martin, and D. Marx. From *ab initio* quantum chemistry to molecular dynamics: The delicate case of hydrogen bonding in ammonia. *Journal of Chemical Physics*, 119:5965–5980, 2003.

[165] A. D. Boese, N. L. Doltsinis, N. C. Handy, and M. Sprik. New generalized gradient approximation functionals. *Journal of Chemical Physics*, 112:1670–1678, 2000.

[166] A. D. Boese and N. C. Handy. New exchange-correlation density functionals: The role of the kinetic-energy density. *Journal of Chemical Physics*, 116:9559–9569, 2002.

[167] A. D. Boese and J. M. L. Martin. Development of density functionals for thermochemical kinetics. *Journal of Chemical Physics*, 121:3405–3416, 2004.

[168] A. Bohm, B. Kendrick, and M. E. Loewe. The Berry phase in molecular physics. *International Journal of Quantum Chemistry*, 41:53–75, 1992.

[169] D. Bohm. *Quantum Theory*. Dover Publications, New York, 1951.

[170] M. C. Böhm, R. Ramírez, and J. Schulte. Electrons and nuclei of C_6H_6 and C_6D_6; a combined Feynman path integral – *ab initio* approach. *Chemical Physics*, 227:271–300, 1998.

[171] M. C. Böhm, J. Schulte, and R. Ramirez. Excited state properties of C_6H_6 and C_6D_6 studied by Feymnan path integral-*ab initio* simulations. *Journal of Physical Chemistry A*, 106:3169–3180, 2002.

[172] M. C. Böhm, J. Schulte, and R. Ramírez. Finite-temperature properties of the muonium substituted ethyl radical CH_2MuCH_2: nuclear degrees of freedom and hyperfine splitting constants. *Molecular Physics*, 103:2407–2436, 2005.

[173] P. G. Bolhuis, D. Chandler, C. Dellago, and P. L. Geissler. Transition path sampling: Throwing ropes over rough mountain passes, in the dark. *Annual Review of Physical Chemistry*, 53:291–318, 2002.

[174] K. Bolton, W. L. Hase, and G. H. Peslherbe. Direct dynamics simulations of reactive systems. In D. L. Thompson, ed., *Modern Methods for Multidimensional Dynamics Computations in Chemistry*, pages 143–189. World Scientific, Singapore, 1998.

[175] K. Bolton, B. H. Schlegel, W. L. Hase, and K. Song. An *ab initio* quasi-classical direct dynamics investigation of the $F+C_2H_4 \rightarrow C_2H_3F+H$ product energy distributions. *Physical Chemistry Chemical Physics*, 1:999–1011, 1999.

[176] V. Bonačić-Koutecký, J. Jellinek, M. Wiechert, and P. Fantucci. *Ab initio* molecular dynamics study of solid-to-liquidlike transitions in Li_9^+, Li_{10}, and Li_{11}^+ clusters. *Journal of Chemical Physics*, 107:6321–6334, 1997.

[177] S. A. Bonev, B. Militzer, and G. Galli. *Ab initio* simulations of dense liquid deuterium: Comparison with gas-gun shock-wave experiments. *Physical Review B*, 69:014101, 2004.

[178] D. W. Boo, Z. F. Liu, A. G. Suits, J. S. Tse, and Y. T. Lee. Dynamics of carbonium ions solvated by molecular hydrogen - $CH_5^+(H_2)_n$ (n = 1, 2, 3). *Science*, 269:57–59, 1995.

[179] M. Born and K. Huang. *Dynamical Theory of Crystal Lattices*. Clarendon Press, Oxford, 1954 and 1988. See in particular Appendix VIII "Elimination of the Electronic Motion".

[180] M. Born and R. Oppenheimer. Zur Quantentheorie der Molekeln. *Annalen der Physik (IV. Folge)*, 84:457–484, 1927.

[181] F. A. Bornemann and C. Schütte. A mathematical investigation of the Car–Parrinello method. *Numerische Mathematik*, 78:359–376, 1998.

[182] F. A. Bornemann and C. Schütte. Adaptive accuracy control for Car–Parrinello simulations. *Numerische Mathematik*, 83:179–186, 1999.

[183] D. R. Bowler, R. Choudhury, M. J. Gillan, and T. Miyazaki. Recent progress with large-scale *ab initio* calculations: the CONQUEST code. *physica status solidi (b)*, 243:989–1000, 2006.

[184] S. F. Boys. Construction of some molecular orbitals to be approximately invariant for changes from one molecule to another. *Reviews of Modern Physics*, 32:296–299, 1960.

[185] S. F. Boys and F. Bernardi. Calculation of small molecular interactions by differences of separate total energies - Some procedures with reduced errors. *Molecular Physics*, 19:553, 1970.

[186] B. J. Braams and D. E. Manolopoulos. On the short-time limit of ring polymer molecular dynamics. *Journal of Chemical Physics*, 125:124105, 2006.

[187] M. Brändle, J. Sauer, R. Dovesi, and N. M. Harrison. Comparison of a combined quantum mechanics/interatomic potential function approach with its periodic quantum-mechanical limit: Proton siting and ammonia adsorption in zeolite chabazite. *Journal of Chemical Physics*, 109:10379–10389, 1998.

[188] E. L. Briggs, D. J. Sullivan, and J. Bernholc. Large-scale electronic-structure calculations with multigrid acceleration. *Physical Review B*, 52:R5471–R5474, 1995.

[189] K. D. Brommer, B. E. Larson, M. Needels, and J. D. Joannopoulos. Implementation of the Car–Parrinello algorithm for *ab initio* total energy calculations on a massively parallel computer. *Computers in Physics*, 7:350–362, 1993.

[190] K. D. Brommer, B. E. Larson, M. Needels, and J. D. Joannopoulos. Modeling large surface reconstructions on the Connection Machine. *Japanese Journal of Applied Physics Part 1*, 32:1360–1367, 1993.

[191] P. Broqvist, I. Panas, and H. Grönbeck. The nature of NO_x species on BaO(100): An *ab initio* molecular dynamics study. *Journal of Physical Chemistry B*, 109:15410–15416, 2005.

[192] J. Broughton and F. Khan. Accuracy of time-dependent properties in electronic structure calculations using a fictitious Lagrangian. *Physical Review B*, 40:12098–12104, 1989.

[193] L. M. Brown, ed. *Feynman's Thesis: A New Approach to Quantum Theory*. World Scientific, Singapore, 2005.

[194] C. G. Broyden. A class of methods for solving nonlinear simultaneous equations. *Mathematics of Computation*, 19:557–593, 1965.

[195] F. Buda, R. Car, and M. Parrinello. Thermal expansion of *c*-Si via *ab initio* molecular dynamics. *Physical Review B*, 41:1680–1683, 1990.

[196] F. Buda, G. L. Chiarotti, R. Car, and M. Parrinello. Proton diffusion in crystalline silicon. *Physical Review Letters*, 63:294–297, 1989.

[197] F. Buda, H. J. M. de Groot, and A. Bifone. Charge localization and dynamics in rhodopsin. *Physical Review Letters*, 77:4474–4477, 1996.

[198] F. Buda and A. Fasolino. Strained semiconductor clusters in sodalite. *Physical Review B*, 60:6131–6136, 1999.

[199] M. Bühl, A. Chaumont, R. Schurhammer, and G. Wipff. *Ab initio* molecular dynamics of liquid 1,3-dimethylimidazolium chloride. *Journal of Physical Chemistry B*, 109:18591–18599, 2005.

[200] M. Bühl, R. Diss, and G. Wipff. Coordination environment of aqueous uranyl(VI) ion. *Journal of the American Chemical Society*, 127:13506–13507, 2005.

[201] M. Bühl, S. Grigoleit, H. Kabrede, and F. T. Mauschick. Simulation of ^{59}Co NMR chemical shifts in aqueous solution. *Chemistry - A European Journal*, 12:477–488, 2006.

[202] M. Bühl and M. Parrinello. Medium effects on ^{51}V NMR chemical shifts: A density functional study. *Chemistry - A European Journal*, 7:4487–4494, 2001.

[203] C. Bungaro, C. Noguera, P. Ballone, and W. Kress. Early oxidation stages of Mg(0001): A density functional study. *Physical Review Letters*, 79:4433–4436, 1997.

[204] J. K. Burdett and T. A. McCormick. Electron localization in molecules and solids: The meaning of ELF. *Journal of Physical Chemistry A*, 102:6366–6372, 1998.

[205] K. Burke, R. Car, and R. Gebauer. Density functional theory of the electrical conductivity of molecular devices. *Physical Review Letters*, 94:146803, 2005. Erratum: [206].

[206] K. Burke, R. Car, and R. Gebauer. Erratum. *Physical Review Letters*, 94:159901, 2005. Original article: [205].

[207] T. Burnus, M. A. L. Marques, and E. K. U. Gross. Time-dependent electron localization function. *Physical Review A*, 71:010501, 2005.

[208] G. Bussi, A. Laio, and M. Parrinello. Equilibrium free energies from nonequilibrium metadynamics. *Physical Review Letters*, 96:090601, 2006.

[209] D. M. Bylander, L. Kleinman, and S. Lee. Self-consistent calculations of the energy bands and bonding properties of $B_{12}C_3$. *Physical Review B*, 42:1394–1403, 1990.

[210] T. Cagin and J. R. Ray. Fundamental treatment of molecular-dynamics ensembles. *Physical Review A*, 37:247–251, 1988.

[211] D. J. E. Callaway and A. Rahman. Micro-canonical ensemble formulation of lattice gauge-theory. *Physical Review Letters*, 49:613–616, 1982.

[212] A. Calzolari, N. Marzari, I. Souza, and M. Buongiorno Nardelli. *Ab initio* transport properties of nanostructures from maximally localized Wannier functions. *Physical Review B*, 69:035108, 2004.

[213] L. Campana, A. Selloni, J. Weber, and A. Goursot. Cation siting and dynamical properties of zeolite offretite from first-principles molecular dynamics. *Journal of Physical Chemistry B*, 101:9932–9939, 1997.

[214] J. Cao and B. J. Berne. A Born–Oppenheimer approximation for path-integrals with an application to electron solvation in polarizable fluids. *Journal of Chemical Physics*, 99:2902–2916, 1993.

[215] J. Cao and G. J. Martyna. Adiabatic path integral molecular dynamics methods. 2. Algorithms. *Journal of Chemical Physics*, 104:2028–2035, 1996.

[216] J. S. Cao and G. A. Voth. A new perspective on quantum time-correlation functions. *Journal of Chemical Physics*, 99:10070–10073, 1993.

[217] J. S. Cao and G. A. Voth. The formulation of quantum-statistical mechanics based on the Feynman path centroid density. 2. Dynamical properties. *Journal of Chemical Physics*, 100:5106–5171, 1994.

[218] J. S. Cao and G. A. Voth. The formulation of quantum-statistical mechanics based on the Feynman path centroid density. 3. Phase-space formalism and analysis of centroid molecular-dynamics. *Journal of Chemical Physics*, 101:6157–6167, 1994.

[219] J. S. Cao and G. A. Voth. The formulation of quantum-statistical mechanics based on the Feynman path centroid density. 4. Algorithms for centroid molecular-dynamics. *Journal of Chemical Physics*, 101:6168–6183, 1994.

[220] R. B. Capaz, L. V. C. Assali, L. C. Kimerling, K. Cho, and J. D. Joannopoulos. Mechanism for hydrogen-enhanced oxygen diffusion in silicon. *Physical Review B*, 59:4898–4900, 1999.

[221] R. Car. Molecular dynamics from first principles. In K. Binder and G. Ciccotti, eds., *Monte Carlo and Molecular Dynamics of Condensed Matter Systems*, chapter 23, pages 601–634. Italian Physical Society SIF, Bologna, 1996.

[222] R. Car and M. Parrinello. Unified approach for molecular dynamics and density-functional theory. *Physical Review Letters*, 55:2471–2474, 1985. See also Refs. [995, 1216].

[223] R. Car and M. Parrinello. The unified approach to density functional and molecular-dynamics in real space. *Solid State Communications*, 62:403–405, 1987.

[224] R. Car, M. Parrinello, and M. Payne. Comment on "Error cancellation in the molecular-dynamics method for total energy calculations". *Journal of Physics: Condensed Matter*, 3:9539–9543, 1991. Comment to Ref. [1121].

[225] M. Cardona. Resonance phenomena. In M. Cardona and G. Güntherodt, eds., *Light Scattering in Solids II*, page 58. Springer, Berlin, 1982.

[226] P. Carloni and F. Alber. Density-functional theory investigations of enzyme–substrate interactions. *Perspectives in Drug Discovery and Design*, 9-11:169–179, 1998.

[227] P. Carloni, W. Andreoni, and M. Parrinello. Self-assembled peptide nanotubes from first principles. *Physical Review Letters*, 79:761–764, 1997.

[228] P. Carloni, U. Röthlisberger, and M. Parrinello. The role and perspective of *ab initio* molecular dynamics in the study of biological systems. *Accounts of Chemical Research*, 35:455–464, 2002.

[229] A. Caro, S. Ramos de Debiaggi, and M. Victoria. Quantum-chemical molecular-dynamics applied to s–p metals. *Physical Review B*, 41:913–919, 1990.

[230] E. A. Carter, G. Ciccotti, J. T. Hynes, and R. Kapral. Constrained reaction coordinate dynamics for the simulation of rare events. *Chemical Physics Letters*, 156:472–477, 1989.

[231] M. Cascella, S. Raugei, and P. Carloni. Formamide hydrolysis investigated by multiple-steering *ab initio* molecular dynamics. *Journal of Physical Chemistry B*, 108:369–375, 2004.

[232] M. E. Casida. Time-dependent density functional response theory for molecules. In D. P. Chong, ed., *Recent Advances in Density Functional Methods (Part I)*, volume 1 of *Recent Advances in Computational Chemistry*, chapter 5, pages 155–192. World Scientific, Singapore, 1995.

[233] M. E. Casida. Time-dependent density functional response theory of molecular systems: Theory, computational methods, and functionals. In J. M. Seminario, ed., *Recent Developments and Applications of Modern Density Functional Theory*, volume 4 of *Theoretical and Computational Chemistry*. Elsevier, Amsterdam, 1996.

[234] CASTEP. Refs. [1123, 1307];
See http://www.tcm.phy.cam.ac.uk/castep/.

[235] A. Castro, M. A. L. Marques, and A. Rubio. Propagators for the time-dependent Kohn–Sham equations. *Journal of Chemical Physics*, 121:3425–3433, 2004.

[236] C. Cattarius, G. A. Worth, H.-D. Meyer, and L. S. Cederbaum. All mode dynamics at the conical intersection of an octa-atomic molecule: Multi-configuration time-dependent Hartree (MCTDH) investigation on the butatriene cation. *Journal of Chemical Physics*, 115:2088–2100, 2001.

[237] M. S. Causo, G. Ciccotti, D. Montemayor, S. Bonella, and D. F. Coker. An adiabatic linearized path integral approach for quantum time functions: Electronic transport in metal–molten salt solutions. *Journal of Physical Chemistry B*, 109:6855–6865, 2005.

[238] M. Cavalleri, M. Odelius, A. Nilsson, and L. G. M. Petterson. X-ray absorption spectra of water within a plane-wave Car–Parrinello molecular dynamics framework. *Journal of Chemical Physics*, 121:10065–10075, 2004.

[239] M. Cavalleri, M. Odelius, D. Nordlund, A. Nilsson, and L. G. M. Petterson. Half or full core hole in density functional theory X-ray absorption spectrum calculations of water? *Physical Chemistry Chemical Physics*, 7:2854–2858, 2005.

[240] C. Cavazzoni. *Large Scale First-Principles Simulations of Water and*

Ammonia at High Pressure and Temperature. PhD thesis, Scuola Internazionale Superiore di Studi Avanzati (SISSA), Trieste, 1998.

[241] C. Cavazzoni, G. L. Chiarotti, S. Scandolo, E. Tosatti, M. Bernasconi, and M. Parrinello. Superionic and metallic states of water and ammonia at giant planet conditions. *Science*, 283:44–46, 1999.

[242] C. Cavazzoni, R. Colle, R. Farchioni, and G. Grosso. Car–Parrinello molecular dynamics study of electronic and structural properties of neutral polyanilines. *Physical Review B*, 66:165110, 2002.

[243] C. Cavazzoni, R. Colle, R. Farchioni, and G. Grosso. *Ab initio* molecular dynamics study of the structure of emeraldine base polymers. *Physical Review B*, 69:115213, 2004.

[244] C. Cavazzoni, R. Colle, R. Farchioni, and G. Grosso. Base and salt 3D forms of Emeraldine II polymers by Car–Parrinello molecular dynamics. *Computer Physics Communications*, 169:135–138, 2005.

[245] D. M. Ceperley. Path-integral calculations of normal liquid ^{3}He. *Physical Review Letters*, 69:331–334, 1992.

[246] D. M. Ceperley. Path integrals in the theory of condensed helium. *Reviews of Modern Physics*, 67:279–355, 1995.

[247] D. M. Ceperley. Path integral Monte Carlo methods for fermions. In K. Binder and G. Ciccotti, eds., *Monte Carlo and Molecular Dynamics of Condensed Matter Systems*, chapter 16, pages 443–482. Italian Physical Society SIF, Bologna, 1996.

[248] C. Ceriani, A. Laio, E. Fois, A. Gamba, R. Martoňák, and M. Parrinello. Molecular dynamics simulation of reconstructive phase transitions on an anhydrous zeolite. *Physical Review B*, 70:113403, 2004.

[249] G. M. Chaban, J. O. Jung, and R. B. Gerber. Ab initio calculation of anharmonic vibrational states of polyatomic systems: Electronic structure combined with vibrational self-consistent field. *Journal of Chemical Physics*, 111:1823–1829, 1999.

[250] D. J. Chadi and M. L. Cohen. Special points in the Brillouin zone. *Physical Review B*, 8:5747–5753, 1973.

[251] Ch. Chakravarty. Path integral simulations of atomic and molecular systems. *International Reviews in Physical Chemistry*, 16:421–444, 1997.

[252] B. Champagne, E. Deumens, and Y. Öhrn. Vibrations and soliton dynamics of positively charged polyacetylene chains. *Journal of Chemical Physics*, 107:5433–5444, 1997.

[253] D. Chandler. Theory of quantum processes in liquids. In J. P. Hansen, D. Levesque, and J. Zinn-Justin, eds., *Liquids, Freezing and Glass Transition (Part I)*, pages 193–285. Elsevier, Amsterdam, 1991.

[254] D. Chandler and P. G. Wolynes. Exploiting the isomorphism between quantum-theory and classical statistical-mechanics of polyatomic fluids. *Journal of Chemical Physics*, 74:4078–4095, 1981.

[255] A. Chandra and D. Marx. Creating interfaces by stretching the solvent is key to metallic ammonia solutions. *Angewandte Chemie International Edition*, 46:3676–3679, 2007.

[256] J.-C. Charlier, A. de Vita, X. Blase, and R. Car. Microscopic growth mechanisms for carbon nanotubes. *Science*, 275:646–649, 1997.

[257] S. Chawla and G. A. Voth. Exact exchange in *ab initio* molecular dynamics: An efficient plane-wave based algorithm. *Journal of Chemical Physics*, 108:4697–4700, 1998.

[258] J. R. Chelikowsky and S. G. Louie. First-principles linear combination of atomic orbitals method for the cohesive and structural properties of solids: Application to diamond. *Physical Review B*, 29:3470–3481, 1984.

[259] J. R. Chelikowsky, N. Troullier, and Y. Saad. Finite-difference-pseudopotential method – Electronic-structure calculations without a basis. *Physical Review Letters*, 72:1240–1243, 1994.

[260] J. R. Chelikowsky, N. Troullier, K. Wu, and Y. Saad. Higher-order finite-difference pseudopotential method – An application to diatomic-molecules. *Physical Review B*, 50:11355–11364, 1994.

[261] B. Chen, I. Ivanov, M. L. Klein, and M. Parrinello. Hydrogen bonding in water. *Physical Review Letters*, 91:215503, 2003.

[262] B. Chen, I. Ivanov, J. M. Park, M. Parrinello, and M. L. Klein. Solvation structure and mobility mechanism of OH^-: A Car–Parrinello molecular dynamics investigation of alkaline solutions. *Journal of Physical Chemistry B*, 106:12006–12016, 2002.

[263] H.-P. Cheng. Water clusters: Fascinating hydrogen-bonding networks, solvation shell structures, and proton motion. *Journal of Physical Chemistry A*, 102:6201–6204, 1998.

[264] H.-P. Cheng, R. N. Barnett, and U. Landman. All-quantum simulations: H_3O^+ and $H_5O_2^+$. *Chemical Physics Letters*, 237:161–170, 1995.

[265] H.-P. Cheng and J. L. Krause. The dynamics of proton transfer in $H_5O_2^+$. *Journal of Chemical Physics*, 107:8461–8468, 1997.

[266] H. Chermette. Chemical reactivity indexes in density functional theory. *Journal of Computational Chemistry*, 20:129–154, 1999.

[267] S. Chiesa, D. M. Ceperley, and S. Zhang. Accurate, efficient, and simple forces computed with quantum Monte Carlo methods. *Physical Review Letters*, 94:036404, 2005.

[268] K. Cho, T. A. Arias, J. D. Joannopoulos, and P. K. Lam. Wavelets in electronic-structure calculations. *Physical Review Letters*, 71:1808–1811, 1993.

[269] S. V. Churakov, M. Iannuzzi, and M. Parrinello. *Ab initio* study of dehydroxylation–carbonation reaction on brucite surface. *Journal of Physical Chemistry B*, 108:11567–11574, 2004.

[270] L. C. Ciacchi and M. C. Payne. "Hot-atom" O_2 dissociation and oxide nucleation on Al(111). *Physical Review Letters*, 92:176104, 2004.

[271] L. C. Ciacchi and M. C. Payne. First-principles molecular-dynamics study of native oxide growth on Si(001). *Physical Review Letters*, 95:196101, 2005.

[272] G. Ciccotti, D. Frenkel, and I. R. McDonald, eds. *Simulation of Liquids and Solids – Molecular Dynamics and Monte Carlo Methods in Statistical Mechanics*. North-Holland, Amsterdam, 1987.

[273] G. Ciccotti and J. P. Ryckaert. Molecular dynamics simulation of rigid molecules. *Computer Physics Reports*, 4:346–392, 1987.

[274] L. J. Clarke, I. Štich, and M. C. Payne. Large-scale ab initio total energy calculations on parallel computers. *Computer Physics Communications*, 72:14–28, 1993.

[275] R. D. Coalson. On the connection between Fourier coefficient and discretized cartesian path integration. *Journal of Chemical Physics*, 85:926–936, 1986.

[276] M. Cococcioni, F. Mauri, G. Ceder, and N. Marzari. Electronic-enthalpy functional for finite systems under pressure. *Physical Review Letters*, 94:145501, 2005.

[277] J. D. Coe and T. J. Martinez. Competitive decay at two- and three-state conical intersections in excited-state intramolecular proton transfer. *Journal of the American Chemical Society*, 127:4560–4561, 2005.

[278] J. D. Coe and T. J. Martinez. *Ab initio* molecular dynamics of excited-state intramolecular proton transfer around a three-state conical intersection in malonaldehyde. *Journal of Physical Chemistry A*, 110:618–630, 2006.

[279] A. J. Cohen and N. C. Handy. Assessment of exchange correlation functionals. *Chemical Physics Letters*, 316:160–166, 2000.

[280] M. H. Cohen and V. Heine. Cancellation of kinetic and potential energy in atoms, molecules, and solids. *Physical Review*, 122:1821–1826, 1961.

[281] R. Colle and A. Curioni. Density-functional theory study of electronic and structural properties of doped polypyrroles. *Journal of the American Chemical Society*, 120:4832–4839, 1998.

[282] CONQUEST: A linear scaling DFT electronic structure code. Ref. [183]; distributed under the terms of the GNU General Public Licence; See http://http://www.conquest.ucl.ac.uk/.

[283] W. D. Cornell, P. Cieplak, C. I. Bayly, I. R. Gould, K. M. Merz Jr., D. M. Ferguson, D. C. Spellmeyer, T. Fox, J. W. Caldwell, and P. A. Kollman. A second generation force field for the simulation of proteins, nucleic acids and organic molecules. *Journal of the American Chemical Society*, 117:5179–5197, 1995.

[284] A. Dal Corso, S. Baroni, and R. Resta. Density-functional theory of the dielectric-constant – Gradient-corrected calculation for silicon. *Physical Review B*, 49:5323–5328, 1994.

[285] COSMO*logic* GmbH & Co. KG, Leverkusen, Germany. See http://www.cosmologic.de/.

[286] B. B. Coussens, F. Buda, H. Oevering, and R. J. Meier. Simulations of ethylene insertion in the Pt^{II}-H bond of $(H)Pt(PX_3)_2^+$ species. *Organometallics*, 17:795–801, 1998.

[287] CP2k: A general program to perform molecular dynamics simulations. Distributed under the terms of the GNU General Public Licence; See http://cp2k.berlios.de/.

[288] CP-PAW. P. E. Blöchl and IBM Zurich Research Laboratory, Refs. [142, 145]; See http://orion.pt.tu-clausthal.de/paw/.

[289] C. F. Craig, W. R. Duncan, and O. V. Prezhdo. Trajectory surface hopping in the time-dependent Kohn–Sham approach for electron-nuclear dynamics. *Physical Review Letters*, 95:163001, 2005.

[290] I. R. Craig and D. E. Manolopoulos. Quantum statistics and classical mechanics: Real time correlation functions from ring polymer molecular dynamics. *Journal of Chemical Physics*, 121:3368–3373, 2004.

[291] I. R. Craig and D. E. Manolopoulos. A refined ring polymer molecular dynamics theory of chemical reaction rates. *Journal of Chemical Physics*, 123:034102, July 2005.

[292] R. Crehuet and M. J. Field. Comment on "Action-derived molecular dynamics in the study of rare events". *Physical Review Letters*, 90:089801, 2003. Comment to Ref. [1116].

[293] C. Császár and P. Pulay. Geometry optimization by direct inversion in the iterative subspace. *Journal of Molecular Structure*, 114:31–34, 1984.

[294] T. R. Cundari, M. T. Benson, M. L. Lutz, and S. O. Sommerer. Effective core potential approaches to the chemistry of the heavier elements. In K. B. Lipkowitz and D. B. Boyd, eds., *Reviews in Computational Chemistry*, volume 8, pages 145–202. VCH, New York, 1996.

[295] A. Curioni, W. Andreoni, J. Hutter, H. Schiffer, and M. Parrinello. Density-functional-theory-based molecular dynamics study of 1,3,5-trioxane and 1,3-dioxolane protolysis. *Journal of the American Chemical Society*, 116:11251–11255, 1994.

[296] A. Curioni, M. Boero, and W. Andreoni. Alq$_3$: *Ab initio* calculations of its structural and electronic properties in neutral and charged states. *Chemical Physics Letters*, 294:263–271, 1998.

[297] A. Curioni, M. Sprik, W. Andreoni, H. Schiffer, J. Hutter, and M. Parrinello. Density functional theory-based molecular dynamics simulation of acid-catalyzed chemical reactions in liquid trioxane. *Journal of the American Chemical Society*, 119:7218–7229, 1997.

[298] M. d' Avezac, M. Calandra, and F. Mauri. Density functional theory description of hole-trapping in SiO$_2$: A self-interaction-corrected approach. *Physical Review B*, 71:205210, 2005.

[299] A. J. R. da Silva, H.-Y. Cheng, D. A. Gibson, K. L. Sorge, Z. Liu, and E. A. Carter. Limitations of *ab initio* molecular dynamics simulations of simple reactions: F+H$_2$ as a prototype. *Spectrochimica Acta A*, 53:1285, 1997.

[300] A. J. R. da Silva, J. W. Pang, E. A. Carter, and D. Neuhauser. Anharmonic vibrations via filter diagonalization of *ab initio* dynamics trajectories. *Journal of Physical Chemistry A*, 102:881–885, 1998.

[301] A. J. R. da Silva, M. R. Radeke, and E. A. Carter. *Ab initio* molecular dynamics of H$_2$ desorption from Si(100)-2×1. *Surface Science*, 381:L628–L635, 1997.

[302] C. R. da Silva and R. M. Wentzcovitch. First principles investigation of the A7 to simple cubic transformation in As. *Computational Materials Science*, 8:219–227, 1997.

[303] DACAPO: An *ab initio* molecular dynamics code based on ultra-soft pseudopotentials.
See http://dcwww.camp.dtu.dk/campos/Dacapo/.

[304] L. Dai, S.-W. Yang, X.-T. Chen, P. Wu, and V. B. C. Tan. Investigation of metal diffusion into polymers by *ab initio* molecular dynamics. *Applied Physics Letters*, 87:032108, 2005.

[305] E. Dalgaard and P. Jorgensen. Optimization of orbitals for multicon-

figurational reference states. *Journal of Chemical Physics*, 69:3833–3844, 1978.

[306] S. Datta. *Electronic Transport in Mesoscopic Systems*. Cambridge University Press, Cambridge, 1995.

[307] C. Daul. Density-functional theory applied to the excited-states of coordination-compounds. *International Journal of Quantum Chemistry*, 52:867–877, 1994.

[308] T. Dauxois, M. Peyrard, and S. Ruffo. The Fermi-Pasta-Ulam 'numerical experiment': history and pedagogical perspectives. *European Journal of Physics*, 26:S3–S11, 2005.

[309] E. R. Davidson. Electronic population analysis of molecular wavefunctions. *Journal of Chemical Physics*, 46:3320–3324, 1967.

[310] E. R. Davidson. Iterative calculation of a few of lowest eigenvalues and corresponding eigenvectors of large real-symmetric matrices. *Journal of Computational Physics*, 17:87–94, 1975.

[311] J. E. Davies, N. L. Doltsinis, A. J. Kirby, C. D. Roussev, and M. Sprik. Estimating pK_a values for pentaoxyphosphoranes. *Journal of the American Chemical Society*, 124:6594–6599, 2002.

[312] I. P. Daykov, T. A. Arias, and T. D. Engeness. Robust *ab initio* calculation of condensed matter: Transparent convergence through semi-cardinal multiresolution analysis. *Physical Review Letters*, 90:216402, 2003.

[313] G. A. de Wijs, G. Kresse, and M. J. Gillan. First-order phase transitions by first-principles free-energy calculations: The melting of Al. *Physical Review B*, 57:8223–8234, 1998.

[314] G. A. de Wijs, G. Kresse, L. Vocadlo, K. Dobson, D. Alfe, M. J. Gillan, and G. D. Price. The viscosity of liquid iron at the physical conditions of the Earth's core. *Nature*, 392:805–807, 1998.

[315] G. A. de Wijs, A. De Vita, and A. Selloni. First-principles study of chlorine adsorption and reactions on Si(100). *Physical Review B*, 57:10021–10029, 1998.

[316] A. Debernardi, M. Bernasconi, M. Cardona, and M. Parrinello. Infrared absorption in amorphous silicon from *ab initio* molecular dynamics. *Applied Physics Letters*, 71:2692–2694, 1997.

[317] R. Declerck, V. Van Speybroeck, and M. Waroquier. First-principles calculation of the EPR g tensor in extended periodic systems. *Physical Review B*, 73:115113, 2006.

[318] P. H. Dederichs and R. Zeller. Self-consistency iterations in electronic-structure calculations. *Physical Review B*, 28:5462–5472, 1983.

[319] P. Deglmann and F. Furche. Efficient characterization of stationary points on potential energy surfaces. *Journal of Chemical Physics*, 117:9535–9538, 2002.

[320] F. Della Sala, R. Rousseau, A. Görling, and D. Marx. Quantum and thermal fluctuation effects on the photoabsorption spectra of clusters. *Physical Review Letters*, 92:183401, 2004.

[321] C. Dellago, P. G. Bolhuis, and P. L. Geissler. Transition path sampling. *Advances in Chemical Physics*, 123:1–78, 2002.

[322] C. Dellago, P. L. Geissler, D. Chandler, J. Hutter, and M. Parrinello. Comment on "Dissociation of water under pressure". *Physical Review Letters*, 89:199601, 2002. Comment to Ref. [1300].

[323] C. Dellago and M. M. Naor. Dipole moment of water molecules in narrow pores. *Computer Physics Communications*, 169:36–39, 2005.

[324] C. Dellago, M. M. Naor, and G. Hummer. Erratum: Proton transport through water-filled carbon nanotubes. *Physical Review Letters*, 91:139902, 2003. Original article: [325].

[325] C. Dellago, M. M. Naor, and G. Hummer. Proton transport through water-filled carbon nanotubes. *Physical Review Letters*, 90:105902, 2003. Erratum: [324].

[326] L. Delle Site, C. E. Abrams, A. Alavi, and K. Kremer. Polymers near metal surfaces: Selective adsorption and global conformations. *Physical Review Letters*, 89:156103, 2002.

[327] L. Delle Site, A. Alavi, and R. M. Lynden-Bell. The electrostatic properties of water molecules in condensed phases: An *ab initio* study. *Molecular Physics*, 96:1683–1693, 1999.

[328] B. Delley. An all-electron numerical method for solving the local density functional for poly-atomic molecules. *Journal of Chemical Physics*, 92:508–517, 1990.

[329] J. B. Delos. Theory of electronic-transitions in slow atomic-collisions. *Reviews of Modern Physics*, 53:287–357, 1981.

[330] J. B. Delos, W. R. Thorson, and S. K. Knudson. Semiclassical theory of inelastic collisions. I. classical picture and semiclassical formulation. *Physical Review A*, 6:709, 1972. See also the two papers that follow in that issue on pp. 720-727 and pp. 728-745.

[331] Z. Deng, G. J. Martyna, and M. L. Klein. Structure and dynamics of bipolarons in liquid-ammonia. *Physical Review Letters*, 68:2496–2499, 1992.

[332] Z. Deng, G. J. Martyna, and M. L. Klein. Electronic states in metal–ammonia solutions. *Physical Review Letters*, 71:267–270, 1993.

[333] Z. Deng, G. J. Martyna, and M. L. Klein. Quantum simulation studies of metal–ammonia solutions. *Journal of Chemical Physics*, 100:7590–7601, 1994.

[334] Z. H. Deng, M. K. Klein, and G. J. Martyna. Electronic states and the metal-insulator transition in cesium ammonia solutions. *Journal of the Chemical Society, Faraday Transactions*, 90:2009–2013, 1994.

[335] P. J. H. Denteneer and W. van Haeringen. The pseudopotential-density-functional method in momentum space – Details and test cases. *Journal of Physics C: Solid State Physics*, 18:4127–4142, 1985.

[336] M. D. Deshpande, D. G. Kanhere, P. V. Panat, I. Vasiliev, and R. M. Martin. Ground-state geometries and optical properties of $Na_{8-x}Li_x$ (x=0–8) clusters. *Physical Review A*, 65:053204, 2002.

[337] M. D. Deshpande, D. G. Kanhere, I. Vasiliev, and R. M. Martin. Density-functional study of structural and electronic properties of Na_nLi and Li_nNa ($1 \leq n \leq 12$) clusters. *Physical Review A*, 65:033202, 2002.

[338] E. Deumens, A. Diz, R. Longo, and Y. Öhrn. Time-dependent theoretical treatments of the dynamics of electrons and nuclei in molecular-systems. *Reviews of Modern Physics*, 66:917–983, 1994.

[339] A. Devenyi, K. Cho, T. A. Arias, and J. D. Joannopoulos. Adaptive Riemannian metric for all-electron calculations. *Physical Review B*, 49:13373–13376, 1994.

[340] B. K. Dey, A. Askar, and H. Rabitz. Multidimensional wave packet dynamics within the fluid dynamical formulation of the Schrödinger equation. *Journal of Chemical Physics*, 109:8770–8782, 1998.

[341] A. Dhavale, D. G. Kanhere, C. Majumder, and G. P. Das. Ground-state geometries and stability of Na_nMg (n=1-12) clusters using *ab initio* molecular dynamics method. *European Physical Journal D*, 6:495–500, 1999.

[342] P. Dieska, I. Štich, and R. Perez. Covalent and reversible short-range electrostatic imaging in noncontact atomic force microscopy. *Physical Review Letters*, 91:216401, 2003.

[343] P. Dieska, I. Štich, and R. Perez. Nanomanipulation using only mechanical energy. *Physical Review Letters*, 95:126103, 2005.

[344] P. A. M. Dirac. Note on exchange phenomena in the Thomas atom. *Proceedings of the Cambridge Philosophical Society*, 26:376–385, 1930.

[345] P. A. M. Dirac. *The Principles of Quantum Mechanics*. Oxford University Press, Oxford, 3rd edition, 1947.

[346] M. Diraison, G. J. Martyna, and M. E. Tuckerman. Simulation studies of liquid ammonia by classical *ab initio*, classical, and path-integral molecular dynamics. *Journal of Chemical Physics*, 111:1096–1103, 1999.

[347] R. Ditchfield. Molecular-orbital theory of magnetic shielding and magnetic susceptibility. *Journal of Chemical Physics*, 56:5688, 1972.

[348] K. Doclo and U. Röthlisberger. Ab initio molecular dynamics simulations of the gas-phase reaction of hydroxyl radical with nitrogen dioxide radical. *Chemical Physics Letters*, 297:205–210, 1998.

[349] M. Dolg and H. Stoll. Electronic structure calculations for molecules containing lanthanide atoms. In K. A. Gschneidner Jr. and L. Eyring, eds., *Handbook on the Physics and Chemistry of Rare Earths Vol. 22*, pages 607–729. Elsevier, Amsterdam, 1996.

[350] N. L. Doltsinis. Nonadiabatic dynamics: Mean-field and surface hopping. In J. Grotendorst, D. Marx, and A. Muramatsu, eds., *Quantum Simulations of Complex Many-Body Systems: From Theory to Algorithms*, pages 377–397. John von Neumann Institute for Computing, Forschungszentrum Jülich, Jülich, 2002. See http://www.theochem.rub.de/go/cprev.html.

[351] N. L. Doltsinis. Ab initio surface hopping study of internal conversion of uridine. *Faraday Discussions*, 127:231–233, 2004.

[352] N. L. Doltsinis. Excited state proton transfer and internal conversion in o-hydroxybenzaldehyde: new insights from non-adiabatic *ab initio* molecular dynamics. *Molecular Physics*, 102:499–506, 2004.

[353] N. L. Doltsinis and K. Fink. Comment on "Excitations in photoactive molecules from quantum Monte Carlo", [*Journal of Chemical Physics*, **121**, 5836 (2004)]. *Journal of Chemical Physics*, 122:087101, 2005. Comment to Ref. [1281], and Reply: Ref. [433].

[354] N. L. Doltsinis and D. S. Kosov. Plane wave/pseudopotential implementation of excited state gradients in density functional linear response theory: A new route via implicit differentiation. *Journal of Chemical Physics*, 122:144101, 2005.

[355] N. L. Doltsinis and D. Marx. First principles molecular dynamics involving excited states and nonadiabatic transitions. *Journal of Theoretical and Computational Chemistry*, 1:319–349, 2002.

[356] N. L. Doltsinis and D. Marx. Nonadiabatic Car–Parrinello molecular dynamics. *Physical Review Letters*, 88:166402, 2002.

[357] N. L. Doltsinis and M. Sprik. Electronic excitation spectra from time-dependent density functional response theory using plane-wave methods. *Chemical Physics Letters*, 330:563–569, 2000.

[358] N. L. Doltsinis and M. Sprik. Theoretical pK_a estimates for solvated $P(OH)_5$ from coordination constrained Car–Parrinello molecular dynamics. *Physical Chemistry Chemical Physics*, 5:2612–2618, 2003.

[359] W. Domcke and G. Stock. Theory of ultrafast nonadiabatic excited-state processes and their spectroscopic detection in real time. *Advances in Chemical Physics*, 100:1–168, 1997.

[360] D. Donadio and M. Bernasconi. *Ab initio* simulation of photoinduced transformation of small rings in amorphous silica. *Physical Review B*, 71:073307, 2005.

[361] D. Donadio, M. Bernasconi, and M. Boero. *Ab initio* simulations of photoinduced interconversions of oxygen deficient centers in amorphous silica. *Physical Review Letters*, 87:93–96, 2001.

[362] J. Douady, Y. Ellinger, R. Subra, and B. Levy. Exponential transformation of molecular-orbitals – Quadratically convergent SCF procedure 1. General formulation and application to closed-shell ground-states. *Journal of Chemical Physics*, 72:1452–1462, 1980.

[363] R. M. Dreizler and E. K. U. Gross. *Density-Functional Theory*. Springer, Berlin, 1990.

[364] A. Dreuw and M. Head-Gordon. Single-reference ab initio methods for the calculation of excited states of large molecules. *Chemical Reviews*, 105:4009–4037, 2005.

[365] X. Duan, M. Peressi, and S. Baroni. Characterization of Si-doped GaAs cross-sectional surfaces via *ab initio* simulations. *Physical Review B*, 72:085341, 2005.

[366] V. Dubois and A. Pasquarello. *Ab initio* molecular dynamics of liquid hydrogen chloride. *Journal of Chemical Physics*, 122:114512, 2005.

[367] V. Dubois, P. Umari, and A. Pasquarello. Dielectric susceptibility of dipolar molecular liquids by *ab initio* molecular dynamics: application to liquid HCl. *Chemical Physics Letters*, 390:193–198, 2004.

[368] O. Dulub, B. Meyer, and U. Diebold. Observation of the dynamical change in a water monolayer adsorbed on a ZnO surface. *Physical Review Letters*, 95:136101, 2005.

[369] W. R. Duncan and O. V. Prezhdo. Nonadiabatic molecular dynamics study of electron transfer from alizarin to the hydrated Ti^{4+} ion. *Journal of Physical Chemistry B*, 109:17998–18002, 2005.

[370] W. R. Duncan, W. M. Stier, and O. V. Prezhdo. *Ab initio* nonadiabatic molecular dynamics of the ultrafast electron injection across the alizarin-TiO_2 interface. *Journal of the American Chemical Society*, 127:7941–7951, 2005.

[371] B. I. Dunlap. Quantum chemical molecular dynamics. *International Journal of Quantum Chemistry*, 69:317–325, 1998.

[372] B. I. Dunlap, J. W. D. Connolly, and J. R. Sabin. Some approximations in applications of $X\alpha$ theory. *Journal of Chemical Physics*, 71:3396–3402, 1979.

[373] B. I. Dunlap and R. W. Warren. Quantum chemical molecular dynamics. *Advances in Quantum Chemistry*, 33:167–187, 1998.

[374] M. Durandurdu. *Ab initio* simulation of polyamorphic phase transition in hydrogenated silicon. *Physical Review B*, 73:035209, 2005.

[375] M. Dyer, C. Zhang, and A. Alavi. Quantum diffusion of hydrogen and isotopes in metals. *ChemPhysChem*, 6:1711–1715, 2005.

[376] Weinan E, Weiqing Ren, and Eric Vanden-Eijnden. String method for the study of rare events. *Physical Review B*, 66:052301, 2002.

[377] A. L. L. East, Z. F. Liu, C. McCague, K. Cheng, and J. S. Tse. The three isomers of protonated ethane, $C_2H_7^+$. *Journal of Physical Chemistry A*, 102:10903–10911, 1998.

[378] J. W. Eastwood and D. R. K. Brownrigg. Remarks on the solution of Poisson's equation for isolated systems. *Journal of Computational Physics*, 32:24–38, 1979.

[379] P. J. Eberlein and H. Park. Efficient implementation of Jacobi algorithms and Jacobi sets on distributed memory architectures. *Journal of Parallel and Distributed Computing*, 8:358–366, 1990.

[380] C. Edmiston and K. Ruedenberg. Localized atomic and molecular orbitals. *Reviews of Modern Physics*, 35:457–464, 1963.

[381] D. Egli and S. Billeter. Analytic second variational derivative of the exchange-correlation functional. *Physical Review B*, 69:115106, 2004.

[382] P. Ehrenfest. Bemerkung über die angenäherte Gültigkeit der klassischen Mechanik innerhalb der Quantenmechanik. *Zeitschrift für Physik*, 45:455–457, 1927.

[383] M. Eichinger. *Berechnung molekularer Eigenschaften in komplexer Lösungsmittelumgebung: Dichtefunktionaltheorie kombiniert mit einem Molekularmechanik-Kraftfeld.* PhD thesis, Ludwig-Maximilians-Universität, München, 1999.

[384] M. Eichinger, H. Grubmüller, and H. Heller. EGO-VIII: A parallel program for molecular dynamics simulations of biomolecules. See http://www.mpibpc.mpg.de/groups/grubmueller/.

[385] M. Eichinger, H. Grubmüller, and H. Heller. EGO-VIII: *User Manual for EGO_VIII, Release 2.0.* Theoretische Biophysik, Institut für Medizinische Optik, Ludwig-Maximilians-Universität München.

[386] M. Eichinger, H. Grubmüller, H. Heller, and P. Tavan. FAMUSAMM: An algorithm for rapid evaluation of electrostatic interactions in molecular dynamics simulations. *Journal of Computational Chemistry*, 18:1729–1749, 1997.

[387] M. Eichinger, H. Heller, and H. Grubmüller. EGO - an efficient molecular dynamics program and its application to protein dynamics simulations. In R. Esser, P. Grassberger, J. Grotendorst, and M. Lewerenz, eds., *Molecular Dynamics on Parallel Computers*, pages 154–174. World Scientific, Singapore, 2000.

[388] M. Eichinger, P. Tavan, J. Hutter, and M. Parrinello. A hybrid method for solutes in complex solvents: Density functional theory combined with empirical force fields. *Journal of Chemical Physics*, 110:10452–10467, 1999.

[389] U. Eichler, C. M. Kölmel, and J. Sauer. Combining *ab initio* techniques with analytical potential functions for structure predictions of large systems: Method and application to crystalline silica polymorphs. *Journal of Computational Chemistry*, 18:463–477, 1997.

[390] T. D. Engeness and T. A. Arias. Multiresolution analysis for efficient, high precision all-electron density-functional calculations. *Physical Review B*, 65:165106, 2002.

[391] B. Ensing and M. L. Klein. Perspective on the reactions between F^- and CH_3CH_2F: The free energy landscape of the E2 and S_N2 reaction channels. *Proceedings of the National Academy of Sciences of the United States of America*, 102:6755–6759, 2005.

[392] B. Ensing, A. Laio, M. Parrinello, and M. L. Klein. A recipe for the computation of the free energy barrier and the lowest free energy path of concerted reactions. *Journal of Physical Chemistry B*, 109:6676–6687, 2005.

[393] B. Ensing, M. De Vivo, Z. W. Liu, P. Moore, and M. L. Klein. Metadynamics as a tool for exploring free energy landscapes of chemical reactions. *Accounts of Chemical Research*, 39:73–81, 2006.

[394] M. Erdmann, E. K. U. Gross, and V. Engel. Time-dependent electron localization functions for coupled nuclear–electronic motion. *Journal of Chemical Physics*, 121:9666–9670, 2004.

[395] E. Ermakova, J. Solca, H. Huber, and D. Marx. Many-body and quantum effects in the radial-distribution function of liquid neon and argon. *Chemical Physics Letters*, 246:204–208, 1995.

[396] M. Ernzerhof and G. E. Scuseria. Assessment of the Perdew–Burke–Ernzerhof exchange–correlation functional. *Journal of Chemical Physics*, 110:5029–5036, 1999.

[397] H. Eschrig. *The Fundamentals of Density Functional Theory*. Teubner, Stuttgart, 1996.

[398] R. Esser, P. Grassberger, J. Grotendorst, and M. Lewerenz, eds. *Molecular Dynamics on Parallel Computers*. World Scientific, Singapore, 2000.

[399] S. K. Estreicher, J. L. Hastings, and P. A. Fedders. Radiation-induced formation of H_2^* in silicon. *Physical Review Letters*, 82:815–818, 1999.

[400] R. A. Evarestov and V. P. Smirnov. Special points of the Brillouin zone and their use in the solid state theory. *physica status solidi (b)*, 119:9–40, 1983.

[401] G. Fabricius, E. Artacho, D. Sánchez-Portal, P. Ordejón, D. A. Drabold, and J. M. Soler. Atomic layering at the liquid silicon surface: A first-principles simulation. *Physical Review B*, 1999:R16283–R16286, 1999.

[402] M. Fähnle, C. Elsässer, and H. Krimmel. The basic strategy behind the derivation of various ab-initio force formulas. *physica status solidi (b)*, 191:9–19, 1995.

[403] A. A. Farajian, M. Mikami, P. Ordejón, and K. Tanabe. Ring closure in dioxin formation process: An *ab initio* molecular dynamics study. *Journal of Chemical Physics*, 115:6401–6405, 2001.

[404] J. L. Fattebert and F. Gygi. First-principles molecular dynamics simulations in a continuum solvent. *International Journal of Quantum Chemistry*, 93:139–147, 2003.

[405] P. A. Fedders and D. A. Drabold. Theory of boron doping in a-Si:H. *Physical Review B*, 56:1864–1867, 1997.

[406] A. Fedorowicz, J. Mavri, P. Bala, and A. Koll. Molecular dynamics study of the tautomeric equilibrium in the Mannich base. *Chemical Physics Letters*, 289:457–462, 1998.

[407] M. D. Feit, J. A. Fleck Jr., and A. Steiger. Solution of the Schrödinger-equation by a spectral method. *Journal of Computational Physics*, 47:412–433, 1982.

[408] H. Feldmeier and J. Schnack. Molecular dynamics for fermions. *Reviews of Modern Physics*, 72:655–688, 2000.

[409] E. Fermi, J. Pasta, S. Ulam, and M. Tsingou. Studies of nonlinear problems. I. Technical Report # LA-1940 (20 pages and 9 figures), Los Alamos Scientific Laboratory of the University of California, May 1955. See Refs. [33, 272, 308, 453, 652] for discussions of this investigation and its implications.

[410] P. Fernandez, A. Dal Corso, A. Baldereschi, and F. Mauri. First-principles Wannier functions of silicon and gallium arsenide. *Physical Review B*, 55:R1909–R1913, 1997.

[411] M. V. Fernández-Serra and E. Artacho. Network equilibration and first-principles liquid water. *Journal of Chemical Physics*, 121:11136–11144, 2004.

[412] M. V. Fernández-Serra and E. Artacho. Electrons and hydrogen-bond connectivity in liquid water. *Physical Review Letters*, 96:016404, 2006.

[413] M. V. Fernández-Serra, G. Ferlat, and E. Artacho. Two exchange-correlation functionals compared for first-principles liquid water. *Molecular Simulation*, 31:361–366, 2005.

[414] G. W. Fernando, G.-X. Qian, M. Weinert, and J. W. Davenport. First-principles molecular dynamics for metals. *Physical Review B*, 40:7985–7988, 1989.

[415] A. L. Fetter and J. D. Walecka. *Quantum Theory of Many-Particle Systems*. McGraw-Hill, New York, 1971.

[416] R. P. Feynman. Forces in molecules. *Physical Review*, 56:340–343, 1939. See in particular equation (2).

[417] R. P. Feynman. *The principle of least action in quantum mechanics*. PhD thesis, Princeton University, 1942. For reprint see Ref. [193].

[418] R. P. Feynman. Space-time approach to non-relativistic quantum mechanics. *Reviews of Modern Physics*, 20:367–387, 1948. See also Refs. [193, 417].

[419] R. P. Feynman. *Statistical Mechanics*. Addison-Wesley, Redwood City, 1972.

[420] R. P. Feynman and A. R. Hibbs. *Quantum Mechanics and Path Integrals*. McGraw-Hill, New York, 1965.

[421] FHI98md. Ref. [154]; See http://www.fhi-berlin.mpg.de/th/fhimd/.

[422] M. J. Field. Simulated annealing, classical molecular-dynamics and the Hartree–Fock method – The NDDO approximation. *Chemical Physics Letters*, 172:83–88, 1990.

[423] M. J. Field. Constrained optimization of *ab initio* and semiempirical Hartree–Fock wave-functions using direct minimization or simulated annealing. *Journal of Physical Chemistry*, 95:5104–5108, 1991.

[424] M. J. Field, P. A. Bash, and M. Karplus. A combined quantum mechanical and molecular mechanical potential for molecular dynamics simulation. *Journal of Computational Chemistry*, 11:700–733, 1990.

[425] M. Filatov and S. Shaik. Spin-restricted density functional approach

to the open-shell problem. *Chemical Physics Letters*, 288:689–697, 1998.

[426] M. Filatov and S. Shaik. Application of spin-restricted open-shell Kohn–Sham method to atomic and molecular multiplet states. *Journal of Chemical Physics*, 110:116–125, 1999.

[427] M. Filatov and S. Shaik. A spin-restricted ensemble-referenced Kohn–Sham method and its application to diradicaloid situations. *Chemical Physics Letters*, 304:429–437, 1999.

[428] M. Filatov and S. Shaik. Artificial symmetry breaking in radicals is avoided by the use of the ensemble-referenced Kohn–Sham (REKS) method. *Chemical Physics Letters*, 332:409–419, 2000.

[429] M. Filatov and W. Thiel. A new gradient-corrected exchange-correlation density functional. *Molecular Physics*, 91:847–859, 1997.

[430] M. Filatov and W. Thiel. Exchange-correlation density functional beyond the gradient approximation. *Physical Review A*, 57:189–199, 1998.

[431] A. Filippetti, A. Satta, D. Vanderbilt, and W. Zhong. Hardness conservation as a new transferability criterion: Application to fully nonlocal pseudopotentials. *International Journal of Quantum Chemistry*, 61:421–427, 1997.

[432] A. Filippetti, D. Vanderbilt, W. Zhong, Y. Cai, and G. B. Bachelet. Chemical hardness, linear response, and pseudopotential transferability. *Physical Review B*, 52:11793–11804, 1995.

[433] C. Filippi and F. Buda. Response to "Comment on 'Excitations in photoactive molecules from quantum Monte Carlo' ". *Journal of Chemical Physics*, 122:087102, 2005. Reply to Comment: Ref. [353].

[434] F. Filippone and F. A. Gianturco. Screening ionic motion in sodalite cages: A dynamical study. *Journal of Chemical Physics*, 111:2761–2769, 1999.

[435] F. Filippone, S. Meloni, and M. Parrinello. A novel implicit Newton-Raphson geometry optimization method for density functional theory calculations. *Journal of Chemical Physics*, 115:636–642, 2001.

[436] F. Filippone and M. Parrinello. Vibrational analysis from linear response theory. *Chemical Physics Letters*, 345:179–182, 2001.

[437] F. Finocchi and C. Noguera. Metal segregation and electronic properties of lithium suboxide clusters. *Physical Review B*, 57:14646–14649, 1998.

[438] D. Fischer, A. Curioni, S. Billeter, and W. Andreoni. Effects of nitridation on the characteristics of silicon dioxide: Dielectric and

structural properties from *ab initio* calculations. *Physical Review Letters*, 92:236405, 2004.

[439] A. J. Fisher, W. Hayes, and F. L. Pratt. Theory of positive muons in polyacetylene. *Journal of Physics: Condensed Matter*, 3:9823–9829, 1991.

[440] P. Focher. *First-principle studies of structural phase transformations*. PhD thesis, Scuola Internazionale Superiore di Studi Avanzati (SISSA), Trieste, 1994.

[441] P. Focher, G. L. Chiarotti, M. Bernasconi, E. Tosatti, and M. Parrinello. Structural phase transformations via first-principles simulation. *Europhysics Letters*, 26:345–351, 1994.

[442] E. Fois and A. Gamba. Host/guest interactions and femtosecond scale proton exchange in a zeolitic cage. *Journal of Physical Chemistry B*, 101:4487, 1997.

[443] E. Fois and A. Gamba. Dynamical host/guest interactions in zeolites: Framework isotope effects on proton transfer studied by Car–Parrinello molecular dynamics. *Journal of Physical Chemistry B*, 103:1794–1799, 1999.

[444] E. Fois, A. Gamba, C. Medici, and G. Tabacchi. Intermolecular electronic excitation transfer in a confined space: A first-principles study. *ChemPhysChem*, 6:1917–1922, 2005.

[445] E. Fois, A. Gamba, E. Spano, and G. Tabacchi. Rotation of molecules and ions in confined spaces: A first-principles simulation study. *Journal of Molecular Structure*, 644:55–66, 2003.

[446] E. Fois, A. Gamba, and G. Tabacchi. Structure and dynamics of a Bronsted acid site in a zeolite: An *ab initio* study of hydrogen sodalite. *Journal of Physical Chemistry B*, 102:3974–3979, 1998.

[447] E. Fois, A. Gamba, and G. Tabacchi. Ab initio molecular dynamics study of the Bronsted acid site in a gallium zeolite. *Physical Chemistry Chemical Physics*, 1:531–536, 1999.

[448] E. Fois, A. Gamba, and G. Tabacchi. First-principles simulation of the intracage oxidation of nitrite to nitrate sodalite. *Chemical Physics Letters*, 329:1–6, 2000.

[449] E. Fois, A. Selloni, and M. Parrinello. Approach to metallic behavior in metal–molten–salt solutions. *Physical Review B*, 39:4812–4815, 1989.

[450] E. S. Fois, J. I. Penman, and P. A. Madden. Control of the adiabatic electronic state in *ab initio* molecular-dynamics. *Journal of Chemical Physics*, 98:6361–6368, 1993.

[451] E. S. Fois, A. Selloni, M. Parrinello, and R. Car. Bipolarons in metal metal halide solutions. *Journal of Physical Chemistry*, 92:3268–3273, 1988.

[452] E. S. Fois, M. Sprik, and M. Parrinello. Properties of supercritical water – An ab-initio simulation. *Chemical Physics Letters*, 223:411–415, 1994.

[453] J. Ford. The Fermi–Pasta–Ulam problem – Paradox turns discovery. *Physics Reports*, 213:271–310, 1992.

[454] L. D. Fosdick. Numerical estimation of the partition function in quantum statistics. *Journal of Mathematical Physics*, 3:1251–1264, 1962.

[455] L. D. Fosdick and H. F. Jordan. Path-integral calculation of the two-particle Slater sum for He^4. *Physical Review*, 143:58–66, 1966.

[456] J. M. Foster and S. F. Boys. Canonical configuration interaction procedure. *Reviews of Modern Physics*, 32:300–302, 1960.

[457] G. P. Francis and M. C. Payne. Finite basis set corrections to total energy pseudopotential calculations. *Journal of Physics: Condensed Matter*, 2:4395–4404, 1990.

[458] I. Frank. *Ab initio* simulation of radical reactions in the atmosphere. *Journal of Information Recording*, 25:137–145, 2000.

[459] I. Frank. Chemical reactions "on the fly". *Angewandte Chemie International Edition*, 42:1569–1571, 2003.

[460] I. Frank. Chemische Reaktionen "on the fly". *Angewandte Chemie*, 115:1607, 2003. See also [459].

[461] I. Frank. Mechanically induced chemistry: New perspectives on the nanoscale. *Angewandte Chemie International Edition*, 45:852–854, 2006.

[462] I. Frank, J. Hutter, D. Marx, and M. Parrinello. Molecular dynamics in low-spin excited states. *Journal of Chemical Physics*, 108:4060–4069, 1998.

[463] I. Frank, E. M. Lupton, C. Nonnenberg, F. Achenbach, J. Weis, and C. Bräuchle. Stretching siloxanes: An *ab initio* molecular dynamics study. *Chemical Physics Letters*, 414:132–137, 2005.

[464] I. Frank, D. Marx, and M. Parrinello. First principles investigation of quinizarin chemisorbed on α-Al_2O_3. *Journal of the American Chemical Society*, 117:8037–8038, 1995.

[465] I. Frank, D. Marx, and M. Parrinello. Structure and electronic properties of quinizarin chemisorbed on alumina. *Journal of Chemical Physics*, 104:8143–8150, 1996.

[466] I. Frank, D. Marx, and M. Parrinello. First-principles molecular dynamics study of a photochromic molecular crystal. *Journal of Physical Chemistry A*, 103:7341–7344, 1999.

[467] I. Frank, M. Parrinello, and A. Klamt. Insight into chemical reactions from first-principles simulations: The mechanism of the gas-phase reaction of OH radicals with ketones. *Journal of Physical Chemistry A*, 102:3614–3617, 1998.

[468] D. Frenkel and B. Smit. *Understanding Molecular Simulation - From Algorithms to Applications*. Academic Press, San Diego, 1996 and 2002.

[469] J. Fritsch and U. Schröder. Density functional calculation of semiconductor surface phonons. *Physics Reports*, 309:209–331, 1999.

[470] S. Froyen and M. L. Cohen. Structural properties of NaCl and KCl under pressure. *Journal of Physics C*, 19:2623–2632, 1986.

[471] C.-L. Fu and K.-M. Ho. First-principles calculation of the equilibrium ground-state properties of transition-metals - Applications to Nb and Mo. *Physical Review B*, 28:5480–5486, 1983.

[472] M. Fuchs and M. Scheffler. Ab initio pseudopotentials for electronic structure calculations of poly-atomic systems using density-functional theory. *Computer Physics Communications*, 119:67–98, 1999. For FHIPP package; See http://www.fhi-berlin.mpg.de/th/fhi98md/fhi98PP/.

[473] K. Fukui, J. I. Cline, and J. H. Frederick. Canonical sampling of classical phase space: Application to molecular vibration-rotation dynamics. *Journal of Chemical Physics*, 107:4551–4563, 1997.

[474] F. Furche and R. Ahlrichs. Adiabatic time-dependent density functional methods for excited state properties. *Journal of Chemical Physics*, 117:7433–7447, 2002. Erratum: [475].

[475] F. Furche and R. Ahlrichs. Erratum: "Time-dependent density functional methods for excited state properties" [*J. Chem. Phys.* **117**, 7433 (2002)]. *Journal of Chemical Physics*, 121:12772–12773, 2004. Original article: [474].

[476] M.-P. Gaigeot and M. Sprik. Ab initio molecular dynamics computation of the infrared spectrum of aqueous uracil. *Journal of Physical Chemistry B*, 107:10344–10358, 2003.

[477] M.-P. Gaigeot and M. Sprik. *Ab initio* molecular dynamics study of uracil in aqueous solution. *Journal of Physical Chemistry B*, 108:7458–7467, 2004.

[478] M.-P. Gaigeot, R. Vuilleumier, M. Sprik, and D. Borgis. Infrared spectroscopy of N-methylacetamide revisited by *ab initio* molecular

dynamics simulations. *Journal of Chemical Theory and Computation*, 1:772–789, 2005.

[479] J. D. Gale, R. Shah, M. C. Payne, I. Štich, and K. Terakura. Methanol in microporous materials from first principles. *Catalysis Today*, 50:525–532, 1999.

[480] G. Galli, A. Catellani, and F. Gygi. Wetting silicon carbide with nitrogen: A theoretical study. *Physical Review Letters*, 83:2006–2009, 1999.

[481] G. Galli, F. Gygi, and A. Catellani. Quantum mechanical simulations of microfracture in a complex material. *Physical Review Letters*, 82:3476–3479, 1999.

[482] G. Galli, R. Q. Hood, A. U. Hazi, and F. Gygi. *Ab initio* simulations of compressed liquid deuterium. *Physical Review B*, 61:909–912, 2000.

[483] G. Galli, R. M. Martin, R. Car, and M. Parrinello. *Ab initio* calculation of properties of carbon in the amorphous and liquid states. *Physical Review B*, 39:7470–7482, 1990.

[484] G. Galli, R. M. Martin, R. Car, and M. Parrinello. Melting of diamond at high pressure. *Science*, 250:1547–1549, 1990.

[485] G. Galli and M. Parrinello. *Ab initio* molecular dynamics: principles and practical implementation. In M. Meyer and V. Pontikis, eds., *Computer Simulations in Materials Science*, pages 283–304. Kluwer, Dordrecht, 1991.

[486] G. Galli and A. Pasquarello. First-principles molecular dynamics. In M. P. Allen and D. J. Tildesley, eds., *Computer Simulation in Chemical Physics*, pages 261–313. Kluwer, Dordrecht, 1993.

[487] A. Gambirasio and M. Bernasconi. *Ab initio* study of boron doping in tetrahedral amorphous carbon. *Physical Review B*, 60:12007–12014, 1999.

[488] A. Gambirasio, M. Bernasconi, G. Benedek, and P. L. Silvestrelli. *Ab initio* simulation of laser-induced transformations in fullerite. *Physical Review B*, 62:12644–12647, 2000.

[489] F. R. Gantmacher. *Matrizenrechnung*. VEB Deutscher Verlag der Wissenschaften, Berlin, 1970.

[490] J. R. Gardner and N. A. W. Holzwarth. Pseudopotential inversion scheme. *Physical Review B*, 33:7139–7143, 1986.

[491] R. Gaudoin and K. Burke. Lack of Hohenberg–Kohn theorem for excited states. *Physical Review Letters*, 93:173001, 2004. Publisher's Note: *Phys. Rev. Lett.* **94**, 029901 (2005).

[492] N. Gayathri, S. Izvekov, and G. A. Voth. *Ab initio* molecular dynamics simulation of the H/InP(100)–water interface. *Journal of Chemical Physics*, 117:872–884, 2002.

[493] R. Gebauer and R. Car. Current in open quantum systems. *Physical Review Letters*, 93:160404, 2004.

[494] R. Gebauer, S. Piccinin, and R. Car. Quantum collision current in electronic circuits. *ChemPhysChem*, 6:1727–1730, 2005.

[495] P. Geerlings, F. De Proft, and W. Langenaeker. Conceptual density functional theory. *Chemical Reviews*, 103:1793–1873, 2003.

[496] P. L. Geissler, C. Dellago, D. Chandler, J. Hutter, and M. Parrinello. *Ab initio* analysis of proton transfer dynamics in $(H_2O)_3H^+$. *Chemical Physics Letters*, 321:225–230, 2000.

[497] P. L. Geissler, C. Dellago, D. Chandler, J. Hutter, and M. Parrinello. Autoionization in liquid water. *Science*, 291:2121–2124, 2001. See also Ref. [751].

[498] F. L. Gervasio, P. Carloni, and M. Parrinello. Electronic structure of wet DNA. *Physical Review Letters*, 89:108102, 2002.

[499] F. L. Gervasio, A. Laio, M. Iannuzzi, and M. Parrinello. Influence of DNA structure on the reactivity of the guanine radical cation. *Chemistry - A European Journal*, 10:4846–4852, 2004.

[500] F. L. Gervasio, A. Laio, M. Parrinello, and M. Boero. Charge localization in DNA fibers. *Physical Review Letters*, 94:158103, 2005.

[501] T. V. Gerya, W. V. Maresch, M. Burchard, V. Zakhartchouk, N. L. Doltsinis, and T. Fockenberg. Thermodynamic modeling of quartz solubility and speciation of silica in aqueous fluid up to 1300 °C and 20 kbar based on the chain reaction formalism. *European Journal of Mineralogy*, 17:269–283, 2005.

[502] L. M. Ghiringhelli and E. J. Meijer. Phosphorus: First principle simulation of a liquid–liquid phase transition. *Journal of Chemical Physics*, 122:184510, 2005.

[503] D. A. Gibson and E. A. Carter. Time-reversible multiple timescale ab-initio molecular-dynamics. *Journal of Physical Chemistry*, 97:13429–13434, 1993.

[504] D. A. Gibson and E. A. Carter. Generalized valence bond molecular dynamics at constant temperature. *Molecular Physics*, 89:1265–1276, 1996.

[505] D. A. Gibson and E. A. Carter. *Ab initio* molecular dynamics of pseudorotating Li_5. *Chemical Physics Letters*, 271:266–272, 1997.

[506] D. A. Gibson, I. V. Ionova, and E. A. Carter. A comparison of Car–Parrinello and Born–Oppenheimer generalized valence bond molecular dynamics. *Chemical Physics Letters*, 240:261–267, 1995.

[507] N. I. Gidopoulos, P. G. Papaconstantinou, and E. K. U. Gross. Spurious interactions, and their correction, in the ensemble-Kohn–Sham scheme for excited states. *Physical Review Letters*, 88:033003, 2002.

[508] M. J. Gillan. Calculation of the vacancy formation energy in aluminium. *Journal of Physics: Condensed Matter*, 1:689–711, 1989.

[509] M. J. Gillan. The path-integral simulation of quantum systems. In C. R. A. Catlow, S. C. Parker, and M. P. Allen, eds., *Computer Modelling of Fluids, Polymers and Solids*, pages 155–188. Kluwer, Dordrecht, 1990.

[510] M. J. Gillan. The virtual matter laboratory. *Contemporary Physics*, 38:115–130, 1997.

[511] M. J. Gillan, P. J. D. Lindan, L. N. Kantorovich, and S. P. Bates. Molecular processes on oxide surfaces studied by first-principles calculations. *Mineralogial Magazine*, 62:669–685, 1998.

[512] E. Gindensperger, C. Meier, and J. A. Beswick. Mixing quantum and classical dynamics using Bohmian trajectories. *Journal of Chemical Physics*, 113:9369–9372, 2000.

[513] E. Gindensperger, C. Meier, and J. A. Beswick. Quantum-classical dynamics including continuum states using quantum trajectories. *Journal of Chemical Physics*, 116:8–13, 2002.

[514] D. Gleich and J. Hutter. Computational approaches to activity in rhodium-catalysed hydroformylation. *Chemistry - A European Journal*, 10:2435–2444, 2004.

[515] V. V. Godlevsky, J. J. Derby, and J. R. Chelikowsky. *Ab initio* molecular dynamics simulation of liquid CdTe and GaAs: Semiconducting versus metallic behavior. *Physical Review Letters*, 81:4959–4962, 1998.

[516] V. V. Godlevsky, M. Jain, J. J. Derby, and J. R. Chelikovsky. First-principles calculations of liquid CdTe at temperatures above and below the melting point. *Physical Review B*, 60:8640, 1999.

[517] S. Goedecker. *Wavelets and their Application for the Solution of Partial Differential Equations in Physics*. Presses Polytechniques et Universitaires Romandes, Lausanne, 1998.

[518] S. Goedecker. Linear scaling electronic structure methods. *Reviews of Modern Physics*, 71:1085–1123, 1999.

[519] S. Goedecker and C. Chauvin. Combining multigrid and wavelet

ideas to construct more efficient multiscale algorithms. *Journal of Theoretical and Computational Chemistry*, 2:483–495, 2003.

[520] S. Goedecker and O. V. Ivanov. Linear scaling solution of the Coulomb problem using wavelets. *Solid State Communications*, 105:665–669, 1998.

[521] S. Goedecker and O. V. Ivanov. Solution of multiscale partial differential equations using wavelets. *Computers in Physics*, 12:548–555, 1998.

[522] S. Goedecker and O. V. Ivanov. Frequency localization properties of the density matrix and its resulting hypersparsity in a wavelet representation. *Physical Review B*, 59:7270–7273, 1999.

[523] S. Goedecker and K. Maschke. Transferability of pseudopotentials. *Physical Review A*, 45:88–93, 1992.

[524] S. Goedecker and G. E. Scuseria. Linear scaling electronic structure methods in chemistry and physics. *Computing in Science & Engineering*, 5:14–21, 2003.

[525] S. Goedecker, M. Teter, and J. Hutter. Separable dual-space Gaussian pseudopotentials. *Physical Review B*, 54:1703–1710, 1996.

[526] S. Goedecker and C. J. Umrigar. Critical assessment of the self-interaction-corrected local-density-functional method and its algorithmic implementation. *Physical Review A*, 55:1765–1771, 1997.

[527] N. Goldman, L. E. Fried, I-F. W. Kuo, and C. J. Mundy. Bonding in the superionic phase of water. *Physical Review Letters*, 94:217801, 2005.

[528] H. Goldstein, C. P. Poole, and J. L. Safko. *Classical Mechanics*. Addison-Wesley, San Francisco, 3rd edition, 2002. For errors see http://astro.physics.sc.edu/goldstein/.

[529] G. H. Golub and C. F. Van Loan. *Matrix Computations*. The John Hopkins University Press, Baltimore, MD, 2nd edition, 1989.

[530] P. Gomes Dacosta, O. H. Nielsen, and K. Kunc. Stress theorem in the determination of static equilibrium by the density functional method. *Journal of Physics C*, 19:3163–3172, 1986.

[531] A. F. Goncharov, N. Goldman, L. E. Fried, J. C. Crowhurst, I-F. W. Kuo, C. J. Mundy, and J. M. Zaug. Dynamic ionization of water under extreme conditions. *Physical Review Letters*, 94:125508, 2005.

[532] X. G. Gong, D. Guenzburger, and E. B. Saitovitch. Structure and dynamic properties of neutral and ionized SiH_5 and Si_2H_3. *Chemical Physics Letters*, 275:392–398, 1997.

[533] D. J. Gonzalez, L. E. Gonzalez, and M. J. Stott. Surface structure

of liquid Li and Na: an *ab initio* molecular dynamics study. *Physical Review Letters*, 92:085501, 2004.

[534] D. J. Gonzalez, L. E. Gonzalez, and M. J. Stott. Liquid–vapor interface in liquid binary alloys: An *ab initio* molecular dynamics study. *Physical Review Letters*, 94:077801, 2005.

[535] X. Gonze. Adiabatic density-functional perturbation-theory. *Physical Review A*, 52:1096–1114, 1995.

[536] X. Gonze. Perturbation expansion of variational-principles at arbitrary order. *Physical Review A*, 52:1086–1095, 1995.

[537] X. Gonze. First-principles responses of solids to atomic displacements and homogeneous electric fields: Implementation of a conjugate-gradient algorithm. *Physical Review B*, 55:10337–10354, 1997.

[538] X. Gonze, P. Käckell, and M. Scheffler. Ghost states for separable, norm-conserving, 'ab initio' pseudopotentials. *Physical Review B*, 41:12264–12267, 1990.

[539] X. Gonze, G. M. Rignanese, M. Verstraete, J. M. Beuken, Y. Pouillon, R. Caracas, F. Jollet, M. Torrent, G. Zerah, M. Mikami, P. Ghosez, M. Veithen, J. Y. Raty, V. Olevanov, F. Bruneval, L. Reining, R. Godby, G. Onida, D. R. Hamann, and D. C. Allan. A brief introduction to the ABINIT software package. *Zeitschrift für Kristallographie*, 220:558–562, 2005.

[540] X. Gonze, R. Stumpf, and M. Scheffler. Analysis of separable potentials. *Physical Review B*, 41:8503–8513, 1991.

[541] L. Gorb and J. Leszczynski. Current trends in modeling interactions of DNA fragments with polar solvents. In J. Leszczynski, ed., *Computational Molecular Biology*, page 179. Elsevier, Amsterdam, 1999.

[542] R. G. Gordon. Molecular motion in infrared and Raman spectra. *Journal of Chemical Physics*, 43:1307–1312, 1965.

[543] C. M. Goringe, L. J. Clark, M. H. Lee, M. C. Payne, I. Štich, J. A. White, M. J. Gillan, and A. P. Sutton. The GaAs(001)-(2×4) surface: Structure, chemistry, and adsorbates. *Journal of Physical Chemistry B*, 101:1498–1509, 1997.

[544] A. Görling. Exact treatment of exchange in Kohn–Sham band-structure schemes. *Physical Review B*, 53:7024–7029, 1996.

[545] A. Görling. Proper treatment of symmetries and excited states in a computationally tractable Kohn–Sham method. *Physical Review Letters*, 85:4229–4232, 2000.

[546] H. W. Graben and J. R. Ray. Eight physical systems of thermodynamics, statistical mechanics, and computer simulations. *Molecular Physics*, 80:1183–1193, 1993.

[547] T. Grabo, E. K. U. Gross, and M. Lüders. Orbital functionals in density functional theory: The optimized effective potential method. See `http://psi-k.dl.ac.uk/index.html?highlights` and `http://psi-k.dl.ac.uk/newsletters/News_16/Highlight_16.pdf`.

[548] J. Gräfenstein and D. Cremer. Can density functional theory describe multi-reference systems? Investigation of carbenes and organic biradicals. *Physical Chemistry Chemical Physics*, 2:2091–2103, 2000.

[549] J. Gräfenstein, E. Kraka, and D. Cremer. Density functional theory for open-shell singlet biradicals. *Chemical Physics Letters*, 288:593–602, 1998.

[550] C. G. Gray and K. E. Gubbins. *Theory of Molecular Fluids Vol. 1.* Clarendon, Oxford, 1984.

[551] J. C. Greer, R. Ahlrichs, and I. V. Hertel. Proton-transfer in ammonia cluster cations – Molecular-dynamics in a self-consistent field. *Zeitschrift für Physik D*, 18:413–426, 1991. See in particular Section 3.

[552] T. Gregor, F. Mauri, and R. Car. A comparison of methods for the calculation of NMR chemical shifts. *Journal of Chemical Physics*, 111:1815–1822, 1999.

[553] S. Grimm, C. Bräuchle, and I. Frank. Light-driven unidirectional rotation in a molecule: ROKS simulation. *ChemPhysChem*, 6:1943–1947, 2005.

[554] S. Grimm, C. Nonnenberg, and I. Frank. Restricted open-shell Kohn–Sham theory for π-π^* transitions. I. Polyenes, cyanines and protonated imines. *Journal of Chemical Physics*, 119:11574–11584, 2003.

[555] H. Grönbeck. First principles studies of metal-oxide surfaces. *Topics in Catalysis*, 28:59–69, 2004.

[556] A. Gross. Reactions at surfaces studied by *ab initio* dynamics calculations. *Surface Science Reports*, 32:291–340, 1998.

[557] A. Gross, M. Bockstedte, and M. Scheffler. Ab initio molecular dynamics study of the desorption of D_2 from Si(100). *Physical Review Letters*, 79:701–704, 1997.

[558] A. Gross, C.-M. Wei, and M. Scheffler. Poisoning of hydrogen dissociation at Pd (100) by adsorbed sulfur studied by *ab initio* quantum dynamics and *ab initio* molecular dynamics. *Surface Science*, 416:L1095–L1100, 1998.

[559] E. K. U. Gross and W. Kohn. Time-dependent density-functional theory. *Advances in Quantum Chemistry*, 21:255–291, 1990.

[560] J. C. Grossman and L. Mitas. Efficient quantum Monte Carlo energies for molecular dynamics simulations. *Physical Review Letters*, 94:056403, 2005.

[561] J. C. Grossman, E. Schwegler, E. W. Draeger, F. Gygi, and G. Galli. Towards an assessment of the accuracy of functional theory for first principles simulations of water. *Journal of Chemical Physics*, 120:300–311, 2004.

[562] F. Grossmann. A hierarchy of semiclassical approximations based on Gaussian wavepackets. *Comments on Atomic and Molecular Physics*, 34:141, 1999.

[563] J. Grotendorst, S. Blügel, and D. Marx, eds. *Computational Nanoscience: Do It Yourself!* John von Neumann Institute for Computing, Forschungszentrum Jülich, Jülich, 2006.
Hardcover Version: ISBN 3-00-017350-1,
Electronic Version: http://www.theochem.rub.de/go/cprev.html,
http://www.fz-juelich.de/nic-series/volume31/.

[564] J. Grotendorst, D. Marx, and A. Muramatsu, eds. *Quantum Simulations of Complex Many-Body Systems: From Theory to Algorithms*. John von Neumann Institute for Computing, Forschungszentrum Jülich, Jülich, 2002.
Hardcover Version: ISBN 3-00-009057-6,
Electronic Version: http://www.theochem.rub.de/go/cprev.html,
Audio-Vis. Lect. Notes: http://www.fz-juelich.de/video/wsqs/.

[565] H. Grubmüller. Predicting slow structural transitions in macromolecular systems – Conformational flooding. *Physical Review E*, 52:2893–2906, 1995.

[566] M. P. Grumbach, D. Hohl, R. M. Martin, and R. Car. *Ab initio* molecular-dynamics with a finite-temperature density-functional. *Journal of Physics: Condensed Matter*, 6:1999–2014, 1994.

[567] M. P. Grumbach and R. M. Martin. Phase diagram of carbon at high pressures and temperatures. *Physical Review B*, 54:15730–15741, 1996.

[568] F. Gygi. Adaptive Riemannian metric for plane-wave electronic-structure calculations. *Europhysics Letters*, 19:617–622, 1992.

[569] F. Gygi. Electronic-structure calculations in adaptive coordinates. *Physical Review B*, 48:11692–11700, 1993.

[570] F. Gygi. *Ab initio* molecular-dynamics in adaptive coordinates. *Physical Review B*, 51:11190–11193, 1995.

[571] F. Gygi. First-principles simulations of organic compounds: Solid

CO_2 under pressure. *Computational Materials Science*, 10:63–66, 1998.

[572] F. Gygi and A. Baldereschi. Self-consistent Hartree–Fock and screened-exchange calculations in solids: Application to silicon. *Physical Review B*, 34:4405–4408, 1986.

[573] F. Gygi, J.-L. Fattebert, and E. Schwegler. Computation of maximally localized Wannier functions using a simultaneous diagonalization algorithm. *Computer Physics Communications*, 155:1–6, 2003.

[574] F. Gygi and G. Galli. Real-space adaptive-coordinate electronic-structure calculations. *Physical Review B*, 52:R2229–R2232, 1995.

[575] F. Haase and J. Sauer. The surface structure of sulfated zirconia: Periodic *ab initio* study of sulfuric acid adsorbed on $ZrO_2(101)$ and $ZrO_2(001)$. *Journal of the American Chemical Society*, 120:13503–13512, 1998.

[576] F. Haase, J. Sauer, and J. Hutter. *Ab initio* molecular dynamics simulation of methanol adsorbed in chabazite. *Chemical Physics Letters*, 266:397–402, 1997.

[577] R. Haberlandt, S. Fritzsche, G. Peinel, and K. Heinzinger. *Molekulardynamik - Grundlagen und Anwendungen*. Vieweg, Braunschweig, 1995.

[578] M. D. Hack, A. M. Wensmann, D. G. Truhlar, M. Ben-Nun, and T. J. Martinez. Comparison of full multiple spawning, trajectory surface hopping, and converged quantum mechanics for electronically nonadiabatic dynamics. *Journal of Chemical Physics*, 115:1172–1186, 2001.

[579] J. C. L. Hageman, R. J. Meier, M. Heinemann, and R. A. de Groot. Young modulus of crystalline polyethylene from *ab initio* molecular dynamics. *Macromolecules*, 30:5953–5957, 1997.

[580] P. H. Hahn, W. G. Schmidt, K. Seino, M. Preuss, F. Bechstedt, and J. Bernholc. Optical absorption of water: Coulomb effects versus hydrogen bonding. *Physical Review Letters*, 94:037404, 2005.

[581] H. Häkkinen, S. Abbet, A. Sanchez, U. Heiz, and U. Landman. Structural, electronic, and impurity-doping effects in nanoscale chemistry: Supported gold nanoclusters. *Angewandte Chemie International Edition*, 42:1297–1300, 2003.

[582] H. Häkkinen, R. N. Barnett, and U. Landman. Gold nanowires and their chemical modifications. *Journal of Physical Chemistry B*, 103:8814–8816, 1999.

[583] H. Häkkinen, M. Moseler, and U. Landman. Bonding in Cu, Ag,

and Au clusters: Relativistic effects, trends, and surprises. *Physical Review Letters*, 89:033401, 2002.

[584] R. W. Hall. Simulation of electronic and geometric degrees of freedom using a kink-based path integral formulation: Application to molecular systems. *Journal of Chemical Physics*, 122:164112, 2005.

[585] R. W. Hall and B. J. Berne. Nonergodicity in path integral molecular-dynamics. *Journal of Chemical Physics*, 81:3641–3643, 1984.

[586] D. R. Hamann. Generalized norm-conserving pseudopotentials. *Physical Review B*, 40:2980–2987, 1989.

[587] D. R. Hamann. Application of adaptive curvilinear coordinates to the electronic structure of solids. *Physical Review B*, 51:7337–7340, 1995.

[588] D. R. Hamann. Band structure in adaptive curvilinear coordinates. *Physical Review B*, 51:9508–9514, 1995.

[589] D. R. Hamann. Generalized-gradient functionals in adaptive curvilinear coordinates. *Physical Review B*, 54:1568–1574, 1996.

[590] D. R. Hamann. Adaptive-coordinate electronic structure of $3d$ bands: TiO_2. *Physical Review B*, 56:14979–14984, 1997.

[591] D. R. Hamann. H_2O hydrogen bonding in density-functional theory. *Physical Review B*, 55:R10157–R10160, 1997.

[592] D. R. Hamann. Diffusion of atomic oxygen in SiO_2. *Physical Review B*, 81:3447–3450, 1998.

[593] D. R. Hamann. Comparison of global and local adaptive coordinates for density-functional calculations. *Physical Review B*, 63:075107, 2001.

[594] D. R. Hamann, M. Schlüter, and C. Chiang. Norm-conserving pseudopotentials. *Physical Review Letters*, 43:1494–1497, 1979.

[595] B. Hammer, L. B. Hansen, and J. K. Nørskov. Improved adsorption energetics within density-functional theory using revised Perdew-Burke-Ernzerhof functionals. *Physical Review B*, 59:7413–7421, 1999.

[596] S. Hammes-Schiffer and H. C. Andersen. *Ab initio* and semiempirical methods for molecular-dynamics simulations based on general Hartree–Fock theory. *Journal of Chemical Physics*, 99:523–532, 1993.

[597] S. Hammes-Schiffer and J. C. Tully. Proton-transfer in solution – Molecular-dynamics with quantum transitions. *Journal of Chemical Physics*, 101:4657–4667, 1994.

[598] J.-W. Handgraaf, T. S. van Erp, and E. J. Meijer. *Ab initio* molecular dynamics study of liquid methanol. *Chemical Physics Letters*, 367:617–624, 2003.

[599] N. C. Handy and H. F. Schaefer III. On the evaluation of analytic energy derivatives for correlated wave-functions. *Journal of Chemical Physics*, 81:5031–5033, 1984.

[600] S. Hannongbua. The best structural data of liquid ammonia based on the pair approximation: First-principles Monte Carlo simulation. *Journal of Chemical Physics*, 113:4707–4712, 2000.

[601] D. J. Harris, J. P. Brodholt, and D. M. Sherman. Zinc complexation in hydrothermal chloride brines: Results from *ab initio* molecular dynamics calculations. *Journal of Physical Chemistry A*, 107:1050–1054, 2003.

[602] J. F. Harrison. On the role of the electron density difference in the interpretation of molecular properties. *Journal of Chemical Physics*, 119:8763–8764, 2003.

[603] B. Hartke. Private communication (1999).

[604] B. Hartke and E. A. Carter. *Ab initio* molecular-dynamics with correlated molecular wave-functions – Generalized valence bond molecular-dynamics and simulated annealing. *Journal of Chemical Physics*, 97:6569–6578, 1992.

[605] B. Hartke and E. A. Carter. Spin eigenstate-dependent Hartree–Fock molecular-dynamics. *Chemical Physics Letters*, 189:358–362, 1992.

[606] B. Hartke and E. A. Carter. *Ab initio* molecular-dynamics simulated annealing at the generalized valence-bond level – Application to a small nickel cluster. *Chemical Physics Letters*, 216:324–328, 1993.

[607] B. Hartke, D. A. Gibson, and E. A. Carter. Multiple time scale Hartree–Fock molecular-dynamics. *International Journal of Quantum Chemistry*, 45:59–70, 1993.

[608] C. Hartwigsen, S. Goedecker, and J. Hutter. Relativistic separable dual-space Gaussian pseudopotentials from H to Rn. *Physical Review B*, 58:3641–3662, 1998.

[609] C. Hartwigsen, W. Witschel, and E. Spohr. *Ab initio* study of structural properties of stage-1 alkali graphite intercalation compounds. *Berichte der Bunsengesellschaft für Physikalische Chemie*, 101:859–862, 1997.

[610] C. Hartwigsen, W. Witschel, and E. Spohr. Charge density and charge transfer in stage-1 alkali–graphite intercalation compounds. *Physical Review B*, 55:4953–4959, 1997.

[611] K. C. Hass, W. F. Schneider, A. Curioni, and W. Andreoni. The chemistry of water on alumina surfaces: Reaction dynamics from first principles. *Science*, 282:265–268, 1998. See also *Science* **282**, 882 (1998).

[612] A. Hayashi, M. Shiga, and M. Tachikawa. *Ab initio* path integral molecular dynamics simulation study on the dihydrogen bond of $NH_4^+ \cdots BeH_2$. *Chemical Physics Letters*, 410:54–58, 2005.

[613] S. Hayashi and I. Ohmine. Proton transfer in bacteriorhodopsin: Structure, excitation, IR spectra and potential energy surface analyses by an *ab initio* QM/MM method. *Journal of Physical Chemistry B*, 104:10678–10691, 2000.

[614] S. Hayashi, E. Tajkhorshid, and K. Schulten. Structural changes during the formation of early intermediates in the bacteriorhodopsin photocycle. *Biophysical Journal*, 83:1281–1297, 2002.

[615] S. Hayashi, E. Tajkhorshid, and K. Schulten. Molecular dynamics simulation of bacteriorhodopsin's photoisomerization using *ab initio* forces for the excited chromophore. *Biophysical Journal*, 85:1440–1449, 2003.

[616] P. D. Haynes, C.-K. Skylaris, A. A. Mostofi, and M. C. Payne. ONETEP: linear-scaling density-functional theory with local orbitals and plane waves. *physica status solidi (b)*, 243:2489–2499, 2006.

[617] M. Head-Gordon and J. A. Pople. Optimization of wave-function and geometry in the finite basis Hartree–Fock method. *Journal of Physical Chemistry*, 92:3063–3069, 1988.

[618] F. Hedman and A. Laaksonen. A parallel quantum mechanical MD simulation of liquids. *Molecular Simululation*, 20:265–284, 1998.

[619] W. J. Hehre, L. Radom, P. v. R. Schleyer, and J. A. Pople. Ab Initio *Molecular Orbital Theory*. John Wiley & Sons, New York, 1986.

[620] A. Heidenreich and J. Sauer. *Ab initio* molecular dynamics simulations of the Li_4F_4 cluster. *Zeitschrift für Physik D*, 35:279–283, 1995.

[621] V. Heine and D. Weaire. Pseudopotential theory of cohesion and structure. In H. Ehrenreich, F. Seitz, and D. Turnbull, eds., *Solid State Physics Vol. 24*, page 249. Academic Press, New York, 1970.

[622] R. Heinzmann and R. Ahlrichs. Population analysis based on occupation numbers of modified atomic orbitals (MAOs). *Theoretica Chimica Acta*, 42:33–45, 1976.

[623] T. Helgaker, M. Jaszunski, and K. Ruud. *Ab initio* methods for the calculation of NMR shielding and indirect spin–spin coupling constants. *Chemical Reviews*, 99:293–352, 1999.

[624] T. Helgaker and P. Jørgensen. Configuration–interaction energy derivatives in a fully variational formulation. *Theoretica Chimica Acta*, 75:111–127, 1989.

[625] T. Helgaker, P. Jørgensen, and J. Olsen. *Molecular Electronic Structure Theory*. John Wiley & Sons, Chichester, 2000.

[626] A. Hellmann, B. Razaznejad, and B. I. Lundqvist. Potential-energy surfaces for excited states in extended systems. *Journal of Chemical Physics*, 120:4593–4602, 2004.

[627] H. Hellmann. Zur Rolle der kinetischen Elektronenenergie für die zwischenatomaren Kräfte. *Zeitschrift für Physik*, 85:180–190, 1933. See equation (3b); note that Hellmann already connects the "Hellmann–Feynman theorem" to first-order perturbation theory for cases where it does not hold rigorously, for instance when the wavefunction used to evaluate the expectation value is not the exact one.

[628] H. Hellmann. A new approximation method in the problem of many electrons. *Journal of Chemical Physics (Letters to the Editor)*, 3:61–61, 1935.

[629] H. Hellmann and W. Kassatotschkin. Metallic binding according to the combined approximation procedure. *Journal of Chemical Physics (Letters to the Editor)*, 4:324–325, 1936.

[630] G. Henkelman, A. Arnaldsson, and H. Jónsson. A fast and robust algorithm for Bader decomposition of charge density. *Computational Materials Science*, 36:354–360, 2006. For download see http://theory.cm.utexas.edu/henkelman/research/bader/.

[631] G. Henkelman and H. Jónsson. Improved tangent estimate in the nudged elastic band method for finding minimum energy paths and saddle points. *Journal of Chemical Physics*, 113:9978–9985, 2000.

[632] G. Henkelman, B. P. Uberuaga, and H. Jónsson. A climbing image nudged elastic band method for finding saddle points and minimum energy paths. *Journal of Chemical Physics*, 113:9901–9904, 2000.

[633] J. M. Herbert and M. Head-Gordon. Curvy-steps approach to constraint-free extended-Lagrangian *ab initio* molecular dynamics, using atom-centered basis functions: Convergence toward Born–Oppenheimer trajectories. *Journal of Chemical Physics*, 121:11542–11556, 2004.

[634] J. M. Herbert and M. Head-Gordon. Accelerated, energy-conserving Born–Oppenheimer molecular dynamics via Fock matrix extrapolation. *Physical Chemistry Chemical Physics*, 7:3269–3275, 2005.

[635] J. M. Herbert and M. Head-Gordon. Response to "Comment on 'Curvy-steps approach to constraint-free extended-Lagrangian *ab initio* molecular dynamics, using atom-centered basis functions: Convergence toward Born–Oppenheimer trajectories' " [*J. Chem. Phys.* **123**, 027101 (2005)]. *Journal of Chemical Physics*, 123:027102, 2005.

[636] C. Herring. A new method for calculating wave functions in crystals. *Physical Review*, 57:1169–1177, 1940.

[637] B. Hetényi, F. De Angelis, P. Giannozzi, and R. Car. Reconstruction of frozen-core all-electron orbitals from pseudo-orbitals. *Journal of Chemical Physics*, 115:5791–5795, 2001.

[638] B. Hetényi, F. De Angelis, P. Giannozzi, and R. Car. Calculation of near-edge X-ray-absorption fine structure at finite temperatures: Spectral signatures of hydrogen bond breaking in liquid water. *Journal of Chemical Physics*, 120:8632–8637, 2004.

[639] S. Hirata and M. Head-Gordon. Time-dependent density functional theory for radicals – An improved description of excited states with substantial double excitation character. *Chemical Physics Letters*, 302:375–382, 1999.

[640] S. Hirata and M. Head-Gordon. Time-dependent density functional theory within the Tamm-Dancoff approximation. *Chemical Physics Letters*, 314:291–299, 1999.

[641] K.-M. Ho, J. Ihm, and J.D. Joannopoulos. Dielectric matrix scheme for fast convergence in self-consistent electronic-structure calculations. *Physical Review B*, 25:4260–4262, 1982.

[642] E. Hobi, A. J. R. da Silva, F. D. Novaes, E. Z. da Silva, and A. Fazzio. Comment on "Contaminants in suspended gold chains: An *ab initio* molecular dynamics study". *Physical Review Letters*, 95:169601, 2005. Comment to Ref. [855], Reply: Ref. [856].

[643] R. W. Hockney. The potential calculation and some applications. *Methods in Computational Physics*, 9:135–211, 1970.

[644] W. A. Hofer, A. S. Foster, and A. L. Shluger. Theories of scanning probe microscopes at the atomic scale. *Reviews of Modern Physics*, 75:1287–1331, 2003.

[645] P. Hohenberg and W. Kohn. Inhomogeneous Electron Gas. *Physical Review*, 136:B864–B871, 1964.

[646] D. Hohl and R. O. Jones. First-principles molecular-dynamics simulation of liquid and amorphous selenium. *Physical Review B*, 43:3856–3870, 1991.

[647] D. Hohl, R. O. Jones, R. Car, and M. Parrinello. The structure of selenium clusters: Se_3 to Se_8. *Chemical Physics Letters*, 139:540–545, 1987.

[648] D. Hohl, R. O. Jones, R. Car, and M. Parrinello. Structure of sulfur clusters using simulated annealing: S_2 to S_{13}. *Journal of Chemical Physics*, 89:6823–6835, 1988.

[649] D. Hohl, V. Natoli, D. M. Ceperley, and R. M. Martin. Molecular-dynamics in dense hydrogen. *Physical Review Letters*, 71:541–544, 1993.

[650] M. Holzmann, C. Pierleoni, and D. M. Ceperley. Coupled electron-ion Monte Carlo calculations of atomic hydrogen. *Computer Physics Communications*, 169:421–425, 2005.

[651] W. G. Hoover. Canonical dynamics – Equilibrium phase-space distributions. *Physical Review A*, 31:1695–1697, 1985.

[652] W. G. Hoover and W. T. Ashurst. Nonequilibrium molecular dynamics. In H. Eyring and D. Henderson, eds., *Theoretical Chemistry - Advances and Perspectives Volume 1*, pages 1–51. Academic Press, New York, 1975.

[653] K. Hoshino, F. Shimojo, and T. Nishida. The photo-induced structural change in a Se chain and a Se_8 ring: An *ab initio* molecular-dynamics simulation. *Journal of the Physical Society of Japan*, 68:1907–1911, 1999.

[654] P. H. Hünenberger. Optimal charge-shaping functions for the particle–particle–particle–mesh (P^3M) method for computing electrostatic interactions in molecular simulations. *Journal of Chemical Physics*, 113:10464–10476, 2000.

[655] P. Hunt and M. Sprik. On the position of the highest occupied molecular orbital in aqueous solutions of simple ions. *ChemPhysChem*, 6:1805–1808, 2005.

[656] P. Hunt, M. Sprik, and R. Vuilleumier. Thermal versus electronic broadening in the density of states of liquid water. *Chemical Physics Letters*, 376:68–74, 2003.

[657] P. A. Hunt and M. A. Robb. Systematic control of photochemistry: The dynamics of photoisomerization of a model cyanine dye. *Journal of the American Chemical Society*, 127:5720–5726, 2005.

[658] G. Hura, D. Russo, R. M. Glaeser, T. Head-Gordon, M. Krack, and M. Parrinello. Water structure as a function of temperature from X-ray scattering experiments and *ab initio* molecular dynamics. *Physical Chemistry Chemical Physics*, 10:1981–1991, 2003.

[659] J. Hutter. Excited state nuclear forces from the Tamm-Dancoff approximation to time-dependent density functional theory within the plane wave basis set framework. *Journal of Chemical Physics*, 118:3928–3934, 2003.

[660] J. Hutter, P. Carloni, and M. Parrinello. Nonempirical calculations of a hydrated RNA duplex. *Journal of the American Chemical Society*, 118:8710–8712, 1996.

[661] J. Hutter and A. Curioni. Car–Parrinello molecular dynamics on massively parallel computers. *ChemPhysChem*, 6:1788–1793, 2005.

[662] J. Hutter and A. Curioni. Dual-level parallelism for *ab initio* molecular dynamics: Reaching teraflop performance with the CPMD code. *Parallel Computing*, 31:1–17, 2005.

[663] J. Hutter, H. P. Lüthi, and M. Parrinello. Electronic structure optimisation in plane-wave-based density functional calculations by direct inversion in the iterative subspace. *Computational Materials Science*, 2:244–248, 1994.

[664] J. Hutter, M. Parrinello, and S. Vogel. Exponential transformation of molecular orbitals. *Journal of Chemical Physics*, 101:3862–3865, 1994.

[665] J. Hutter, M. E. Tuckerman, and M. Parrinello. Integrating the Car–Parrinello equations. III. Techniques for ultrasoft pseudopotentials. *Journal of Chemical Physics*, 102:859–871, 1995.

[666] J. T. Hynes. Physical chemistry – The protean proton in water. *Nature*, 397:565–567, 1999.

[667] M. Iannuzzi, A. Laio, and M. Parrinello. Efficient exploration of reactive potential energy surfaces using Car–Parrinello molecular dynamics. *Physical Review Letters*, 90:238302, 2003.

[668] M. Iannuzzi and M. Parrinello. Proton transfer in heterocycle crystals. *Physical Review Letters*, 93:025901, 2004.

[669] R. Iftimie, P. Minary, and M. E. Tuckerman. *Ab initio* molecular dynamics: Concepts, recent developments, and future trends. *Proceedings of the National Academy of Sciences of the United States of America*, 102:6654–6659, 2005.

[670] R. Iftimie, J. W. Thomas, and M. E. Tuckerman. On-the-fly localization of electronic orbitals in Car–Parrinello molecular dynamics. *Journal of Chemical Physics*, 120:2169–2181, 2004.

[671] R. Iftimie and M. E. Tuckerman. Decomposing total IR spectra of aqueous systems into solute and solvent contributions: A computational approach using maximally localized wannier orbitals. *Journal of Chemical Physics*, 122:214508, 2005.

[672] R. Iftimie and M. E. Tuckerman. The molecular origin of the "continuous" infrared absorption in aqueous solutions of acids: A computational approach. *Angewandte Chemie International Edition*, 45:1144–1147, 2006.

[673] J. Ihm, A. Zunger, and M. L. Cohen. Momentum-space formalism for the total energy of solids. *Journal of Physics C*, 12:4409–4422, 1979.

[674] T. Ikeda, M. Hirata, and T. Kimura. *Ab initio* molecular dynamics study of polarization effects on ionic hydration in aqueous $AlCl_3$ solution. *Journal of Chemical Physics*, 119:12386–12392, 2003.

[675] T. Ikeda, M. Hirata, and T. Kimura. Hydration structure of Y^{3+} and La^{3+} compared: An application of metadynamics. *Journal of Chemical Physics*, 122:244507, 2005.

[676] T. Ikeda, M. Sprik, K. Terakura, and M. Parrinello. Pressure effects on hydrogen bonding in the disordered phase of solid HBr. *Physical Review Letters*, 81:4416–4419, 1998.

[677] F. Illas, I. de P. R. Moreira, C. de Graaf, and V. Barone. Magnetic coupling in biradicals, binuclear complexes and wide-gap insulators: A survey of *ab initio* wave function and density functional theory approaches. *Theoretical Chemistry Accounts*, 104:265–272, 2000.

[678] Y. Imamura, N. Matsui, Y. Morikawa, M. Hada, T. Kubo, M. Nishijima, and H. Nakatsuji. First-principles molecular dynamics study of CO adsorption on the Si(001) surface. *Chemical Physics Letters*, 287:131–136, 1998.

[679] Y. Ishikawa and R. C. Binning Jr. Direct *ab initio* molecular dynamics study of the protonation of nitric acid. *Chemical Physics Letters*, 338:353–360, 2001.

[680] S. Ismail-Beigi and S. G. Louie. Excited-state forces within a first-principles Green's function formalism. *Physical Review Letters*, 90:076401, 2003.

[681] S. Ispas, M. Benoit, P. Jund, and R. Jullien. Structural and electronic properties of the sodium tetrasilicate glass $Na_2Si_4O_9$ from classical and *ab initio* molecular dynamics simulations. *Physical Review B*, 64:214206, 2001.

[682] S. Ispas, M. Benoit, P. Jund, and R. Jullien. Structural properties of glassy and liquid sodium tetrasilicate: Comparison between *ab initio* and classical molecular dynamics simulations. *Journal of Non-Crystalline Solids*, 307-310:946–955, 2002.

[683] I. Ivanov and M. L. Klein. Deprotonation of a histidine residue in aqueous solution using constrained *ab initio* molecular dynamics. *Journal of the American Chemical Society*, 124:13380–13381, 2002.

[684] I. Ivanov and M. L. Klein. First principles computational study of the active site of arginase. *Proteins - Structure, Function and Genetics*, 54:1–7, 2004.

[685] I. Ivanov and M. L. Klein. Dynamical flexibility and proton transfer in the arginase active site probed by *ab initio* molecular dynamics. *Journal of the American Chemical Society*, 127:4010–4020, 2005.

[686] S. S. Iyengar, G. E. Scuseria H. B. Schlegel, J. M. Millam, and M. J. Frisch. Comment on "Curvy-steps approach to constraint-free extended-Lagrangian *ab initio* molecular dynamics, using atom-centered basis functions: Convergence toward Born–Oppenheimer trajectories" [*J. Chem. Phys.* **121**, 11542 (2004)]. *Journal of Chemical Physics*, 123:027101, 2005.

[687] S. S. Iyengar and J. Jakowski. Quantum wave packet *ab initio* molecular dynamics: An approach to study quantum dynamics in large systems. *Journal of Chemical Physics*, 122:114105, 2005.

[688] S. S. Iyengar, H. B. Schlegel, J. M. Millam, G. A. Voth, G. E. Scuseria, and M. J. Frisch. *Ab initio* molecular dynamics: Propagating the density matrix with Gaussian orbitals. II. Generalizations based on mass-weighting, idempotency, energy conservation and choice of initial conditions. *Journal of Chemical Physics*, 115:10291–10302, 2001.

[689] S. S. Iyengar, H. B. Schlegel, and G. A. Voth. Atom-centered density matrix propagation (ADMP): Generalizations using Bohmian mechanics. *Journal of Physical Chemistry A*, 107:7269–7277, 2003.

[690] S. S. Iyengar, H. B. Schlegel, G. A. Voth, J. M. Millam, G. E. Scuseria, and M. J. Frisch. *Ab initio* molecular dynamics: Propagating the density matrix with Gaussian orbitals. IV. Formal analysis of the deviations from Born–Oppenheimer dynamics. *Israel Journal of Chemistry*, 42:191–202, 2002.

[691] S. Izvekov, A. Mazzolo, K. VanOpdorp, and G. A. Voth. *Ab initio* molecular dynamics simulation of the Cu(110)–water interface. *Journal of Chemical Physics*, 114:3248–3257, 2001.

[692] S. Izvekov and G. A. Voth. *Ab initio* molecular dynamics simulation of the Ag(111)–water interface. *Journal of Chemical Physics*, 115:7196–7206, 2001.

[693] S. Izvekov and G. A. Voth. Car–Parrinello molecular dynamics simulation of liquid water: New results. *Journal of Chemical Physics*, 116:10372–10376, 2002.

[694] S. Izvekov and G. A. Voth. *Ab initio* molecular-dynamics simulation of aqueous proton solvation and transport revisited. *Journal of Chemical Physics*, 123:044505, 2005. Erratum: [695].

[695] S. Izvekov and G. A. Voth. Erratum: *Ab initio* molecular-dynamics simulation of aqueous proton solvation and transport revisited. *Journal of Chemical Physics*, 124:039901, 2006. Original article: [694].

[696] J. Hutter *et al.* CPMD (Car–Parrinello Molecular Dynamics):

An *ab initio* electronic structure and molecular dynamics program; IBM Zurich Research Laboratory (1990–2007) and Max–Planck–Institut für Festkörperforschung Stuttgart (1997–2001); See http://www.cpmd.org/.

[697] M. Jacoby. Catalysis by the numbers. *Chemistry and Engineering News*, November 29:25–28, 2004.

[698] N. Jakse, O. Lebacq, and A. Pasturel. *Ab initio* molecular-dynamics simulations of short-range order in liquid $Al_{80}Mn_{20}$ and $Al_{80}Ni_{20}$ alloys. *Physical Review Letters*, 93:207801, 2004.

[699] N. Jakse and A. Pasturel. Local order of liquid and supercooled zirconium by *ab initio* molecular dynamics. *Physical Review Letters*, 91:195501, 2003.

[700] N. Jakse and A. Pasturel. *Ab initio* molecular dynamics simulations of local structure of supercooled Ni. *Journal of Chemical Physics*, 120:6124–6127, 2004.

[701] J. F. Janak. Proof that $\partial E / \partial n_i = \epsilon_i$ in density-functional theory. *Physical Review B*, 18:7165–7168, 1978.

[702] S. Jang and G. A. Voth. Path integral centroid variables and the formulation of their exact real time dynamics. *Journal of Chemical Physics*, 111:2357–2370, 1999.

[703] E. A. A. Jarvis, E. Fattal, J. R. da Silva, and E. A. Carter. Characterization of photoionization intermediates via *ab initio* molecular dynamics. *Journal of Physical Chemistry A*, 104:2333–2340, 2000.

[704] M. R. Jarvis, R. Perez, and M. C. Payne. Can atomic force microscopy achieve atomic resolution in contact mode? *Physical Review Letters*, 86:1287–1290, 2001.

[705] Y. Jeanvoine, J. G. Ángyán, G. Kresse, and J. Hafner. On the nature of water interacting with Bronsted acidic sites. *Ab initio* molecular dynamics study of hydrated HSAPO-34. *Journal of Physical Chemistry B*, 102:7307–7310, 1998.

[706] P. Jelinek, R. Perez, J. Ortega, and F. Flores. First-principles simulations of the stretching and final breaking of Al nanowires: Mechanical properties and electrical conductance. *Physical Review B*, 68:085403, 2003.

[707] J. Jellinek, V. Bonačić-Koutecký, P. Fantucci, and M. Wiechert. *Ab initio* Hartree–Fock self-consistent-field molecular-dynamics study of structure and dynamics of Li_8. *Journal of Chemical Physics*, 101:10092–10100, 1994.

[708] J. Jellinek, S. Srinivas, and P. Fantucci. *Ab initio* Monte Carlo: Application to Li_8. *Chemical Physics Letters*, 288:705–713, 1998.

[709] M. O. Jensen, U. Röthlisberger, and C. Rovira. Hydroxide and proton migration in aquaporins. *Biophysical Journal*, 89:1744–1759, 2005.

[710] V. R. Jensen, M. Graf, and W. Thiel. Unusual temperature effects in propene polymerization using stereorigid zirconocene catalysts. *ChemPhysChem*, 6:1929–1933, 2005.

[711] B. J. Jesson and P. A. Madden. Structure and dynamics at the aluminum solid–liquid interface: An *ab initio* simulation. *Journal of Chemical Physics*, 113:5935–5946, 2000.

[712] B. G. Johnson, P. M. W. Gill, and J. A. Pople. The performance of a family of density functional methods. *Journal of Chemical Physics*, 98:5612–5626, 1993.

[713] D. D. Johnson. Modified Broyden method for accelerating convergence in self-consistent calculations. *Physical Review B*, 38:12807–12813, 1988.

[714] G. A. Jones, B. K. Carpenter, and M. N. Paddon-Row. Application of trajectory surface hopping to the study of intramolecular electron transfer in polyatomic organic systems. *Journal of the American Chemical Society*, 120:5499–5508, 1998.

[715] R. O. Jones. Molecular-structures from density functional calculations with simulated annealing. *Angewandte Chemie International Edition*, 30:630–640, 1991.

[716] R. O. Jones and O. Gunnarsson. The density functional formalism, its applications and prospects. *Reviews of Modern Physics*, 61:689–746, 1989.

[717] H. F. Jordan and L. D. Fosdick. Three-particle effects in the pair distribution function for He^4 Gas. *Physical Review*, 171:128–149, 1968.

[718] P. Jorgensen and J. Simons, eds. *Geometrical Derivatives of Energy Surfaces and Molecular Properties*. NATO ASI Series, Vol. 166, Series C, Mathematical and Physical Sciences, Reidel, Dordrecht, 1985.

[719] Y.-M. Juan and E. Kaxiras. Application of gradient corrections to density-functional theory for atoms and solids. *Physical Review B*, 48:14944–14952, 1993.

[720] K. Jug, N. N. Nair, and T. Bredow. Molecular dynamics investigation of oxygen vacancy diffusion in rutile. *Physical Chemistry Chemical Physics*, 7:2616–2621, 2005.

[721] K. Jug, N. N. Nair, and T. Bredow. Reaction of surface hydroxyl groups with VO_4H_3 on anatase surfaces. *Surface Science*, 596:108–116, 2005.

[722] P. Jungwirth and D. J. Tobias. Chloride anion on aqueous clusters, at the air–water interface, and in liquid water: Solvent effects on Cl^- polarizability. *Journal of Physical Chemistry A*, 106:379–383, 2002.

[723] J. Junquera, O. Paz, D. Sanchez-Portal, and E. Artacho. Numerical atomic orbital for linear scaling calculations. *Physical Review B*, 64:235111, 2001.

[724] M. Kac. On distributions of certain Wiener functionals. *Transactions of the American Mathematical Society*, 65:1–13, 1949.

[725] Y. Kanai, A. Tilocca, A. Selloni, and R. Car. First-principles string molecular dynamics: An efficient approach for finding chemical reaction pathways. *Journal of Chemical Physics*, 121:3359–3367, 2004.

[726] M. Karplus. Molecular dynamics simulations of biomolecules. *Accounts of Chemical Research*, 35:321–323, 2002. Guest Editorial to: *Special Issue on Molecular Dynamics Simulations of Biomolecules*, *Accounts of Chemical Research* **35**, 321-489 (2002).

[727] R. Kaschner and D. Hohl. Density functional theory and biomolecules: A study of glycine, alanine, and their oligopeptides. *Journal of Physical Chemistry A*, 102:5111–5116, 1998.

[728] J. Kästner and P. E. Blöchl. Towards an understanding of the workings of nitrogenase from DFT calculations. *ChemPhysChem*, 6:1724–1726, 2005.

[729] C. Katan. First-principles study of the structures and vibrational frequencies for tetrathiafulvalene TTF and TTF-d(4) in different oxidation states. *Journal of Physical Chemistry A*, 103:1407–1413, 1999.

[730] M. I. Katsnelson, M. van Schilfgaarde, V. P. Antropov, and B. N. Harmon. *Ab initio* instanton molecular dynamics for the description of tunneling phenomena. *Physical Review A*, 54:4802–4809, 1996.

[731] M. Kaupp, M. Bühl, and V. G. Malkin, eds. *Calculation of NMR and EPR Parameters*. Wiley-VCH, Weinheim, 2004.

[732] S. H. Ke, T. Uda, I. Štich, and K. Terakura. First-principles simulation of atomic force microscopy image formation on a GaAs(110) surface: Effect of tip morphology. *Physical Review B*, 63:245323, 2001.

[733] T. A. Keith and R. F. W. Bader. Calculation of magnetic response properties using a continuous set of gauge transformations. *Chemical Physics Letters*, 210:223–231, 1993.

[734] R. A. Kendall, E. Apra, D. E. Bernholdt, E. J. Bylaska, M. Dupuis, G. I. Fann, R. J. Harrison, J. L. Ju, J. A. Nichols, J. Nieplocha, T. P. Straatsma, T. L. Windus, and A. T. Wong. High performance computational chemistry: An overview of NWChem a distributed

parallel application. *Computer Physics Communications*, 128:260–283, 2000.

[735] G. P. Kerker. Non-singular atomic pseudopotentials for solid-state applications. *Journal of Physics C*, 13:L189–L194, 1980.

[736] G. Kern and J. Hafner. *Ab initio* molecular-dynamics studies of the graphitization of flat and stepped diamond (111) surfaces. *Physical Review B*, 58:13167–13175, 1998.

[737] V. Keshari and Y. Ishikawa. *Ab initio* Monte-Carlo simulated annealing method. *Chemical Physics Letters*, 218:406–412, 1994.

[738] E. Kim, Y. H. Lee, J. J. Lee, and Y. G. Hwang. Defects in amorphous $Si_{1-x}Ge_x$ alloys: An explanation of electron spin resonance signals. *Europhysics Letters*, 40:147–152, 1997.

[739] E. Kim, C. W. Oh, and Y. H. Lee. Diffusion mechanism of Si adatoms on a double-layer stepped Si(001) surface. *Physical Review Letters*, 79:4621–4624, 1997.

[740] R. D. King-Smith and D. Vanderbilt. Theory of polarization of crystalline solids. *Physical Review B*, 47:1651–1654, 1993.

[741] F. Kirchhoff, G. Kresse, and M. J. Gillan. Structure and dynamics of liquid selenium. *Physical Review B*, 57:10482–10495, 1998.

[742] B. Kirchner and J. Hutter. The structure of a DMSO–water mixture from Car–Parrinello simulations. *Chemical Physics Letters*, 364:497–502, 2002.

[743] B. Kirchner and J. Hutter. Solvent effects on electronic properties from Wannier functions in a dimethyl sulfoxide/water mixture. *Journal of Chemical Physics*, 121:5133–5142, 2004.

[744] B. Kirchner, M. Reiher, A. Hille, J. Hutter, and B. A. Hess. Car–Parrinello molecular dynamics study of the initial dinitrogen reduction step in Sellmann-type nitrogenase model complexes. *Chemistry - A European Journal*, 11:574–583, 2005.

[745] B. Kirchner, J. Stubbs, and D. Marx. Fast anomalous diffusion of small hydrophobic species in water. *Physical Review Letters*, 89:215901, 2002.

[746] S. Kirkpatrick, C. D. Gelatt Jr., and M. P. Vecchi. Optimization by simulated annealing. *Science*, 220:671–680, 1983.

[747] H. Kitamura, S. Tsuneyuki, T. Ogitsu, and T. Miyake. Quantum distribution of protons in solid molecular hydrogen at megabar pressures. *Nature*, 404:259–262, 2000.

[748] M. Klähn, G. Mathias, C. Kotting, M. Nonella, J. Schlitter, K. Gerwert, and P. Tavan. IR spectra of phosphate ions in aqueous solution:

Predictions of a DFT/MM approach compared with observations. *Journal of Physical Chemistry A*, 108:6186–6194, 2004.

[749] A. Klamt. Conductor-like screening model for real solvents – A new approach to the quantitative calculation of solvation phenomena. *Journal of Physical Chemistry*, 99:2224–2235, 1995.

[750] A. Klamt and G. Schüürmann. COSMO: A new approach to dielectric screening in solvents with explicit expressions for the screening energy and its gradient. *Journal of the Chemical Society, Perkin Transactions 2*, pages 799–805, 1993.

[751] M. L. Klein. Chemistry - Water on the move. *Science*, 291:2106–2107, 2001.

[752] S. Klein, M. J. Bearpark, B. R. Smith, M. A. Robb, M. Olivucci, and F. Bernardi. Mixed state 'on the fly' non-adiabatic dynamics: The role of the conical intersection topology. *Chemical Physics Letters*, 292:259–266, 1998.

[753] H. Kleinert. *Path Integrals in Quantum Mechanics, Statistics, Polymer Physics, and Financial Markets*. World Scientific, Singapore, 3rd edition, 2004.

[754] L. Kleinman and D. M. Bylander. Efficacious form for model pseudopotentials. *Physical Review Letters*, 48:1425–1428, 1982. Fully non-local form for pseudo potentials.

[755] A. Klesing, D. Labrenz, and R. A. van Santen. *Ab initio* simulation of 'liquid' water on a Pd surface. *Journal of the Chemical Society, Faraday Transactions*, 94:3229–3235, 1998.

[756] M. Klessinger and J. Michl. *Excited States and Photochemistry of Organic Molecules*. VCH, New York, 1995.

[757] D. D. Klug, R. Rousseau, K. Uehara, M. Bernasconi, Y. Le Page, and J. S. Tse. *Ab initio* molecular dynamics study of the pressure-induced phase transformations in cristobalite. *Physical Review B*, 63:104106, 2001.

[758] L. Knoll, Z. Vager, and D. Marx. Experimental versus simulated Coulomb-explosion images of flexible molecules: Structure of protonated acetylene $C_2H_3^+$. *Physical Review A*, 67:022506, 2003.

[759] E. Ko, M. M. G. Alemany, and J. R. Chelikowsky. Viscosities of liquid CdTe near melting point from *ab initio* molecular-dynamics calculations. *Journal of Chemical Physics*, 121:942–945, 2004.

[760] W. Koch and M. C. Holthausen. *A Chemist's Guide to Density Functional Theory*. Wiley-VCH, Weinheim, 2001.

[761] J. Kohanoff. Phonon spectra from short non-thermally equilibrated

molecular dynamics simulations. *Computational Materials Science*, 2:221–232, 1994.

[762] J. Kohanoff. *Electronic Structure Calculations for Solids and Molecules – Theory and Computational Methods*. Cambridge University Press, Cambridge, 2006.

[763] J. Kohanoff, S. Scandolo, G. L. Chiarotti, and E. Tosatti. Solid molecular hydrogen: The broken symmetry phase. *Physical Review Letters*, 78:2783–2786, 1997.

[764] W. Kohn. Analytic properties of Bloch waves and Wannier functions. *Physical Review*, 115:809–821, 1959.

[765] W. Kohn. Density-functional Wannier function-theory for systems of very many atoms. *Chemical Physics Letters*, 208:167–172, 1993.

[766] W. Kohn. Nobel Lecture: Electronic structure of matter-wave functions and density functionals. *Reviews of Modern Physics*, 71:1253–1266, 1999. Nobel Prize in Chemistry awarded to Walter Kohn "for his development of the density-functional theory" and to John Pople "for his development of computational methods in quantum chemistry" in 1998; See http://www.nobel.se/chemistry/laureates/1998/.

[767] W. Kohn and L. J. Sham. Self-consistent equations including exchange and correlation effects. *Physical Review*, 140:A1133–A1138, 1965.

[768] M. Kohout and A. Savin. Atomic shell structure and electron numbers. *International Journal of Quantum Chemistry*, 60:875–882, 1996.

[769] M. Kohout, F. R. Wagner, and Y. Grin. Electron localization function for transition-metal compounds. *Theoretical Chemistry Accounts*, 108:150–156, 2002.

[770] M. Kolaski, H. M. Lee, C. Pak, M. Dupuis, and K. S. Kim. *Ab initio* molecular dynamics simulations of an excited state of $X–(H_2O)_3$ (X = Cl, I) complex. *Journal of Physical Chemistry A*, 109:9419–9423, 2005.

[771] W. Kołos. Adiabatic approximation and its accuracy. *Advances in Quantum Chemistry*, 5:99–133, 1970.

[772] M. Konôpka, R. Rousseau, I. Štich, and D. Marx. Detaching thiolates from copper and gold clusters: Which bonds to break? *Journal of the American Chemical Society*, 126:12103–12111, 2004.

[773] M. Konôpka, R. Rousseau, I. Štich, and D. Marx. Electronic origin of disorder and diffusion at a molecule–metal interface: Self-assembled

monolayers of CH_3S on Cu(111). *Physical Review Letters*, 95:096102, 2005.

[774] H. Köppel, W. Domcke, and L. S. Cederbaum. Multimode molecular-dynamics beyond the Born–Oppenheimer approximation. *Advances in Chemical Physics*, 57:59–246, 1984.

[775] R. Kosloff. Time-dependent quantum-mechanical methods for molecular-dynamics. *Journal of Physical Chemistry*, 92:2087–2100, 1988.

[776] R. Kosloff. Propagation methods for quantum molecular-dynamics. *Annual Review of Physical Chemistry*, 45:145–178, 1994.

[777] S. Koval, J. Kohanoff, J. Lasave, G. Colizzi, and R. L. Migoni. First-principles study of ferroelectricity and isotope effects in H-bonded KH_2PO_4 crystals. *Physical Review B*, 71:184102, 2005.

[778] S. Koval, J. Kohanoff, R. L. Migoni, and E. Tosatti. Ferroelectricity and isotope effects in hydrogen-bonded KDP crystals. *Physical Review Letters*, 89:187602, 2002.

[779] R. Kováčik, B. Meyer, and D. Marx. *F* Centers versus dimer vacancies on ZnO surfaces: Characterization by STM and STS calculations. *Angewandte Chemie International Edition*, 46:4894–4897, 2007.

[780] M. Krack, A. Gambirasio, and M. Parrinello. *Ab initio* X-ray scattering of liquid water. *Journal of Chemical Physics*, 117:9409–9412, 2002.

[781] M. Kreitmeir, H. Bertagnolli, J. J. Mortensen, and M. Parrinello. *Ab initio* molecular dynamics simulation of hydrogen fluoride at several thermodynamic states. *Journal of Chemical Physics*, 118:3639–3645, 2003.

[782] M. Kreitmeir, G. Heusel, H. Bertagnolli, K. Todheide, C. J. Mundy, and G. J. Cuello. Structure of dense hydrogen fluoride gas from neutron diffraction and molecular dynamics simulations. *Journal of Chemical Physics*, 122:154511, 2005.

[783] V. V. Kresin. Comment on "Shape phase transitions in the absorption spectra of atomic clusters". *Physical Review Letters*, 81:5702, 1998. Comment to Ref. [1091], see also Ref. [320].

[784] G. Kresse. *Ab initio* molecular dynamics: recent progresses and limitations. *Journal of Non-Crystalline Solids*, 312:52–59, 2002.

[785] G. Kresse and J. Furthmüller. Efficiency of ab-initio total energy calculations for metals and semiconductors using a plane-wave basis set. *Computational Materials Science*, 6:15–50, 1996.

[786] G. Kresse and J. Furthmüller. Efficient iterative schemes for *ab initio* total-energy calculations using a plane-wave basis set. *Physical Review B*, 54:11169–11186, 1996.

[787] G. Kresse and J. Hafner. *Ab initio* molecular dynamics for liquid metals. *Physical Review B*, 47:558–561, 1993.

[788] G. Kresse and J. Hafner. Norm-conserving and ultrasoft pseudopotentials for first-row and transition-elements. *Journal of Physics: Condensed Matter*, 6:8245–8257, 1994.

[789] G. Kresse and J. Hafner. *Ab initio* simulation of the metal/nonmetal transition in expanded fluid mercury. *Physical Review B*, 55:7539–7548, 1997.

[790] G. Kresse and D. Joubert. From ultrasoft pseudopotentials to the projector augmented-wave method. *Physical Review B*, 59:1758–1775, 1999.

[791] S. Krishnamurty, K. Joshi, D. G. Kanhere, and S. A. Blundell. Finite-temperature behavior of small silicon and tin clusters: An *ab initio* molecular dynamics study. *Physical Review B*, 73:045419, 2005.

[792] D. Krüger, H. Fuchs, R. Rousseau, D. Marx, and M. Parrinello. Pulling monatomic gold wires with single molecules: An *ab initio* simulation. *Physical Review Letters*, 89:186402, 2002. See animation of the pulling process http://www.theochem.rub.de/go/afm.html.

[793] D. Krüger, R. Rousseau, H. Fuchs, and D. Marx. Towards "mechanochemistry": Mechanically induced isomerizations of thiolate–gold clusters. *Angewandte Chemie International Edition*, 42:2251–2253, 2003. See animation of the pulling process http://www.theochem.rub.de/go/afm.html.

[794] R. Kubo, M. Toda, and N. Hashitsume. *Statistical Physics II*. Springer, Berlin, 1991.

[795] T. D. Kühne, M. Krack, F. R. Mohamed, and M. Parrinello. Efficient and accurate Car–Parrinello-like approach to Born–Oppenheimer molecular dynamics. *Physical Review Letters*, 98:066401, 2007.

[796] P. Kumar and D. Marx. Understanding hydrogen scrambling and infrared spectrum of bare CH_5^+ based on *ab initio* simulations. *Physical Chemistry Chemical Physics*, 8:573–586, 2006.

[797] V. Kumar. $Al_{10}Li_8$: A magic compound cluster. *Physical Review B*, 60:2916–2920, 1999.

[798] V. Kumar and V. Sundararajan. *Ab initio* molecular-dynamics studies of doped magic clusters and their interaction with atoms. *Physical Review B*, 57:4939–4942, 1998.

[799] K. Kunc. Density functional method in physics of solids. *Journal de chimie physique et de physico-chimie biologique*, 86:647–669, 1989.

[800] T. Kunert and R. Schmidt. Excitation and fragmentation mechanisms in ion–fullerene collisions. *Physical Review Letters*, 86:5258–5261, 2001.

[801] T. Kunert and R. Schmidt. Non-adiabatic quantum molecular dynamics: General formalism and case study H_2^+ in strong laser fields. *European Physical Journal D*, 25:15–24, 2003.

[802] I-F. W. Kuo and C. J. Mundy. An *ab initio* molecular dynamics study of the aqueous liquid-vapor interface. *Science*, 303:658–660, 2004. For a *Perspective Article* see Ref. [935].

[803] I-F. W. Kuo, C. J. Mundy, M. J. McGrath, and J. I. Siepmann. Time-dependent properties of liquid water: A comparison of Car–Parrinello and Born–Oppenheimer molecular dynamics simulations. *Journal of Chemical Theory and Computation*, 2:1274–1281, 2006.

[804] I-F. W. Kuo, C. J. Mundy, M. J. McGrath, J. I. Siepmann, J. VandeVondele, M. Sprik, J. Hutter, B. Chen, M. L. Klein, F. Mohamed, M. Krack, and M. Parrinello. Liquid water from first principles: Investigation of different sampling approaches. *Journal of Physical Chemistry B*, 108:12990–12998, 2004.

[805] I-F. W. Kuo and D. J. Tobias. Thermal fluctuations of the unusually symmetric and stable superoxide tetrahydrate complex: An *ab initio* molecular dynamics study. *Journal of Physical Chemistry A*, 106:10969–10976, 2002.

[806] Y. Kurosaki, K. Yokoyama, and Y. Teranishi. Direct *ab initio* molecular dynamics study of the two photodissociation channels of formic acid. *Chemical Physics*, 308:325–334, 2005.

[807] W. Kutzelnigg. The physical mechanism of the chemical bond. *Angewandte Chemie international Edition*, 12:546–562, 1973.

[808] W. Kutzelnigg. Theory of magnetic-susceptibilities and NMR chemical-shifts in terms of localized quantities. *Israel Journal of Chemistry*, 19:193–200, 1980.

[809] W. Kutzelnigg. Chemical bonding in higher main group elements. *Angewandte Chemie International Edition in English*, 23:272–295, 1984.

[810] W. Kutzelnigg. Stationary perturbation-theory 1. Survey of basic concepts. *Theoretica Chimica Acta*, 83:263–312, 1992.

[811] W. Kutzelnigg. The adiabatic approximation 1. The physical background of the Born-Handy ansatz. *Molecular Physics*, 90:909–916, 1997.

[812] W. Kutzelnigg. *Einführung in die Theoretische Chemie.* Wiley-VCH, Weinheim, 2002.

[813] K. Laasonen and M. L. Klein. *Ab initio* molecular-dynamics study of hydrochloric-acid in water. *Journal of the American Chemical Society*, 116:11620–11621, 1994.

[814] K. Laasonen and M. L. Klein. Structural study of $(H_2O)_{20}$ and $(H_2O)_{21}H^+$ using density-functional methods. *Journal of Physical Chemistry*, 98:10079, 1994.

[815] K. Laasonen, R. M. Nieminen, and M. J. Puska. First-principles study of fully relaxed vacancies in GaAs. *Physical Review B*, 45:4122–4130, 1992.

[816] K. Laasonen, A. Pasquarello, R. Car, C. Lee, and D. Vanderbilt. Car–Parrinello molecular dynamics with Vanderbilt ultrasoft pseudopotentials. *Physical Review B*, 47:10142–10153, 1993.

[817] K. Laasonen, M. Sprik, M. Parrinello, and R. Car. "Ab initio" liquid water. *Journal of Chemical Physics*, 99:9080–9089, 1993.

[818] T. Laino, F. Mohamed, A. Laio, and M. Parrinello. An efficient real space multigrid QM/MM electrostatic coupling. *Journal of Chemical Theory and Computation*, 1:1176–1184, 2005.

[819] T. Laino, F. Mohamed, A. Laio, and M. Parrinello. An efficient linear-scaling electrostatic coupling for treating periodic boundary conditions in QM/MM simulations. *Journal of Chemical Theory and Computation*, 2:1370–1378, 2006.

[820] T. Laino and D. Passerone. Pseudo-dynamics and band optimizations: Shedding light into conical intersection seams. *Chemical Physics Letters*, 389:1–6, 2004.

[821] A. Laio and M. Parrinello. Escaping free-energy minima. *Proceedings of the National Academy of Sciences of the United States of America*, 99:12562–12566, 2002.

[822] A. Laio, A. Rodriguez-Fortea, F. L. Gervasio, M. Ceccarelli, and M. Parrinello. Assessing the accuracy of metadynamics. *Journal of Physical Chemistry B*, 109:6714–6721, 2005.

[823] A. Laio, J. VandeVondele, and U. Röthlisberger. A Hamiltonian electrostatic coupling scheme for hybrid Car–Parrinello molecular dynamics simulations. *Journal of Chemical Physics*, 116:6941–6947, 2002.

[824] A. Laio, J. VandeVondele, and U. Röthlisberger. D-RESP: Dynamically generated electrostatic potential derived charges from quantum mechanics/molecular mechanics simulations. *Journal of Physical Chemistry B*, 106:7300–7307, 2002.

[825] D. Lamoen, P. Ballone, and M. Parrinello. Electronic structure, screening, and charging effects at a metal/organic tunneling junction: A first-principles study. *Physical Review B*, 54:5097–5102, 1996.

[826] D. Lamoen and M. Parrinello. Geometry and electronic structure of porphyrins and porphyrazines. *Chemical Physics Letters*, 248:309–315, 1996.

[827] C. Lanczos. An iteration method for the solution of the eigenvalue problem of linear differential and integral operators. *Journal of Research of the National Bureau of Standards*, 45:255–282, 1950.

[828] W. Langel. Car–Parrinello simulation of NH_3 adsorbed on the MgO(100) surface. *Chemical Physics Letters*, 259:7–14, 1996.

[829] W. Langel. Car–Parrinello simulation of H_2O dissociation on rutile. *Surface Science*, 496:141–150, 2002.

[830] W. Langel and L. Menken. Simulation of the interface between titanium oxide and amino acids in solution by first principles MD. *Surface Science*, 538:1–9, 2003.

[831] W. Langel and M. Parrinello. Hydrolysis at stepped MgO surfaces. *Physical Review Letters*, 73:504–507, 1994.

[832] W. Langel and M. Parrinello. *Ab initio* molecular-dynamics of H_2O adsorbed on solid MgO. *Journal of Chemical Physics*, 103:3240–3252, 1995.

[833] H. Langer and N. L. Doltsinis. Excited state tautomerism of the DNA base guanine: A restricted open-shell Kohn–Sham study. *Journal of Chemical Physics*, 118:5400–5407, 2003.

[834] H. Langer and N. L. Doltsinis. Selective photostabilisation of guanine by methylation. *Physical Chemistry Chemical Physics*, 5:4516–4518, 2003.

[835] H. Langer and N. L. Doltsinis. Nonradiative decay of photoexcited methylated guanine. *Physical Chemistry Chemical Physics*, 6:2742–2748, 2004.

[836] H. Langer, N. L. Doltsinis, and D. Marx. Excited state dynamics and coupled proton–electron transfer of guanine: from the gas phase via microsolvation to aqueous solution. *ChemPhysChem*, 6:1734–1737, 2005.

[837] P. W. Langhoff, S. T. Epstein, and M. Karplus. Aspects of time-dependent perturbation theory. *Reviews of Modern Physics*, 44:602–644, 1972.

[838] H. Lapid, N. Agmon, M. K. Petersen, and G. A. Voth. A bond-order analysis of the mechanism for hydrated proton mobility in liquid water. *Journal of Chemical Physics*, 122:014506, 2005.

[839] Y. Laudernet, J. Clerouin, and S. Mazevet. *Ab initio* simulations of the electrical and optical properties of shock-compressed SiO_2. *Physical Review B*, 70:165108, 2004.

[840] G. Lauritsch and P.-G. Reinhard. An FFT solver for the Coulomb problem. *International Journal of Modern Physics C*, 5:65–75, 1994.

[841] C. P. Lawrence, N. Nakayama, N. Makri, and J. L. Skinner. Quantum dynamics in simple fluids. *Journal of Chemical Physics*, 120:6621–6624, 2004.

[842] M. Lax. The Franck–Condon principle and its application to crystals. *Journal of Chemical Physics*, 20:1752–1760, 1952.

[843] C. Lee and D. Vanderbilt. Erratum: Proton transfer in ice. *Chemical Physics Letters*, 226:610, 1993. Original article: [844].

[844] C. Lee and D. Vanderbilt. Proton transfer in ice. *Chemical Physics Letters*, 210:279–284, 1993. Erratum: [843].

[845] C. Lee, D. Vanderbilt, K. Laasonen, R. Car, and M. Parrinello. *Ab initio* studies on high pressure phases of ice. *Physical Review Letters*, 69:462–465, 1992.

[846] C. Lee, D. Vanderbilt, K. Laasonen, R. Car, and M. Parrinello. *Ab initio* studies on the structural and dynamic properties of ice. *Physical Review B*, 47:4863–4872, 1993.

[847] C. Lee, W. Yang, and R. G. Parr. Development of the Colle–Salvetti correlation-energy formula into a functional of the electron density. *Physical Review B*, 37:785–789, 1988.

[848] H. S. Lee and M. E. Tuckerman. *Ab initio* molecular dynamics with discrete variable representation basis sets: Techniques and application to liquid water. *Journal of Physical Chemistry A*, 110:5549–5560, 2006.

[849] H. S. Lee and M. E. Tuckerman. Structure of liquid water at ambient temperature from *ab initio* molecular dynamics performed in the complete basis set limit. *Journal of Chemical Physics*, 125:154507, 2006.

[850] J. G. Lee, E. Asciutto, V. Babin, C. Sagui, T. Darden, and C. Roland. Deprotonation of solvated formic acid: Car–Parrinello and metadynamics simulations. *Journal of Physical Chemistry B*, 110:2325–2331, 2006.

[851] Y.-S. Lee and M. Krauss. Dynamics of proton transfer in bacteriorhodopsin. *Journal of the American Chemical Society*, 126:2225–2230, 2004.

[852] Y.-S. Lee, M. Buongiorno Nardelli, and N. Marzari. Band structure

and quantum conductance of nanostructures from maximally localized Wannier functions: The case of functionalized carbon nanotubes. *Physical Review Letters*, 95:076804, 2005.

[853] C. Leforestier. Classical trajectories using full *ab initio* potential-energy surface $H^- + CH_4 \rightarrow CH_4 + H^-$. *Journal of Chemical Physics*, 68:4406–4410, 1978.

[854] C. Leforestier, R. H. Bisseling, C. Cerjan, M. D. Feit, R. Friesner, A. Guldberg, A. Hammerich, G. Jolicard, W. Karrlein, H.-D. Meyer, N. Lipkin, O. Roncero, and R. Kosloff. A comparison of different propagation schemes for the time-dependent Schrödinger-equation. *Journal of Computational Physics*, 94:59–80, 1991.

[855] S. B. Legoas, V. Rodrigues, D. Ugarte, and D. S. Galvao. Contaminants in suspended gold chains: An *ab initio* molecular dynamics study. *Physical Review Letters*, 93:216103, 2004. Comment: Ref. [642] and Reply: Ref. [856].

[856] S. B. Legoas, V. Rodrigues, D. Ugarte, and D. S. Galvao. Reply to Comment on "Contaminants in suspended gold chains: An *ab initio* molecular dynamics study". *Physical Review Letters*, 95:169602, 2005. Reply to Comment: Ref. [642].

[857] J. G. LePage, M. Alouani, D. L. Dorsey, J. W. Wilkins, and P. E. Blöchl. *Ab initio* calculation of binding and diffusion of a Ga adatom on the GaAs(001)-c(4×4) surface. *Physical Review B*, 58:1499–1505, 1998.

[858] K. Leung and S. B. Rempe. *Ab initio* molecular dynamics study of glycine intramolecular proton transfer in water. *Journal of Chemical Physics*, 122:184506, 2005.

[859] I. N. Levine. *Quantum Chemistry*. Allyn and Bacon, Boston, 1983.

[860] J. Li and D. A. Drabold. First-principles molecular-dynamics study of glassy As_2Se_3. *Physical Review B*, 61:11998–12004, 2000.

[861] J. Li, G. Speyer, and O. F. Sankey. Conduction switching of photochromic molecules. *Physical Review Letters*, 93:248302, 2004.

[862] X. Li, J. C. Tully, H. B. Schlegel, and M. J. Frisch. *Ab initio* Ehrenfest dynamics. *Journal of Chemical Physics*, 123:084106, 2005.

[863] X. P. Li, W. Nunes, and D. Vanderbilt. Density-matrix electronic-structure method with linear system-size scaling. *Physical Review B*, 47:10891–10894, 1993.

[864] J. C. Light and T. Carrington. Discrete-variable representations and their utilization. *Advances in Chemical Physics*, 114:263–310, 2000.

[865] J. M. Lighthill. *Introduction to Fourier Analysis and Generalized Functions*. Cambridge University Press, Cambridge, 1958.

[866] F. C. Lightstone, E. Schwegler, M. Allesch, F. Gygi, and G. Galli. A first-principles molecular dynamics study of calcium in water. *ChemPhysChem*, 6:1745–1749, 2005.

[867] F. C. Lightstone, E. Schwegler, R. Q. Hood, F. Gygi, and G. Galli. A first principles molecular dynamics simulation of the hydrated magnesium ion. *Chemical Physics Letters*, 343:549–555, 2001.

[868] J. S. Lin, A. Qteish, M. C. Payne, and V. Heine. Optimized and transferable nonlocal separable *ab initio* pseudopotentials. *Physical Review B*, 47:4174–4180, 1993.

[869] G. Lippert, J. Hutter, and M. Parrinello. A hybrid Gaussian and plane wave density functional scheme. *Molecular Physics*, 92:477–488, 1997.

[870] G. Lippert, J. Hutter, and M. Parrinello. The Gaussian and augmented-plane-wave density functional method for *ab initio* molecular dynamics simulations. *Theoretical Chemistry Accounts*, 103:124–140, 1999.

[871] Yi Liu, D. A. Yarne, and M. E. Tuckerman. *Ab initio* molecular dynamics calculations with simple, localized, orthonormal real-space basis sets. *Physical Review B*, 68:125110, 2003.

[872] Z. Liu, L. E. Carter, and E. A. Carter. Full configuration-interaction molecular-dynamics of Na_2 and Na_3. *Journal of Physical Chemistry*, 99:4355–4359, 1995.

[873] Z. F. Liu, C. K. Siu, and J. S. Tse. *Ab initio* molecular dynamics study on the hydrolysis of molecular chlorine. *Chemical Physics Letters*, 311:93–101, 1999.

[874] Z. F. Liu, C. K. Siu, and J. S. Tse. Catalysis of the reaction hcl + hocl $\rightarrow$ H_2O + Cl_2 on an ice surface. *Chemical Physics Letters*, 309:335–343, 1999.

[875] Z. F. Liu, C. K. Siu, and J. S. Tse. Catalysis of the reaction HCl+HOCl $\rightarrow$ H_2O+Cl_2 on an ice surface. *Chemical Physics Letters*, 314:317–325, 1999.

[876] Z.-P. Liu, X.-Q. Gong, J. Kohanoff, C. Sanchez, and P. Hu. Catalytic role of metal oxides in gold-based catalysts: A first principles study of CO oxidation on TiO_2 supported Au. *Physical Review Letters*, 91:266102, 2003.

[877] T. López-Ciudad, R. Ramírez, J. Schulte, and M. C. Böhm. Anharmonic effects on the structural and vibrational properties of the ethyl radical: A path integral Monte Carlo study. *Journal of Chemical Physics*, 119:4328–4338, August 2003.

[878] C. L. Lopreore and R. E. Wyatt. Quantum wave packet dynamics with trajectories. *Physical Review Letters*, 82:5190–5193, 1999.

[879] S. G. Louie, S. Froyen, and M. L. Cohen. Non-linear ionic pseudopotentials in spin-density-functional calculations. *Physical Review B*, 26:1738–1742, 1982.

[880] S. G. Louie, K.-M. Ho, and M. L. Cohen. Self-consistent mixed-basis approach to the electronic structure of solids. *Physical Review B*, 19:1774–1782, 1979.

[881] P. O. Löwdin. On the non-orthogonality problem. *Advances in Quantum Chemistry*, 5:185–199, 1970.

[882] H. Löwen and I. D'Amico. Testing of pseudopotentials used in classical Car–Parrinello simulations. *Journal of Physics: Condensed Matter*, 9:8879–8892, 1997.

[883] H. Löwen, J.-P. Hansen, and P. A. Madden. Nonlinear counterion screening in colloidal suspensions. *Journal of Chemical Physics*, 98:3275–3289, 1993.

[884] H. Löwen, P. A. Madden, and J.-P. Hansen. *Ab initio* description of counterion screening in colloidal suspensions. *Physical Review Letters*, 68:1081–1084, 1992. See also Ref. [882].

[885] A. Y. Lozovoi, A. Alavi, J. Kohanoff, and R. M. Lynden-Bell. *Ab initio* simulation of charged slabs at constant chemical potential. *Journal of Chemical Physics*, 115:1661–1669, 2001.

[886] M. I. Lubin, E. J. Bylaska, and J. H. Weare. *Ab initio* molecular dynamics simulations of aluminum ion solvation in water clusters. *Chemical Physics Letters*, 322:447–453, 2000.

[887] A. P. Lyubartsev, K. Laasonen, and A. Laaksonen. Hydration of Li^+ ion. An *ab initio* molecular dynamics simulation. *Journal of Chemical Physics*, 114:3120–3126, 2001.

[888] A. D. MacKerell, D. Bashford, M. Bellott, R. L. Dunbrack, J. D. Evanseck, M. J. Field, S. Fischer, J. Gao, H. Guo, S. Ha, D. Joseph-McCarthy, L. Kuchnir, K. Kuczera, F. T. K. Lau, C. Mattos, S. Michnick, T. Ngo, D. T. Nguyen, B. Prodhom, W. E. Reiher, B. Roux, M. Schlenkrich, J. C. Smith, B. Stote, J. Straub, M. Watanabe, J. Wiorkiewicz-Kuczera, D. Yin, and M. Karplus. All-atom empirical potential for molecular modeling and dynamics studies of proteins. *Journal of Physical Chemistry B*, 102:3586–3616, 1998.

[889] A. C. Maggs and V. Rossetto. Local simulation algorithms for Coulomb interactions. *Physical Review Letters*, 88:196402, 2002.

[890] A. C. Maggs and J. Rottler. Auxiliary field simulation and Coulomb's law. *Computer Physics Communications*, 169:160–165, 2005.

[891] A. Magistrato, W. F. DeGrado, A. Laio, U. Röthlisberger, J. VandeVondele, and M. L. Klein. Characterization of the dizinc analogue of the synthetic diiron protein DF1 using *ab initio* and hybrid quantum/classical molecular dynamics simulations. *Journal of Physical Chemistry B*, 107:4182–4188, 2003.

[892] A. Magistrato, A. Togni, and U. Röthlisberger. Enantioselective palladium-catalyzed hydrosilylation of styrene: Influence of electronic and steric effects on enantioselectivity and catalyst design via hybrid QM/MM molecular dynamics simulations. *Organometallics*, 25:1151–1157, 2006.

[893] J.-B. Maillet, E. Bourasseau, and V. Recoules. *Ab initio* molecular dynamic study of liquid hydrogen fluorine under pressure. *Physical Review B*, 72:224103, 2005.

[894] N. T. Maitra. On correlated electron-nuclear dynamics using time-dependent density functional theory. *Journal of Chemical Physics*, 125:014110, 2006.

[895] G. Makov and M. C. Payne. Periodic boundary conditions in *ab initio* calculations. *Physical Review B*, 51:4014–4022, 1995.

[896] S. A. Maluendes and M. Dupuis. *Ab initio* SCF molecular-dynamics: Exploring the potential-energy surface of small silicon clusters. *International Journal of Quantum Chemistry*, 42:1327–1338, 1992.

[897] D. J. Mann and M. D. Halls. *Ab initio* simulations of oxygen atom insertion and substitutional doping of carbon nanotubes. *Journal of Chemical Physics*, 116:9014–9020, 2002.

[898] D. J. Mann and M. D. Halls. Water alignment and proton conduction inside carbon nanotubes. *Physical Review Letters*, 90:195503, 2003.

[899] Y. A. Mantz, B. Chen, and G. J. Martyna. Temperature-dependent water structure: *Ab initio* and empirical model predictions. *Chemical Physics Letters*, 405:294–299, 2005.

[900] Y. A. Mantz, B. Chen, and G. J. Martyna. Structural correlations and motifs in liquid water at selected temperatures: Ab initio and empirical model predictions. *Journal of Physical Chemistry B*, 110:3540–3554, 2006.

[901] Y. A. Mantz, F. M. Geiger, L. T. Molina, M. J. Molina, and B. L. Trout. First-principles molecular-dynamics study of surface disordering of the (0001) face of hexagonal ice. *Journal of Chemical Physics*, 113:10733–10743, 2000.

[902] Y. A. Mantz, F. M. Geiger, L. T. Molina, M. J. Molina, and B. L. Trout. The interaction of HCl with the (0001) face of hexagonal ice

studied theoretically via Car–Parrinello molecular dynamics. *Chemical Physics Letters*, 348:285–292, 2001.

[903] A. Marcellini, C. A. Pignedoli, M. Ferrario, and C. M. Bertoni. Interaction of Cl_2 molecules with GaAs(110) surface. *Surface Science*, 402:47–51, 1998.

[904] M. Marchi, J. Hutter, and M. Parrinello. A first principles investigation of the structure of a bacteriochlorophyll crystal. *Journal of the American Chemical Society*, 118:7847–7848, 1996.

[905] P. Margl, K. Schwarz, and P. E. Blöchl. Finite-temperature characterization of ferrocene from first-principles molecular-dynamics simulations. *Journal of Chemical Physics*, 100:8194–8203, 1994.

[906] P. Margl, T. Ziegler, and P. E. Blöchl. Reaction of methane with $Rh(PH_3)_2Cl$: A dynamical density functional study. *Journal of the American Chemical Society*, 117:12625–12634, 1995.

[907] P. M. Margl, T. K. Woo, P. E. Blöchl, and T. Ziegler. Evidence for a stable Ti(IV) metallocene dihydrogen complex from *ab initio* molecular dynamics. *Journal of the American Chemical Society*, 120:2174–2175, 1998.

[908] N. A. Marks, D. R. McKenzie, M. Bernasconi B. A. Pailthorpe, and M. Parrinello. Microscopic structure of tetrahedral amorphous carbon. *Physical Review Letters*, 76:768–771, 1996.

[909] P. R. L. Markwick, N. L. Doltsinis, and D. Marx. Targeted Car–Parrinello molecular dynamics: Elucidating double proton transfer in formic acid dimer. *Journal of Chemical Physics*, 122:054112, 2005.

[910] M. A. L. Marques and E. K. U. Gross. Time-dependent density functional theory. *Annual Review of Physical Chemistry*, 55:427–455, 2004.

[911] M. A. L. Marques, X. López, D. Varsano, A. Castro, and A. Rubio. Time-dependent density-functional approach for biological chromophores: The case of the green fluorescent protein. *Physical Review Letters*, 90:258101, 2003.

[912] M. A. L. Marques, C. A. Ullrich, F. Nogueira, A. Rubio, K. Burke, and E. K. U. Gross, eds. *Time-Dependent Density Functional Theory*. Springer, Berlin, 2006.

[913] R. M. Martin. *Electronic Structure: Basic Theory and Practical Methods*. Cambridge University Press, Cambridge, 2004. For additional information, errata, web resources, etc. see `http://electronicstructure.org/`.

[914] T. J. Martinez. *Ab initio* molecular dynamics around a conical intersection: Li(2p) + H_2. *Chemical Physics Letters*, 272:139–147, 1997.

[915] T. J. Martinez. Insights for light-driven molecular devices from *ab initio* multiple spawning excited-state dynamics of organic and biological chromophores. *Accounts of Chemical Research*, 39:119–126, 2006.

[916] T. J. Martinez, M. Ben-Nun, and G. Ashkenazi. Classical quantal method for multistate dynamics: A computational study. *Journal of Chemical Physics*, 104:2847–2856, 1996.

[917] T. J. Martinez, M. Ben-Nun, and R. D. Levine. Multi-electronic-state molecular dynamics: A wave function approach with applications. *Journal of Physical Chemistry*, 100:7884–7895, 1996.

[918] R. Martoňák, L. Colombo, C. Molteni, and M. Parrinello. Pressure-induced structural transformations in a medium-sized silicon nanocrystal by tight-binding molecular dynamics. *Journal of Chemical Physics*, 117:11329–11335, 2002.

[919] R. Martoňák, D. Donadio, and M. Parrinello. Polyamorphism of ice at low temperatures from constant-pressure simulations. *Physical Review Letters*, 92:225702, 2004.

[920] R. Martoňák, A. Laio, M. Bernasconi, C. Ceriani, P. Raiteri, F. Zipoli, and M. Parrinello. Simulation of structural phase transitions by metadynamics. *Zeitschrift für Kristallographie*, 220:489–498, 2005.

[921] R. Martoňák, A. Laio, and M. Parrinello. Predicting crystal structures: The Parrinello–Rahman method revisited. *Physical Review Letters*, 90:075503, 2003.

[922] R. Martoňák, C. Molteni, and M. Parrinello. *Ab Initio* molecular dynamics with a classical pressure reservoir: Simulation of pressure-induced amorphization in a $Si_{35}H_{36}$ cluster. *Physical Review Letters*, 84:682–685, 2000.

[923] R. Martoňák, C. Molteni, and M. Parrinello. A new constant-pressure *ab initio*/classical molecular dynamics method: Simulation of pressure-induced amorphization in a $Si_{35}H_{36}$ cluster. *Computational Materials Science*, 20:293–299, 2001.

[924] G. Martyna, C. Cheng, and M. L. Klein. Electronic states and dynamical behavior of $LiXe_n$ and $CsXe_n$ clusters. *Journal of Chemical Physics*, 95:1318–1336, 1991.

[925] G. J. Martyna. Adiabatic path integral molecular dynamics methods. 1. Theory. *Journal of Chemical Physics*, 104:2018–2027, 1996.

[926] G. J. Martyna, A. Hughes, and M. E. Tuckerman. Molecular dynamics algorithms for path integrals at constant pressure. *Journal of Chemical Physics*, 110:3275–3290, 1999.

[927] G. J. Martyna, M. L. Klein, and M. Tuckerman. Nosé–Hoover chains – The canonical ensemble via continuous dynamics. *Journal of Chemical Physics*, 97:2635–2643, 1992.

[928] G. J. Martyna, D. J. Tobias, and M. L. Klein. Constant-pressure molecular-dynamics algorithms. *Journal of Chemical Physics*, 101:4177–4189, 1994.

[929] G. J. Martyna and M. E. Tuckerman. A reciprocal space based method for treating long range interactions in *ab initio* and force-field-based calculations in clusters. *Journal of Chemical Physics*, 110:2810–2821, 1999.

[930] G. J. Martyna, M. E. Tuckerman, D. J. Tobias, and M. L. Klein. Explicit reversible integrators for extended systems dynamics. *Molecular Physics*, 87:1117–1157, 1996.

[931] D. Marx. Proton transfer in ice. In B. J. Berne, G. Ciccotti, and D. F. Coker, eds., *Classical and Quantum Dynamics in Condensed Phase Simulations*, chapter 15, pages 359–383. World Scientific, Singapore, 1998.

[932] D. Marx. *Ab initio* liquids: Simulating liquids based on first principles. In C. Caccamo, J.-P. Hansen, and G. Stell, eds., *New Approaches to Problems in Liquid State Theory*, pages 439–457. Kluwer, Dordrecht, 1999.

[933] D. Marx. Theoretical Chemistry 1998. *Nachrichten aus Chemie, Technik und Laboratorium*, 47:186–195, 1999.

[934] D. Marx. Theoretical chemistry in the 21st century: The "virtual lab". In W. Greiner and J. Reinhardt, eds., *Proceedings of the "Idea-Finding Symposium: Frankfurt Institute for Advanced Studies"*, pages 139–153. EP Systema Bt., Debrecen, 2004. For download see http://www.theochem.rub.de/go/cprev.html.

[935] D. Marx. Throwing tetrahedral dice. *Science*, 303:634–636, 2004.

[936] D. Marx. Wasser, Eis und Protonen: Mit Quantensimulationen zum molekularen Verständnis von Wasserstoffbrücken, Protonentransfer und Phasenübergängen. *Physik Journal*, 3:33–39, 2004.

[937] D. Marx. Advanced Car–Parrinello techniques: Path integrals and nonadiabaticity in condensed matter simulations. In M. Ferrario, K. Binder, and G. Ciccotti, eds., *Computer Simulations in Condensed Matter: From Materials to Chemical Biology (Lecture Notes in Physics 704)*, volume 2, pages 507–539. Springer, Berlin, 2006.

[938] D. Marx. An introduction to ab initio molecular dynamics simulations. In J. Grotendorst, S. Blügel, and D. Marx, eds., *Computational*

Nanoscience: Do It Yourself!, pages 195–244. John von Neumann Institute for Computing, Forschungszentrum Jülich, Jülich, 2006.

[939] D. Marx. Proton transfer 200 years after von Grotthuss: Insights from ab initio simulations. *ChemPhysChem*, 7:1848–1870, 2006. Addendum: [940].

[940] D. Marx. Addendum to: Proton Transfer 200 Years after von Grotthuss: Insights from Ab Initio Simulations. *ChemPhysChem*, 8:209–210, 2007. Original article: [939].

[941] D. Marx and coworkers. Animation of thermally induced hydrogen scrambling of a protonated methane molecule (CH_5^+) in vacuum. See http://www.theochem.rub.de/go/ch5p.html.

[942] D. Marx, E. Fois, and M. Parrinello. Static and Dynamic Density Functional Investigation of Hydrated Beryllium Dications. *International Journal of Quantum Chemistry*, 57:655–662, 1996.

[943] D. Marx and J. Hutter. *Ab Initio* molecular dynamics: Theory and implementation. In J. Grotendorst, ed., *Modern Methods and Algorithms of Quantum Chemistry*, pages 301–449 (first edition, paperback, ISBN 3--00--005618--1) or 329–477 (second edition, hardcover, ISBN 3--00--005834--6). John von Neumann Institute for Computing, Forschungszentrum Jülich, Jülich, 2000. See http://www.theochem.rub.de/go/cprev.html.

[944] D. Marx, J. Hutter, and M. Parrinello. Density-functional study of small aqueous Be^{2+} clusters. *Chemical Physics Letters*, 241:457–462, 1995.

[945] D. Marx and M. H. Müser. Path integral simulations of rotors: Theory and applications. *Journal of Physics: Condensed Matter*, 11:R117–R155, 1999.

[946] D. Marx and M. Parrinello. *Ab initio* path-integral molecular dynamics. *Zeitschrift für Physik B (Rapid Note)*, 95:143–144, 1994. Note: A misprinted sign in the Lagrangian is correctly given in Eq. (5.208) in this book. However, the correct definition was implemented in the CPMD package [696] so that all data reported in the paper are unaffected.

[947] D. Marx and M. Parrinello. Structural quantum effects and three-centre two-electron bonding in CH_5^+. *Nature*, 375:216–218, 1995.

[948] D. Marx and M. Parrinello. *Ab initio* path integral molecular dynamics: Basic ideas. *Journal of Chemical Physics*, 104:4077–4082, 1996.

[949] D. Marx and M. Parrinello. The effect of quantum and thermal

fluctuations on the structure of the floppy molecule $C_2H_3^+$. *Science*, 271:179–181, 1996.

[950] D. Marx and M. Parrinello. Structure and dynamics of protonated methane: CH_5^+ at finite temperatures. *Zeitschrift für Physik D*, 41:253–260, 1997.

[951] D. Marx and M. Parrinello. CH_5^+ Stability and mass spectrometry. *Science*, 286:1051, 1999. DOI:10.1126/science.286.5442.1051a; See `http://www.sciencemag.org/cgi/content/full/286/5442/1051a`.

[952] D. Marx and M. Parrinello. Molecular spectroscopy – CH_5^+: The Cheshire Cat smiles. *Science*, 284:59–61, 1999. See also Ref. [951].

[953] D. Marx and A. Savin. Topological bifurcation analysis: Electronic structure of CH_5^+. *Angewandte Chemie International Edition*, 36:2077–2080, 1997.

[954] D. Marx, M. Sprik, and M. Parrinello. Ab initio molecular dynamics of ion solvation. The case of Be_2^+ in water. *Chemical Physics Letters*, 272:360–366, 1997. See also Refs. [942, 944].

[955] D. Marx, M. E. Tuckerman, J. Hutter, and M. Parrinello. The nature of the hydrated excess proton in water. *Nature*, 397:601–604, 1999. See also Ref. [666].

[956] D. Marx, M. E. Tuckerman, and G. J. Martyna. Quantum dynamics via adiabatic *ab initio* centroid molecular dynamics. *Computer Physics Communications*, 118:166–184, 1999. Note: The misprinted definition of the fictitious normal mode masses in Eq. (2.51) is corrected in Eq. (5.230) in this book. However, the correct definition was implemented in the CPMD package [696] so that all data reported in the paper are unaffected.

[957] D. Marx, M. E. Tuckerman, and M. Parrinello. Solvated excess protons in water: Quantum effects on the hydration structure. *Journal of Physics: Condensed Matter*, 12:A153–A159, 2000.

[958] N. Marzari and R. Car. Extended space Car–Parrinello molecular dynamics. *Abstracts of Papers of the American Chemical Society*, 222:195, August 2001. Private communication.

[959] N. Marzari, I. Souza, and D. Vanderbilt. An introduction to maximally-localized Wannier functions.
See `http://psi-k.dl.ac.uk/index.html?highlights` and `http://psi-k.dl.ac.uk/newsletters/News_57/Highlight_57.pdf`.

[960] N. Marzari and D. Vanderbilt. Maximally localized generalized Wannier functions for composite energy bands. *Physical Review B*, 56:12847–12865, 1997. See Ref. [959] for a review.

[961] N. Marzari, D. Vanderbilt, and M. C. Payne. Ensemble density-functional theory for ab initio molecular dynamics of metals and finite-temperature insulators. *Physical Review Letters*, 79:1337–1340, 1997.

[962] N. Marzari, D. Vanderbilt, A. De Vita, and M. C. Payne. Thermal contraction and disordering of the Al(110) surface. *Physical Review Letters*, 82:3296–3299, 1999.

[963] F. Maseras and K. Morokuma. IMOMM – A new integrated ab-initio plus molecular mechanics geometry optimization scheme of equilibrium structures and transition-states. *Journal of Computational Chemistry*, 16:1170–1179, 1995.

[964] P. Masini and M. Bernasconi. *Ab initio* simulations of hydroxylation and dehydroxylation reactions at surfaces: Amorphous silica and brucite. *Journal of Physics: Condensed Matter*, 14:4133–4144, 2002.

[965] H. S. W. Massey. Collisions between atoms and molecules at ordinary temperatures. *Reports on Progress in Physics*, 12:248–269, 1948.

[966] C. Massobrio, A. Pasquarello, and R. Car. Microscopic structure of liquid $GeSe_2$: The problem of concentration fluctuations over intermediate range distances. *Physical Review Letters*, 80:2342–2345, 1998.

[967] C. Massobrio, A. Pasquarello, and R. Car. Short- and intermediate-range structure of liquid $GeSe_2$. *Physical Review B*, 64:144205, 2001.

[968] G. Mathias. *Elektrostatische Wechselwirkungen in komplexen Flüssigkeiten und ihre Beschreibung mit Molekulardynamiksimulationen.* PhD thesis, Ludwig-Maximilians-Universität, München, 2004.

[969] G. Mathias, B. Egwolf, M. Nonella, and P. Tavan. A fast multipole method combined with a reaction field for long-range electrostatics in molecular dynamics simulations: The effects of truncation on the properties of water. *Journal of Chemical Physics*, 118:10847–10860, 2003.

[970] G. Mathias and D. Marx. Structures and spectral signatures of protonated water networks in bacteriorhodopsin. *Proceedings of the National Academy of Sciences of the United States of America*, 104:6980–6985, 2007.

[971] W. D. Mattson, D. Sánchez-Portal, S. Chiesa, and R. M. Martin. Prediction of new phases of nitrogen at high pressure from first-principles simulations. *Physical Review Letters*, 93:125501, 2004.

[972] A. E. Mattsson. In pursuit of the "divine" functional (perspectives: Density functional theory). *Science*, 298:759–760, 2002.

[973] T. R. Mattsson and S. J. Paddison. Methanol at the water–platinum interface studied by *ab initio* molecular dynamics. *Surface Science*, 544:L697–L702, 2003.

[974] F. Mauri, G. Galli, and R. Car. Orbital formulation for electronic-structure calculations with linear system-size scaling. *Physical Review B*, 47:9973–9976, 1993.

[975] F. Mauri and S. Louie. Magnetic susceptibility of insulators from first principles. *Physical Review Letters*, 76:4246–4249, 1996.

[976] F. Mauri, B. Pfrommer, and S. Louie. Ab initio theory of NMR chemical shifts in solids and liquids. *Physical Review Letters*, 77:5300–5303, 1996.

[977] F. Mauri, B. G. Pfrommer, and S. G. Louie. *Ab initio* NMR chemical shift of diamond, chemical-vapor-deposited diamond, and amorphous carbon. *Physical Review Letters*, 79:2340–2343, 1997.

[978] I. Mayer. Charge, bond order and valence in the *ab initio* SCF theory. *Chemical Physics Letters*, 97:270–274, 1984.

[979] S. Mazevet, J. Clerouin, V. Recoules, P. M. Anglade, and G. Zerah. *Ab initio* simulations of the optical properties of warm dense gold. *Physical Review Letters*, 95:085002, 2005.

[980] D. G. McCulloch, D. R. McKenzie, and C. M. Goringe. *Ab initio* simulations of the structure of amorphous carbon. *Physical Review B*, 61:2349–2355, 2000.

[981] M. J. McGrath, J. I. Siepmann, I-F. W. Kuo, C. J. Mundy, J. Vande-Vondele, J. Hutter, F. Mohamed, and M. Krack. Isobaric-isothermal Monte Carlo simulations from first principles: Application to liquid water at ambient conditions. *ChemPhysChem*, 6:1894–1901, 2005.

[982] M. J. McGrath, J. I. Siepmann, I-F. W. Kuo, C. J. Mundy, J. VandeVondele, J. Hutter, F. Mohamed, and M. Krack. Simulating fluid-phase equilibria of water from first principles. *Journal of Physical Chemistry A*, 110:640–646, 2006.

[983] M. J. McGrath, J. I. Siepmann, I-F. W. Kuo, C. J. Mundy, J. Vande-Vondele, M. Sprik, J. Hutter, F. Mohamed, M. Krack, and M. Parrinello. Toward a Monte Carlo program for simulating vapor–liquid phase equilibria from first principles. *Computer Physics Communications*, 169:289–294, 2005.

[984] D. A. McQuarrie. *Statistical Mechanics*. Harper & Row Publishers, New York, 1976.

[985] R. McWeeny. *Methods of Molecular Quantum Mechanics*. Academic Press, London, 1992.

[986] C. A. Mead. The geometric phase in molecular systems. *Reviews of Modern Physics*, 64:51–85, 1992.

[987] H. S. Mei, M. E. Tuckerman, D. E. Sagnella, and M. L. Klein. Quantum nuclear *ab initio* molecular dynamics study of water wires. *Journal of Physical Chemistry B*, 102:10446–10458, 1998.

[988] E. J. Meijer and M. Sprik. A density-functional study of the intermolecular interactions of benzene. *Journal of Chemical Physics*, 105:8684–8689, 1996.

[989] E. J. Meijer and M. Sprik. *Ab initio* molecular dynamics study of the reaction of water with formaldehyde in sulfuric acid solution. *Journal of the American Chemical Society*, 120:6345–6355, 1998.

[990] E. J. Meijer and M. Sprik. A density functional study of the addition of water to SO_3 in the gas phase and in aqueous solution. *Journal of Physical Chemistry A*, 102:2893–2898, 1998.

[991] B. Meng, D. Maroudas, and W. H. Weinberg. Structure of chemisorbed acetylene on the Si(001)-(2×1) surface and the effect of coadsorbed atomic hydrogen. *Chemical Physics Letters*, 278:97–101, 1997.

[992] S. Meng, L. f. Xu, E. G. Wang, and S. Gao. Vibrational recognition of hydrogen-bonded water networks on a metal surface. *Physical Review Letters*, 89:176104, 2002.

[993] F. Mercuri, C. J. Mundy, and M. Parrinello. Formation of a reactive intermediate in molecular beam chemistry of sodium and water. *Journal of Physical Chemistry A*, 105:8423–8427, 2001.

[994] N. D. Mermin. Thermal properties of the inhomogeneous electron gas. *Physical Review*, 137:A1441–A1443, 1965.

[995] N. D. Mermin. What's wrong with this Lagrangean? *Physics Today*, page 9, April 1988. Note: the correct spelling "Lagrangian" has been copy-edited into "Lagrangean" also in Ref. [222].

[996] A. Messiah. *Quantum Mechanics*. North-Holland, Amsterdam, 1964.

[997] B. Meyer, K. Hummler, C. Elsässer, and M. Fähnle. Reconstruction of the true wavefunctions from the pseudowavefunctions in a crystal and calculation of electric field gradients. *Journal of Physics: Condensed Matter*, 7:9201–9217, 1995.

[998] B. Meyer, D. Marx, O. Dulub, U. Diebold, M. Kunat, D. Langenberg, and Ch. Wöll. Partial dissociation of water leads to stable superstructures on the surface of zinc oxide. *Angewandte Chemie International Edition*, 43:6641–6645, 2005.

[999] B. Meyer, H. Raaba, and D. Marx. Water adsorption on $ZnO(10\bar{1}0)$:

from single molecules to partially dissociated monolayers. *Physical Chemistry Chemical Physics*, 8:1513–1520, 2006.

[1000] H.-D. Meyer, U. Manthe, and L. S. Cederbaum. The multi-configurational time-dependent Hartree approach. *Chemical Physics Letters*, 165:73–78, 1990.

[1001] H.-D. Meyer and W. H. Miller. Classical analog for electronic degrees of freedom in non-adiabatic collision processes. *Journal of Chemical Physics*, 70:3214–3223, 1979.

[1002] H.-D. Meyer and G. A. Worth. Quantum molecular dynamics: Propagating wavepackets and density operators using the multiconfiguration time-dependent Hartree method. *Theoretical Chemistry Accounts*, 109:251–267, 2003.

[1003] C. Micheletti, A. Laio, and M. Parrinello. Reconstructing the density of states by history-dependent metadynamics. *Physical Review Letters*, 92:170601, 2004.

[1004] B. Militzer, D. M. Ceperley, J. D. Kress, J. D. Johnson, L. A. Collins, and S. Mazevet. Calculation of a deuterium double shock Hugoniot from *ab initio* simulations. *Physical Review Letters*, 87:275502, 2001.

[1005] B. Militzer, F. Gygi, and G. Galli. Structure and bonding of dense liquid oxygen from first principles simulations. *Physical Review Letters*, 91:265503, 2003.

[1006] T. F. Miller and D. E. Manolopoulos. Quantum diffusion in liquid para-hydrogen from ring-polymer molecular dynamics. *Journal of Chemical Physics*, 122:184503, May 2005.

[1007] P. Minary, G. J. Martyna, and M. E. Tuckerman. Algorithms and novel applications based on the isokinetic ensemble. II. *Ab initio* molecular dynamics. *Journal of Chemical Physics*, 118:2527–2538, 2003.

[1008] P. Minary, J. A. Morrone, D. A. Yarne, M. E. Tuckerman, and G. J. Martyna. Long range interactions on wires: A reciprocal space based formalism. *Journal of Chemical Physics*, 121:11949–11956, 2004.

[1009] P. Minary, M. E. Tuckerman, K. A. Pihakari, and G. J. Martyna. A new reciprocal space based treatment of long range interactions on surfaces. *Journal of Chemical Physics*, 116:5351–5362, 2002.

[1010] C. Mischler, J. Horbach, W. Kob, and K. Binder. Water adsorption on amorphous silica surfaces: A Car–Parrinello simulation study. *Journal of Physics: Condensed Matter*, 17:4005–4013, 2005.

[1011] C. Mischler, W. Kob, and K. Binder. Classical and *ab initio* molecular dynamic simulation of an amorphous silica surface. *Computer Physics Communications*, 147:222–225, 2002.

[1012] S. Miura and S. Okazaki. Path integral molecular dynamics for Bose-Einstein and Fermi-Dirac statistics. *Journal of Chemical Physics*, 112:10116–10124, 2000.

[1013] S. Miura, M. E. Tuckerman, and M. L. Klein. An *ab initio* path integral molecular dynamics study of double proton transfer in the formic acid dimer. *Journal of Chemical Physics*, 109:5290–5299, 1998.

[1014] T. Miyake, T. Ogitsu, and S. Tsuneyuki. Quantum distributions of muonium and hydrogen in crystalline silicon. *Physical Review Letters*, 81:1873–1876, 1998.

[1015] T. Miyake, T. Ogitsu, and S. Tsuneyuki. First-principles study of the quantum states of muonium and hydrogen in crystalline silicon. *Physical Review B*, 60:14197–14204, 1999.

[1016] Y. Miyamoto, O. Sugino, and Y. Mochizuki. Real-time electron-ion dynamics for photoinduced reactivation of hydrogen-passivated donors in GaAs. *Applied Physics Letters*, 75:2915–2917, 1999.

[1017] N. A. Modine, G. Zumbach, and E. Kaxiras. Adaptive-coordinate real-space electronic-structure calculations for atoms, molecules, and solids. *Physical Review B*, 55:10289–10301, 1997.

[1018] M. Mohr, D. Marx, M. Parrinello, and H. Zipse. Solvation of radical cations in water – Reactive or unreactive solvation? *Chemistry - A European Journal*, 6:4009–4015, 2000.

[1019] C. Molteni, I. Frank, and M. Parrinello. An excited state density functional theory study of the rhodopsin chromophore. *Journal of the American Chemical Society*, 121:12177–12183, 1999.

[1020] C. Molteni, I. Frank, and M. Parrinello. Modelling photoreactions in proteins by density functional theory. *Computational Materials Science*, 20:311–317, 2001.

[1021] C. Molteni and R. Martoňák. Polymorphism in silicon nanocrystals under pressure. *ChemPhysChem*, 6:1765–1768, 2005.

[1022] C. Molteni, R. Martoňák, and M. Parrinello. First principles molecular dynamics simulations of pressure-induced structural transformations in silicon clusters. *Journal of Chemical Physics*, 114:5358–5365, 2001.

[1023] C. Molteni, N. Marzari, M. C. Payne, and V. Heine. Sliding mechanisms in aluminum grain boundaries. *Physical Review Letters*, 79:869–872, 1997.

[1024] C. Molteni and M. Parrinello. Condensed matter effects on the structure of crystalline glucose. *Chemical Physics Letters*, 275:409–413, 1998.

[1025] C. Molteni and M. Parrinello. Glucose in aqueous solution by first principles molecular dynamics. *Journal of the American Chemical Society*, 120:2168–2171, 1998.

[1026] H. J. Monkhorst and J. D. Pack. Special points for Brillouin-zone integrations. *Physical Review B*, 13:5188–5192, 1976.

[1027] T. Z. Mordasini and W. Thiel. Combined quantum mechanical and molecular mechanical approaches. *Chimia*, 52:288–291, 1998.

[1028] T. Morishita. Liquid–liquid phase transitions of phosphorus via constant-pressure first-principles molecular dynamics simulations. *Physical Review Letters*, 87:105701, 2001.

[1029] T. Morishita. Polymeric liquid of phosphorus at high pressure: First-principles molecular-dynamics simulations. *Physical Review B*, 66:054204, 2002.

[1030] T. Morishita and S. Nosé. Momentum conservation law in the Car–Parrinello method. *Physical Review B*, 59:15126–15132, 1999.

[1031] T. Morita. Solution of the Bloch equation for many-particle systems in terms of the path integral. *Journal of the Physical Society of Japan*, 35:980–984, 1973.

[1032] J. A. Morrone and M. E. Tuckerman. *Ab initio* molecular dynamics study of proton mobility in liquid methanol. *Journal of Chemical Physics*, 117:4403–4413, 2002.

[1033] J. A. Morrone and M. E. Tuckerman. A simple quantum mechanical/molecular mechanical (QM/MM) model for methanol. *Chemical Physics Letters*, 370:406–411, 2003.

[1034] J. J. Mortensen and M. Parrinello. A density functional theory study of a silica-supported zirconium monohydride catalyst for depolymerization of polyethylene. *Journal of Physical Chemistry B*, 104:2901–2907, 2000. See reference [23].

[1035] M. Moseler, H. Häkkinen, and U. Landman. Photoabsorption spectra of Na_n^+ clusters: Thermal line-broadening mechanisms. *Physical Review Letters*, 87:053401, 2001.

[1036] N. J. Mosey, A. G. Hu, and T. K. Woo. *Ab initio* molecular dynamics simulations with a HOMO-LUMO gap biasing potential to accelerate rare reaction events. *Chemical Physics Letters*, 373:498–505, 2003.

[1037] N. J. Mosey, M. H. Müser, and T. K. Woo. Molecular mechanisms for the functionality of lubricant additives. *Science*, 307:1612–1615, 2005.

[1038] N. J. Mosey, T. K. Woo, and M. H. Müser. Energy dissipation via quantum chemical hysteresis during high-pressure compression: A first-principles molecular dynamics study of phosphates. *Physical Review B*, 72:054124, 2005.

[1039] M. Mugnai, G. Cardini, and V. Schettino. High pressure reactivity of propene by first principles molecular dynamics calculations. *Journal of Chemical Physics*, 120:5327–5333, 2004.

[1040] E. M. Müller, A. de Meijere, and H. Grubmüller. Predicting unimolecular chemical reactions: Chemical flooding. *Journal of Chemical Physics*, 116:897–905, 2002.

[1041] R. S. Mulliken. Criteria for the construction of good self-consistent-field molecular orbital wave functions, and the significance of LCAO-MO population analysis. *Journal of Chemical Physics*, 36:3428–3439, 1962.

[1042] W. Münch, K.-D. Kreuer, W. Silvestri, J. Maier, and G. Seifert. The diffusion mechanism of an excess proton in imidazole molecule chains: First results of an *ab initio* molecular dynamics study. *Solid State Ionics*, 145:437–443, 2001.

[1043] C. J. Mundy, M. E. Colvin, and A. A. Quong. Irradiated guanine: A Car–Parrinello molecular dynamics study of dehydrogenation in the presence of an OH radical. *Journal of Physical Chemistry A*, 106:10063–10071, 2002.

[1044] C. J. Mundy, J. Hutter, and M. Parrinello. Microsolvation and chemical reactivity of sodium and water clusters. *Journal of the American Chemical Society*, 122:4837–4838, 2000.

[1045] V. Musolino, A. Selloni, and R. Car. Structure and dynamics of small metallic clusters on an insulating metal-oxide surface: Copper on MgO(100). *Physical Review Letters*, 83:3242–3245, 1999.

[1046] Á. Nagy. Kohn–Sham equations for multiplets. *Physical Review A*, 57:1672–1677, 1998.

[1047] N. N. Nair, E. Schreiner, and D. Marx. Glycine at the pyrite-water interface: The role of surface defects. *Journal of the American Chemical Society*, 128:13815–13826, 2006.

[1048] H. Nakamura. *Nonadiabatic Transition: Concepts, Basic Theories and Applications*. World Scientific, Singapore, 2002.

[1049] M. M. Naor, K. Van Nostrand, and C. Dellago. Car–Parrinello molecular dynamics simulation of the calcium ion in liquid water. *Chemical Physics Letters*, 369:159–164, 2003.

[1050] V. Natoli, R. M. Martin, and D. M. Ceperley. Crystal-structure of atomic hydrogen. *Physical Review Letters*, 70:1952–1955, 1993.

[1051] R. Neumann and N. C. Handy. Higher-order gradient corrections for exchange-correlation functionals. *Chemical Physics Letters*, 266:16–22, 1997.

[1052] R. Neumann, R. H. Nobes, and N. C. Handy. Exchange functionals and potentials. *Molecular Physics*, 87:1–36, 1996.

[1053] C. Niedermeier and P. Tavan. A structure adapted multipole method for electrostatic interactions in protein dynamics. *Journal of Chemical Physics*, 101:734–748, 1994.

[1054] C. Niedermeier and P. Tavan. Fast version of the structure adapted multipole method-efficient calculation of electrostatic forces in protein dynamics. *Molecular Simululation*, 17:57–66, 1996.

[1055] O. H. Nielsen and R. M. Martin. First-principles calculation of stress. *Physical Review Letters*, 50:697–700, 1983.

[1056] O. H. Nielsen and R. M. Martin. Quantum-mechanical theory of stress and force. *Physical Review B*, 32:3780–3791, 1985.

[1057] O. H. Nielsen and R. M. Martin. Stresses in semiconductors: *Ab initio* calculations on Si, Ge and GaAs. *Physical Review B*, 32:3792–3805, 1985.

[1058] A. Nitzan and M. A. Ratner. Electron transport in molecular wire junctions. *Science*, 300:1384–1389, 2003.

[1059] M. Nonella, G. Mathias, M. Eichinger, and P. Tavan. Structures and vibrational frequencies of the quinones in *Rb. sphaeroides* derived by a combined density functional/molecular mechanics approach. *Journal of Physical Chemistry B*, 107:316–322, 2003.

[1060] M. Nonella, G. Mathias, and P. Tavan. Infrared spectrum of p-benzoquinone in water obtained from a QM/MM hybrid molecular dynamics simulation. *Journal of Physical Chemistry A*, 107:8638–8647, 2003.

[1061] C. Nonnenberg, S. Grimm, and I. Frank. Restricted open-shell Kohn–Sham theory for π-π^* transitions. II. Simulation of photochemical reactions. *Journal of Chemical Physics*, 119:11585–11590, 2003.

[1062] L. Noodleman. Valence bond description of anti-ferromagnetic coupling in transition-metal dimers. *Journal of Chemical Physics*, 74:5737–5743, 1981.

[1063] S. Nosé. A molecular-dynamics method for simulations in the canonical ensemble. *Molecular Physics*, 52:255–268, 1984.

[1064] S. Nosé. A unified formulation of the constant temperature molecular-dynamics methods. *Journal of Chemical Physics*, 81:511–519, 1984.

[1065] S. Nosé. Constant temperature molecular-dynamics methods. *Progress of Theoretical Physics Supplement*, 103:1–46, 1991.

[1066] S. Nosé and M. L. Klein. Constant pressure molecular-dynamics for molecular-systems. *Molecular Physics*, 50:1055–1076, 1983.

[1067] F. D. Novaes, A. J. R. da Silva, E. Z. da Silva, and A. Fazzio. Effect of impurities in the large Au–Au distances in gold nanowires. *Physical Review Letters*, 90:036101, 2003.

[1068] F. D. Novaes, A. J. R. da Silva, E. Z. da Silva, and A. Fazzio. Oxygen clamps in gold nanowires. *Physical Review Letters*, 96:016104, 2006.

[1069] NWChem: PSPW module. Ref. [734]; developed and distributed by Pacific Northwest National Laboratory, USA; See http://www.emsl.pnl.gov/docs/nwchem/.

[1070] T. Oda and A. Pasquarello. Ab initio molecular dynamics investigation of the structure and the noncollinear magnetism in liquid oxygen: Occurrence of O_4 molecular units. *Physical Review Letters*, 89:197204, 2002.

[1071] T. Oda and A. Pasquarello. Noncollinear magnetism in liquid oxygen: A first-principles molecular dynamics study. *Physical Review B*, 70:134402, 2004.

[1072] M. Odelius. Mixed molecular and dissociative water adsorption on MgO[100]. *Physical Review Letters*, 82:3919–3922, 1999.

[1073] M. Odelius, M. Bernasconi, and M. Parrinello. Two dimensional ice adsorbed on mica surface. *Physical Review Letters*, 78:2855–2858, 1997.

[1074] M. Odelius, M. Cavalleri, A. Nilsson, and L. G. M. Pettersson. X-ray absorption spectrum of liquid water from molecular dynamics simulations: Asymmetric model. *Physical Review B*, 73:024205, 2006.

[1075] M. Odelius, M. Kadi, J. Davidsson, and A. N. Tarnovsky. Photodissociation of diiodomethane in acetonitrile solution and fragment recombination into iso-diiodomethane studied with *ab initio* molecular dynamics simulations. *Journal of Chemical Physics*, 121:2208–2214, 2004.

[1076] M. Odelius, B. Kirchner, and J. Hutter. s-Tetrazine in aqueous solution: A density functional study of hydrogen bonding and electronic excitations. *Journal of Physical Chemistry A*, 108:2044–2052, 2004.

[1077] M. Odelius, D. Laikov, and J. Hutter. Excited state geometries within time-dependent and restricted open-shell density functional theories. *Journal of Molecular Structure-THEOCHEM*, 630:163–175, 2003.

[1078] A. R. Oganov, R. Martoňák, A. Laio, P. Raiteri, and M. Parrinello. Anisotropy of Earth's D'' layer and stacking faults in the $MgSiO_3$ post-perovskite phase. *Nature*, 438:1142–1144, 2005.

[1079] K.-D. Oh and P. A. Deymier. *Ab initio* molecular-dynamics method based on the restricted path integral: Application to the electron plasma and liquid alkali metal. *Physical Review B*, 58:7577–7584, 1998.

[1080] K.-D. Oh and P. A. Deymier. Path-integral molecular dynamics calculations of electron plasma. *Physical Review Letters*, 81:3104–3107, 1998.

[1081] Y. Ohta, K. Ohta, and K. Kinugawa. Unified quantum molecular dynamics method based on centroid molecular dynamics and semiempirical molecular orbital theory. *International Journal of Quantum Chemistry*, 95:372–379, 2003.

[1082] Y. Ohta, K. Ohta, and K. Kinugawa. *Ab initio* centroid path integral molecular dynamics: Application to vibrational dynamics of diatomic molecular systems. *Journal of Chemical Physics*, 120:312–320, 2004.

[1083] Y. Ohta, K. Ohta, and K. Kinugawa. Quantum effect on the internal proton transfer and structural fluctuation in the H_5^+ cluster. *Journal of Chemical Physics*, 121:10991–10999, 2004.

[1084] T. Ohtsuki, K. Ohno, K. Shiga, Y. Kawazoe, Y. Maruyama, and K. Masumoto. Insertion of Xe and Kr atoms into C_{60} and C_{70} fullerenes and the formation of dimers. *Physical Review Letters*, 81:967–970, 1998.

[1085] ONETEP (Order-N Electronic Total Energy Package): Linear-scaling density-functional theory with plane waves. Ref. [616]; See http://www.tcm.phy.cam.ac.uk/onetep/.

[1086] L. Onsager. Electric moments of molecules in liquids. *Journal of the American Chemical Society*, 58:1486–1493, 1936.

[1087] P. Ordejón. Order-N tight-binding methods for electronic-structure and molecular dynamics. *Computational Materials Science*, 12:157–191, 1998.

[1088] P. Ordejón, D. Drabold, M. Grumbach, and R. M. Martin. Unconstrained minimization approach for electronic computations that scales linearly with system size. *Physical Review B*, 48:14646–14649, 1993.

[1089] J. Ortega, J. P. Lewis, and O. F. Sankey. Simplified electronic-structure model for hydrogen-bonded systems – water. *Physical Review B*, 50:10516–10530, 1994.

[1090] J. Ortega, J. P. Lewis, and O. F. Sankey. First principles simulations of fluid water: The radial distribution functions. *Journal of Chemical Physics*, 106:3696–3702, 1997.

[1091] J. M. Pacheco and W.-D. Schöne. Shape phase transitions in the absorption spectra of atomic clusters. *Physical Review Letters*, 79:4986–4989, 1997. Comment: Ref. [783] and Reply: Ref. [1092], see also Ref. [320].

[1092] J. M. Pacheco and W.-D. Schöne. Comment on "Shape phase transitions in the absorption spectra of atomic clusters" - Reply. *Physical Review Letters*, 81:5703, 1998. Reply to Comment: Ref. [783], see also Ref. [320].

[1093] PACS (Physics and Astronomy Classification Scheme). "71.15.Pd - Electronic Structure: Molecular dynamics calculations (Car–Parrinello) and other numerical simulations"; See http://publish.aps.org/PACS/.

[1094] M. Pagliai, M. Iannuzzi, G. Cardini, M. Parrinello, and V. Schettino. Lithium hydroxide phase transition under high pressure: An *ab initio* molecular dynamics study. *ChemPhysChem*, 7:141–147, 2006.

[1095] M. Pagliai, S. Raugei, G. Cardini, and V. Schettino. Thermal effects on the $Cl^- + ClCH_2CN$ reaction by Car–Parrinello molecular dynamics. *Journal of Chemical Physics*, 117:2199–2204, 2003.

[1096] J. Paier, R. Hirschl, M. Marsman, and G. Kresse. The Perdew-Burke-Ernzerhof exchange-correlation functional applied to the G2-1 test set using a plane-wave basis set. *Journal of Chemical Physics*, 122:234102, 2005.

[1097] J. Paier, M. Marsman, K. Hummer, G. Kresse, I. C. Gerber, and J. G. Ángyán. Screened hybrid density functionals applied to solids. *Journal of Chemical Physics*, 124:154709, 2006.

[1098] A. Palma, A. Pasquarello, G. Ciccotti, and R. Car. Cu^{++} and Li^+ interaction with polyethylene oxide by *ab initio* molecular dynamics. *Journal of Chemical Physics*, 108:9933–9936, 1998.

[1099] M. Palummo, L. Reining, and P. Ballone. First principles simulations. *Journal de physique IV (Paris)*, 3:(C7):1955–1964, 1993.

[1100] G. A. Papoian, W. F. DeGrado, and M. L. Klein. Probing the configurational space of a metalloprotein core: An *ab initio* molecular dynamics study of Duo Ferro 1 Binuclear Zn Cofactor. *Journal of the American Chemical Society*, 125:560–569, 2003.

[1101] P. V. Parandekar and J. C. Tully. Mixed quantum–classical equilibrium. *Journal of Chemical Physics*, 122:094102, 2005.

[1102] R. G. Parr and W. Yang. *Density-Functional Theory of Atoms and Molecules*. Oxford University Press, Oxford, 1989.

[1103] M. Parrinello. From silicon to RNA: The coming of age of *ab ini-*

tio molecular dynamics. *Solid State Communications*, 102:107–120, 1997.

[1104] M. Parrinello. Simulating complex systems without adjustable parameters. *Computing in Science & Engineering*, 2:22–27, 2000.

[1105] M. Parrinello and A. Rahman. Crystal structure and pair potentials: A molecular-dynamics study. *Physical Review Letters*, 45:1196–1199, 1980.

[1106] M. Parrinello and A. Rahman. Polymorphic transitions in single-crystals – A new molecular-dynamics method. *Journal of Applied Physics*, 52:7182–7190, 1981.

[1107] M. Parrinello and A. Rahman. Strain fluctuations and elastic-constants. *Journal of Chemical Physics*, 76:2662–2666, 1982.

[1108] M. Parrinello and A. Rahman. Study of an F-center in molten KCl. *Journal of Chemical Physics*, 80:860–867, 1984.

[1109] I. Pasichnyk and B. Dünweg. Coulomb interactions via local dynamics: A molecular-dynamics algorithm. *Journal of Physics: Condensed Matter*, 16:S3999–S4020, 2004.

[1110] A. Pasquarello and R. Car. Identification of Raman defect lines as signatures of ring structures in vitreous silica. *Physical Review Letters*, 80:5145–5147, 1998.

[1111] A. Pasquarello, K. Laasonen, R. Car, Ch. Lee, and D. Vanderbilt. *Ab initio* molecular dynamics for d-electron systems: Liquid copper at 1500 K. *Physical Review Letters*, 69:1982–1985, 1992.

[1112] A. Pasquarello, I. Petri, P. S. Salmon, O. Parisel, R. Car, E. Tóth, D. H. Powell, H. E. Fischer, L. Helm, and A. E. Merbach. First solvation shell of the Cu(II) aqua ion: Evidence for fivefold coordination. *Science*, 291:856–859, 2001.

[1113] A. Pasquarello, J. Sarnthein, and R. Car. Dynamic structure factor of vitreous silica from first principles: Comparison to neutron-inelastic-scattering experiments. *Physical Review B*, 57:14133–14140, 1998.

[1114] D. Passerone, M. Ceccarelli, and M. Parrinello. A concerted variational strategy for investigating rare events. *Journal of Chemical Physics*, 118:2025–2032, 2003.

[1115] D. Passerone and M. Parrinello. Action-derived molecular dynamics in the study of rare events. *Physical Review Letters*, 87:108302, September 2001.

[1116] D. Passerone and M. Parrinello. Reply to comment on "Action-derived molecular dynamics in the study of rare events" – Reply. *Physical Review Letters*, 90:089802, 2003. Reply to Comment: Ref. [292].

[1117] G. Pastore. Car–Parrinello method and adiabatic invariants. In K. Binder and G. Ciccotti, eds., *Monte Carlo and Molecular Dynamics of Condensed Matter Systems*, chapter 24, pages 635–647. Italian Physical Society SIF, Bologna, 1996.

[1118] G. Pastore, E. Smargiassi, and F. Buda. Theory of *ab initio* molecular-dynamics calculations. *Physical Review A*, 44:6334–6347, 1991.

[1119] M. Pavese, D. R. Berard, and G. A. Voth. *Ab initio* centroid molecular dynamics: A fully quantum method for condensed-phase dynamics simulations. *Chemical Physics Letters*, 300:93–98, 1999.

[1120] M. Pavone, V. Barone, I. Ciofini, and C. Adamo. First-principle molecular dynamics of the Berry pseudorotation: Insights on ^{19}F NMR in SF_4. *Journal of Chemical Physics*, 120:9167–9174, 2004.

[1121] M. C. Payne. Error cancellation in the molecular-dynamics method for total energy calculations. *Journal of Physics: Condensed Matter*, 1:2199–2210, 1989. Comment: Ref. [224].

[1122] M. C. Payne, J. D. Joannopoulos, D. C. Allan, M. P. Teter, and D. Vanderbilt. Molecular dynamics and *ab initio* total energy calculations. *Physical Review Letters*, 56:2656, 1986.

[1123] M. C. Payne, M. P. Teter, D. C. Allan, T. A. Arias, and J. D. Joannopoulos. Iterative minimization techniques for *ab initio* total-energy calculations: molecular dynamics and conjugate gradients. *Reviews of Modern Physics*, 64:1045–1097, 1992.

[1124] M. Pearson, E. Smargiassi, and P. Madden. *Ab initio* molecular-dynamics with an orbital-free density-functional. *Journal of Physics: Condensed Matter*, 5:3221–3240, 1993.

[1125] G. La Penna, F. Buda, A. Bifone, and H. J. M. de Groot. The transition state in the isomerization of rhodopsin. *Chemical Physics Letters*, 294:447–453, 1998.

[1126] M. Dal Peraro, F. Alber, and P. Carloni. Ser133 phosphate–KIX interactions in the CREB–CBP complex: An *ab initio* molecular dynamics study. *European Biophysics Journal*, 30:75–81, 2001.

[1127] M. Dal Peraro, L. L. Llarrull, U. Röthlisberger, A. J. Vila, and P. Carloni. Water-assisted reaction mechanism of monozinc beta-lactamases. *Journal of the American Chemical Society*, 126:12661–12668, 2004.

[1128] M. Dal Peraro, S. Raugei, P. Carloni, and M. L. Klein. Solute-solvent charge transfer in aqueous solution. *ChemPhysChem*, 6:1715–1718, 2005.

[1129] J. P. Perdew. Correction. *Physical Review B*, 34:7406, 1986. Original article: [1130].

[1130] J. P. Perdew. Density-functional approximation for the correlation-energy of the inhomogeneous electron-gas. *Physical Review B*, 33:8822–8824, 1986. Erratum: [1129].

[1131] J. P. Perdew, K. Burke, and M. Ernzerhof. Generalized gradient approximation made simple. *Physical Review Letters*, 77:3865–3868, 1996. Erratum: Ref. [1132]; Comment: Ref. [1657] and Reply: Ref. [1133]; two different "revised versions" of the original PBE parameterization were proposed: revPBE [1657] and RPBE [595].

[1132] J. P. Perdew, K. Burke, and M. Ernzerhof. Erratum: Generalized gradient approximation made simple. *Physical Review Letters*, 78:1396, 1997. Original article: [1131].

[1133] J. P. Perdew, K. Burke, and M. Ernzerhof. Reply to comment on "Generalized gradient approximation made simple". *Physical Review Letters*, 80:891, 1998. Comment: Ref. [1657] and original article: Ref. [1131].

[1134] J. P. Perdew, J. A. Chevary, S. H. Vosko, K. A. Jackson, M. R. Pederson, D. J. Singh, and C. Fiolhais. Atoms, molecules, solids, and surfaces: Applications of the generalized gradient approximation for exchange and correlation. *Physical Review B*, 46:6671–6687, 1992. Erratum: [1135].

[1135] J. P. Perdew, J. A. Chevary, S. H. Vosko, K. A. Jackson, M. R. Pederson, D. J. Singh, and C. Fiolhais. Erratum: Atoms, molecules, solids, and surfaces: Applications of the generalized gradient approximation for exchange and correlation. *Physical Review B*, 48:4978, 1993. Original article: [1134].

[1136] J. P. Perdew, A. Ruzsinszky, J. Tao, V. N. Staroverov, G. E. Scuseria, and G. I. Csonka. Prescription for the design and selection of density functional approximations: More constraint satisfaction with fewer fits. *Journal of Chemical Physics*, 123:062201, 2005.

[1137] J. P. Perdew and Y. Wang. Accurate and simple analytic representation of the electron-gas correlation energy. *Physical Review B*, 45:13244–13249, 1992.

[1138] J. P. Perdew and A. Zunger. Self-interaction correction to density-functional approximations for many-electron systems. *Physical Review B*, 23:5048–5079, 1981.

[1139] R. Perez, M. C. Payne, I. Štich, and K. Terakura. Role of covalent tip-surface interactions in noncontact atomic force microscopy on reactive surfaces. *Physical Review Letters*, 78:678–681, 1997.

[1140] R. Perez, M. C. Payne, I. Štich, and K. Terakura. Contrast mechanism in non-contact AFM on reactive surfaces. *Applied Surface Science*, 123/124:249–254, 1998.

[1141] R. Pérez, I. Štich, M. C. Payne, and K. Terakura. Surface–tip interactions in noncontact atomic-force microscopy on reactive surfaces: Si(111). *Physical Review B*, 58:10835–10849, 1998.

[1142] R. Perez, I. Štich, M. C. Payne, and K. Terakura. Chemical interactions in noncontact AFM on semiconductor surfaces: Si(111), Si(100) and GaAs(110). *Applied Surface Science*, 140:320–326, 1999.

[1143] Á. J. Pérez-Jiménez, J. M. Pérez-Jordá, and F. Illas. Density functional theory with alternative spin densities: Application to magnetic systems with localized spins. *Journal of Chemical Physics*, 120:18–25, 2004.

[1144] B. G. Pfrommer, J. Demmel, and H. Simon. Unconstrained energy functionals for electronic structure calculations. *Journal of Computational Physics*, 150:287–298, 1999.

[1145] J. C. Phillips. Energy-band interpolation scheme based on a pseudopotential. *Physical Review*, 112:685–695, 1958.

[1146] J. C. Phillips and L. Kleinman. New method for calculating wave functions in crystals and molecules. *Physical Review*, 116:287–294, 1959.

[1147] S. Piana, D. Bucher, P. Carloni, and U. Röthlisberger. Reaction mechanism of HIV-1 protease by hybrid Car–Parrinello/classical MD simulations. *Journal of Physical Chemistry B*, 108:11139–11149, 2004.

[1148] C. J. Pickard and F. Mauri. All-electron magnetic response with pseudopotentials: NMR chemical shifts. *Physical Review B*, 63:245101, 2001.

[1149] W. E. Pickett. Pseudopotential methods in condensed matter applications. *Computer Physics Reports*, 9:115–197, 1989.

[1150] C. Pierleoni, D. M. Ceperley, B. Bernu, and W. R. Magro. Equation of state of the hydrogen plasma by path-integral Monte-Carlo simulation. *Physical Review Letters*, 73:2145–2149, 1994.

[1151] C. Pierleoni, D. M. Ceperley, and M. Holzmann. Coupled electron-ion Monte Carlo calculations of dense metallic hydrogen. *Physical Review Letters*, 93:146402, 2004.

[1152] C. A. Pignedoli, A. Curioni, and W. Andreoni. Disproving a silicon analog of an alkyne with the aid of topological analyses of the electronic structure and *ab initio* molecular dynamics calculations. *ChemPhysChem*, 6:1795–1799, 2005.

[1153] PINY. The PINY_MD(c) simulation package; Principle authors: G. J. Martyna and M. E. Tuckerman; Other authors: D. A. Yarne, S. O. Samuelson, A. L. Hughes, Y. Liu, Z. Zhu, M. Diraison, K. Pihakari; See Ref. [1513]; OSI Certified Open Source Software (Common Public License); See http://homepages.nyu.edu/~mt33/PINY_MD/PINY.html.

[1154] J. Pipek and P. G. Mezey. A fast intrinsic localization procedure applicable for *ab initio* and semiempirical linear combination of atomic orbital wave functions. *Journal of Chemical Physics*, 90:4916–4926, 1989.

[1155] W. T. Pollard and R. A. Friesner. Efficient Fock matrix diagonalization by a Krylov-space method. *Journal of Chemical Physics*, 99:6742–6750, 1993.

[1156] R. Pollet, C. Boehme, and D. Marx. *Ab initio* simulations of desorption and reactivity of glycine at a water-pyrite interface at "Iron-Sulfur World" prebiotic conditions. *Origins of Life and Evolution of Biospheres*, 36:363–379, 2006.

[1157] J. Polonyi and H. W. Wyld. Microcanonical simulation of fermionic systems. *Physical Review Letters*, 51:2257–2260, 1983.

[1158] V. Pophristic, V. S. K. Balagurusamy, and M. L. Klein. Structure and dynamics of the aluminum chlorohydrate polymer $Al_{13}O_4(OH)_{24}(H_2O)_{12}Cl_7$. *Physical Chemistry Chemical Physics*, 6:919–923, 2004.

[1159] V. Pophristic, M. L. Klein, and M. N. Holerca. Modeling small aluminum chlorohydrate polymers. *Journal of Physical Chemistry A*, 108:113–120, 2004.

[1160] D. Prendergast, J. C. Grossman, and G. Galli. The electronic structure of liquid water within density-functional theory. *Journal of Chemical Physics*, 123:014501, 2005.

[1161] W. H. Press, S. A. Teukolsky, W. T. Vetterling, and B. P. Flannery. *Numerical Recipes - The Art of Scientific Computing*. Cambridge University Press, Cambridge, 1992.

[1162] E. Prodan and W. Kohn. Nearsightedness of electronic matter. *Proceedings of the National Academy of Sciences of the United States of America*, 102:11635–11638, 2005.

[1163] E. I. Proynov, S. Sirois, and D. R. Salahub. Extension of the LAP functional to include parallel spin correlation. *International Journal of Quantum Chemistry*, 64:427–446, 1997.

[1164] P. Pulay. *Ab initio* calculation of force constants and equilibrium geometries in polyatomic molecules. I. Theory. *Molecular Physics*, 17:197–204, 1969.

[1165] P. Pulay. Convergence acceleration of iterative sequences - The case of SCF iteration. *Chemical Physics Letters*, 73:393–398, 1980.

[1166] P. Pulay. Derivative methods in quantum chemistry. *Advances in Chemical Physics*, 69:241–286, 1987.

[1167] P. Pulay and G. Fogarasi. Fock matrix dynamics. *Chemical Physics Letters*, 386:272–278, 2004.

[1168] R. Puthenkovilakam, E. A. Carter, and J. P. Chang. First-principles exploration of alternative gate dielectrics: Electronic structure of ZrO_2/Si and $ZrSiO_4$/Si interfaces. *Physical Review B*, 69:155329, 2004.

[1169] A. Putrino and G. B. Bachelet. Curvilinear coordinates for full-core atoms. In V. Fiorentini and F. Meloni, eds., *Advances in Computational Materials Science II*. Italian Physical Society SIF, Bologna, 1998.

[1170] A. Putrino and M. Parrinello. Anharmonic Raman spectra in high-pressure ice from *ab initio* simulations. *Physical Review Letters*, 88:176401, 2002.

[1171] A. Putrino, D. Sebastiani, and M. Parrinello. Generalized variational density functional perturbation theory. *Journal of Chemical Physics*, 113:7102–7109, 2000.

[1172] PWscf. Plane-Wave Self-Consistent Field is a set of programs for electronic structure calculations within density-functional theory and density-functional perturbation theory, using a plane-wave basis set and pseudopotentials; released under the GNU General Public Licence; See http://www.pwscf.org/.

[1173] M. Qiu, X.-Y. Zhou, M. Jiang, P.-L. Cao, and Z. Zeng. Ammonia adsorption and saturation on small Si cluster surfaces: A full-potential linear-muffin-tin-orbital molecular-dynamics study. *Physics Letters A*, 245:430–434, 1998.

[1174] A. Qteish. Conjugate-gradient methods for metallic systems and band-structure calculations. *Physical Review B*, 52:14497–14504, 1995.

[1175] M. R. Radeke and E. A. Carter. *Ab initio* dynamics of surface chemistry. *Annual Review of Physical Chemistry*, 48:243–270, 1997.

[1176] B. De Raedt, M. Sprik, and M. L. Klein. Computer-simulation of muonium in water. *Journal of Chemical Physics*, 80:5719–5724, 1984.

[1177] C. Rajesh, C. Majumder, M. G. R. Rajan, and S. K. Kulshreshtha.

Isomers of small Pb_n clusters ($n = 2 - 15$): Geometric and electronic structures based on *ab initio* molecular dynamics simulations. *Physical Review B*, 72:235411, 2005.

[1178] L. M. Ramaniah, M. Bernasconi, and M. Parrinello. *Ab initio* molecular-dynamics simulation of K^+ solvation in water. *Journal of Chemical Physics*, 111:1587–1591, 1999.

[1179] R. Ramírez, E. Hernandez, J. Schulte, and M. C. Böhm. Nuclear quantum effects in the electronic structure of C_2H_4: A combined Feynman path integral *ab initio* approach. *Chemical Physics Letters*, 291:44–50, 1998.

[1180] R. Ramírez and T. López-Ciudad. Phase-space formulation of thermodynamic and dynamical properties of quantum particles. *Physical Review Letters*, 83:4456–4459, 1999.

[1181] R. Ramírez and T. López-Ciudad. The Schrödinger formulation of the Feynman path centroid density. *Journal of Chemical Physics*, 111:3339–3348, 1999.

[1182] R. Ramírez and T. López-Ciudad. Dynamic properties via fixed centroid path integrals. In J. Grotendorst, D. Marx, and A. Muramatsu, eds., *Quantum Simulations of Complex Many-Body Systems: From Theory to Algorithms*, pages 325–360. John von Neumann Institute for Computing, Forschungszentrum Jülich, Jülich, 2002. See http://www.theochem.rub.de/go/cprev.html.

[1183] R. Ramírez, T. López-Ciudad, P. Kumar, and D. Marx. Quantum corrections to classical time-correlation functions: Hydrogen bonding and anharmonic floppy modes. *Journal of Chemical Physics*, 121:3973–3983, 2004.

[1184] R. Ramírez, T. López-Ciudad, and J. C. Noya. Feynman effective classical potential in the Schrödinger formulation. *Physical Review Letters*, 81:3303–3306, 1998.

[1185] R. Ramírez, J. Schulte, and M. C. Böhm. All-quantum description of molecules: Electrons and nuclei of C_6H_6. *Chemical Physics Letters*, 275:377–385, 1997.

[1186] R. Ramírez, J. Schulte, and M. C. Böhm. Ground state and excited state properties of ethylene isomers studied by a combined Feynman path integral – *ab initio* approach. *Molecular Physics*, 99:1249–1273, August 2001.

[1187] R. Ramírez, J. Schulte, and M. C. Böhm. Feynman path integral – *ab initio* study of the isotropic hyperfine coupling constants of the muonium substituted ethyl radical CH_2MuCH_2. *Chemical Physics Letters*, 402:346–351, February 2005.

[1188] G. Ranghino, A. Anselmino, L. Meda, C. Tonini, and G. F. Cerofolini. Theoretical analysis of CO_2 addition to ion-bombarded porous silica. *Journal of Physical Chemistry B*, 101:7723–7726, 1997.

[1189] D. C. Rapaport. *The Art of Molecular Dynamics Simulation*. Cambridge University Press, Cambridge, 2001 and 2005.

[1190] A. M. Rappe, J. D. Joannopoulos, and P. A. Bash. A test of the utility of plane-waves for the study of molecules from first principles. *Journal of the American Chemical Society*, 114:6466–6469, 1992.

[1191] A. M. Rappe, K. M. Rabe, E. Kaxiras, and J. D. Joannopoulos. Optimized pseudopotentials. *Physical Review B*, 41:1227–1230, 1990.

[1192] J.-Y. Raty, F. Gygi, and G. Galli. Growth of carbon nanotubes on metal nanoparticles: A microscopic mechanism from *ab initio* molecular dynamics simulations. *Physical Review Letters*, 95:096103, 2005.

[1193] S. Raugei, G. Cardini, and V. Schettino. An *ab initio* molecular dynamics study of the S_N2 reaction $Cl^-+CH_3Br \rightarrow CH_3Cl^+Br^-$. *Journal of Chemical Physics*, 111:10887–10894, 1999.

[1194] S. Raugei, M. Cascella, and P. Carloni. A proficient enzyme: Insights on the mechanism of orotidine monophosphate decarboxylase from computer simulations. *Journal of the American Chemical Society*, 126:15730–15737, 2004.

[1195] S. Raugei and M. L. Klein. Dynamics of water molecules in the Br^- solvation shell: An *ab initio* molecular dynamics study. *Journal of the American Chemical Society*, 123:9484–9485, 2001.

[1196] S. Raugei and M. L. Klein. An *ab initio* study of water molecules in the bromide ion solvation shell. *Journal of Chemical Physics*, 116:196–202, 2002.

[1197] S. Raugei and M. L. Klein. Hydrocarbon reactivity in the superacid SbF_5/HF: An *ab initio* molecular dynamics study. *Journal of Physical Chemistry B*, 106:11596–11605, 2002.

[1198] S. Raugei and M. L. Klein. Structure of the strongly associated liquid antimony pentafluoride: An *ab initio* molecular dynamics study. *Journal of Chemical Physics*, 116:7087–7093, 2002.

[1199] S. Raugei and M. L. Klein. Nuclear quantum effects and hydrogen bonding in liquids. *Journal of the American Chemical Society*, 125:8992–8993, 2003.

[1200] S. Raugei and M. L. Klein. On the quantum nature of an excess proton in liquid hydrogen fluoride. *ChemPhysChem*, 5:1569–1576, 2004.

[1201] S. Raugei, P. L. Silvestrelli, and M. Parrinello. Pressure-induced frustration and disorder in $Mg(OH)_2$ and $Ca(OH)_2$. *Physical Review Letters*, 83:2222–2225, 1999.

[1202] P. Raybaud, J. Hafner, G. Kresse, and H. Toulhoat. Adsorption of thiophene on the catalytically active surface of MoS_2: An *ab initio* local-density-functional study. *Physical Review Letters*, 80:1481–1484, 1998.

[1203] C. Raynaud, L. Maron, J.-P. Daudey, and F. Jolibois. Reconsidering Car–Parrinello molecular dynamics using direct propagation of molecular orbitals developed upon Gaussian type atomic orbitals. *Physical Chemistry Chemical Physics*, 6:4226–4232, 2004.

[1204] N. Rega, S. S. Iyengar, G. A. Voth, T. Vreven H. B. Schlegel, and M. J. Frisch. Hybrid *ab-initio*/empirical molecular dynamics: Combining the ONIOM scheme with the atom-centered density matrix propagation (ADMP) approach. *Journal of Physical Chemistry B*, 108:4210–4220, 2004.

[1205] C. Reichardt. *Solvents and Solvent Effects in Organic Chemistry*. Wiley-VCH, Weinheim, 2003.

[1206] M. Reiher, B. Kirchner, J. Hutter, D. Sellmann, and B. A. Hess. A photochemical activation scheme of inert dinitrogen by dinuclear Ru^{II} and Fe^{II} complexes. *Chemistry - A European Journal*, 10:4443–4453, 2004.

[1207] M. Reiher and J. Neugebauer. A mode-selective quantum chemical method for tracking molecular vibrations applied to functionalized carbon nanotubes. *Journal of Chemical Physics*, 118:1634–1641, 2003.

[1208] S. Reinhardt, C. M. Marian, and I. Frank. The influence of excess ammonia on the mechanism of the reaction of boron trichloride with ammonia – An *ab initio* molecular dynamics study. *Angewandte Chemie International Edition*, 40:3683–3685, 2001.

[1209] D. K. Remler and P. A. Madden. Molecular-dynamics without effective potentials via the Car–Parrinello approach. *Molecular Physics*, 70:921–966, 1990.

[1210] R. Resta. Macroscopic polarization in crystalline dielectrics: The geometric phase approach. *Reviews of Modern Physics*, 66:899–915, 1994.

[1211] R. Resta. Quantum-mechanical position operator in extended systems. *Physical Review Letters*, 80:1800–1803, 1998.

[1212] R. Resta and S. Sorella. Electron localization in the insulating state. *Physical Review Letters*, 82:370–373, 1999.

[1213] S. L. Richardson and J. L. Martins. *Ab initio* studies of the structural and electronic properties of solid cubane. *Physical Review B*, 58:15307–15309, 1998.

[1214] M. C. Righi, C. A. Pignedoli, R. Di Felice, C. M. Bertoni, and A. Catellani. *Ab initio* simulations of homoepitaxial SiC growth. *Physical Review Letters*, 91:136101, 2003.

[1215] G.-M. Rignanese, A. De Vita, J.-C. Charlier, X. Gonze, and R. Car. First-principles molecular-dynamics study of the (0001) alpha-quartz surface. *Physical Review B*, 61:13250–13255, 2000.

[1216] J. Riordon. Physical Review Letters' Top Ten: Number Five. *APS News*, 12:3, April 2003.
The Car–Parrinello article [222] is "Number 5" of the *"Physical Review Letters'* Top 10" most cited papers with 2819 citations at that time,
see http://www.aps.org/publications/apsnews/200304/upload/apr03.pdf.

[1217] M. A. Robb, M. Garavelli, M. Olivucci, and F. Bernardi. A computational strategy for organic photochemistry. *Reviews in Computational Chemistry*, 15:87–146, 2000.

[1218] K. R. Roby. Quantum theory of chemical valence concepts I. Definition of the charge on an atom in a molecule and of occupation numbers for electron density shared between atoms. *Molecular Physics*, 27:81–104, 1974.

[1219] P. Rodziewicz, S. M. Melikova, K. S. Rutkowski, and F. Buda. Car–Parrinello molecular dynamics study of a blue-shifted intermolecular weak-hydrogen-bond system. *ChemPhysChem*, 6:1719–1724, 2005.

[1220] U. F. Röhrig and I. Frank. First-principles molecular dynamics study of a polymer under tensile stress. *Journal of Chemical Physics*, 115:8670–8674, 2001.

[1221] U. F. Röhrig, I. Frank, J. Hutter, A. Laio, J. VandeVondele, and U. Röthlisberger. A QM/MM Car–Parrinello molecular dynamics study of the solvent effects on the ground state and on the first excited singlet state of acetone in water. *ChemPhysChem*, 4:1177–1182, 2003.

[1222] U. F. Röhrig, L. Guidoni, A. Laio, I. Frank, and U. Röthlisberger. A molecular spring for vision. *Journal of the American Chemical Society*, 126:15328–15329, 2004.

[1223] U. F. Röhrig, L. Guidoni, and U. Röthlisberger. Solvent and protein effects on the structure and dynamics of the rhodopsin chromophore. *ChemPhysChem*, 6:1836–1847, 2005.

[1224] A. H. Romero, C. Sbraccia, P. L. Silvestrelli, and F. Ancilotto. Adsorption of methylchloride on Si(100) from first principles. *Journal of Chemical Physics*, 119:1085–1092, 2003.

[1225] A. H. Romero, P. L. Silvestrelli, and M. Parrinello. Compton anisotropy from Wannier functions in the case of ice I_h. *physica status solidi (b)*, 220:703–708, 2000.

[1226] A. H. Romero, P. L. Silvestrelli, and M. Parrinello. Compton scattering and the character of the hydrogen bond in ice I_h. *Journal of Chemical Physics*, 115:115–123, 2001.

[1227] C. C. J. Roothaan. Self-consistent field theory for open shells of electronic systems. *Reviews of Modern Physics*, 32:179–185, 1960.

[1228] J. Rosen, K. Larsson, and J. M. Schneider. *Ab initio* molecular dynamics study of hydrogen removal by ion–surface interactions. *Journal of Physics: Condensed Matter*, 17:L137–L142, 2005.

[1229] J. Rosen, J. M. Schneider, and K. Larsson. *Ab initio* molecular dynamics study of ion–surface interactions. *Solid State Communications*, 134:333–336, 2005.

[1230] L. Rosso, P. Minary, Z. Zhu, and M. E. Tuckerman. On the use of the adiabatic molecular dynamics technique in the calculation of free energy profiles. *Journal of Chemical Physics*, 116:4389–4402, 2002.

[1231] U. Röthlisberger and W. Andreoni. Structural and electronic properties of sodium microclusters ($n = 2 - -20$) at low and high temperatures: New insights from *ab initio* molecular dynamics studies. *Journal of Chemical Physics*, 94:8129–8151, 1991.

[1232] U. Röthlisberger and W. Andreoni. Metal clusters with impurities: $Na_n Mg$ (n=6–9, 18). *Chemical Physics Letters*, 198:478–482, 1992.

[1233] U. Röthlisberger, W. Andreoni, and M. Parrinello. Structure of nanoscale silicon clusters. *Physical Review Letters*, 72:665–668, 1994.

[1234] U. Röthlisberger and P. Carloni. *Ab initio* molecular dynamics studies of a synthetic biomimetic model of galactose oxidase. *International Journal of Quantum Chemistry*, 73:209–218, 1999.

[1235] U. Röthlisberger and M. L. Klein. *Ab initio* molecular dynamics investigation of singlet $C_2H_2Li_2$: Determination of the ground state structure and observation of lih intermediates. *Journal of the American Chemical Society*, 117:42–48, 1995.

[1236] U. Röthlisberger and M. Parrinello. *Ab initio* molecular dynamics simulation of liquid hydrogen fluoride. *Journal of Chemical Physics*, 106:4658–4664, 1997.

[1237] U. Röthlisberger, M. Sprik, and M. L. Klein. Living polymers – *ab initio* molecular dynamics study of the initiation step in the polymerization of isoprene induced by ethyl lithium. *Journal of the Chemical Society, Faraday Transactions*, 94:501–508, 1998.

[1238] J. Rottler and A. C. Maggs. Local molecular dynamics with Coulombic interactions. *Physical Review Letters*, 93:170201, 2004.

[1239] R. Rousseau, M. Boero, M. Bernasconi, M. Parrinello, and K. Terakura. *Ab initio* simulation of phase transitions and dissociation of H_2S at high pressure. *Physical Review Letters*, 85:1254–1257, 2000.

[1240] R. Rousseau, G. Dietrich, S. Krückeberg, K. Lützenkirchen, D. Marx, L. Schweikhard, and C. Walther. Probing cluster structures with sensor molecules: Methanol adsorbed onto gold clusters. *Chemical Physics Letters*, 295:41–46, 1998.

[1241] R. Rousseau, V. Kleinschmidt, U. W. Schmitt, and D. Marx. Assigning protonation patterns in water networks in bacteriorhodopsin based on computed IR spectra. *Angewandte Chemie International Edition*, 43:4804–4807, 2004.

[1242] R. Rousseau, V. Kleinschmidt, U. W. Schmitt, and D. Marx. Modeling protonated water networks in bacteriorhodopsin. *Physical Chemistry Chemical Physics*, 6:1848–1859, 2004.

[1243] R. Rousseau and D. Marx. Fluctuations and bonding in lithium clusters. *Physical Review Letters*, 80:2574–2577, 1998.

[1244] R. Rousseau and D. Marx. The role of quantum and thermal fluctuations upon properties of lithium clusters. *Journal of Chemical Physics*, 111:5091–5101, 1999.

[1245] R. Rousseau and D. Marx. Exploring the electronic structure of elemental lithium: From small molecules to nanoclusters, bulk metal, and surfaces. *Chemistry - A European Journal*, 6:2982–2993, 2000.

[1246] R. Rousseau and D. Marx. The interaction of gold clusters with methanol molecules: Ab initio molecular dynamics of $Au_n^+CH_3OH$ and Au_nCH_3OH. *Journal of Chemical Physics*, 112:761–769, 2000.

[1247] C. Rovira. Structure, protonation state and dynamics of catalase compound II. *ChemPhysChem*, 6:1820–1826, 2005.

[1248] C. Rovira, K. Kunc, J. Hutter, P. Ballone, and M. Parrinello. A comparative study of O_2, CO, and NO binding to iron-porphyrin. *International Journal of Quantum Chemistry*, 69:31–35, 1998.

[1249] C. Rovira and M. Parrinello. Oxygen binding to iron-porphyrin: A density functional study using both LSD and LSD+GC schemes. *International Journal of Quantum Chemistry*, 70:387–394, 1998.

[1250] C. Rovira and M. Parrinello. Factors influencing ligand-binding properties of heme models: A first principles study of picket-fence and protoheme complexes. *Chemistry - A European Journal*, 5:250–262, 1999.

[1251] C. Rovira and M. Parrinello. First-principles molecular dynamics simulations of models for the myoglobin active center. *International Journal of Quantum Chemistry*, 80:1172–1180, 2000.

[1252] C. Rovira and M. Parrinello. Harmonic and anharmonic dynamics of Fe–CO and Fe–O_2 in heme models. *Biophysical Journal*, 78:93–100, 2000.

[1253] C. Rovira, B. Schulze, M. Eichinger, J. D. Evanseck, and M. Parrinello. Influence of the heme pocket conformation on the structure and vibrations of the Fe–CO bond in myoglobin: A QM/MM density functional study. *Biophysical Journal*, 81:435–445, 2001.

[1254] E. Ruiz and M. C. Payne. One-dimensional intercalation compound $2HgS \cdot SnBr_2$: Ab initio electronic structure calculations and molecular dynamics simulations. *Chemistry - A European Journal*, 4:2485–2492, 1998.

[1255] E. Runge and E. K. U. Gross. Density-functional theory for time-dependent systems. *Physical Review Letters*, 52:997–1000, 1984.

[1256] A. Rytkönen, H. Häkkinen, and M. Manninen. Melting and octupole deformation of Na_{40}. *Physical Review Letters*, 80:3940–3943, 1998.

[1257] U. Saalmann. *Nicht-adiabatische Quantenmolekulardynamik - Ein neuer Zugang zur Dynamik atomarer Vielteilchensysteme*. PhD thesis, Technische Universität Dresden, Dresden, 1997.

[1258] U. Saalmann and R. Schmidt. Non-adiabatic quantum molecular dynamics: Basic formalism and case study. *Zeitschrift für Physik D*, 38:153–163, 1996.

[1259] U. Saalmann and R. Schmidt. Excitation and relaxation in atom-cluster collisions. *Physical Review Letters*, 80:3213–3216, 1998.

[1260] R. R. Sadeghi and H.-P. Cheng. The dynamics of proton transfer in a water chain. *Journal of Chemical Physics*, 111:2086–2094, 1999.

[1261] D. E. Sagnella, K. Laasonen, and M. L. Klein. *Ab initio* molecular dynamics study of proton transfer in a polyglycine analog of the ion channel gramicidin A. *Biophysical Journal*, 71:1172–1178, 1996.

[1262] C. Sagui, P. Pomorski, T. A. Darden, and C. Roland. *Ab initio* calculation of electrostatic multipoles with Wannier functions for large-scale biomolecular simulations. *Journal of Chemical Physics*, 120:4530–4544, 2004.

[1263] M. Saharay and S. Balasubramanian. *Ab initio* molecular-dynamics study of supercritical carbon dioxide. *Journal of Chemical Physics*, 120:9694–9702, 2004.

[1264] B. Sahli and W. Fichtner. *Ab initio* molecular dynamics simulation of self-interstitial diffusion in silicon. *Physical Review B*, 72:245210, 2005.

[1265] A. M. Saitta and M. L. Klein. Polyethylene under tensil load: Strain energy storage and breaking of linear and knotted alkanes probed by first-principles molecular dynamics calculations. *Journal of Chemical Physics*, 111:9434–9440, 1999.

[1266] A. M. Saitta and M. L. Klein. First-principles molecular dynamics study of the rupture processes of a bulklike polyethylene knot. *Journal of Physical Chemistry B*, 105:6495–6499, 2001.

[1267] A. M. Saitta, P. D. Soper, E. Wasserman, and M. L. Klein. Influence of a knot on the strength of a polymer strand. *Nature*, 399:46–48, 1999.

[1268] J. J. Sakurai. *Modern Quantum Mechanics*. Addison-Wesley, Redwood City, 1985.

[1269] H. Sambe and R. H. Felton. New computational approach to Slater's SCF-$X\alpha$ equation. *Journal of Chemical Physics*, 62:1122–1126, 1975.

[1270] D. Sánchez-Portal, E. Artacho, and J. M. Soler. Projection of plane-wave calculations into atomic orbitals. *Solid State Communications*, 95:685–690, 1995.

[1271] D. Sánchez-Portal, E. Artacho, and J. M. Soler. Analysis of atomic orbital basis sets from the projection of plane-wave results. *Journal of Physics: Condensed Matter*, 8:3859–3880, 1996.

[1272] E. Sandré and A. Pasturel. An introduction to *ab initio* molecular dynamics schemes. *Molecular Simulations*, 20:63–77, 1997.

[1273] J. F. Sanz and N. C. Hernández. Mechanism of Cu deposition on the α-Al_2O_3 (0001) surface. *Physical Review Letters*, 94:016104, 2005.

[1274] J. Sauer and J. Döbler. Gas-phase infrared spectrum of the protonated water dimer: Molecular dynamics simulation and accuracy of the potential energy surface. *ChemPhysChem*, 6:1706–1710, 2005.

[1275] J. Sauer and M. Sierka. Combining quantum mechanics and interatomic potential functions in *ab initio* studies of extended systems. *Journal of Computational Chemistry*, 21:1470–1493, December 2000.

[1276] A. Savin, A. D. Becke, J. Flad, R. Nesper, H. Preuss, and H. G. von Schnering. A new look at electron localization. *Angewandte Chemie International Edition in English*, 30:409–412, 1991.

[1277] A. Savin, O. Jepsen, J. Flad, O. K. Andersen, H. Preuß, and H. G. von Schnering. Electron localization in solid-state structures of the elements - The diamond structure. *Angewandte Chemie International Edition in English*, 31:187–188, 1992.

[1278] A. Savin, R. Nesper, S. Wengert, and T. F. Fässler. ELF: The electron localization function. *Angewandte Chemie International Edition in English*, 36:1809–1832, 1997. See http://www.cpfs.mpg.de/ELF/.

[1279] C. Sbraccia. *Computer Simulation of Thermally Activated Processes*. PhD thesis, Scuola Internazionale Superiore di Studi Avanzati (SISSA), Trieste, 2005.
See http://www.sissa.it/cm/thesis/2005/sbraccia.pdf.

[1280] G. C. Schatz. The analytical representation of electronic potential-energy surfaces. *Reviews of Modern Physics*, 61:669–688, 1989.

[1281] F. Schautz, F. Buda, and C. Filippi. Excitations in photoactive molecules from quantum Monte Carlo. *Journal of Chemical Physics*, 121:5836–5844, 2004. Comment: Ref. [353] and Reply: Ref. [433].

[1282] F. Scheck. *Mechanics: From Newton's Laws to Deterministic Chaos*. Springer, Berlin, 4th edition, 2005.

[1283] D. A. Scherlis, J. L. Fattebert, F. Gygi, M. Cococcioni, and N. Marzari. A unified electrostatic and cavitation model for first-principles molecular dynamics in solution. *Journal of Chemical Physics*, 124:074103, 2006.

[1284] R. Schinke. *Photodissociation Dynamics*. Cambridge University Press, Cambridge, 1995.

[1285] H. B. Schlegel, S. S. Iyengar, X. Li, J. M. Millam, G. A. Voth, G. E. Scuseria, and M. J. Frisch. *Ab initio* molecular dynamics: Propagating the density matrix with Gaussian orbitals. III. Comparison with Born–Oppenheimer dynamics. *Journal of Chemical Physics*, 117:8694–8704, 2002.

[1286] H. B. Schlegel, J. M. Millam, S. S. Iyengar, G. A. Voth, A. D. Daniels, G. E. Scuseria, and M. J. Frisch. *Ab initio* molecular dynamics: Propagating the density matrix with Gaussian orbitals. *Journal of Chemical Physics*, 114:9758, 2001.

[1287] J. Schlitter, M. Engels, P. Krüger, E. Jacoby, and A. Wollmer. Targeted molecular dynamics simulation of conformational change – Application to the T↔R transition in insulin. *Molecular Simulations*, 10:291–308, 1993.

[1288] R. Schmid. Car–Parrinello simulations with a real space method. *Journal of Computational Chemistry*, 25:799–812, April 2004.

[1289] R. Schmid, M. Tafipolski, P. H. König, and H. Köstler. Car–Parrinello molecular dynamics using real space wavefunctions. *physica status solidi (b)*, 243:1001–1015, 2006.

[1290] K. E. Schmidt and D. M. Ceperley. Monte Carlo techniques for quantum fluids, solids and droplets. In K. Binder, ed., *The Monte Carlo*

Method in Condensed Matter Physics, page 205. Springer, Berlin, 1992.

[1291] M. Schmitz and P. Tavan. Vibrational spectra from atomic fluctuations in dynamics simulations I. Theory, limitations and a sample application. *Journal of Chemical Physics*, 121:12233–12246, 2004.

[1292] M. Schmitz and P. Tavan. Vibrational spectra from atomic fluctuations in dynamics simulations II. Solvent-induced frequency fluctuations at femtosecond time resolution. *Journal of Chemical Physics*, 121:12247–12258, 2004.

[1293] P. Schravendijk, N. van der Vegt, L. Delle Site, and K. Kremer. Dual-scale modeling of benzene adsorption onto Ni(111) and Au(111) surfaces in explicit water. *ChemPhysChem*, 6:1866–1871, 2005.

[1294] J. Schulte, M. C. Böhm, and R. Ramírez. The isotope effect in electronic expectation values: an all-quantum study of C_6H_6 and C_6D_6. *Molecular Physics*, 93:801–807, 1998.

[1295] J. Schütt, M. C. Böhm, and R. Ramírez. Quantum delocalization of nuclei and electrons: Cyclobutadiene. *Chemical Physics Letters*, 248:379–385, 1996.

[1296] K. Schwarz, E. Nusterer, and P. E. Blöchl. First-principles molecular dynamics study of small molecules in zeolites. *Catalysis Today*, 50:501–509, 1999.

[1297] K. Schwarz, E. Nusterer, and P. Margl. *Ab initio* molecular dynamics calculations to study catalysis. *International Journal of Quantum Chemistry*, 61:369–380, 1997.

[1298] E. Schwegler, G. Galli, and F. Gygi. Water under pressure. *Physical Review Letters*, 84:2429–2432, 2000.

[1299] E. Schwegler, G. Galli, F. Gygi, and R. Q. Hood. Dissociation of water under pressure. *Physical Review Letters*, 87:265501, 2001. Comment: Ref. [322] and Reply: Ref. [1300].

[1300] E. Schwegler, G. Galli, F. Gygi, and R. Q. Hood. Reply to comment. *Physical Review Letters*, 89:199602, 2002. Comment: Ref. [322].

[1301] E. Schwegler, J. C. Grossman, F. Gygi, and G. Galli. Towards an assessment of the accuracy of density functional theory for first principles simulations of water. II. *Journal of Chemical Physics*, 121:5400–5409, September 2004.

[1302] W. R. P. Scott, P. H. Hünenberger, I. G. Tironi, A. E. Mark, S. R. Billeter, J. Fennen, A. E. Torda, T. Huber, P. Krüger, and W. F. van Gunsteren. The GROMOS biomolecular simulation program package. *Journal of Physical Chemistry A*, 103:3596–3607, 1999.

[1303] D. Sebastiani, G. Goward, I. Schnell, and M. Parrinello. NMR chemical shifts in periodic systems from first principles. *Computer Physics Communications*, 147:707–710, 2002.

[1304] D. Sebastiani and M. Parrinello. A new *ab initio* approach for NMR chemical shifts in periodic systems. *Journal of Physical Chemistry A*, 105:1951–1958, 2001.

[1305] D. Sebastiani and U. Röthlisberger. Nuclear magnetic resonance chemical shifts from hybrid DFT QM/MM calculations. *Journal of Physical Chemistry B*, 108:2807–2815, 2004.

[1306] M. D. Segall. Applications of *ab initio* atomistic simulations to biology. *Journal of Physics: Condensed Matter*, 14:2957–2973, 2002.

[1307] M. D. Segall, P. J. D. Lindan, M. J. Probert, C. J. Pickard, P.J. Hasnip, S. J. Clark, and M. C. Payne. First-principles simulation: ideas, illustrations and the CASTEP code. *Journal of Physics: Condensed Matter*, 14:2717–2744, 2002.

[1308] M. D. Segall, M. C. Payne, S. W. Ellis, G. T. Tucker, and R. N. Boyes. *Ab initio* molecular modeling in the study of drug metabolism. *Eur. J. Drug Metab. Pharmacokinet.*, 22:283–289, 1997.

[1309] M. D. Segall, M. C. Payne, S. W. Ellis, G. T. Tucker, and R. N. Boyes. First principles calculation of the activity of cytochrome P450. *Physical Review E*, 57:4618–4621, 1998.

[1310] M. D. Segall, M. C. Payne, S. W. Ellis, G. T. Tucker, and P. J. Eddershaw. First principles investigation of singly reduced cytochrome P450. *Xenobiotica*, 29:561–571, 1999.

[1311] M. D. Segall, C. J. Pickard, R. Shah, and M. C. Payne. Population analysis in plane wave electronic structure calculations. *Molecular Physics*, 89:571–577, 1996.

[1312] M. D. Segall, R. Shah, C. J. Pickard, and M. C. Payne. Population analysis of plane-wave electronic structure calculations of bulk materials. *Physical Review B*, 54:16317–16320, 1996.

[1313] G. Seifert and R. O. Jones. Geometric and electronic structure of clusters. *Zeitschrift für Physik D*, 20:77–80, 1991.

[1314] G. Seifert, R. Kaschner, M. Schöne, and G. Pastore. Density functional calculations for Zintl systems: structure, electronic structure and electrical conductivity of liquid NaSn alloys. *Journal of Physics: Condensed Matter*, 10:1175–1198, 1998.

[1315] G. Seifert, D. Porezag, and Th. Frauenheim. Calculations of molecules, clusters, and solids with a simplified LCAO-DFT-LDA scheme. *International Journal of Quantum Chemistry*, 58:185, 1996.

[1316] A. Selloni, P. Carnevali, R. Car, and M. Parrinello. Localization, hopping, and diffusion of electrons in molten-salts. *Physical Review Letters*, 59:823–826, 1987.

[1317] A. Selloni, A. Vittadini, and M. Grätzel. The adsorption of small molecules on the TiO_2 anatase(101) surface by first-principles molecular dynamics. *Surface Science*, 402-404:219–222, 1998.

[1318] S. Sen and J. E. Dickinson. *Ab initio* molecular dynamics simulation of femtosecond laser-induced structural modification in vitreous silica. *Physical Review B*, 68:214204, 2003.

[1319] Y. Senda, F. Shimojo, and K. Hoshino. Composition dependence of the structure and the electronic states of liquid K-Pb alloys: *Ab initio* molecular-dynamics simulations. *Journal of Physics: Condensed Matter*, 11:5387–5398, 1999.

[1320] Y. Senda, F. Shimojo, and K. Hoshino. The origin of the first sharp diffraction peak in liquid Na-Pb alloys: *ab initio* molecular-dynamics simulations. *Journal of Physics: Condensed Matter*, 11:2199–2210, 1999.

[1321] Y. Senda, F. Shimojo, and K. Hoshino. The metal–nonmetal transition of liquid phosphorus by *ab initio* molecular-dynamics simulations. *Journal of Physics: Condensed Matter*, 14:3715–3723, 2002.

[1322] H. M. Senn, P. M. Margl, R. Schmid, T. Ziegler, and P. E. Blöchl. *Ab initio* molecular dynamics with a continuum solvation model. *Journal of Chemical Physics*, 118:1089–1100, 2003.

[1323] A. Sergi, M. Ferrario, F. Buda, and I. R. McDonald. Structure of phosphorus–selenium glasses: Results from *ab initio* molecular dynamics simulations. *Molecular Physics*, 98:701–707, 2000.

[1324] S. Serra, C. Cavazzoni, G. L. Chiarotti, S. Scandolo, and E. Tosatti. Pressure-induced solid carbonates from molecular CO_2 by computer simulation. *Science*, 284:788–790, 1999.

[1325] S. Serra, G. Chiarotti, S. Scandolo, and E. Tosatti. Pressure-induced magnetic collapse and metallization of molecular oxygen: The ζ-O_2 phase. *Physical Review Letters*, 80:5160–5163, 1998.

[1326] S. Serra, S. Iarlori, E. Tosatti, S. Scandolo, and G. Santoro. Dynamical and thermal properties of polyethylene by *ab initio* simulation. *Chemical Physics Letters*, 331:339–345, 2001.

[1327] R. Shah, M. C. Payne, and J. D. Gale. Acid–base catalysis in zeolites from first principles. *International Journal of Quantum Chemistry*, 61:393–398, 1997.

[1328] R. Shah, M. C. Payne, M.-H. Lee, and J. D. Gale. Understanding the catalytic behavior of zeolites: A first-principles study of the adsorption of methanol. *Science*, 271:1395–1397, 1996.

[1329] A. Shapere and F. Wilczek, eds. *Geometric Phases in Physics*. World Scientific, Singapore, 1989.

[1330] M. Sharma, R. Resta, and R. Car. Intermolecular dynamical charge fluctuations in water: A signature of the H-bond network. *Physical Review Letters*, 95:187401, 2005.

[1331] M. Sheng, E. G. Wang, and G. Shiwu. The pressure induced phase transition of confined water from *ab initio* molecular dynamics simulation. *Journal of Physics: Condensed Matter*, 16:8851–8859, 2004.

[1332] P. Sherwood. Hybrid quantum mechanics/molecular mechanics approaches. In J. Grotendorst, ed., *Modern Methods and Algorithms of Quantum Chemistry*, pages 257–277 (first edition, paperback, ISBN 3--00--005618--1), pages 285–305 (second edition, hardcover, ISBN 3--00--005834--6). John von Neumann Institute for Computing, Forschungszentrum Jülich, Jülich, 2000.

[1333] M. Shiga and M. Tachikawa. *Ab initio* path integral study of isotope effect of hydronium ion. *Chemical Physics Letters*, 374:229–234, 2003.

[1334] M. Shiga, M. Tachikawa, and S. Miura. *Ab initio* molecular orbital calculation considering the quantum mechanical effect of nuclei by path integral molecular dynamics. *Chemical Physics Letters*, 332:396–402, 2000.

[1335] M. Shiga, M. Tachikawa, and S. Miura. A unified scheme for *ab initio* molecular orbital theory and path integral molecular dynamics. *Journal of Chemical Physics*, 115:9149–9159, 2001.

[1336] M. Shiga and T. Takayanagi. Quantum path-integral molecular dynamics calculations of the dipole-bound state of the water dimer anion. *Chemical Physics Letters*, 378:539–547, 2003.

[1337] F. Shimojo and M. Aniya. Diffusion of mobile ions and bond fluctuations in superionic conductor CuI from *ab initio* molecular-dynamics simulations. *Journal of the Physical Society of Japan*, 72:2702–2705, 2003.

[1338] F. Shimojo and M. Aniya. Diffusion mechanism of Ag ions in superionic conductor Ag_2Se from *ab initio* molecular-dynamics simulations. *Journal of the Physical Society of Japan*, 74:1224–1230, 2005.

[1339] F. Shimojo, K. Hoshino, and H. Okazaki. First-principles molecular-dynamics simulation of proton diffusion in perovskite oxides. *Solid State Ionics*, 113:319–323, 1998.

[1340] F. Shimojo, K. Hoshino, and Y. Zempo. Photo-induced bond breaking in the S_8 ring: An *ab initio* molecular-dynamics simulation. *Journal of Physics: Condensed Matter*, 10:L177–L182, 1998.

[1341] F. Shimojo, S. Munejiri, K. Hoshino, and Y. Zempo. The microscopic mechanism of the semiconductor–metal transition in liquid arsenic triselenide. *Journal of Physics: Condensed Matter*, 11:L153–L158, 1999.

[1342] E. L. Shirley, D. C. Allan, R. M. Martin, and J. D. Joannopoulos. Extended norm-conserving pseudopotentials. *Physical Review B*, 40:3652–3660, 1989.

[1343] SIESTA: Spanish Initiative for Electronic Simulations with Thousands of Atoms. Ref. [1370]; See http://www.uam.es/siesta/.

[1344] A. J. Sillanpää, M. L. Klein, and K. Laasonen. Structural and spectral properties of aqueous hydrogen fluoride studied using *ab initio* molecular dynamics. *Journal of Physical Chemistry B*, 106:11315–11322, 2002.

[1345] A. J. Sillanpää and K. Laasonen. Structure and dynamics of concentrated hydrochloric acid solutions. A first principles molecular dynamics study. *Physical Chemistry Chemical Physics*, 6:555–565, 2004.

[1346] A. J. Sillanpää and K. Laasonen. Car–Parrinello molecular dynamics study of DCl hydrate crystals. *ChemPhysChem*, 6:1879–1883, 2005.

[1347] P. L. Silvestrelli. Maximally localized Wannier functions for simulations with supercells of general symmetry. *Physical Review B*, 59:9703–9706, 1999.

[1348] P. L. Silvestrelli, A. Alavi, and M. Parrinello. Electrical-conductivity calculation in *ab initio* simulations of metals: Application to liquid sodium. *Physical Review B*, 55:15515–15522, 1997.

[1349] P. L. Silvestrelli, A. Alavi, M. Parrinello, and D. Frenkel. *Ab initio* molecular dynamics simulation of laser melting of silicon. *Physical Review Letters*, 77:3149–3152, 1996.

[1350] P. L. Silvestrelli, A. Alavi, M. Parrinello, and D. Frenkel. Hot electrons and the approach to metallic behaviour in $K_x(KCl)_{1-x}$. *Europhysics Letters*, 33:551–556, 1996.

[1351] P. L Silvestrelli, A. Alavi, M. Parrinello, and D. Frenkel. Nonmetal–metal transition in metal–molten–salt solutions. *Physical Review B*, 53:12750–12760, 1996.

[1352] P. L. Silvestrelli, A. Alavi, M. Parrinello, and D. Frenkel. Structural, dynamical, electronic, and bonding properties of laser-heated silicon: An *ab initio* molecular-dynamics study. *Physical Review B*, 56:3806–3812, 1997.

[1353] P. L. Silvestrelli, F. Ancilotto, F. Toigo, C. Sbraccia, T. Ikeda, and M. Boero. Hydrophobic–hydrophilic interactions of water with alka-nethiolate chains from first-principles calculations. *ChemPhysChem*, 6:1889–1893, 2005.

[1354] P. L. Silvestrelli, M. Bernasconi, and M. Parrinello. *Ab initio* infrared spectrum of liquid water. *Chemical Physics Letters*, 277:478–482, 1997.

[1355] P. L. Silvestrelli, N. Marzari, D. Vanderbilt, and M. Parrinello. Maximally-localized Wannier functions for disordered systems: Application to amorphous silicon. *Solid State Communications*, 107:7–11, 1998.

[1356] P. L. Silvestrelli and M. Parrinello. *Ab initio* molecular dynamics simulation of laser melting of graphite. *Journal of Applied Physics*, 83:2478–2483, 1998.

[1357] P. L. Silvestrelli and M. Parrinello. Structural, electronic, and bonding properties of liquid water from first principles. *Journal of Chemical Physics*, 111:3572–3580, 1999.

[1358] P. L. Silvestrelli and M. Parrinello. Water molecule dipole in the gas and in the liquid phase. *Physical Review Letters*, 82:3308–3311, 1999. Erratum: [1359].

[1359] P. L. Silvestrelli and M. Parrinello. Water molecule dipole in the gas and in the liquid phase – erratum. *Physical Review Letters*, 82:5415, 1999.

[1360] B. Silvi and A. Savin. Classification of chemical-bonds based on topological analysis of electron localization functions. *Nature*, 371:683–686, 1994. See http://www.cpfs.mpg.de/ELF/.

[1361] F. Sim, A. St.-Amant, I. Papai, and D. R. Salahub. Gaussian density functional calculations on hydrogen-bonded systems. *Journal of the American Chemical Society*, 114:4391–4400, 1992.

[1362] C. Simon and M. L. Klein. *Ab initio* molecular dynamics simulation of a water–hydrogen fluoride equimolar mixture. *ChemPhysChem*, 6:148–153, 2005.

[1363] O. Sinanoglu. Relation of perturbation theory to variation method. *Journal of Chemical Physics*, 34:1237–1240, 1961.

[1364] S. J. Singer, J.-L. Kuo, T. K. Hirsch, C. Knight, L. Ojamäe, and M. L. Klein. Hydrogen-bond topology and the ice VII/VIII and ice Ih/XI proton-ordering phase transitions. *Physical Review Letters*, 94:135701, 2005.

[1365] D. J. Singh. *Planewaves, Pseudopotentials and the LAPW Method*. Kluwer, Dordrecht, 1994.

[1366] U. C. Singh and P. A. Kollman. A combined *ab initio* quantum-mechanical and molecular mechanical method for carrying out simulations on complex molecular-systems – Applications to the CH_3Cl + Cl^- exchange-reaction and gas-phase protonation of polyethers. *Journal of Computational Chemistry*, 7:718–730, 1986.

[1367] P. H.-L. Sit and N. Marzari. Static and dynamical properties of heavy water at ambient conditions from first-principles molecular dynamics. *Journal of Chemical Physics*, 122:204510, 2005.

[1368] E. Smargiassi. Self-diffusion in sodium via *ab initio* molecular dynamics. *Physical Review B*, 65:012301, 2002.

[1369] V. Smelyansky, J. Hafner, and G. Kresse. Adsorption of thiophene on RuS_2: An *ab initio* density-functional study. *Physical Review B*, 58:R1782–R1785, 1998.

[1370] J. M. Soler, E. Artacho, J. D. Gale, A. Garcia, J. Junquera, P. Ordejón, and D. Sanchez-Portal. The SIESTA method for *ab initio* order-N materials simulation. *Journal of Physics: Condensed Matter*, 14:2745–2779, 2002.

[1371] A. K. Soper. The quest for the structure of water and aqueous solutions. *Journal of Physics: Condensed Matter*, 9:2717–2730, 1997.

[1372] I. Souza, J. Iniguez, and D. Vanderbilt. First-principles approach to insulators in finite electric fields. *Physical Review Letters*, 89:117602, 2002.

[1373] I. Souza and R. M. Martin. Polarization and strong infrared activity in compressed solid hydrogen. *Physical Review Letters*, 81:4452–4455, 1998.

[1374] I. Souza, N. Marzari, and D. Vanderbilt. Maximally localized Wannier functions for entangled energy bands. *Physical Review B*, 65:035106, 2001.

[1375] I. Souza, T. Wilkens, and R. M. Martin. Polarization and localization in insulators: Generating function approach. *Physical Review B*, 62:1666–1683, 2000.

[1376] R. Spezia, C. Nicolas, A. Boutin, and R. Vuilleumier. Molecular dynamics simulations of a silver atom in water: Evidence for a dipolar excitonic state. *Physical Review Letters*, 91:208304, 2003.

[1377] S/PHI/nX (formerly SFHIngX). Written by S. Boeck, J. Neugebauer *et al.*; See http://www.sxlib.de/.

[1378] K. Spiegel, U. Röthlisberger, and P. Carloni. Cisplatin binding to DNA oligomers from hybrid Car–Parrinello/molecular dynamics simulations. *Journal of Physical Chemistry B*, 108:2699–2707, 2004.

[1379] M. Sprik. Computer simulation of the dynamics of induced polarization fluctuations in water. *Journal of Physical Chemistry*, 95:2283–2291, March 1991.

[1380] M. Sprik. Effective pair potentials and beyond. In M. P. Allen and D. J. Tildesley, eds., *Computer Simulation in Chemical Physics*. Kluwer, Dordrecht, 1993.

[1381] M. Sprik. *Ab initio* molecular dynamics simulation of liquids and solutions. *Journal of Physics: Condensed Matter*, 8:9405–9409, 1996.

[1382] M. Sprik. Introduction to molecular dynamics methods. In K. Binder and G. Ciccotti, eds., *Monte Carlo and Molecular Dynamics of Condensed Matter Systems*, chapter 2, pages 43–88. Italian Physical Society SIF, Bologna, 1996.

[1383] M. Sprik. Density functional techniques for the simulation of chemical reactions. In B. J. Berne, G. Ciccotti, and D. F. Coker, eds., *Classical and Quantum Dynamics in Condensed Phase Simulations*, chapter 13, pages 285–309. World Scientific, Singapore, 1998.

[1384] M. Sprik. *Ab initio* molecular dynamics simulation of liquids and solutions. *Journal of Physics: Condensed Matter*, 12:A161–A163, 2000.

[1385] M. Sprik. Computation of the pK of liquid water using coordination constraints. *Chemical Physics*, 258:139–150, August 2000.

[1386] M. Sprik and G. Ciccotti. Free energy from constrained molecular dynamics. *Journal of Chemical Physics*, 109:7737–7744, 1998.

[1387] M. Sprik, J. Hutter, and M. Parrinello. *Ab initio* molecular dynamics simulation of liquid water: Comparison of three gradient-corrected density functionals. *Journal of Chemical Physics*, 105:1142–1152, 1996.

[1388] M. Sprik and M. L. Klein. Optimization of a distributed Gaussian-basis set using simulated annealing – Application to the adiabatic dynamics of the solvated electron. *Journal of Chemical Physics*, 89:1592–1607, 1988. Erratum: [1389].

[1389] M. Sprik and M. L. Klein. Erratum. *Journal of Chemical Physics*, 90:7614, 1989. Original article: [1388].

[1390] M. Springborg, ed. *Density-Functional Methods in Chemistry and Materials Science*. John Wiley & Sons, New York, 1997.

[1391] G. P. Srivastava and D. Weaire. The theory of the cohesive energies of solids. *Advances in Physics*, 36:463–517, 1987.

[1392] R. Stadler, A. Alfe, G. Kresse, G. A. de Wijs, and M. J. Gillan. Transport coefficients of liquids from first principles. *Journal of Non-Crystalline Solids*, 250–252:82–90, 1999.

[1393] R. Stadler and M. J. Gillan. First-principles molecular dynamics studies of liquid tellurium. *Journal of Physics: Condensed Matter*, 12:6053–6061, 2000.

[1394] T. Starkloff and J.D. Joannopoulos. Local pseudopotential theory for transition-metals. *Physical Review B*, 16:5212–5215, 1977.

[1395] V. N. Staroverov, G. E. Scuseria, J. Tao, and J. P. Perdew. Comparative assessment of a new nonempirical density functional: Molecules and hydrogen-bonded complexes. *Journal of Chemical Physics*, 119:12129–12137, 2003. Erratum [1396].

[1396] V. N. Staroverov, G. E. Scuseria, J. Tao, and J. P. Perdew. Erratum: "Comparative assessment of a new nonempirical density functional: Molecules and hydrogen-bonded complexes" (vol. 119, pg 12129, 2003). *Journal of Chemical Physics*, 121:11507, 2004. Original article: [1395].

[1397] V. N. Staroverov, G. E. Scuseria, J. Tao, and J. P. Perdew. Tests of a ladder of density functionals for bulk solids and surfaces. *Physics Review B*, 69:075102, 2004.

[1398] M. Stengel and A. De Vita. First-principles molecular dynamics of metals: A Lagrangian formulation. *Physical Review B*, 62:15283–15286, 2000.

[1399] R. M. Sternheimer. Electronic polarizabilities of ions from the Hartree–Fock wave functions. *Physical Review*, 96:951–968, 1954.

[1400] W. Stier and O. V. Prezhdo. Nonadiabatic molecular dynamics simulation of light-induced, electron transfer from an anchored molecular electron donor to a semiconductor acceptor. *Journal of Physical Chemistry B*, 106:8047–8054, 2002.

[1401] A. Stirling, M. Bernasconi, and M. Parrinello. *Ab initio* simulation of H_2S adsorption on the (100) surface of pyrite. *Journal of Chemical Physics*, 119:4934–4939, 2003.

[1402] A. Stirling, M. Bernasconi, and M. Parrinello. *Ab initio* simulation of water interaction with the (100) surface of pyrite. *Journal of Chemical Physics*, 118:8917–8926, 2003.

[1403] A. Stirling, M. Iannuzzi, A. Laio, and M. Parrinello. Azulene-to-naphthalene rearrangement: The Car–Parrinello metadynamics method explores various reaction mechanisms. *ChemPhysChem*, 5:1558–1568, 2004.

[1404] A. Stirling, M. Iannuzzi, M. Parrinello, F. Molnar, V. Bernhart, and G. A. Luinstra. β-Lactone synthesis from epoxide and CO: Reaction mechanism revisited. *Organometallics*, 24:2533–2537, 2005.

[1405] A. J. Stone. *The Theory of Intermolecular Forces*. Clarendon Press, Oxford, 1996 and 2002.

[1406] J. M. Stubbs and D. Marx. Glycosidic bond formation in aqueous solution: On the oxocarbenium intermediate. *Journal of the American Chemical Society*, 125:10960–10962, 2003.

[1407] J. M. Stubbs and D. Marx. Aspects of glycosidic bond formation in aqueous solution: Chemical bonding and the role of water. *Chemistry - A European Journal*, 11:2651–2659, 2005.

[1408] A. C. Stückl, C. A. Daul, and H. U. Güdel. Density functional calculations of optical excitation energies by a transition-state method. *International Journal of Quantum Chemistry*, 61:579–588, 1997.

[1409] P. Stumm and D. A. Drabold. Can amorphous GaN serve as a useful electronic material? *Physical Review Letters*, 79:677–680, 1997.

[1410] Y.-S. Su and S. T. Pantelides. Diffusion mechanism of hydrogen in amorphous silicon: Ab initio molecular dynamics simulation. *Physical Review Letters*, 88:165503, 2002.

[1411] O. Sugino and Y. Miyamoto. Density-functional approach to electron dynamics: Stable simulation under a self-consistent field. *Physical Review B*, 59:2579–2586, 1999. Erratum: [1412].

[1412] O. Sugino and Y. Miyamoto. Erratum. *Physical Review B*, 66:089901, 2002. Original article: [1411].

[1413] D. M. Sullivan, K. Bagchi, M. E. Tuckerman, and M. L. Klein. *Ab initio* molecular dynamics study of crystalline nitric acid trihydrate. *Journal of Physical Chemistry A*, 103:8678–8683, 1999.

[1414] M. Sulpizi, P. Carloni, J. Hutter, and U. Röthlisberger. A hybrid TDDFT/MM investigation of the optical properties of aminocoumarins in water and acetonitrile solution. *Physical Chemistry Chemical Physics*, 5:4798–4805, 2003.

[1415] M. Sulpizi, A. Laio, J. VandeVondele, A. Cattaneo, U. Röthlisberger, and P Carloni. Reaction mechanism of caspases: Insights from QM/MM Car–Parrinello simulations. *Proteins - Structure, Function and Genetics*, 52:212–224, 2003.

[1416] M. Sulpizi, U. F. Röhrig, J. Hutter, and U. Röthlisberger. Optical properties of molecules in solution via hybrid TDDFT/MM simulations. *International Journal of Quantum Chemistry*, 101:671–682, 2005.

[1417] M. Sulpizi, U. Röthlisberger, A. Laio, A. Cattaneo, and P. Carloni. Reaction mechanism of caspases: Insights from QM/MM Car–Parrinello simulations. *Biophysical Journal (Annual Meeting Abstracts)*, 82:359A–360A, 2002.

[1418] D. Y. Sun and X. G. Gong. A new constant-pressure molecular dynamics method for finite systems. *Journal of Physics: Condensed Matter*, 14:L487–L493, 2002.

[1419] G. Y. Sun, J. Kurti, P. Rajczy, M. Kertesz, J. Hafner, and G. Kresse. Performance of the Vienna *ab initio* Simulation Package (VASP) in chemical applications. *Journal of Molecular Structure-THEOCHEM*, 624:37–45, 2003.

[1420] M. P. Surh, T. W. Barbee III, and L. H. Yang. First principles molecular dynamics of dense plasmas. *Physical Review Letters*, 86:5958–5961, 2001.

[1421] T. W. Swaddle, J. Rosenqvist, P. Yu, E. Bylaska, B. L. Phillips, and W. H. Casey. Kinetic evidence for five-coordination in $AlOH(aq)^{2+}$ ion. *Science*, 308:1450–1453, 2005.

[1422] W. C. Swope, H. C. Andersen, P. H. Berens, and K. R. Wilson. A computer-simulation method for the calculation of equilibrium-constants for the formation of physical clusters of molecules – Application to small water clusters. *Journal of Chemical Physics*, 76:637–649, 1982.

[1423] A. Szabo and N. S. Ostlund. *Modern Quantum Chemistry – Introduction to Advanced Electronic Structure Theory*. McGraw-Hill, New York, 1989.

[1424] H. Tachikawa. Temperature dependence of the hyperfine coupling constant of the D_3O radical: A direct *ab initio* molecular dynamics (MD) study. *Chemical Physics*, 276:257–262, 2002.

[1425] H. Tachikawa and M. Igarashi. A direct *ab initio* dynamics study on a gas phase SN_2 reaction $F^- + CH_3Cl \rightarrow CH_3F + Cl^-$: Dynamics of near-collinear collision. *Chemical Physics Letters*, 303:81–86, 1999.

[1426] M. Tachikawa. Multi-component molecular orbital theory for electrons and nuclei including many-body effect with full configuration interaction treatment: Isotope effects on hydrogen molecules. *Chemical Physics Letters*, 360:494–500, 2002.

[1427] M. Tachikawa, T. Ishimoto, H. Tokiwa, H. Kasatani, and K. Deguchi. First-principle calculation on isotope effect in KH_2PO_4 and KD_2PO_4 of hydrogen-bonded dielectric materials. Approach with dynamic extended molecular orbital method. *Ferroelectrics*, 268:423–429, 2002.

[1428] M. Tachikawa, K. Mori, H. Nakai, and K. Iguchi. An extension of *ab initio* molecular orbital theory to nuclear motion. *Chemical Physics Letters*, 290:437–442, 1998.

[1429] M. Tachikawa and M. Shiga. Theoretical study on isotope and temperature effect in hydronium ion using *ab initio* path integral simulation. *Journal of Chemical Physics*, 121:5985–5991, 2004.

[1430] M. Tachikawa and M. Shiga. *Ab initio* path integral simulation study on $^{16}O/^{18}O$ isotope effect in water and hydronium ion. *Chemical Physics Letters*, 407:135–138, 2005.

[1431] M. Tachikawa and M. Shiga. *Ab initio* path integral study on isotope effect of ammonia molecule. *Journal of Theoretical and Compututational Chemistry*, 4:175–181, 2005.

[1432] M. Tachikawa and M. Shiga. Geometrical H/D isotope effects on hydrogen bonds in charged water clusters. *Journal of the American Chemical Society*, 127:11908–11909, 2005.

[1433] M. Takahashi and M. Imada. Monte Carlo calculation of quantum systems. *Journal of the Physical Society of Japan*, 53:963–974, 1984.

[1434] M. Takahashi and M. Imada. Monte Carlo calculation of quantum systems. II. Higher order correction. *Journal of the Physical Society of Japan*, 53:3765–3769, 1984.

[1435] M. Takahashi and M. Imada. Quantum Monte Carlo simulation of a two-dimensional electron system – Melting of Wigner crystal. *Journal of the Physical Society of Japan*, 53:3770–3781, 1984.

[1436] O. Takahashi, M. Odelius, D. Nordlund, A. Nilsson, H. Bluhm, and L. G. M. Pettersson. Auger decay calculations with core–hole excited-state molecular-dynamics simulations of water. *Journal of Chemical Physics*, 124:064307, 2006.

[1437] N. Takeuchi, A. Selloni, and E. Tosatti. Transition from surface vibrations to liquidlike dynamics at an incompletely melted semiconductor surface. *Physical Review B*, 55:15405–15407, 1997.

[1438] M. Tamaoki, Y. Yamauchi, and H. Nakai. Short-time Fourier transform analysis of *ab initio* molecular dynamics simulation: Collision reaction between CN and C_4H_6. *Journal of Computational Chemistry*, 26:436–442, 2005.

[1439] S. Tanaka. Structural optimization in variational quantum Monte-Carlo. *Journal of Chemical Physics*, 100:7416–7420, 1994.

[1440] P. Tangney. On the theory underlying the Car–Parrinello method and the role of the fictitious mass parameter. *Journal of Chemical Physics*, 124:044111, 2006.

[1441] P. Tangney and S. Fahy. Calculations of the a_1 phonon frequency in photoexcited tellurium. *Physical Review Letters*, 82:4340–4343, 1999.

[1442] P. Tangney and S. Scandolo. How well do Car–Parrinello simulations reproduce the Born–Oppenheimer surface? Theory and examples. *Journal of Chemical Physics*, 116:14–24, 2002.

[1443] J. Tao, J. P. Perdew, V .N. Staroverov, and G. E. Scuseria. Climbing the density functional ladder: Nonempirical meta-generalized gradient approximation designed for molecules and solids. *Physical Review Letters*, 91:146401, 2003.

[1444] E. Tapavicza, I. Tavernelli, and U. Röthlisberger. Trajectory surface hopping within linear response time-dependent density-functional theory. *Physical Review Letters*, 98:023001, 2007.

[1445] A. Tarazona, E. Koglin, F. Buda, B. B. Coussens, J. Renkema, S. van Heel, and R. J. Maier. Structure and stability of aluminum alkyl cocatalysts in Ziegler–Natta catalysis. *Journal of Physical Chemistry B*, 101:4370–4378, 1997.

[1446] F. Tassone, G. L. Chiarotti, R. Rousseau, S. Scandolo, and E. Tosatti. Dimerization of CO_2 at high pressure and temperature. *ChemPhysChem*, 6:1752–1756, 2005.

[1447] F. Tassone, F. Mauri, and R. Car. Acceleration schemes for *ab initio* molecular-dynamics simulations and electronic-structure calculations. *Physical Review B*, 50:10561–10573, 1994.

[1448] Y. Tateyama, J. Blumberger, M. Sprik, and I. Tavernelli. Density-functional molecular-dynamics study of the redox reactions of two anionic, aqueous transition-metal complexes. *Journal of Chemical Physics*, 122:234505, 2005.

[1449] P. Tavan, H. Carstens, and G. Mathias. Molecular dynamics simulations of proteins and peptides: Problems, achievements and perspectives. In J. Buchner and T. Kiefhaber, eds., *Protein Folding Handbook, Part I*, pages 1170–1195. Wiley-VCH, Weinheim, 2005.

[1450] P. Tavan and G. Mathias *et al.* EGO-MMII User's Guide. See http://www.bmo.physik.uni-muenchen.de/forschung/tavan/ego/.

[1451] I. Tavernelli, U. T. Röhrig, and U. Röthlisberger. Molecular dynamics in electronically excited states using time-dependent density functional theory. *Molecular Physics*, 103:963–981, 2005.

[1452] I. Tavernelli, E. Tapavicza, and U. Röthlisberger. Photoinduced electron transfer in DNA. Preprint, 2005.

[1453] I. Tavernelli, R. Vuilleumier, and M. Sprik. *Ab Initio* molecular dynamics for molecules with variable numbers of electrons. *Physical Review Letters*, 88:213002, 2002.

[1454] G. te Velde and E. J. Baerends. Slab versus cluster approach for chemisorption studies - CO on Cu(100). *Chemical Physics*, 177:399–406, 1993.

[1455] J. Teixeira. High-pressure physics – The double identity of ice X. *Nature*, 392:232–233, 1998. See also [105].

[1456] K. Terakura, T. Yamasaki, T. Uda, and I. Štich. Atomic and molecular processes on Si(001) and Si(111) surfaces. *Surface Science*, 386:207–215, 1997.

[1457] V. Termath, F. Haase, J. Sauer, J. Hutter, and M. Parrinello. Understanding the nature of water bound to solid acid surfaces. Ab initio simulation on HSAPO-34. *Journal of the American Chemical Society*, 120:8512–8516, 1998.

[1458] V. Termath and J. Sauer. Optimized molecular integration schemes for density functional theory *ab initio* molecular dynamics simulations. *Chemical Physics Letters*, 255:187–194, 1996.

[1459] V. Termath and J. Sauer. *Ab initio* molecular dynamics simulation of $H_5O_2^+$ and $H_7O_3^+$ gas phase clusters based on density functional theory. *Molecular Physics*, 91:963–975, 1997.

[1460] F. Terstegen, E. A. Carter, and V. Buss. Interconversion pathways of the protonated beta-ionone Schiff base: An *ab initio* molecular dynamics study. *International Journal of Quantum Chemistry*, 75:141–145, 1999.

[1461] M. Teter. Additional condition for transferability in pseudopotentials. *Physical Review B*, 48:5031–5041, 1993.

[1462] M. P. Teter, M. C. Payne, and D. C. Allen. Solution of Schrödinger-equation for large systems. *Physical Review B*, 40:12255–12263, 1989.

[1463] J. Theilhaber. *Ab initio* simulations of sodium using time-dependent density-functional theory. *Physical Review B*, 46:12990–13003, 1992.

[1464] J. W. Thomas, R. Iftimie, and M. E. Tuckerman. Field theoretic approach to dynamical orbital localization in *ab initio* molecular dynamics. *Physical Review B*, 69:125105, 2004.

[1465] D. J. Thouless. Stability conditions and nuclear rotations in the Hartree–Fock theory. *Nuclear Physics*, 21:225–232, 1960.

[1466] K. S. Thygesen, L. B. Hansen, and K. W. Jacobsen. Partly occupied wannier functions. *Physical Review Letters*, 94:026405, 2005.

[1467] A. Tilocca and A. Selloni. O_2 and vacancy diffusion on rutile(110): Pathways and electronic properties. *ChemPhysChem*, 6:1911–1916, 2005.

[1468] D. J. Tobias, P. Jungwirth, and M. Parrinello. Surface solvation of halogen anions in water clusters: An *ab initio* molecular dynamics study of the $Cl^-(H_2O)_6$ complex. *Journal of Chemical Physics*, 114:7036–7044, 2001.

[1469] J. Tobik, I. Štich, R. Perez, and K. Terakura. Simulation of tip–surface interactions in atomic force microscopy of an InP(110) surface with a Si tip. *Physical Review B*, 60:11639–11644, 1999.

[1470] T. Todorova, A. P. Seitsonen, J. Hutter, I-F. W. Kuo, and C. J. Mundy. Molecular dynamics simulation of liquid water: Hybrid density functionals. *Journal of Physical Chemistry B*, 110:3685–3691, 2006.

[1471] J. Tomasi, B. Mennucci, and R. Cammi. Quantum mechanical continuum solvation models. *Chemical Reviews*, 105:2999–3093, 2005.

[1472] A. Toniolo, C. Ciminelli, M. Persico, and T. J. Martinez. Simulation of the photodynamics of azobenzene on its first excited state: Comparison of full multiple spawning and surface hopping treatments. *Journal of Chemical Physics*, 123:234308, 2005.

[1473] W.C. Topp and J.J. Hopfield. Chemically motivated pseudopotential for sodium. *Physical Review B*, 7:1295–1303, 1973.

[1474] E. Tornaghi, W. Andreoni, P. Carloni, J. Hutter, and M. Parrinello. Carboplatin versus cisplatin: Density functional approach to their molecular properties. *Chemical Physics Letters*, 246:469–474, 1995.

[1475] G. Toth. Quantum chemical study of the different forms of nitric acid monohydrate. *Journal of Physical Chemistry A*, 101:8871–8876, 1997.

[1476] V. Tozzini, F. Buda, and A. Fasolino. Spontaneous formation and stability of small GaP fullerenes. *Physical Review Letters*, 85:4554–4557, 2000.

[1477] V. Tozzini and P. Giannozzi. Vibrational properties of DsRed model chromophores. *ChemPhysChem*, 6:1786–1788, 2005.

[1478] A. Trave, F. Buda, and A. Fasolino. Band-gap engineering by III–V infill in sodalite. *Physical Review Letters*, 77:5405–5408, 1996.

[1479] A. Trave, P. Tangney, S. Scandolo, A. Pasquarello, and R. Car. Pressure-induced structural changes in liquid SiO_2 from *ab initio* simulations. *Physical Review Letters*, 89:245504, 2002.

[1480] N. Troullier and J. L. Martins. Efficient pseudopotentials for plane-wave calculations. *Physical Review B*, 43:1993–2006, 1991.

[1481] N. Troullier and J. L. Martins. Efficient pseudopotentials for plane-wave calculations. II. Operators for fast iterative diagonalization. *Physical Review B*, 43:8861–8869, 1991.

[1482] B. L. Trout and M. Parrinello. The dissociation mechanism of H_2O in water studied by first-principles molecular dynamics. *Chemical Physics Letters*, 288:343–347, 1998.

[1483] B. L. Trout and M. Parrinello. Analysis of the dissociation of H_2O in water using first-principles molecular dynamics. *Journal of Physical Chemistry B*, 103:7340–7345, 1999.

[1484] J. S. Tse and D. D. Klug. Anomalous isostructural transformation in ice VIII. *Physical Review Letters*, 81:2466–2469, 1998.

[1485] J. S. Tse and D. D. Klug. Structure and dynamics of liquid sulphur. *Physical Review B*, 59:34–37, 1999.

[1486] J. S. Tse, D. D. Klug, and K. Laasonen. Structural dynamics of protonated methane and acetylene. *Physical Review Letters*, 74:876–879, 1995.

[1487] E. Tsuchida. *Ab initio* molecular-dynamics study of liquid formamide. *Journal of Chemical Physics*, 121:4740–4746, 2004.

[1488] E. Tsuchida, Y. Kanada, and M. Tsukada. Density-functional study of liquid methanol. *Chemical Physics Letters*, 311:236–240, 1999.

[1489] E. Tsuchida and M. Tsukada. Electronic-structure calculations based on the finite-element method. *Physical Review B*, 52:5573–5578, 1995.

[1490] E. Tsuchida and M. Tsukada. Real-space approach to electronic-structure calculations. *Solid State Communications*, 94:5–8, 1995.

[1491] E. Tsuchida and M. Tsukada. Large-scale electronic-structure calculations based on the adaptive finite-element method. *Journal of the Physical Society of Japan*, 67:3844–3858, 1998.

[1492] S. Tsuneyuki, H. Kitamura, T. Ogitsu, and T. Miyake. Quantum-mechanical properties of protons in solid molecular hydrogen at megabar pressures. *Journal of Low Temperature Physics*, 122:291–296, February 2001.

[1493] M. Tuckerman, B. J. Berne, and G. J. Martyna. Reversible multiple time scale molecular-dynamics. *Journal of Chemical Physics*, 97:1990–2001, 1992.

[1494] M. Tuckerman, K. Laasonen, M. Sprik, and M. Parrinello. *Ab initio* simulations of water and water ions. *Journal of Physics: Condensed Matter*, 6:A93–A100, 1994.

[1495] M. Tuckerman, K. Laasonen, M. Sprik, and M. Parrinello. *Ab initio* molecular dynamics simulation of the solvation and transport of H_3O^+ and OH^- ions in water. *Journal of Physical Chemistry*, 99:5749–5752, 1995.

[1496] M. Tuckerman, K. Laasonen, M. Sprik, and M. Parrinello. *Ab initio* molecular dynamics simulation of the solvation and transport of hydronium and hydroxyl ions in water. *Journal of Chemical Physics*, 103:150–161, 1995.

[1497] M. E. Tuckerman. Private communication.

[1498] M. E. Tuckerman. *Ab initio* molecular dynamics: basic concepts, current trends and novel applications. *Journal of Physics: Condensed Matter*, 14:R1297–R1355, 2002.

[1499] M. E. Tuckerman. Path integration via molecular dynamics. In J. Grotendorst, D. Marx, and A. Muramatsu, eds., *Quantum Simulations of Complex Many-Body Systems: From Theory to Algorithms*, pages 269–298. John von Neumann Institute for Computing, Forschungszentrum Jülich, Jülich, 2002. See http://www.theochem.rub.de/go/cprev.html.

[1500] M. E. Tuckerman, B. J. Berne, G. J. Martyna, and M. L. Klein. Efficient molecular-dynamics and hybrid Monte-Carlo algorithms for path-integrals. *Journal of Chemical Physics*, 99:2796–2808, 1993.

[1501] M. E. Tuckerman, A. Chandra, and D. Marx. Structure and dynamics of OH$^-$(aq). *Accounts of Chemical Research*, 39:151–158, 2006.

[1502] M. E. Tuckerman and A. Hughes. Path integral molecular dynamics: a computational approach to quantum statistical mechanics. In B. J. Berne, G. Ciccotti, and D. F. Coker, eds., *Classical and Quantum Dynamics in Condensed Phase Simulations*, chapter 14, pages 311–357. World Scientific, Singapore, 1998.

[1503] M. E. Tuckerman and M. L. Klein. *Ab initio* molecular dynamics study of solid nitromethane. *Chemical Physics Letters*, 283:147–151, 1998.

[1504] M. E. Tuckerman and G. J. Martyna. Understanding modern molecular dynamics: Techniques and applications. *Journal of Physical Chemistry B*, 104:159–178, 2000. Additions and Corrections: Ref. [1505].

[1505] M. E. Tuckerman and G. J. Martyna. Additions and corrections: Understanding modern molecular dynamics: Techniques and applications. *Journal of Physical Chemistry B*, 105:7598, 2001. See Ref. [1504] for the original article.

[1506] M. E. Tuckerman and D. Marx. Heavy-atom skeleton quantization and proton tunneling in "intermediate-barrier" hydrogen bonds. *Physical Review Letters*, 86:4946–4949, 2001. See animation of quantum fluctuations and of the tunneling proton http://www.theochem.rub.de/go/malon.html.

[1507] M. E. Tuckerman, D. Marx, M. L. Klein, and M. Parrinello. Efficient and general algorithms for path integral Car–Parrinello molecular dynamics. *Journal of Chemical Physics*, 104:5579–5588, 1996. See Refs. [946, 948] for the introduction and outline, respectively, of the underlying *ab initio* path integral technique.

[1508] M. E. Tuckerman, D. Marx, M. L. Klein, and M. Parrinello. On the quantum nature of the shared proton in hydrogen bonds. *Science*, 275:817–820, 1997.

[1509] M. E. Tuckerman, D. Marx, and M. Parrinello. The nature and transport mechanism of hydrated hydroxide ions in aqueous solution. *Nature*, 417:925–929, 2002.

[1510] M. E. Tuckerman and M. Parrinello. Integrating the Car–Parrinello equations. I. Basic integration techniques. *Journal of Chemical Physics*, 101:1302–1315, 1994. See Refs. [665, 1511] for Parts II and III.

[1511] M. E. Tuckerman and M. Parrinello. Integrating the Car–Parrinello equations. II. Multiple time scale techniques. *Journal of Chemical Physics*, 101:1316–1329, 1994.

[1512] M. E. Tuckerman, P. J. Ungar, T. von Rosenvinge, and M. L. Klein. *Ab initio* molecular dynamics simulations. *Journal of Physical Chemistry*, 100:12878–12887, 1996.

[1513] M. E. Tuckerman, D. A. Yarne, S. O. Samuelson, A. L. Hughes, and G. J. Martyna. Exploiting multiple levels of parallelism in molecular dynamics based calculations via modern techniques and software paradigms on distributed memory computers. *Computer Physics Communications*, 128:333–376, 2000.

[1514] J. C. Tully. Nonadiabatic processes in molecular collisions. In W. H. Miller, ed., *Modern Theoretical Chemistry: Dynamics of Molecular Collisions (Part B)*, page 217. Plenum Press, New York, 1976.

[1515] J. C. Tully. Molecular-dynamics with electronic-transitions. *Journal of Chemical Physics*, 93:1061–1071, 1990.

[1516] J. C. Tully. Mixed quantum-classical dynamics: Mean-field and surface-hopping. In B. J. Berne, G. Ciccotti, and D. F. Coker, eds., *Classical and Quantum Dynamics in Condensed Phase Simulations*, chapter 21, pages 489–538. World Scientific, Singapore, 1998.

[1517] J. C. Tully. Nonadiabatic Dynamics. In D. L. Thompson, ed., *Modern Methods for Multidimensional Dynamics Computations in Chemistry*, pages 34–72. World Scientific, Singapore, 1998.

[1518] J. C. Tully and R. K. Preston. Trajectory surface hopping approach to nonadiabatic molecular collisions – Reaction of H^+ with D_2. *Journal of Chemical Physics*, 55:562–572, 1971.

[1519] D. Tunega, L. Benco, G. Haberhauer, M. H. Gerzabek, and H. Lischka. *Ab initio* molecular dynamics study of adsorption sites on the (001) surfaces of 1 : 1 dioctahedral clay minerals. *Journal of Physical Chemistry B*, 106:11515–11525, 2002.

[1520] C. J. Tymczak and X.-Q. Wang. Orthonormal wavelet bases for quantum molecular dynamics. *Physical Review Letters*, 78:3654–3657, 1997.

[1521] T. Uchiyama, T. Uda, and K. Terakura. Initial oxide-growth process on the Si(100) surface. *Surface Science*, 433–435:896–899, 1999.

[1522] T. Udagawa, T. Ishimoto, H. Tokiwa, M. Tachikawa, and U. Nagashima. The geometrical isotope effect of C–H $\cdots$ O type hydrogen bonds revealed by multi-component molecular orbital calculation. *Chemical Physics Letters*, 389:236–240, 2004.

[1523] K. Uehara, M. Ishitobi, T. Oda, and Y. Hiwatari. First-principles molecular dynamics calculation of selenium clusters. *Molecular Simulations*, 18:385–394, 1997.

[1524] K. Uehara, M. Ishitobi, T. Oda, and Y. Hiwatari. First-principles molecular dynamics simulations for Se_8 and Se_8^+ clusters. *Molecular Simulations*, 19:75–84, 1997.

[1525] M. Uhlmann, T. Kunert, F. Grossmann, and R. Schmidt. Mixed classical-quantum approach to excitation, ionization, and fragmentation of H_2^+ in intense laser fields. *Physical Review A*, 67:013413, 2003.

[1526] P. Ullersma. An exactly solvable model for Brownian motion: I. Derivation of the Langevin equation. *Physica*, 32:27–55, 1966.

[1527] P. Ullersma. An exactly solvable model for Brownian motion: II. Derivation of the Fokker-Planck equation and the master equation. *Physica*, 32:56–73, 1966.

[1528] P. Ullersma. An exactly solvable model for Brownian motion: III. Motion of a heavy mass in a linear chain. *Physica*, 32:74–89, 1966.

[1529] P. Ullersma. An exactly solvable model for Brownian motion: IV. Susceptibility and Nyquist's theorem. *Physica*, 32:90–96, 1966.

[1530] P. Umari and Pasquarello. *Ab initio* molecular dynamics in a finite homogeneous electric field. *Physical Review Letters*, 89:157602, 2002.

[1531] C. P. Ursenbach, A. Calhoun, and G. A. Voth. A first-principles simulation of the semiconductor/water interface. *Journal of Chemical Physics*, 106:2811–2818, 1997.

[1532] H. Ushiyama and K. Takatsuka. Successive mechanism of double-proton transfer in formic acid dimer: A classical study. *Journal of Chemical Physics*, 115:5903–5912, 2001.

[1533] M. Valiev and J. M. Weare. The projector-augmented plane wave method applied to molecular bonding. *Journal of Physical Chemistry A*, 103:10588–10601, 1999.

[1534] R. M. Valladares, A. J. Fisher, S. J. Blundell, and W. Hayes. Studies of implanted muons in organic radicals. *Journal of Physics: Condensed Matter*, 10:10701–10713, 1998.

[1535] R. M. Valladares, A. J. Fisher, and W. Hayes. Path-integral simulations of zero-point effects for implanted muons in benzene. *Chemical Physics Letters*, 242:1–6, 1995.

[1536] C. G. van de Walle and P. E. Blöchl. First-principles calculations of hyperfine parameters. *Physical Review B*, 47:4244–4255, 1993.

[1537] T. S. van Erp and E. J. Meijer. *Ab initio* molecular dynamics study of aqueous solvation of ethanol and ethylene. *Journal of Chemical Physics*, 118:8831–8840, 2003.

[1538] W. F. van Gunsteren, D. Bakowies, R. Baron, I. Chandrasekhar, M. Christen, X. Daura, P. Gee, D. P. Geerke, A. Glättli, P. H. Hünenberger, M. A. Kastenholz, C. Ostenbrink, M. Schenk, D. Trzesniak, N. F. A. van der Vegt, and H. B. Yu. Biomolecular modeling: Goals, problems, perspectives. *Angewandte Chemie International Edition*, 45:4064–4092, 2006.

[1539] W. F. van Gunsteren and H. J. C. Berendsen. Computer simulation of molecular dynamics: Methodology, applications, and perspectives in chemistry. *Angewandte Chemie International Edition*, 29:992–1023, 1990.

[1540] W. F. van Gunsteren, S. R. Billeter, A. A. Eising, P. H. Hünenberger, P. Krüger, A. E. Mark, W. R. P. Scott, and I. G. Tironi. *Biomolecular Simulation:* GROMOS96 *Manual and User Guide.* BIOMOS b.v. ETH, Zürich 1996.

[1541] R. van Leeuwen. Causality and symmetry in time-dependent density-functional theory. *Physical Review Letters*, 80:1280–1283, 1998.

[1542] R. van Leeuwen. Key concepts in time-dependent density-functional theory. *International Journal of Modern Physics B*, 15:1969–2023, 2001.

[1543] R. van Leeuwen. Beyond the Runge-Gross Theorem. In M. A. L. Marques, C. A. Ullrich, F. Nogueira, A. Rubio, K. Burke, and E. K. U. Gross, eds., *Time-Dependent Density Functional Theory*, chapter I.2, pages 17–31. Springer, Berlin, 2006.

[1544] V. van Speybroeck and R. J. Meier. A recent development in computational chemistry: Chemical reactions from first principles molecular dynamics simulations. *Chemical Society Reviews*, 32:151–157, 2003.

[1545] T. Van Voorhis and G. Scuseria. A novel form for the exchange-correlation energy functional. *Journal of Chemical Physics*, 109:400–410, 1998.

[1546] D. Vanderbilt. Optimally smooth norm-conserving pseudopotentials. *Physical Review B*, 32:8412–8415, 1985.

[1547] D. Vanderbilt. Absence of large compressive stress on Si(111). *Physical Review Letters*, 59:1456–1459, 1987.

[1548] D. Vanderbilt. Soft self-consistent pseudopotentials in a generalized eigenvalue formalism. *Physical Review B*, 41:7892–7895, 1990.

[1549] D. Vanderbilt and J. D. Joannopoulos. Off-diagonal occupation numbers in local-density theory. *Physical Review B*, 26:3203–3210, 1982.

[1550] D. Vanderbilt and S. G. Louie. Total energies of diamond (111) surface reconstructions by a linear combination of atomic orbitals method. *Physical Review B*, 30:6118–6130, 1984.

[1551] J. VandeVondele and J. Hutter. An efficient orbital transformation method for electronic structure calculations. *Journal of Chemical Physics*, 118:4365–4369, 2003.

[1552] J. VandeVondele, M. Krack, F. Mohamed, M. Parrinello, T. Chassaing, and J. Hutter. Quickstep: Fast and accurate density functional calculations using a mixed Gaussian and plane waves approach. *Computer Physics Communications*, 167:103–128, 2005.

[1553] J. VandeVondele, F. Mohamed, M. Krack, J. Hutter, M. Sprik, and M. Parrinello. The influence of temperature and density functional models in *ab initio* molecular dynamics simulation of liquid water. *Journal of Chemical Physics*, 122:014515, 2005.

[1554] J. VandeVondele and U. Röthlisberger. Efficient multidimensional free energy calculations for *ab initio* molecular dynamics using classical bias potentials. *Journal of Chemical Physics*, 113:4863–4868, 2000.

[1555] J. VandeVondele and U. Röthlisberger. Accelerating rare reactive events by means of a finite electronic temperature. *Journal of the American Chemical Society*, 124:8163–8171, 2002.

[1556] J. VandeVondele and U. Röthlisberger. Canonical adiabatic free energy sampling (CAFES): A novel method for the exploration of free energy surfaces. *Journal of Physical Chemistry B*, 106:203–208, 2002.

[1557] J. VandeVondele and M. Sprik. A molecular dynamics study of the hydroxyl radical in solution applying self-interaction-corrected density functional methods. *Physical Chemistry Chemical Physics*, 7:1363–1367, 2005.

[1558] J. VandeVondele and A. De Vita. First-principles molecular dynamics of metallic systems. *Physical Review B*, 60:13241–13244, 1999.

[1559] VASP: Vienna ab-initio simulation package. Refs. [785, 786]; See http://cms.mpi.univie.ac.at/vasp/.

[1560] P. Vassilev, C. Hartnig, M. T. M. Koper, F. Frechard, and R. A. van Santen. *Ab initio* molecular dynamics simulation of liquid water and water–vapor interface. *Journal of Chemical Physics*, 115:9815–9820, 2001.

[1561] P. Vassilev, M. T. M. Koper, and R. A. van Santen. *Ab initio* molecular dynamics of hydroxyl–water coadsorption on Rh(111). *Chemical Physics Letters*, 359:337–342, 2002.

[1562] P. Vélez, S. A. Dassie, and E. P. M. Leiva. First principles calculations of mechanical properties of 4,4′-bipyridine attached to au nanowires. *Physical Review Letters*, 95:045503, 2005.

[1563] M. V. Vener, O. Kühn, and J. Sauer. The infrared spectrum of the $O\cdots H\cdots O$ fragment of $H_5O_2^+$: Ab initio classical molecular dynamics and quantum 4D model calculations. *Journal of Chemical Physics*, 114:240–249, 2001.

[1564] M. V. Vener and J. Sauer. Quantum anharmonic frequencies of the $O\cdots H\cdots O$ fragment of the $H_5O_2^+$ ion: A model three-dimensional study. *Chemical Physics Letters*, 312:591–597, 1999.

[1565] M. V. Vener and J. Sauer. Environmental effects on vibrational proton dynamics in $H_5O_2^+$: DFT study on crystalline $H_5O_2^+ClO_4^-$. *Physical Chemistry Chemical Physics*, 7:258–263, 2005.

[1566] M. Di Ventra, S. T. Pantelides, and N. D. Lang. Current-induced forces in molecular wires. *Physical Review Letters*, 88:046801, 2002.

[1567] L. Verlet. Computer "experiments" on classical fluids. I. Thermodynamical properties of Lennard-Jones molecules. *Physical Review*, 159:98–103, 1967.

[1568] P. Vidossich and P. Carloni. Binding of phosphinate and phosphonate inhibitors to aspartic proteases: A first-principles study. *Journal of Physical Chemistry B*, 110:1437–1442, 2006.

[1569] A. Viel, R. P. Krawczyk, U. Manthe, and W. Domcke. Photoinduced dynamics of ethene in the N, V and Z valence states: A six-dimensional nonadiabatic quantum dynamics investigation. *Journal of Chemical Physics*, 120:11000–11010, 2004.

[1570] U. von Barth and L. Hedin. Local exchange-correlation potential for spin polarized case 1. *Journal of Physics C*, 5:1629–1672, 1972.

[1571] O. A. von Lilienfeld, R. D. Lins, and U. Röthlisberger. Variational particle number approach for rational compound design. *Physical Review Letters*, 95:153002, 2005.

[1572] O. A. von Lilienfeld, I. Tavernelli, U. Röthlisberger, and D. Sebastiani. Optimization of effective atom centered potentials for London dispersion forces in density functional theory. *Physical Review Letters*, 93:153004, 2004.

[1573] O. A. von Lilienfeld, I. Tavernelli, U. Röthlisberger, and D. Sebastiani. Variational optimization of effective atom centered potentials for molecular properties. *Journal of Chemical Physics*, 122:014113, 2005.

[1574] T. von Rosenvinge, M. Parrinello, and M. L. Klein. *Ab initio* molecular dynamics study of polyfluoride anions. *Journal of Chemical Physics*, 107:8012–8019, 1997.

[1575] T. von Rosenvinge, M. E. Tuckerman, and M. L. Klein. *Ab initio* molecular dynamics study of crystal hydrates of HCl including path integral results. *Faraday Discussions*, 106:273–289, 1997.

[1576] G. A. Voth. Path-integral centroid methods in quantum statistical mechanics and dynamics. *Advances in Chemical Physics*, 93:135–218, 1996.

[1577] I. Štich. First-principles finite-temperature simulation of surface dynamics: Si(111)-(7×7). *Surface Science*, 368:152–162, 1996.

[1578] I. Štich, R. Car, M. Parrinello, and S. Baroni. Conjugate gradient minimization of the energy functional: A new method for electronic structure calculation. *Physical Review B*, 39:4997–5004, 1989.

[1579] I. Štich, J. D. Gale, K. Terakura, and M. C. Payne. Dynamical observation of the catalytic activation of methanol in zeolites. *Chemical Physics Letters*, 283:402–408, 1998.

[1580] I. Štich, J. D. Gale, K. Terakura, and M. C. Payne. Role of the zeolitic environment in catalytic activation of methanol. *Journal of the American Chemical Society*, 121:3292–3302, 1999.

[1581] I. Štich, D. Marx, M. Parrinello, and K. Terakura. Proton-induced plasticity in hydrogen clusters. *Physical Review Letters*, 78:3669–3672, 1997.

[1582] I. Štich, D. Marx, M. Parrinello, and K. Terakura. Protonated hydrogen clusters. *Journal of Chemical Physics*, 107:9482–9492, 1997.

[1583] R. Vuilleumier and M. Sprik. Electronic properties of hard and soft ions in solution: Aqueous Na^+ and Ag^+ compared. *Journal of Chemical Physics*, 115:3454–3468, 2001.

[1584] R. Vuilleumier and M. Sprik. Electronic control of reactivity using density functional perturbation methods. *Chemical Physics Letters*, 365:305–312, 2002.

[1585] U. V. Waghmare, H. Kim, I. J. Park, N. Modine, P. Maragakis, and E. Kaxiras. HARES: An efficient method for first-principles electronic structure calculations of complex systems. *Computer Physics Communications*, 137:341–360, 2001.

[1586] B. G. Walker, C. Molteni, and N. Marzari. *Ab initio* molecular dynamics of metal surfaces. *Journal of Physics: Condensed Matter*, 16:S2575–S2596, 2004.

[1587] D. S. Wallace, A. M. Stoneham, W. Hayes, A. J. Fisher, and A. H. Harker. Theory of defects in conducting polymers. 1. Theoretical principles and simple applications. *Journal of Physics: Condensed Matter*, 3:3879–3904, 1991.

[1588] D. S. Wallace, A. M. Stoneham, W. Hayes, A. J. Fisher, and A. Testa. Theory of defects in conducting polymers. 2. Application to polyacetylene. *Journal of Physics: Condensed Matter*, 3:3905–3920, 1991.

[1589] C. Wang and Q.-M. Zhang. Amphoteric charge states and diffusion barriers of hydrogen in GaAs. *Physical Review B*, 59:4864–4868, 1999.

[1590] I. S. Y. Wang and M. Karplus. Dynamics of organic reactions. *Journal of the American Chemical Society*, 95:8160–8164, 1973.

[1591] J. Wang, J. D. Gu, and A. M. Tian. The mechanisms of the thermal decomposition of 5-nitro-1-hydrogen-tetrazole: *Ab initio* MD and quantum chemistry studies. *Chemical Physics Letters*, 351:459–468, 2002.

[1592] S. W. Wang, S. J. Mitchell, and P. A. Rikvold. *Ab initio* Monte Carlo simulations for finite-temperature properties: Application to lithium clusters and bulk liquid lithium. *Computational Materials Science*, 29:145–151, 2004.

[1593] X. F. Wang, S. Scandolo, and R. Car. Carbon phase diagram from *ab initio* molecular dynamics. *Physical Review Letters*, 95:185701, 2005.

[1594] G. H. Wannier. The structure of electronic excitation levels in insulating crystals. *Physical Review*, 52:191–197, 1937.

[1595] M. C. Warren, G. J. Ackland, B. B. Karki, and S. J. Clark. Phase transitions in silicate perovskites from first principles. *Mineralogial Magazine*, 62:585–598, 1998.

[1596] A. Warshel and M. Karplus. Semiclassical trajectory approach to photoisomerization. *Chemical Physics Letters*, 32:11–17, 1975.

[1597] A. Warshel and M. Levitt. Theoretical studies of enzymic reactions: Dielectric, electrostatic and steric stabilization of the carbonium ion in the reaction of lysozyme. *Journal of Molecular Biology*, 103:227–249, 1976.

[1598] N. Watanabe and M. Tsukada. Efficient method for simulating quantum electron dynamics under the time-dependent Kohn–Sham equation. *Physical Review E*, 65:036705, 2002.

[1599] R. O. Weht, J. Kohanoff, D. A. Estrin, and C. Chakravarty. An *ab initio* path integral Monte Carlo simulation method for molecules and clusters: Application to Li_4 and Li_5^+. *Journal of Chemical Physics*, 108:8848–8858, 1998.

[1600] D. Wei, H. Guo, and D. R. Salahub. Conformational dynamics of an alanine dipeptide analog: An *ab initio* molecular dynamics study. *Physical Review E*, 64:011907, 2002.

[1601] D. Wei and D. R. Salahub. Hydrated proton clusters: Ab initio molecular dynamics simulation and simulated annealing. *Journal of Chemical Physics*, 106:6086–6094, 1997.

[1602] S. Q. Wei and M. Y. Chou. Wavelets in self-consistent electronic structure calculations. *Physical Review Letters*, 76:2650–2653, 1996.

[1603] F. Weich, J. Widany, and Th. Frauenheim. Paracyanogen-like structures in high-density amorphous carbon nitride. *Carbon*, 37:545–548, 1999.

[1604] M. Weinert and J. W. Davenport. Fractional occupations and density-functional energies and forces. *Physical Review B*, 45:13709–13712, 1992.

[1605] U. Weiss. *Quantum Dissipative Systems*. World Scientific, Singapore, 1999.

[1606] M. Weissbluth. *Atoms and Molecules*. Academic Press, New York, 1978.

[1607] S. Wengert, R. Nesper, W. Andreoni, and M. Parrinello. Ionic diffusion in a ternary superionic conductor: An *ab initio* molecular dynamics study. *Physical Review Letters*, 77:5083–5085, 1996.

[1608] R. M. Wentzcovitch. Invariant molecular-dynamics approach to structural phase-transitions. *Physical Review B*, 44:2358–2361, 1991.

[1609] R. M. Wentzcovitch. Hcp-to-bcc pressure-induced transition in Mg simulated by *ab initio* molecular-dynamics. *Physical Review B*, 50:10358–10361, 1994.

[1610] R. M. Wentzcovitch, C. da Silva, J. R. Chelikowsky, and N. Binggeli. A new phase and pressure induced amorphization in silica. *Physical Review Letters*, 80:2149–2152, 1998.

[1611] R. M. Wentzcovitch and J. L. Martins. First principles molecular dynamics of Li - Test of a new algorithm. *Solid State Communications*, 78:831–834, 1991.

[1612] R. M. Wentzcovitch, J. L. Martins, and P. B. Allen. Energy versus free-energy conservation in first-principles molecular dynamics. *Physical Review B*, 45:11372–11374, 1992.

[1613] R. M. Wentzcovitch, J. L. Martins, and G. D. Price. *Ab initio* molecular dynamics with variable cell shape: Application to $MgSiO_3$. *Physical Review Letters*, 70:3947–3950, 1993.

[1614] R. M. Wentzcovitch and G. D. Price. High pressure studies of mantle minerals by *ab initio* variable cell shape molecular dynamics. In B. Silvi and P. D'Arco, eds., *Molecular Engineering*, pages 39–61. Kluwer, Dordrecht, 1996.

[1615] J. A. White and D. M. Bird. Implementation of gradient-corrected exchange-correlation potentials in Car–Parrinello total-energy calculations. *Physical Review B*, 50:4954–4957, 1994.

[1616] J. A. White, E. Schwegler, G. Galli, and F. Gygi. The solvation of Na^+ in water: First-principles simulations. *Journal of Chemical Physics*, 113:4668–4673, 2000.

[1617] S. R. White, J. W. Wilkins, and M. P. Teter. Finite-element method for electronic-structure. *Physical Review B*, 39:5819–5833, 1989.

[1618] J. Wiggs and H. Jónsson. A parallel implementation of the Car–Parrinello method by orbital decomposition. *Computer Physics Communications*, 81:1–18, 1994.

[1619] J. Wiggs and H. Jónsson. A hybrid decomposition parallel implementation of the Car–Parrinello method. *Computer Physics Communications*, 87:319–340, 1995.

[1620] A. Willetts and N. C. Handy. Dynamic optimization of molecular wave-functions and geometries. *Chemical Physics Letters*, 227:194–200, 1994.

[1621] E. B. Wilson Jr., J. C. Decius, and P. C. Cross. *Molecular Vibrations: The Theory of Infrared and Raman Vibrational Spectra*. Dover, New York, 1980.

[1622] E. Wimmer. Computational materials design with first-principles quantum-mechanics. *Science*, 269:1397–1398, 1995.

[1623] K. Wolf, W. Mikenda, E. Nusterer, K. Schwarz, and C. Ulbricht. Proton transfer in malonaldehyde: An *ab initio* projector augmented wave molecular dynamics study. *Chemistry - A European Journal*, 4:1418–1427, 1998.

[1624] T. K. Woo, P. E. Blöchl, and T. Ziegler. Monomer capture in Brookhart's Ni(II) diimine olefin polymerization catalyst: Static and dynamic quantum mechanics/molecular mechanics study. *Journal of Physical Chemistry A*, 104:121–129, 2000.

[1625] T. K. Woo, P. E. Blöchl, and T. Ziegler. Towards solvation simulations with a combined *ab initio* molecular dynamics and molecular mechanics approach. *Journal of Molecular Structure-THEOCHEM*, 506:313–334, 2000.

[1626] T. K. Woo, L. Cavallo, and T. Ziegler. Implementation of the IMOMM methodology for performing combined QM/MM molecular dynamics simulations and frequency calculations. *Theoretical Chemistry Accounts*, 100:307–313, 1998.

[1627] T. K. Woo, P. Margl, P. E. Blöchl, and T. Ziegler. Sampling phase space by a combined QM/MM *ab initio* Car–Parrinello molecular dynamics method with different (multiple) time steps in the quantum mechanical (QM) and molecular mechanical (MM) domains. *Journal of Physical Chemistry A*, 106:1173–1182, 2002.

[1628] T. K. Woo, P. M. Margl, P. E. Blöchl, and T. Ziegler. A combined Car–Parrinello QM/MM implementation for *ab initio* molecular dynamics simulations of extended systems: Application to transition metal catalysis. *Journal of Physical Chemistry B*, 101:7877–7880, 1997.

[1629] T. K. Woo, P. M. Margl, L. Deng, L. Cavallo, and T. Ziegler. Towards more realistic computational modeling of homogenous catalysis by density functional theory: Combined QM/MM and *ab initio* molecular dynamics. *Catalysis Today*, 50:479–500, 1999.

[1630] T. K. Woo, P. M. Margl, T. Ziegler, and P. E. Blöchl. Static and *ab initio* molecular dynamics study of the titanium(IV)-constrained geometry catalyst $(CpSiH_2NH)Ti-R^+$. 2. Chain termination and long chain branching. *Organometallics*, 16:3454–3468, 1997.

[1631] D. M. Wood and A. Zunger. A new method for diagonalizing large matrices. *Journal of Physics A*, 18:1343–1359, 1985.

[1632] G. A. Worth, P. Hunt, and M. A. Robb. Nonadiabatic dynamics: A comparison of surface hopping direct dynamics with quantum wavepacket calculations. *Journal of Physical Chemistry A*, 107:621–631, 2003.

[1633] G. A. Worth and M. A. Robb. Applying direct molecular dynamics to non-adiabatic systems. *Advances in Chemical Physics*, 124:355–431, 2002.

[1634] K. Wright, I. H. Hillier, M. A. Vincent, and G. Kresse. Dissociation of water on the surface of galena (PbS): A comparison of periodic and cluster models. *Journal of Chemical Physics*, 111:6942–6946, 1999.

[1635] Y. D. Wu, C. J. Mundy, M. E. Colvin, and R. Car. On the mechanisms of OH radical induced DNA-base damage: A comparative quantum chemical and Car–Parrinello molecular dynamics study. *Journal of Physical Chemistry A*, 108:2922–2929, 2004.

[1636] R. E. Wyatt. *Quantum Dynamics with Trajectories - Introduction to Quantum Hydrodynamics*. Springer, Berlin, 2005.

[1637] K. Yabana and G. F. Bertsch. Time-dependent local-density approximation in real time. *Physical Review B*, 54:4484–4487, 1996.

[1638] K. Yabana and G. F. Bertsch. Time-dependent local-density approximation in real time: Application to conjugated molecules. *International Journal of Quantum Chemistry*, 75:55–66, 1999.

[1639] K. Yamaguchi and T. Mukoyama. Wavelet representation for the solution of radial Schrödinger equation. *Journal of Physics B*, 29:4059–4071, 1996.

[1640] H. Yamataka, M. Aida, and M. Dupuis. One transition state leading to two product states: *Ab initio* molecular dynamics simulations of the reaction of formaldehyde radical anion and methyl chloride. *Chemical Physics Letters*, 300:583–587, 1999.

[1641] H. Yamataka, M. Aida, and M. Dupuis. Analysis of borderline substitution/electron transfer pathways from direct *ab initio* MD simulations. *Chemical Physics Letters*, 353:310–316, 2002.

[1642] D. R. Yarkony. Diabolical conical intersections. *Reviews of Modern Physics*, 68:985–1013, 1996.

[1643] D. A. Yarne, M. E. Tuckerman, and M. L. Klein. Structural and dynamical behavior of an azide anion in water from *ab initio* molecular dynamics calculations. *Chemical Physics*, 258:163–169, 2000.

[1644] D. A. Yarne, M. E. Tuckerman, and G. J. Martyna. A dual length scale method for plane-wave-based, simulation studies of chemical systems modeled using mixed *ab initio*/empirical force field descriptions. *Journal of Chemical Physics*, 115:3531–3539, 2001.

[1645] O. V. Yazyev and L. Helm. Hyperfine interactions in aqueous solution of Cr^{3+}: an *ab initio* molecular dynamics study. *Theoretical Chemistry Accounts*, 115:190–195, 2006.

[1646] O. V. Yazyev, I. Tavernelli, L. Helm, and U. Röthlisberger. Core spin-polarization correction in pseudopotential-based electronic structure calculations. *Physical Review B*, 71:115110, 2005.

[1647] L. Ye and H.-P. Cheng. A quantum molecular dynamics study of the properties of NO^+ $(H_2O)_n$ clusters. *Journal of Chemical Physics*, 108:2015–2023, 1998.

[1648] M. T. Yin and M. L. Cohen. Theory of lattice-dynamical properties of solids - Application to Si and Ge. *Physical Review B*, 26:3259–3272, 1982.

[1649] B. Yoon, H. Häkkinen, U. Landman, A. S. Wörz, J.-M. Antonietti, S. Abbet, K. Judai, and U. Heiz. Charging effects on bonding and catalyzed oxidation of CO on Au_8 clusters on MgO. *Science*, 307:403–407, 2005.

[1650] Y.-G. Yoon, B. G. Pfrommer, F. Mauri, and S. G. Louie. NMR chemical shifts in hard carbon nitride compounds. *Physical Review Letters*, 80:3388–3391, 1998.

[1651] A. H. Zewail. *Femtochemistry: Ultrafast Dynamics of the Chemical Bond*. World Scientific, Singapore, 1994.

[1652] C. Zhang, M.-H. Du, H.-P. Cheng, X.-G. Zhang, A. E. Roitberg, and J. L. Krause. Coherent electron transport through an azobenzene molecule: A light-driven molecular switch. *Physical Review Letters*, 92:158301, 2004.

[1653] C. Zhang and P. J. D. Lindan. Multilayer water adsorption on rutile $TiO_2(110)$: A first-principles study. *Journal of Chemical Physics*, 118:4620–4630, 2003.

[1654] C. Zhang and P. J. D. Lindan. Towards a first-principles picture of the oxide–water interface. *Journal of Chemical Physics*, 119:9183–9190, 2003.

[1655] C. J. Zhang, P. Hu, and A. Alavi. A density functional theory study of CO oxidation on Ru(0001) at low coverage. *Journal of Chemical Physics*, 112:10564–10570, 2000.

[1656] C. J. Zhang, P. J. Hu, and A. Alavi. A general mechanism for CO oxidation on close-packed transition metal surfaces. *Journal of the American Chemical Society*, 121:7931–7932, 1999.

[1657] Y. K. Zhang and W. T. Yang. Comment on "Generalized gradient approximation made simple". *Physical Review Letters*, 80:890–890, 1998. Original article: Ref. [1131] and Reply: Ref. [1133].

[1658] G. Zhao, C. S. Liu, and Z. G. Zhu. *Ab initio* molecular-dynamics simulations of the structural properties of liquid $In_{20}Sn_{80}$ in the temperature range 798–1193 K. *Physical Review B*, 73:024201, 2005.

[1659] X. Y. Zhao, D. Ceresoli, and D. Vanderbilt. Structural, electronic, and dielectric properties of amorphous ZrO_2 from *ab initio* molecular dynamics. *Physical Review B*, 71:085107, 2005.

[1660] Y.-J. Zhao, M. Jiang, G. M. Lai, and P.-L. Cao. *Ab initio* molecular dynamics study of sulfur adsorption on Si(100)2×1. *Physics Letters A*, 255:361–363, 1999.

[1661] B. B. Zhou and R. P. Brent. On parallel implementation of the one-sided Jacobi algorithm for singular value decompositions. In *Proceedings of Euromicro Workshop on Parallel and Distributed Processing (San Remo, Italy, January 25–27, 1995)*, page 401. IEEE Computer Society Press, 1995.

[1662] Z. Zhu and M. E. Tuckerman. *Ab initio* molecular dynamics investigation of the concentration dependence of charged defect transport in basic solutions via calculation of the infrared spectrum. *Journal of Physical Chemistry B*, 106:8009–8018, 2002.

[1663] W. Zhuang and C. Dellago. Dissociation of hydrogen chloride and proton transfer in liquid glycerol: An *ab initio* molecular dynamics study. *Journal of Physical Chemistry B*, 108:19647–19656, 2004.

[1664] C. M. Zicovich-Wilson, A. Bert, C. Roetti, R. Dovesi, and V. R. Saunders. Characterization of the electronic structure of crystalline compounds through their localized Wannier functions. *Journal of Chemical Physics*, 116:1120–1127, 2002.

[1665] T. Ziegler, A. Rauk, and E. J. Baerends. Calculation of multiplet energies by Hartree–Fock–Slater method. *Theoretica Chimica Acta*, 43:261–271, 1977.

[1666] F. Zipoli, M. Bernasconi, and A. Laio. *Ab initio* simulations of Lewis-acid-catalyzed hydrosilylation of alkynes. *ChemPhysChem*, 6:1772–1775, 2005.

[1667] F. Zong and D. M. Ceperley. Path integral Monte Carlo calculation of electronic forces. *Physical Review E*, 58:5123–5130, 1998.

[1668] G. Zumbach, N. A. Modine, and E. Kaxiras. Adaptive coordinate, real-space electronic structure calculations on parallel computers. *Solid State Communications*, 99:57–61, 1996.

[1669] E. Zurek, O. Jepsen, and O. K. Andersen. Muffin-tin orbital Wannier-like functions for insulators and metals. *ChemPhysChem*, 6:1934–1942, 2005.

Index

ab initio molecular dynamics (AIMD), *see*
    molecular dynamics: *ab initio* (AIMD)

ABINIT software package, 7, 416
accuracy
    Car–Parrinello propagation, 37, 45, 58, 60,
      66
    density functional, 71, 100
    dipole calculation, 342
    electronic structure, 67
    excited-state forces, 203
    finite differences, 52, 314
    Lagrange multiplier, 126
    link atom, 280
    liquid water, 50, 382
    metadynamics sampling, 193
    non-self-consistency, 54
    plane wave expansion, 88
    Poisson solver, 99
    pseudopotential, 91, 138, 154, 157, 170, 286
    structure optimization, 100
    thermostat integration, 181
    vibrational spectra, 45, 278
    Wannier localization, 340–342
acetic acid, 377
acetone, 385, 387, 402, 403
acetonitrile, 384, 387, 401, 403
acetylene, 376, 394, 399
activation energy/barrier, 189
adiabaticity, 37, 39
    centroid, 255
    controlling, 36, 301
    parameter, 30, 37, 43
    path integral Car–Parrinello, 248
AFM (Atomic Force Microscopy), 380, 381,
    414
Ag, 396
    (111) surface, 379
Ag_2Se, 374
AIMD (*ab initio* molecular dynamics), *see*
    molecular dynamics: *ab initio* (AIMD)
Al
    (110) surface, 217, 376, 377

(111) surface, 379
cluster, 393
Al_2O_3, 376
alanine, 404
ALDA, *see* functional: Adiabatic Local
    Density Approximation (ALDA)
algorithm, *see also* diagonalization,
    propagation
    SHAKE/RATTLE, 120, 125, 182
    Born–Oppenheimer, *see also* dynamics:
      Born–Oppenheimer
    Car–Parrinello, *see* Car–Parrinello
    communication, 350, 353
    Direct Inversion in the Iterative Subspace
      method (DIIS), 116, 118, 126
    dissipation/friction, 127
    Edmiston–Ruedenberg localization, 337
    generalized/quasi-Newton, 118, 126
    Hockney, 96
    load-balancing, 353
    metadynamics, 186, 190
    multiple time step, 57, 122, 273, 278
    orbital-dependent functionals, 201, 202
    preconditioning, 118
    temperature control, 120, 179, 183
    Tully surface-hopping/fewest switches, *see*
      molecular dynamics: surface-hopping,
      molecular dynamics: Tully, transition:
      probability
    velocity Verlet, 120–127, 182, 300, 305
    Wannier localization, 335–340
    Wannier propagation, 340
alkali, 162, 375
alkanes, 397
all-electron treatment, *see* electron:
    all-electron
AMBER software package, 96, 268, 271, 273,
    278, 283, 405
aminocoumarins, 385, 387, 403, 406
ammonia, 58, 377, 386, 391, 395, 397
    dimer, 386
amorphization, 372, 399
AMP, 403

annealing, simulated, **42**, **127**, 389, 394, 395
antimony pentafluoride, 388
approximation
  adiabatic, **13**, 204, 235, 239, 241
  beyond Born–Oppenheimer, 13, 204, 211,
    401, 406
  Born–Oppenheimer, **14**, 28, 204, 234, 235,
    241, 243
  clamped nuclei, 12, 20, 234, 243, 256
  distinguishable particles, 234, 244
  frozen-core, 136, 160, 296, 326, 410
  harmonic, 32, 36
  isolated atom, 48
  Massey criterion, 204
  one-determinant, 17
  pseudopotential, 78, 90
  rigid ion, 48
  semiempirical, 26
aquaporin, 405
argon, 22, 390
Arrhenius, 374
arsenic, 388
ATP, 405
Au, 392
  (111) surface, 375, 380
  cluster, 378
  film, 378
  wire, 381
Aufbau principle, 114
Auger decay, 413
augmentation charge, 286, 294
augmented basis set, *see* basis set:
  augmented/mixed
azobenzene, 382, 400

Bachelet–Hamann–Schlüter–Chiang
  pseudopotential, *see* pseudopotential:
  norm-conserving
bacteriochlorophyll, 396, 403
bacteriorhodopsin, 383, 395, 405
bandwidth, *see* communication: bandwidth
BaO
  (100) surface, 378
barostat, *see* pressure: barostat, pressure:
  control, pressure: stress
basis set
  adiabatic, 20, 204, 236
  augmented/mixed, 82, 235, 236
  Discrete Variable Representation (DVR),
    80, 82
  Gaussian, 56, 75–77, 81, 120
  generalized plane wave, 79
  plane wave, 77, 86, 87
  real space grids, 83
  Slater, 75, 114, 408
  superposition error (BSSE), 54, 81, 186, 187
  Wannier, 83, 226, 327, 408
  wavelet, 80, 81
Basis Set Superposition Error, *see* basis set:
  superposition error (BSSE)
bath, 178, 181, 267

benzene, 266, 372, 375, 382, 395, 415
benzoquinone, 385
benzosemiquinone, 412
Berry phase
  adiabatic approximation, 14
  first-order variation, 227
  formulation of polarization, 222, 226, **317**,
    410
Bessel transform/function, 92, 98, 151, 154,
   161
bias potential sampling, 179, 189
BLAS library, 130–132, 363
Bloch
  function, *see* Kohn–Sham: Bloch orbital,
    orbital: Bloch
  orbital, *see* Kohn–Sham: Bloch orbital,
    orbital: Bloch
  theorem, 77, 87, 327, 329
Blue Moon sampling method, **124**, 179, 189,
   396, 400, *see also* constraint, ensemble,
   free energy calculation, rare event,
   thermodynamic integration
Born–Oppenheimer
  approximation, *see* approximation:
    Born–Oppenheimer
  molecular dynamics, *see* molecular
    dynamics: Born–Oppenheimer
  potential, *see* potential: Born–Oppenheimer
  potential energy surface, 16, 20, 33, 39, 42,
    46
  quantum dynamics, *see* dynamics:
    Born–Oppenheimer quantum
boundary condition
  Born-von Kármán, 331
  Born-von Kármán, *see* boundary condition:
    periodic
  cluster, 96, 98, 188, 272, 376
  cyclic Trotter/imaginary time, 237, 241,
    245, 251
  periodic, 77, **86**, 95, 106, 184, 188, 228, 278,
    328
  periodic DVR, 81
  periodic position operator, 261, 317, 319,
    324, 327–331, 334
  Poisson equation, 96–99, 110
  QM/MM interface, 267–270, 277, 282, 283
  slab, 95, 97, 98, 376, 380
  stochastic droplet, 278
  wire, 97, 98
Bravais lattice, 85, 184
Brillouin theorem, 201
Brillouin zone
  Γ-point approximation, 106, 187, 213, 226,
    317, 330
  integration, **88**, 106, 215
  parallelization, 351
  sampling/special points, 38, 79, **88**, 105,
    186, 215, 331
  size estimate, 186
  vector, 87, 331
broadcast, *see* communication: broadcast

BSSE, *see* basis set: superposition error (BSSE)
butadiene, 201, 398, 401

Ca(OH)$_2$, 372
CADPAC software package, 43
capping QM/MM atom, *see* QM/MM coupling scheme: capping/link atom
Car–Parrinello
  adiabaticity, 29, 31, 34, 37, 46, 59, 64
  conserved energy, **33**, 65, 182
  conserved linear momentum, 183
  constraints, 30, 65
  dissipation/friction, 42, 127
  dynamics, 34, 35, 39, 41, 55–58, 63, 67
  equations of motion, **30**, 40, 47, 120, 180
  Euler–Lagrange equations, **29**, 40, 44, 120
  fictitious electron mass effect, 34–41, 45–50, 64
  fictitious electron mass parameter, 30, 36–43, 45, 48, 57, 64, 120, 123
  fictitious kinetic energy, 33, 42, 46, 55, 63, 64, 181
  forces, 35, 47, 51, 65
  H mass substitution by D, 37
  Hellmann–Feynman force, 52
  Lagrangian, 3, **29**, **30**, 38, 40, 44, 119, 120, 185, 188, 247, 282, 340
  mathematical proof, 31, 39, 46
  metadynamics, *see* metadynamics sampling
  metallic system, 37, 38
  metastability, 31, 39
  non-self-consistency force, 53–55, 105
  PAW equations of motion, 297
  Pulay force, 53
  stability, 34, 38–41, 55, 63
  temperature, *see* temperature: Car–Parrinello renormalization
  time scales, 33, 57, 60
carbon, 138, 167, 168, 170, 171, 277, 393
  amorphous, 389
  phase diagram, 391
carbon monoxide, 391
carbon nitride, 390
carbonylation, 399
carboplatin, 396
carboxyl, 278
CASSCF (Complete Active Space Self-Consistent Field), 67, 74, 194, 396
CASTEP software package, 7, 416
catalysis
  heterogeneous, 177, 376, 399
  homogeneous, 271, 399
CdTe, 387–389
  liquid, 387
cefotaxime, 404
centroid, *see also* path integral
  ab initio molecular dynamics, *see* molecular dynamics: *ab initio* centroid
  adiabaticity, 255
  coordinate, 249

correlation function, 249, 250, 265
  dipole moment, 265
  force, 253
  mass, 251, 252, 255
  mode, 252
  molecular dynamics, *see* molecular dynamics: centroid
  potential, 250, 255
  thermostatting, 254
ceramic, 380
charge
  atomic, 347, 348
  augmentation, 286, 294
  core, 130, 136–161, 296
  Davidson partitioning, 408
  density, **38**, **94**, 184, 213
  density from Kohn–Sham orbitals, 68
  density in DVR basis, 82
  density model, 276
  density, expansion, 87
  density, generalized, 166
  effective point, 272, 273, 279, 280, 284, 345
  excited-state density, 284
  Gaussian smearing, 93, 271, 272, 274, 279, 283
  input/output density, 105
  Löwdin partitioning, 408
  linear response/perturbed density, 223, 312
  Mayer partitioning, 114, 408
  Mulliken partitioning, 114, 408
  multipole moments, 82, 96, 272, 276, 278, 279, 347
  neutralizing background, 95, 272
  nuclear, 93, 137
  periodic system, 95, 272
  projector augmented-wave (PAW) density, 292
  pseudized core density, 161
  pseudo density, 114, 139, 142, 343
  R/ESP, 276, 347
  semi-core, 160
  sloshing, 38
  time-dependent density, 219
  ultrasoft pseudopotential (USPP) density, 293, 294
  valence, 136–161, 343
  Wannier, 345, 347
CHARMM software package, 268, 278, 280
chemical flooding method, 179, 189
chemical potential, *see* potential: chemical
chemical shielding, *see* NMR
chemical shift, *see* NMR
chemisorption, 375, 376, 378, 381
chlorine, 377
chlorophenol, 397
chromophore, 284, 346, 402, 403, 406
cisplatin, 404
Cl$_2$, 377
classical limit, 14–18, 210, 247, 250–263, 267
classical mechanics, 11, 15, 16, 18, 21, 233, 242
clay, 377

cluster, 77
CNO (Constraint Nonorthogonal Orbital)
    method, *see* constraint: nonorthogonal
    orbital (CNO) method
CO, 376–378, 387, 398, 400
    oxidation, 378
collision reaction, 24, 378, 382, 393
communication
    all-to-all, 364–366
    bandwidth, 249, 350, 355, 364
    broadcast, 354, 355, 361
    collective, 354
    dominance, 361
    global, 361
    latency time, 350, 355, 364–366
    library, 350
    Linpack performance, 364
    load-balancing, 351–355, 359, 360, 366
    overhead, 249, 351, 364, 367
    parallelization, *see* parallelization
    scaling, 359, 364, 366
    synchronization, 351, 354
Compton scattering, 414
conductance, 382
conductivity
    electric, 387, 388, 391–393
    electronic, 382
    Green–Kubo relation, 414
    ionic, 374
conductor, protonic/superionic, 374
CONQUEST software package, 7, 416
constraint, *see also* ensemble, free energy
    calculation, metadynamics sampling, rare
    event, thermodynamic integration
    Blue Moon sampling method, 124, 179, 189,
        396, 400
    geometrical, 124
    holonomic, 30, 189, 190
    nonorthogonal, 65, 301
    nonorthogonal orbital (CNO) method,
        300–307
    orbitals, 25, 26, 29, 38, 65, 120, 301
    orthonormality, 25, 29, 117, 120, 122, 126,
        131, 201, 230, 231
    position-dependent, 56, 65, 301
    position-independent, 30, 66, 120
    reaction coordinate dynamics/ensemble,
        124, 179, 189, 396, 400
    symmetry, 196, 201
    velocity, 122, 126
Constraint Nonorthogonal Orbital (CNO)
    method, *see* constraint: nonorthogonal
    orbital (CNO) method
continuum solvation, *see* reaction field,
    QM/MM, embedding
coordinate
    adaptive, 79
    centroid, 249, 253
    collective, 124, 190
    curvilinear, 79
    cyclic, 319, 321

extended/auxiliary/fictitious, 29, 44, 181,
    185, 191, 192, 246, 247, 254, 282
    normal mode, 248, 251, 253
    primitive, 246, 251, 252
    reaction, 124, 190, 397
    scaled, 86, 184, 332
    staging, 248
    transformation, 79, 80, 86, 184, 251, 252
correlation function
    centroid, 249, 250, 265
    dipole, 45, 49, 262, 263, 265, 346, 410
    Kubo, 250, 262
    quantum, 250, 262
    quantum correction, 263, 264
    velocity, 32, 45, 49, 410
COSMO solvation/electrostatics, 281, 282
Coulomb operator, 74
coupling
    adiabatic/diagonal, 13
    neglect, 14, 24, 204, 232, 233, 401
    nonadiabatic/off-diagonal, 13, 20, 23, 195,
        204–210, 220, 224, 232, 400–402, 406,
        412
CP-PAW software package, 7, 96, 271, 281,
    283, 416
CP2k software package, 7, 271, 416
CPMD software package, 7, 416
cross section, *see* spectra
CSGT method, *see* gauge
Cu, 372, 380
    (100) surface, 376
    (110) surface, 379
    (111) surface, 380
    cluster, 375
cubane, 373
CuCl, 372
CuI, 374
cutoff
    constant-pressure/variable-cell, 53, 186
    density, 88, 133
    effective/smooth plane wave, 187, 188
    Ewald real space, 89
    generalized plane waves, 80
    Green's function, 97, 98
    kinetic energy, 37, 53, 77, 88
    parallelization, 353
    plane wave, 79, 87, 88, 137, 286
    pseudopotential, 137, 160, 167, 169, 170, 286
    pseudopotential generation, 145
    system size scaling, 129, 133
    ultrasoft pseudopotential, 289, 297
cyanines, 202, 401
cyclization, 398
cycloaddition, 398
cyclohexadiene, 201, 401
cysteine, 377
cytochrom P450, 403

DACAPO software package, 7, 416
Davidson diagonalization, *see* diagonalization:
    Davidson

decarboxylase, 405
decoupling
  adiabatic, 31, 46
  periodic images, 96–98, 272, 274
  total wave function, 13, 16, 236
defect, 375, 377–379, 385, 386, 389, 391, 392, 406
dehydrogenation, 377
dehydroxylation, 377
delta-SCF/ΔSCF method, 197
density, *see* charge density
density functional, *see* functional
Density Functional Perturbation Theory (DFPT), 310, 325
Density Functional Theory (DFT), *see* Kohn–Sham
detailed balance condition, 260, 263, 264
determinant
  broken symmetry, 207
  factorization, 330
  Fixman, 124
  fractional occupation, 38, 195, 196, 207, 213, 215–217, 408, *see also* occupation: fractional
  Hartree–Fock, 25, 73
  Kohn–Sham, 68
  microstate, 197–199
  mixed spin, 198, 204
  occupation number, *see* occupation: number
  overlap, 330
  ROKS, 199, 204
  single/mean-field, 17, 25, 73
  Slater, 25, 29, 52, 68, 73, 329, 330
  Slater transition state, 207
  spin density, 199
  spin-restricted, 196, 197
  triplet, 198
deuterium
  high-pressure, 391
  liquid, 391
  substitution for H, 37
DFT (Density Functional Theory), *see* Kohn–Sham
diabatic, 209, 210
diagonal/off-diagonal correction, *see* approximation, coupling, dynamics, gradient, molecular dynamics, TDDFT
diagonalization
  Davidson, 118, 316
  in fix-point optimization, 111, 118
  iterative, 118, 316
  Jacobi, 336, 340
  Lanczos, 118, 215, 316
diamagnetic current, 321, 322
diamond, 138, 170–172, 391
  (111) surface, 379
dielectric continuum solvation, *see* reaction field, QM/MM, embedding
diffusion
  adatom, 377
  confined, 393

defect, 374, 377, 379, 383
  Grotthuss, 379, 383, 398
  hydrogen, 374, 384
  hydroxyl, 383, 398
  ionic, 374, 375
  mechanism, 374, 375, 378, 384
  oxygen, 374
  proton, 374, 383, 398
  self, 374
  surface, 377, 379
diiodomethane, 401
diiron, 405
DIIS (Direct Inversion in the Iterative Subspace) method, *see* algorithm: Direct Inversion in the Iterative Subspace method (DIIS)
dimethyl sulfoxide, 385
dimethylacetylene, 378
dipole, *see also* accuracy: dipole calculation, centroid: dipole moment, correlation function: dipole, infrared (IR) spectra: dipole (E1) approximation, operator: dipole moment, transition: dipole approximation, 218, 222, 260, 261, 317
  moment, **318**, 328, 344–346
  textbf, 330
  moment, molecular, 342, 344, 345, 384, 411
Direct Inversion in the Iterative Subspace (DIIS) method, *see* algorithm: Direct Inversion in the Iterative Subspace method (DIIS)
Discrete Variable Representation (DVR) basis set, *see* basis set: Discrete Variable Representation (DVR)
dissipation, 43, 181
dissociation, 376–378, 381–383, 386, 391, 396–398
  O_2, 379
dithienylethene, 382
DNA, 403–406
doping, 375, 389, 393
drag effect, *see* Car–Parrinello: fictitious electron mass effect
DVR, *see* basis set: Discrete Variable Representation (DVR)
dynamical matrix, *see* Hessian/Hesse matrix
dynamics, *see also* molecular dynamics
  adiabatic quantum, 13
  Born–Oppenheimer, *see* molecular dynamics: Born–Oppenheimer
  Born–Oppenheimer quantum, 14
  Car–Parrinello, *see* Car–Parrinello, molecular dynamics: Car–Parrinello
  classical, 15
  classical path, 19
  classical trajectory, 21
  dissipative, 42, 181, 382, 415
  Ehrenfest, *see* molecular dynamics: Ehrenfest
  Hamilton–Jacobi formulation, 15
  instanton, 266, 395

mean-field, 17, 207, 210
Mott, 19
Newtonian formulation, 16
non-Markovian/history-dependent, 191
nonadiabatic quantum, 13, 177
quantum, 13, 18
quantum corrected, 263
Quantum Molecular Dynamics (QMD), 2
quasiclassical, 234, 249, 263, 411
semiclassical, 14, 15, 266, 395
time-dependent self-consistent field
(TDSCF), 17–19, 22, 23

ECP (Effective Core Potential), *see*
pseudopotential
Effective Core Potential (ECP), *see*
pseudopotential
EGO software package, 275, 278
Ehrenfest
molecular dynamics, *see* molecular
dynamics: Ehrenfest
potential, *see* potential: Ehrenfest
Eigen complex, 383
eigenmode, 251, 315, 316
elastomer, 380
electric field
periodic boundary condition, 414
electrochemistry, 413, 414
electrode, 382, 415
electroluminescence, 396
electron
all-electron, 78, 80, 136, 139, 164, 167,
286–296, 410, 413
core, 78, 90, 136–138, 160, 168, 286
frozen-core, 136, 160, 294, 296, 326, 410
non-interacting, 69, 211, 212, 214, 329, 330
nonlinear core, *see* Nonlinear Core
Correction (NLCC)
pseudization, 78, 90, 136–138, 168, 286, 287
semi-core, 78, 160
valence, 78, 91, 136–138, 160, 168, 286
electron density, *see* charge: density
ELF (Electron Localization Function), 372,
384, 404, 407, 409
ELMD (Extended Lagrangian Molecular
Dynamics), 2, 44
embedding, *see also* QM/MM, reaction field
additive scheme, 267
continuum environment, 270, 271, 281
electron spill-out, 269, 275
electrostatic, 269, 274, 276, 279, 284
excited state, 271, 283–285
general concept, 177, 267, 271
implementation approach, 270
link atom, 267, 268, 270, 277, 278, 280, 281
mechanical, 188, 268, 271, 277
polarized, 269
subtractive scheme, 267
energy
Born–Oppenheimer potential, 16, 20
computation, 109

conservation, 58, 60, 61, 63–66
conserved, 15, 33, 63, 182
Coulomb, 69, 93
cutoff, 37, 53, 77, 88
dissipation, 42, 43, 127, 181
drift, *see* energy: conservation
Ehrenfest potential, 18, 20
electronic, 33, 68, 73, 104
electrostatic, 69, 93, 98, 104
exact exchange, 72, 74, 103, 104
exchange–correlation, 69, 71, 72, 99, 100,
104
excited state, 197, 198, 202, 229, 284
fictitious kinetic, 33, 59, 63
Hartree, 69, 93, 98, 99
Hartree–Fock, 73
kinetic, 104
kinetic energy density, 101
Kohn–Sham, 68, 70
non-interacting kinetic, 69
PAW expression, 296
perturbed, 310, 315, 317, 325
physical total, 33, 34, 59, 63, 64
plane wave expression, 99, 103, 104, 109
pseudopotential, 91, 92, 104, 143, 152
QM/MM continuum embedding, 282
QM/MM electrostatic, 269, 273–277, 279,
284
QM/MM link atom correction, 280, 281
QM/MM partitioning, 267, 268, 274
ripples, 100
self-energy, 94
total, 104, 105
ultrasoft pseudopotential, 293, 294
ensemble, *see also* constraint, free energy
calculation, rare event, thermodynamic
integration
Blue Moon, 124, 179, 189, 396, 400
Boltzmann statistics, 234, 241, 244, 256, 258
Bose–Einstein statistics, 242
canonical, 43, 44, 46, 179, 181, 183, 193,
211, 236, 241, 262, 383
centroid dynamics, 265
constrained, 124, 193
constrained reaction coordinate dynamics,
124, 179, 189, 396, 400
density functional theory, 195–197, 201, 207,
210, 216, 217, 376
Fermi–Dirac statistics, 38, 195, 211, 213,
215, 217, 242, 244
Gibbs, 383
grand canonical, 178, 211, 212, 413
independent trajectories, 208, 233
isobaric, 178, 233
isobaric-isoenthalpic, 186
isobaric-isothermal, 186, 248, 373, 383
isothermal, 44, 178, 183, 233
microcanonical, 34, 41, 44, 177–179, 182,
183, 247, 277
quantum-statistical, 235, 236, 241
transformation, 178

enzyme, 403, 404
epoxide, 393, 398, 400
EPR (Electron Paramagnetic Resonance)
  g-tensor, 412
  hyperfine splitting/coupling, 410, 412
  spectroscopy, 411
ESR (Electron Paramagnetic Resonance)
  hyperfine splitting/coupling, 412
ESR (Electron Spin Resonance), *see* EPR
ethane, protonated, 394
ethanol, 384
ethene, 399
ethylene, 266, 384, 395, 398–400
Ewald summation, 88, 93, 107
exact exchange, *see* functional: exact
  exchange (EXX)
exchange operator, 74
excited state, *see* charge, coupling, delta-SCF
  method, dynamics, embedding, energy,
  molecular dynamics, nonadiabatic, photo,
  potential, QM/MM coupling scheme,
  ROKS, ROSS, TDA, TDDFT
Excited State Proton Transfer (ESPT), 195,
  401
EXX, *see* functional: exact exchange (EXX)

F-center, 58, 378, 382
FCI (Full Configuration Interaction), 67, 74,
  194
Fermi–Pasta–Ulam nonergodicity, 3, 247
Feynman–Kac path integral, 235
Feynman–Vernon influence functional, 235,
  242–244
FFT library, 363
FHI98md software package, 7, 416
fictitious electron mass, *see* Car–Parrinello:
  fictitious electron mass
Fixman determinant, 124
fluctuation-dissipation theorem, 264
fluoroform, 387
Fock operator, 73
force, *see* gradient
force field, 21, 22, 267, 268, 271
formaldehyde, 385, 396, 397
formaldimine, 203, 400, 401
formamide, 386, 402, 412
formic acid, 384, 386, 397, 398
Fourier transform, 134, 357
FPMD (First Principles Molecular Dynamics),
  2
free energy calculation, 124, 179, 189, 192,
  396, 400, *see also* constraint, ensemble,
  metadynamics sampling, rare event,
  thermodynamic integration
friction, 43, 127, 179, 181, 186
frozen-core, *see* approximation: frozen-core,
  electron: frozen-core
Fukui function, 165, 407, 409, 410, *see also*
  hardness
fullerene, 393
functional

Adiabatic Local Density Approximation
  (ALDA), 219, 222
asymptotic behavior, 93, 219
exact exchange (EXX), 72, 103, 104, 203,
  219, 226
finite-temperature, *see* functional: Mermin
Generalized Gradient Approximation
  (GGA), 71, 72, 101, 219, 386
hybrid, 72, 103, 104, 226, 386, 402
hyper-GGA, 72
influence, 235, 242–244
Jacob's Ladder, 72, 101
Janak, 165
Laplacian, 72
Local Density Approximation (LDA), 71,
  72, 101, 144, 313
Local Spin Density (LSD), 71
memory effect/time-nonlocality, 218, 219,
  243
Mermin, 214, 216, 217
meta-GGA, 72, 101, 386
Optimized Effective Potential (OEP), 72,
  217
Optimized Potential Method (OPM), 72
Restricted Open-Shell Kohn–Sham (ROKS),
  see ROKS
Restricted Open-Shell Singlet (ROSS), *see*
  ROSS
Self-Interaction Correction (SIC), 71, 384

Ga, 376, 381
  adatom, 377
GaAs, 373, 374, 388
  (001) surface, 376
  (100) surface, 377
  (110) surface, 377, 379, 381
GaN, 390
GaP
  fullerene, 393
gasoline, 400
gauge
  canonical, 311
  choice, 319, 321, 324, 326
  Continuous Set of Gauge Transformations
    (CSGT) method, 319, 321, 323
  Dirac method, 340
  field theoretic Car–Parrinello Lagrangian,
    340
  Gauge Including Atomic Orbitals (GIAO)
    method, 319, 326
  Gauge Including Projector Augmented
    Wave (GIPAW) method, 326
  Individual Gauges for Localized Orbitals
    (IGLO) method, 319
  invariance, 328
  origin, 321–324, 326
  parallel transport, 311
  periodic, 330
  Wannier, 340, 341
Gaussian basis set, *see* basis set: Gaussian

Gaussian core charge, *see* charge: Gaussian smearing, charge: core
Ge
    (111) surface, 377
generalized plane wave basis set, *see* basis set: generalized plane wave
GeSe$_2$
    liquid, 387
GFP (Green Fluorescent Protein), 402, 403, 406
GGA, *see* functional: Generalized Gradient Approximation (GGA)
GIAO method, *see* gauge
gibbsite
    surface, 378
GIPAW method, *see* gauge
glycerol, 386
glycine, 377, 379, 392, 398, 404
Goedecker pseudopotential, *see* pseudopotential: dual-space Gaussian
gold, 381, 382, 415, *see* Au
gradient
    Basis Set Superposition Error (BSSE), *see* basis set: superposition error
    conjugate (CG) method, 115, 118
    constraint, 30, 65, 124–126, 304
    Constraint Nonorthogonal Orbital (CNO) method, 304
    corrected (GC) density functional, *see* functional: Generalized Gradient Approximation (GGA)
    density, 71, 82, 99, 101, 313
    DVR basis set, 82
    excited state, 195, 207, 211, 221, 224, **229**, 402, 412
    excited state in QM/MM, 284
    excited state in ROKS/ROSS, 202
    force field, 21, 119
    friction, 127
    Hartree–Fock/exact exchange, 72, 103, 104
    Hellmann–Feynman contribution, *see* Hellmann–Feynman
    linear response, 224, 229, 232, 309, 313
    local kinetic energy density (meta-GGA), 72, 101
    long-range force, 95, 96, 99, *see also* Ewald summation
    LR-TDDFT, 195, 221, 224, 229, 232, 402, 412, *see also* TDDFT: linear response gradient and dynamics
    LR-TDDFT in QM/MM, 284
    non-self-consistency force, *see* non-self-consistency force
    nonadiabatic coupling, *see* coupling: nonadiabatic/off-diagonal
    Nonlinear Core Correction (NLCC), 162, *see also* Nonlinear Core Correction (NLCC)
    nonorthogonal wave function, 303
    normal modes/centroid, 253
    nuclear, 52, 53, 106–111, 121
    nuclear for NLCC, 162

    nuclear for ultrasoft pseudopotentials, 295, 304
    nuclear in ROKS/ROSS, 202
    optimization, 115, 118, 127, 339, 414
    perturbation theory, 75, 229, 314–316
    preconditioned, 123
    Pulay contribution, *see* Pulay force/stress
    QM/MM coupling, 277, 280, 281, 284
    self energy contribution, 107
    ultrasoft pseudopotential (USPP), 65, 295–297, 302–304
    ultrasoft pseudopotential wave function, 295
    Wannier localization, 338–342
    wave function, 70, 74, 103–121, 123
grain boundary, 373
gramicidin, 405
graphite
    intercalation, 372
    laser heating, 216, 392
grid
    k-point, 79, 87, 88, 105, 106, 186
        parallelization, 351
    artifact, 78, 84, 100
    augmentation charge, 287
    auxiliary/DVR, 80, 81
    commensurate, 100, 270
    Fourier space, 90
        parallelization, 351
    interpolation, 100, 270
    Laplacian, 83
    Mehrstellen scheme, 84
    multiscale, 79, 82, 84, 270, 287
    non-uniform/curvilinear, 79
    P^{3}M method, 274
    PAW decomposition, 296
    Pulay correction, 79, 84, 186
    radial/angular, 154, 155, 296
    real space, 89, 100, 287
        parallelization, 351, 358
    reciprocal lattice vector, **86**
    resolution, 78, 83, 89, 100, 269, 270
    spacing, 78, 83, 89, 100, 269
    vector, **85**, 327, 332
    vector in Parrinello–Rahman molecular dynamics, 184, 186
GROMOS software package, 268, 274, 278, 404
Grotthuss mechanism, 379, 383, 398
guanine, 405, 406
GVB (Generalized Valence Bond), 65, 67, 74, 194

H, *see* hydrogen
H$_2$O
    adsorbate, 377, 378
    cluster, 383, 395
    dipole moment, 411
    microsolvated cluster, 400
    molecule, 395, 409
    protonated cluster, 395
H$_2$S
    adsorbate, 379

crystal, 372
Hamann-Schlüter-Chiang pseudopotential, *see*
    Pseudopotential: norm-conserving
Hamilton function, 15, 272
Hamiltonian, 18, 25, 28, 52, 124, 137
    effective one-particle, 30, 52, 53, 70, 71, 73
    electronic, 11, 12, 56, 236–238
    first-order perturbed, 312, 321, 325
    Hartree–Fock, 73
    Kohn–Sham, 26, 36, 70, 201, 202, 212
    nuclear, 11, 13, 236–238
    one-particle, 211
    second-order perturbed, 321
    standard molecular, 11, 236
    time-dependent Kohn–Sham, 218, 220, 221
    ultrasoft pseudopotential, 295, 302, 304
hardness, 165–167, 409, *see also* Fukui
    function
harmonic
    analysis, 36, 49, 315, 316, 411, *see also*
        Hessian/Hesse matrix, phonon
    approximation, 32, 36, 280
    constraint potential, 191
    coupling, 240, 245, 247, 254
    diagonalized coupling, 248, 251
    force constant, 310, 315, 316
    frequency, 36, 49, 100, 123, 310, 315, 316,
        411
    frequency mass renormalization, 49
    oscillator model, 49, 180, 250, 280
    oscillator nonergodicity, 180, 247
    perturbation, 222
    QM/MM link atom correction, 280
    Quantum Correction Factor (QCF), 264,
        265
    quantum Correction Factor (QCF), 263
    reference system integration, 123, 127
Hartree potential, *see* potential: Hartree,
    potential: electrostatic
Hartree–Fock
    approximation, **25**, 41, **73**
    canonical equations, 26, 73
    canonical orbital, 73
    Car–Parrinello molecular dynamics, 40
    Coulomb operator, 74
    coupled-perturbed, 309
    delta-SCF/ΔSCF method, 197
    determinant, *see* determinant: Hartree–Fock
    energy, 73
    equations, 26, 41, 73
    exchange, *see* functional: exact exchange
        (EXX), energy: exact exchange
    Exchange operator, 74
    exchange operator, 74
    Fock operator, 73
    force, 74
    open-shell, 202
    orbital, 73
    post-HF, 74, 75
    Restricted Open-Shell (ROHF), 196
HBr

crystal, disordered, 372
HBr crystal, 372
HCl
    adsorption, 377
    hydrate, 373, 374
    liquid, 385, 414
    molecule, 377
    solution, 383, 397
Hellmann–Feynman
    force, 23, 51, 52, 54, 77, 214, 296
    force correction, 52–56, 65, 77, 80, 105, 272,
        296, 297, 304
    generalized theorem, 315
    theorem, 52, 54, 55, 165, 214
Hellmann–Feynman potential, *see*
    Hellmann–Feynman
Hessian/Hesse matrix, 49, 315, 316, 414, *see*
    also harmonic, phonon
HF
    liquid, 385
    molecule, 411
    solution, 383, 384
Hg, 376
histidine, 404
HIV, 404, 405
Hockney potential, *see* potential: electrostatic
    Hockney
Hohenberg–Kohn–Sham theory, *see*
    Kohn–Sham
hybrid functional, *see* functional: hybrid
hydrocarbon, 266, 393, 395, 397
hydroformylation, 399
hydrogen
    β-elimination, 399
    adsorbed, 376, 378
    chemical shift, 326
    detachment, 406
    diffusion, 374, 384, 394
    fluid, 391
    high-pressure, 216, 391
    impurity, 374
    solid, 138, 391
    sublattice, 372
hydrogen bond(ed), 71, 177, 234, 282, 374,
    379, 382, 384–387, 395, 397, 405, 412, 414
hydrophilic, 384
hydrophobic, 384
hydrosilylation, 398–400
hydroxyl
    adsorbed, 377–379, 381
    hydrated, 383, 398
    radical, 384, 396
hydroxylation, 377, 406
hyper-GGA, *see* Functional: hyper-GGA
hyperfine splitting/coupling, *see* EPR:
    hyperfine splitting/coupling

IBM
    Blue Gene/L, 361, 362
    pSeries 690 (Power4), 364–366
    RS6000/Model 390 (Power2), 61

ice
  electronic properties, 383
  high-pressure, 391, 411, 413
  infrared spectrum, 410
  melting, 379
  optical properties, 413
  phase transition, 379, 391
  surface, 377
  two-dimensional, 379
  X-ray diffraction, 413
IGLO method, *see* gauge
imine, 202, 401
incomplete-basis-set force, *see* Pulay
    force/stress, Car–Parrinello: Pulay force
Inelastic Neutron Scattering (INS), 373
influence functional, *see* Feynman–Vernon
    influence functional
infrared (IR) spectra
  absorption coefficient/cross-section,
      260–263, 265
  analysis, 315–317, 327, 343–347, 411
  computation, 260, 264, 265, 315–317, 327,
      343, 346, 410, 411
  cross-correlation term, 347
  dipole (E1) approximation, 218, 260, 261,
      317
  electronic polarization, 262
  harmonic, *see* harmonic
  intensity, 262, 264, 265, 310, 347, 411
  lineshape function/spectral density,
      262–265, 411
  oscillator strength, *see* oscillator strength
  QM/MM framework, 277, 281
  quantum correction, 234, 260, 263–265, 347,
      411
  quasiclassical approximation, 265, 411
  solvent shift, 277, 281, 346
  specific systems, 373, 383, 390, 395, 405,
      406, 410, 411
  susceptibility tensor, 262, 263
initial guess
  extrapolation, 59
  wave function, **112**, 119
initialization
  orthogonality constraint, 122
  wave function, 112, 119
InP, 379, 380
instanton dynamics, 266, 395
integration
  Brillouin zone, *see* Brillouin zone
  Car–Parrinello equations, 119, 120, 125,
      273, 305
  Chebyshev, 221
  equations of motion, 119–123, **125**, **305**
  equations of motion (NpT), 186
  equations of motion (thermostats), 181, 182
  Gauss–Hermite, 153, 154, 172
  harmonic reference system, 123, 127
  multiple time step, 57, 122, 123, 273, 278
  path integral, *see* path integral
  radial/angular, 147, 154, 155

Suzuki–Yoshida, 181, 182
  thermodynamic, *see* Thermodynamic
      Integration (TI)
  time-reversible, 122, 186, 273
  with dissipation/friction, 127
interface
  growth process, 379
  liquid–solid, 379, 404, 410
  liquid–vapor, 380
  liquid-vapor, 362
  metal–ceramic, 380
  metal–organic, 380
  solid–solid, 373, 379
Intrinsic Reaction Coordinate (IRC), 397, *see
    also* coordinate: reaction, constraint:
    reaction coordinate dynamics/ensemble,
    ensemble: constrained reaction coordinate
    dynamics
iron, 391–393
  liquid, 391
isomerization
  pathway, 195, 196, 375, 401
  photo-induced, 195, 196, 396, 399–401, 406
isotope effect, 37, 45, 49, 374, 394, 395
iterative diagonalization, *see* diagonalization:
    iterative

Jacob's Ladder concept, *see* functional:
    Jacob's Ladder
Jacobi diagonalization, *see* diagonalization:
    Jacobi
Jacobi optimization, 340–342
Jacobian, 79, 126, 252

kaolinite, 377
KCl, 58, 387
KDP crystal, 374
KH_2PO_4, 374
KOH, 383
Kohn–Sham
  Bloch orbital, 87, 345
  canonical equations, 70
  canonical orbital, 70
  coupled-perturbed, 223, 310
  determinant, *see* determinant: Kohn–Sham
  energy, 68
  equations, 26, 70
  exchange–correlation energy, 69, 71, 72
  force, 70
  non-interacting reference system, 69, 211,
      212, 214, 218
  orbital, 68, 87
  orbital plane wave expansion, 87
  potential, *see* potential: Kohn–Sham
  TDDFT, *see* TDDFT
  theory, 68
  time-dependent theory, 71, 195, 209, 217,
      218, 221

Lagrange multiplier, 202
  SHAKE/RATTLE algorithm, 121, 125

constraint nonorthogonal orbital (CNO)
    method, 304, 305
gauge, 311
generalized orthogonality condition, 291,
    293, 295, 298–302, 304, 305
geometrical constraint, 124–126
numerical accuracy, 122, 126
orthonormality constraint, 26, 30, 57,
    120–122, 223, 310
stationarity condition, 230, 231
velocity Verlet algorithm, 121, 125
Lagrangean, 3, 29, 496
Lagrangian
    ab initio centroid, 252, 253, 255
    ab initio path integral, 247, 252
    Car–Parrinello, *see* Car–Parrinello:
        Lagrangian
    constant-pressure (anisotropic), 185
    constant-pressure (isotropic), 188
    constraint nonorthogonal orbital (CNO),
        304
    covariant, 340
    Euclidean, 242
    extended, 38, 44, 120, 184, 246, 282
    Hartree–Fock, 26, 41
    metadynamics, 179, 190, 191
    method for derivatives, 229
    mixed quantum-classical/QM/MM, 272, 282
    Parrinello–Rahman, 185
    path integral, 246
    plane wave formulation, 120
    unitary transformation, 339
    Wannier gauge, 340–342
Lanczos diagonalization, *see* diagonalization:
    Lanczos
Landau–Zener transition probability, 209, 210,
    221
Landau–Zener–Stückelberg theory, 209
Laplacian functional, *see* functional: Laplacian
Laplacian, lattice-discretized, 83
laser heating, 196, 216, 389, 392, 393
latency, *see* communication: latency
lattice, *see* grid
LDA, *see* functional: Local Density
    Approximation (LDA)
Li, 260, 372, 380, 394
    cluster, 388, 392, 413
    liquid, 388
LiH, 138, 394
linear response theory, 309–326, *see also*
    TDDFT: linear response approximation
lineshape function, *see* infrared (IR) spectra:
    lineshape function
link QM/MM atom, *see* QM/MM coupling
    scheme: capping/link atom
Linpack performance, 364
LiOH, 373
liquid, 22, 188, 344
    ammonia, 58, 386, 397
    metals, 376, 388
    Na, 376

polarizable, 43
selenium, 58
silicon, 376
sodium, 58
water, 50, 62, 72, 104, 285, 383, 384
water autodissociation, 383
water solution, 410
LMTO (Linear Muffin Tin Orbital), 391
LSD, *see* functional: Local Spin Density
    (LSD)

malonaldehyde, 395, 398, 400
Mannich base, 398
mass
    centroid, 251, 252, 255
    fictitious barostat, 185
    fictitious collective variable, 191
    fictitious electron, **30**, 39, 46, 120
    fictitious normal mode, 251, 252, 255
    fictitious nuclear, 57, 58, 246
    fictitious thermostat, 181, 254
    general fictitious, 44
    inertia effect, *see* Car–Parrinello: fictitious
        electron mass effect
    isotope substitution, 37, 45
    physical electron, 11
    physical nuclear, 11, 246, 255
    preconditioned fictitious electron, 48, 123,
        127
    renormalization, **45**, 48–50, 123
    rescaling, *see* mass: renormalization
Massey parameter, 204
Maximally-Localized Wannier Functions
    (MLWF), *see* orbital: Wannier
mechanochemistry, 380, 381
Mehrstellen scheme, 84
Mermin functional, *see* functional: Mermin
mesh, *see* grid
meta-GGA, *see* Functional: meta-GGA
metadynamics sampling, 179, 186, 190, 373,
    384, 397–399, 401, 404
metallic system, 37, 38, 195
metallocence, 399
metathesis, 399
methane, 391, 399
methane, protonated, 394
methanol, 362, 377, 386, 400
methionine, 377
methyl, 278
Mg
    (0001) surface, 377
Mg(OH)$_2$, 372, 377
MgO, 49, 377, 380, 391, 414
    (001) surface, 378
    (100) surface, 377, 393
mica, 379
microsolvation, 402
microstate, *see* determinant: microstate
mixed basis set, *see* basis set:
    augmented/mixed
molecular dynamics, *see also* dynamics

ab initio (AIMD), 2, 54, 66
ab initio centroid, 234, 250, 252, 253, 255, 265
ab initio path integral, 233–266
   application, 374, 383, 385, 393, 395, 398, 405
adiabatic, 24
Born–Oppenheimer, 16, 20, **24**, 27, 119
Car–Parrinello, 30, **30**, 31, 36, 39, 45, 47, 119, *see also* Car–Parrinello
CASSCF, 396
centroid, 234, **249**, 253, 411
classical, 11, 15, 16, 18, 119
comparison
   Born–Oppenheimer/Car–Parrinello/ Ehrenfest, 60
comparison Born–Oppenheimer/Car– Parrinello/Ehrenfest, 40, 54, 57–66
constant-pressure, 107, 184–186, 387, 391
constant-temperature, 180
Ehrenfest, 18, 19, 22–24, 196, 204, 207–208, 221, 402
EOM-CCSD, 400
excited state, 196–209, 214, 217, 218, 229, 283–285, 402, 406
Extended Lagrangian (ELMD), 2, 44
First Principles (FPMD), 2
force field, 21, 119
instanton, 266, 395
mixed quantum-classical/QM/MM, 267–285
non-Markovian/history-dependent, 191
nonadiabatic, 23, 204, 207, 208, 219, 284
Parrinello–Rahman, 107, 184–186, 387, 391
path integral, 246
QM(LR-TDDFT)/MM, 284
quantum (QMD), 2
quantum corrected, 263
quasiclassical, 234, 249, 250, 253, 263, 411
Ring Polymer (RPMD), 249, 250
surface-hopping, 207, 209, 210, 221, 232, 283, 284, 400, 401, 406
TDDFT, 218, 219, 229, 232, 402, 406
Tully, 207, 208, 232, 283, 284, 400, 401, 406
Wannier function, 340, 346
Monte Carlo method, 43, 75, 136, 246, 265, 266
ab initio, 383, 386, 388
MoS_2
   (010) surface, 376
MP2 (Møller-Plesset second-order perturbation theory), 67, 74, 75, 266, 395
MRCI (Multi-Reference Configuration Interaction, 75, 401
multiple steering sampling, 179, 189
multiscale, *see also* grid: multiscale approach, 80, 375
muonium, 177, 265, 374, 375, 395
muscovite, 379

Na, 58, 374, 380, 389

liquid, 376
nano
   cluster, 378, 392, 393
   contact, 381
   friction, 381
   hook, 381
   manipulation, 381, 414
   structure, 415
   tube, 382, 383, 393, 395, 403, 405, 415
   wire, 381, 382, 393
NaSn, 387
Ne, 139
NEXAFS, 413
NH_4F, solid, 375
Ni
   (111) surface, 375
nickel, 389
   surface, 375
nitrate, 378
nitric acid, 398
nitric acid trihydrate, 372
nitride, 375
nitrite, 378
nitrogen, 375, 391
nitrogen fixation, 403
nitrogenase, 403
nitromethane, 372
NMR
   chemical shift, 318, 319, 323, 326, 410–412
   diamagnetic contribution, 321, 322
   magnetic susceptibility, 319, 320, 411
   nuclear magnetic moment, 310, 387
   paramagnetic contribution, 321
   periodic boundary condition, 319, 411
   QM/MM framework, 275, 278, 412
   shielding, 310, 326
   spectroscopy, 318, 411
non-self-consistency force, 53–55, 105
nonadiabatic, *see* approximation, coupling, dynamics, molecular dynamics, TDDFT
nondiagonal/diagonal correction, 25, 243, *see also* approximation, coupling, dynamics, gradient, molecular dynamics, TDDFT
Nonlinear Core Correction (NLCC), 160, 162, 168, 294, *see also* gradient: Nonlinear Core Correction (NLCC)
normal mode transformation, *see* coordinate: normal mode
Nuclear Magnetic Resonance (NMR), *see* NMR
nucleobase, 401, 405, 406
Nudged Elastic Band (NEB) method, 179, 189
NWChem software package, 7, 416

occupation
   fractional, 38, 165, 195, 196, 207, 213, 215, 217, 408, *see also* determinant: fractional occupation
   matrix, 38, 166, 216, 217
   number, 20, 29, 38, 68, 87, 114, 165, 204, 207, 208, 216, 217, 301

number transformation, 216

OEP, *see* functional: Optimized Effective Potential (OEP)

off-diagonal/diagonal correction, *see* approximation, coupling, dynamics, gradient, molecular dynamics, TDDFT

olefine, 399

ONETEP software package, 7, 416

OpenMP standard, 363

operator
  angular momentum, 148
  Boltzmann, 215
  Coulomb, 74
  density, 38, 258, 262
  derivative, 84
  dipole moment, 256, 261, 262, 265
  electron kinetic energy, 139
  exchange, 74
  Fermi, 216
  Fock, 73
  Heisenberg, 250, 260, 262
  metric, 88
  momentum, 321
  nonadiabatic/off-diagonal coupling, 13
  nuclear kinetic energy, 13, 236, 238, 239
  overlap, 291, 293, 294
  perturbation, 319, 322, 325
  position, 18, 261, 319, 324, 325, 327–335, 345
  spin, 148, 198
  time evolution, 220
  time-ordering, 56

OPM, *see* functional: Optimized Potential Method (OPM)

orbital
  Bloch, 83, 87, 327, 329–331, 336, 340
  Boys, 83, 327
  canonical, *see* Hartree–Fock: canonical orbital, Kohn–Sham: canonical orbital
  crystal momentum part, 87, 327
  Discrete Variable Representation (DVR) expansion, 80
  Gaussian function expansion, 75
  generalized plane wave expansion, 79
  Hartree–Fock, *see* Hartree–Fock: orbital
  Kohn–Sham, *see* Kohn–Sham: orbital
  Linear Combination of Atomic Orbitals (LCAO) expansion, 51, 76, 136
  localized, 83, 327
  mixed/augmented basis set expansion, 82
  periodic part, 77, 87, 327
  plane wave expansion, 51, 77, 87, 136
  Slater function expansion, 75
  spin orbital, 73
  valence, 139
  Wannier, 226, 324, 325, 327–347
    application of, 382, 385, 408, 410, 414, 415
    basis set, 83
    NMR, 319
  wavelet expansion, 80

orthogonalization, 117, 131

Car–Parrinello molecular dynamics, 57
  Cholesky, 117
  Gram-Schmidt, 117
  Löwdin, 117
  parallel, 363

oscillator strength, 227, 228, 233, 258, 260, 412

oxygen, 49, 391, 412

parallelization, 104, 350–367
  communication routines, 354
  CPMD, 351
  data structures, 355
  Jacobi rotation, 340
  OpenMP, 363
  over orbitals, 351
  over plane waves, 351
  path integral, 248, 249, 351
  taskgroups, 360
  Wannier function, 340

paramagnetic current, 321

Parrinello-Rahman method, *see* molecular dynamics: Parrinello-Rahman, Lagrangian: Parrinello-Rahman

path integral
  (non)adiabatic correction, 239–241, 243
  ab initio, 233–266
    application, 374, 383, 385, 393, 395, 398, 405
  ab initio molecular dynamics, *see* molecular dynamics: *ab initio* path integral
  adiabatic approximation, 239–241, 243
  Born–Oppenheimer approximation, 241–244
  centroid, *see* centroid
  ergodic sampling, 182, 247
  Euclidean Lagrangian/action, 242, 243, 259
  Fermi–Dirac/Bose–Einstein quantum statistics, 242, 244, 266
  Feynman–Kac, *see* Feynman–Kac path integral
  Feynman–Vernon, *see* Feynman–Vernon influence functional
  high-temperature approximation, 247, 259, 267
  Lagrangian, 246, 247, 252
  Maxwell-Boltzmann quantum statistics, 234, 241, 244, 256, 258
  molecular dynamics, *see* molecular dynamics: path integral
  MP2 (Møller-Plesset second-order perturbation theory), 265
  normal mode/staging transformation, 248, 251–253
  QM/MM coupling, 247, 267
  quasiclassical dynamics, 234, 246, 249, 250, 253, 255, 265
  quenched average, 243, 244
  ring polymer isomorphism, 245, 246, 249
  sampling (Monte Carlo/molecular dynamics), 182, 235, 246, 247, 249, 255, 266
  thermostatting, *see* thermostat

Trotter discretization, *see* Trotter
Pauli repulsion, 78, 136, 268, 275, 282
PAW, 78, 287–308
  continuum models, 281
  exact exchange, 104
  NMR and EPR, 326, 410
  QM/MM, 271, 399
Pd, 380
  (100) surface, 378, 379
penicillopepsin, 405
peptide, 403, 404
perovskite, 373
perturbation theory
  $2n + 1$ theorem, 310
  Hylleras variational technique, 309
  Sternheimer technique, 309, 312
  Sum-Over-States technique, 309
perylene, 380
pH, 383
phenylacetylene, 398, 400
phonon, 32, 33, 36–38, 49, 57, 182, 217, 248,
    266, 376, *see also* harmonic,
    Hessian/Hesse matrix
phosphate, 374, 381, 385, 405
phosphorus, 378, 388, 389
photo
  biochemistry, 406
  chemistry, 203, 209, 278, 284, 401, 403, 406
  chromic crystal, 375
  chromic molecule, 382, 415
  dissociation, 397, 401
  excitation, 177, 195, 217, 400
  isomerization, *see* isomerization:
    photo-induced
  physics, 203, 209, 401
photoabsorption spectra, 232–234, 255, 258,
    284, 402, 412
photoelectron spectra, 393, 413
physisorption, 376
pinacolyl alcohol, 397
PINY software package, 7, 269, 271, 416
plane wave basis set, *see* basis set: plane wave
plasma, 57, 392
Poisson equation, 69, 94, 96, 98, 136
polarizability, 227, 310, 317, 318
polarization, 222, 261, 262, 317, 318, 327, 328,
    330, 402, *see also* dipole
polaron, 58, 375
polyacetylene, 375
polyaniline, 375
polycarbonate, 375
polyenes, 202, 401
polyethylene, 375
polyethylene oxide, 375
polyfluoride, 387
polyglycine, 405
polymer, 375, 380, 399, 400
  conducting, 375
polymerization, 397, 399
porphyrazine, 396
porphyrin, 380, 396

potential
  Born–Oppenheimer, 16, 20, 22, 39, 46
  centroid, 250, 255
  chemical, 178, 212, 217, 414
  clamped nuclei, 20
  classical force field, 21, 99, 119, 267, 268
  Coulomb, 69, 73
  effective, 16–18, 137
  Effective Core Potential (ECP), *see*
    pseudopotential
  effective path integral, 246, 250
  Ehrenfest, 18, 20
  electrostatic, 69, 73, 96, 268, 272, 275–277,
    279–281
  electrostatic: Hockney method, 96
  electrostatic: Martyna-Tuckerman method,
    98
  electrostatic: Mortensen-Parrinello method,
    98
  electrostatic: Wannier representation, 347,
    348
  energy surface, 20, 22, 39, 46, 204, 207, 208,
    210, 243, 257, 396
  energy surface, diabatic, 209
  energy surface, excited-state, 23, 194–210,
    229, 232
  exchange–correlation, 70, 99, 102, 105, 313
  external, 69, 211, 218
  first-order perturbation, 222
  global fit, 21–23
  grand canonical, 211, 212
  harmonic, *see* harmonic
  Hartree, 69, 74, 80, 96
  Kohn–Sham, 70, 86, 88, 222
  Lennard-Jones, 268
  many-body interactions, 21, 22
  matter-field interaction, 222
  non-Markovian/history-dependent, 191
  nonbonded, 268
  of Mean Force (PMF), 124, 396, 398, 400,
    404
  pair, 21, 22, 188, 189, 386
  periodic, 77, 86, 330
  pseudo, *see* pseudopotential
  restoring, 196
  thermodynamic, 211
  vector, 320, 321
power spectrum, 31, 32, 38, 49, 50, 181, 182,
    187, 248, 410
preconditioning, 67, 114
pressure, *see also* stress
  barostat, 44, 178, 183, 185, 186, 188
  control, 44, 53, 177, 178, 183–186, 373, 390,
    391, 393, 398
  finite system, 188, 189, 390
  hysteresis, 186
  induced transformation, 189, 372, 373, 379,
    381, 383, 385, 387, 388, 390–393, 398,
    410, 411, 413
  internal/external, 186
  isotropic stress, 183, 185, 371

non-isotropic stress, 184
Projector Augmented Wave (PAW) method,
    see PAW
propagation
    Born–Oppenheimer, 25, 66, *see also*
        molecular dynamics:
        Born–Oppenheimer
    Car–Parrinello, 36, 66, *see also*
        Car–Parrinello, molecular dynamics:
        Car–Parrinello
    Ehrenfest, 66, *see also* molecular dynamics:
        Ehrenfest
    unitary, 56
    wave function, 24, 25
propene, 399
protein, 383, 402–406
proton
    excited state transfer, 400, *see* Excited
        State Proton Transfer (ESPT)
    Grotthuss/structural diffusion, *see*
        Grotthuss mechanism
    hydrated, 373, 374, 379, 382–384, 395, 398,
        401, 404, 405
    quantum effects, 398
    solvated, 384–386, 397, 398
    transfer, 177, 373, 379, 381–385, 395, 398,
        404, 405
pseudopotential, 78, 85, 91
    Bachelet–Hamann–Schlüter–Chiang, *see*
        pseudopotential: norm-conserving
    dual-space Gaussian, 157
    empirical, 139, 275, 278
    for link atoms in QM/MM, 267, 277
    Gauss–Hermite integration, 172
    Ghost states, 157
    Goedecker, *see* pseudopotential: dual-space
        Gaussian
    Hamann–Schlüter–Chiang, *see*
        pseudopotential: norm-conserving
    Hamiltonian, 156
    Kerker, 161
    Kleinman-Bylander form, 155
    local, 92
    Nonlinear Core Correction (NLCC), *see*
        Nonlinear Core Correction (NLCC)
    nonlocal, 105, 134
    norm-conserving, 32, 78, 90–92, 118, 128,
        136–165, 343, 344
    Projector Augmented Wave (PAW), *see*
        PAW
    relativistic, 137, 147, 148
    separable nonlocal, 156
    spin–orbit coupling, 144, 147
    transferability, 137, 139, 143, 157–170
    transferable, 91, 137
    Troullier–Martins, 149–151, 167–173
    ultrasoft, 78, 104, 287, 344
    Vanderbilt, *see* pseudopotential: ultrasoft
Pt
    (111) surface, 377, 379

Pulay force/stress, 53–55, 65, 76, 77, 79–81,
    84, 162, 186, 187, 272, 296
PWscf software package, 7, 416
pyridine, 375
pyrite
    (100) surface, 379

QM/MM, *see also* embedding
    biochemistry, 278, 403–405
    biophysics, 278, 403–405
    general concept, 247, 267, 385
    NMR chemical shift, 275, 278, 412
QM/MM coupling scheme, *see also* embedding
    capping/link atom, 188, 270, 272, 273, 277,
        278, 281
    constant pressure, 188
    CP-PAW/AMBER interface, 96, 271, 273,
        283
    CP-PAW/COSMO interface, 271, 281
    CPMD/GROMOS interface, 271, 274, 278,
        283
    EGO/CPMD interface, 271, 278
    excited state, 283–285, 402, 406
    reaction field, *see* reaction field
    surface-hopping/Tully, 283, 284, 402
    TDDFT, *see* TDDFT: QM/MM coupling
        scheme
quantum
    classical mixed/hybrid, *see* QM/MM
    conductance/current, 382, 415
    confinement/size effect, 393
    Correction Factor (QCF), 234, 260,
        263–265, 347, 411
    effects of heavy atoms, 245, 395
    effects on dynamics, 13, 18, 234, 249–255,
        260, 262, 265, 266, 347, 394, 395, 398
    effects on electronic structure, 18, 234, 255,
        256, 258, 259, 266, 347, 394, 395, 400,
        402, 409, 412
    effects on spectra, 234, 255, 256, 258–263,
        265, 266, 347, 402, 411, 412
    effects on structure, 234, 256, 259, 347, 383,
        385, 394, 395, 398, 402, 409
    high-temperature/classical limit, 14–18, 247,
        250, 257, 259, 260, 262, 263, 265, 267
    Monte Carlo (QMC), *see* Monte Carlo
        method
    path integral, *see* path integral
    subsystem/bath, 19, 267
    tunneling, 233, 234, 256, 382, 394, 395, 402,
        414
    zero-point motion, 233, 234, 256, 394, 402
quantum mechanics
    Bohm's formulation, 15, 267
    Feynman's formulation, 234, 235, 267
    hydrodynamic formulation, 15, 267
    non-relativistic, 11
quartz, 373, 389, 391
    (0001) surface, 376
quasiclassical, *see* dynamics: quasiclassical,
    infrared (IR) spectra: quasiclassical

approximation, molecular dynamics:
quasiclassical
quinizarin, 376

radial distribution function, 50, 51
radiationless decay, 195, 401, 406
Raman spectra, 310, 389, 406, 410, 411
rare event, 124, 179, 189, 396, 400, *see also*
constraint, ensemble, free energy
calculation, metadynamics sampling,
thermodynamic integration
reaction field, 270, 271, 278, 281, *see also*
embedding, QM/MM
real space basis set, *see* basis set: real space
redox reaction, 178, 384, 413
relativistic effect, 137, 147, 148
Reppe carbonylation, 399
Restricted Open-Shell Hartree–Fock (ROHF),
196, 202
Restricted Open-Shell Kohn–Sham (ROKS)
method, *see* ROKS
Restricted Open-Shell Singlet (ROSS)
method, *see* ROSS
Rh
(111) surface, 379
Rhodobacter sphaeroides, 405
rhodopsin, 403, 406
Ring Polymer Molecular Dynamics (RPMD),
249, 250
ripples, 100
RNA, 395
ROKS
application, 400–402, 406
ROKS/ROSS, 196–203
nonadiabatic dynamics, 204, 205, 207
QM/MM, 283–285
ROSS, 401
Ru
(0001) surface, 378
rutile
surface, 377, 378

S/PHI/nX software package, 7, 416
Sakakura–Tanaka functionalization, 399
scanning probe microscopy, *see also* STM,
AFM, 414
SCF, *see* Hartree–Fock
Schrödinger equation
stationary, 20, 21, 24, 25, 28, 218
time-dependent, 11, 19, 28
segregation, 374
selenium, 58, 389, 393
amorphous, 389
Self-Consistent Field (SCF), *see* Hartree–Fock
Self-Interaction Correction (SIC), *see*
functional: Self-Interaction Correction
(SIC)
serine, 377
Si, 160, 164, 376, 379
(001) surface, 376–379
(100) surface, 377

(111) surface, 376, 379, 380
cluster, 393
crystal, 49, 59, 101, 128, 361
tip, 381
SIC, *see* functional: Self-Interaction
Correction (SIC)
SiC, 364, 365, 373, 375, 379
(001) surface, 375
SIESTA software package, 7, 416
silanol group, 377
silica
amorphization, 389
amorphous, 389, 401
aqueous, 373
dielectric constant, 414
diffusion, 374
impurity, 373
laser heating, 389
melt, 389
optical properties, 391
surface, 377, 389
silicate, 373, 389
silicon, 140, 141, 372, 374, 380, 389, *see also* Si
amorphous, 374, 389, 390
carbide, *see* SiC
cluster, 390, 392
laser heating, 216, 392
liquid, 376
p-doped, 374
siloxane, 380
silver, *see* Ag
simulated annealing, *see* annealing, simulated
Slater basis set, *see* basis set: Slater
Slater transition state, 207
SMP (symmetric multiprocessing), 364–366
soliton, 375
solvation
continuum, *see* reaction field, QM/MM,
embedding
explicit, 270, 281
free energy, 270
implicit, *see* reaction field, QM/MM,
embedding
model, 270
sparsity, 134
spectra/spectroscopy
Auger, *see* Auger decay
EPR, *see* EPR
ESR, *see* EPR
infrared, *see* infrared (IR) spectra
optical, *see* photoabsorption spectra
oscillator strength, *see* oscillator strength
photoabsorption, *see* photoabsorption
spectra
photoelectron, *see* photoelectron spectra
Raman, *see* Raman spectra
spin
contamination, 198
density, 196, 199, 200, 203, 285
local spin density (LSD), *see* functional:
Local Spin Density (LSD)

noncollinear magnetism, 412
nuclear, 326
operator, 148, 198
orbital, *see* orbital: spin orbital
polarization effect, 412
polarized/unrestricted, 71, 345, 347, 409
spin–orbit effect, 144, 147
state, 196–200
unpolarized/restricted, 68, 73, 85, 196–200, 205, 226, 344, 345, 347
$SrCeO_3$, 374
$SrTiO_3$, 374
staging transformation, *see* coordinate: staging
stishovite, 373
STM (Scanning Tunneling Microscopy), 381, 382, 414
strain, 184, *see also* pressure
stress, *see also* pressure, Pulay force/stress
tensile, 380, 382
tensor, 107, 108
string molecular dynamics, 179, 189, 378
structure factor, 92, 94, 113, 129, 130, 132, 161, 387, 412
dynamic, 389
X-ray scattering, 413
STS (Scanning Tunneling Spectroscopy), 382
styrene, 399
sulfur, 376, 378
liquid, 387
sulfur–gold chemical bond, 381
sulfuric acid, 377, 397
supercell, 85, 226
decoupling, 95, 98, 272
symmetry, 333, 335
volume/shape change, 107
volume/shape changes, 183
supercritical
CO_2, 387
hydrated electron, 384
water, 383
surface-hopping, *see* Landau–Zener, molecular dynamics: surface-hopping, molecular dynamics: Tully, transition: probability, QM/MM coupling scheme: surface-hopping/Tully
susceptibility, 262, 263, 319, 320, 385, 409, 414

Ta
cluster, 375
Targeted Molecular Dynamics (TMD), 179, 189, 398
taskgroup, 362
TDA (Tamm–Dancoff Approximation), 224, 225, 227–230
TDDFT, *see also* molecular dynamics: TDDFT, Kohn–Sham: time-dependent theory
electronic excitation, 221, 230, 232, 233, 255
general concept, 23, 39, 195, 207, 217, 221
hybrid functional, 402

Kohn–Sham equations, 218, 221, 223
linear response approximation, 195, 196, 209, 221–233, 401, 402, 412
path integrals, 234, 256
QM/MM, 283, 284
linear response charge density, 223, 284
linear response gradient and dynamics, 195, 209, 224, 229, 232, 400, 402, 406
linear response oscillator strength, 227, 228, 233, 258, 260, 412
nonadiabatic coupling vector, 220, 224, 232, 402
protein effect, 402, 403, 405
QM/MM coupling scheme, 283, 284, 402, 403, 405, 406
real-time propagation, 23, 195, 196, 217, 220
solvent effect, 402, 403
TDSCF (Time-Dependent Self-Consistent Field, *see* dynamics: TDSCF
tellurium, 217, 388
temperature, *see also* thermostat
adiabatic decoupling, 31, 38, 180, 248, 272
Car–Parrinello renormalization, 50
control, 38, 120, 272
equipartition theorem/kinetic energy, 50, 179, 182
fictitious electronic, 31, 33, 42, 55
physical, 31, 181
QM/MM coupling scheme, 272
real electronic, 195, 211–213, 215–217
tetrathiafulvalene, 396
Thermodynamic Integration (TI), 124, 179, 189, 271, 273, 396, 400, *see also* constraint, ensemble, free energy calculation, metadynamics sampling, rare event
thermostat, 37, 38, 44, 65, 178–180, 186, 302
adiabatic decoupling, 31, 38, 180, 248, 272, 301
conserved energy, 182
conserved linear momentum, 183
ergodicity, 180, 182, 247, 254
mass and parameters, 181, 182, 254, 255
massive Nosé–Hoover chain, 182, 248, 254
nonconserved linear momentum, 183
Nosé–Hoover, 180, 272, 301
Nosé–Hoover chain, 180, 247, 301
QM/MM coupling scheme, 272
with barostat, 186
thiolate, 380–382, 384, 413, 415
thiophene, 376
thymine, 406
time scale
centroid motion, 255
comparison Born–Oppenheimer/Car–Parrinello/Ehrenfest, 57–59, 62
computer simulation, 189, 190, 273, 361, 389, 404
electronic motion, 28, 57, 195, 203, 208

fictitious electronic motion, 28, 31, 33–37, 57, 59
geometrical constraints, 124
integration time step, 28, 37, 57, 58, 60, 123, 124, 208, 248
multiple, 57, 122, 273, 278
nonadiabatic transition, 203, 204, 208, *see also* Massey parameter
nuclear motion, 28, 31, 34, 35, 47, 48, 57, 58, 61, 195, 204, 211, 255
separation/decoupling, 28, 31, 34–36, 47, 57–60, 123, 189, 190, 195, 248, 255, 273, 404
Time-Dependent Density Functional Theory (TDDFT), *see* TDDFT
Time-Dependent Self-Consistent Field (TDSCF), *see* dynamics: TDSCF
TiO_2
  (110) surface, 378, 379
  surface, 376
transferability, *see* pseudopotential: transferability
  pseudopotential, 286
Transition
  Path Sampling (TPS), 179, 189
transition
  coherent/incoherent, 211
  dipole, 228, 229, 232, 256, 413
  electronic $\pi - \pi^\star$, 201, 202
  electronic $n - \pi^\star$, 202
  metal, 78, 162
  nonadiabatic, *see* Coupling: nonadiabatic/ off-diagonal, dynamics: nonadiabatic quantum, molecular dynamics: nonadiabatic, time scale: nonadiabatic transition, transition: probability
  oscillator strength, *see* oscillator strength
  probability: Landau–Zener, 209, 210, 221
  probability: Tully/fewest-switches, 208, 221
  state, 190, 207, 397, 398
  state Slater's, *see* Slater transition state
trimethylaluminum, 396
trimethylgallium, 376
Trotter
  cyclic boundary condition, 237, 241, 245, 251
  discretization, 214, 234–237, 244, 245, 259
  parallelization, 351, 352
  parameter/imaginary time slice, 215, 237, 244, 259, 351, 352
  product, 241, 244
Troullier–Martins pseudopotential, *see* pseudopotential: Troullier–Martins
Tully
  molecular dynamics, *see* molecular dynamics: surface-hopping, molecular dynamics: Tully
  transition probability, *see* transition: probability
Tunneling, *see* quantum: tunneling

ultrasoft pseudopotential, *see* pseudopotential: ultrasoft
uracil, 411
uranium, 80
uridine, 405
USPP, *see* pseudopotential: ultrasoft

Vanderbilt pseudopotential, *see* pseudopotential: ultrasoft
VASP software package, 7, 416

Wannier
  basis set, *see* basis set: Wannier
  charge, 345, 347
  function, *see* orbital: Wannier
  function analysis, 340, 343, 345–347, 385, 386, 404, 407, 408, 411
  Function Center (WFC), 340–342, 345–347, 411
  orbital, *see* orbital: Wannier
water, *see* H_2O, liquid: water
  adsorption, 379
  liquid, 382
  supercritical, 383
Wavelet basis set, *see* basis set: wavelet

X-ray absorption, 383
XAS, 413

zeolite, 400, 401
zero-point motion, *see* quantum: zero-point motion
Ziegler–Natta polymerization, 399
Zintl-alloys, 387
ZnO
  surface, 381
$ZrO_2/ZrSiO_4$, 373, 377, 390
Zundel complex, 373, 383
Zundel continua, 411
zwitterionic, 398, 404